Lecture Notes in Computer Science 5756

Commenced Publication in 1973
Founding and Former Series Editors:
Gerhard Goos, Juris Hartmanis, and Jan van Leeuwen

Kathleen Stewart Hornsby
Christophe Claramunt
Michel Denis
Gérard Ligozat (Eds.)

Spatial
Information Theory

9th International Conference, COSIT 2009
Aber Wrac'h, France, September 21-25, 2009
Proceedings

 Springer

Volume Editors

Kathleen Stewart Hornsby
The University of Iowa
Iowa City, IA 52242, USA
E-mail: kathleen-stewart@uiowa.edu

Christophe Claramunt
Naval Academy Research Institute
29240 Brest Naval, France
E-mail: claramunt@ecole-navale.fr

Michel Denis
Gérard Ligozat
Université de Paris-Sud
91403 Orsay, France
E-mail: {denis, ligozat}@limsi.fr

Library of Congress Control Number: 2009932257

CR Subject Classification (1998): E.1, H.2.8, J.2, I.5.3, I.2, F.1

LNCS Sublibrary: SL 1 – Theoretical Computer Science and General Issues

ISSN   0302-9743
ISBN-10   3-642-03831-X Springer Berlin Heidelberg New York
ISBN-13   978-3-642-03831-0 Springer Berlin Heidelberg New York

springer.com

© Springer-Verlag Berlin Heidelberg 2009
Printed in Germany

Typesetting: Camera-ready by author, data conversion by Scientific Publishing Services, Chennai, India
Printed on acid-free paper   SPIN: 12738410   06/3180   5 4 3 2 1 0

Preface

First established in 1993 with a conference in Elba, Italy, COSIT (the International COnference on Spatial Information Theory) is widely acknowledged as one of the most important conferences for the field of spatial information theory. This conference series brings together researchers from a wide range of disciplines for intensive scientific exchanges centered on spatial information theory. COSIT submissions typically address research questions drawn from cognitive, perceptual, and environmental psychology, geography, spatial information science, computer science, artificial intelligence, cognitive science, engineering, cognitive anthropology, linguistics, ontology, architecture, planning, and environmental design. Some of the topical areas include, for example, the cognitive structure of spatial knowledge; events and processes in geographic space; incomplete or imprecise spatial knowledge; languages of spatial relations; navigation by organisms and robots; ontology of space; communication of spatial information; and the social and cultural organization of space to name a few. This volume contains the papers presented at the 9th International Conference on Spatial Information Theory, COSIT 2009, held in Aber Wrac'h, France, September 21–25, 2009.

For COSIT 2009, 70 full paper submissions were received. These papers were carefully reviewed by an international Program Committee based on relevance to the conference, intellectual quality, scientific significance, novelty, relation to previously published literature, and clarity of presentation. After reviewing was completed, 30 papers were selected for presentation at the conference and appear in this volume. This number of papers reflects the high quality of submissions to COSIT this year. These papers were presented in sessions held September 22–24, 2009 at the Centre de la Mer, Aber Wrac'h, France.

The conference began with a day of tutorials and workshops. Two tutorials were offered for COSIT registrants. They included a half-day tutorial on "Processes and Events in Geographical Space" (Antony Galton) and a full-day tutorial on "Perspectives on Semantic Similarity for the Spatial Sciences" (Martin Raubal and Alex Klippel). Full-day workshops were offered on "Spatial and Temporal Reasoning for Ambient Intelligence Systems" (Mehul Bhatt and Hans Guesgen) and "Presenting Spatial Information: Granularity, Relevance and Integration" (Thora Tenbrink and Stephan Winter). Based on the outcome of the workshops, special issues of journals are planned that will feature the research presented at these workshops. The final day of the conference is traditionally reserved for the COSIT doctoral colloquium. The COSIT 2009 doctoral colloquium was organized and chaired by Clare Davies of the Ordnance Survey, UK. This colloquium provides an opportunity for PhD students to present their research in a friendly atmosphere to an international audience of fellow students, researchers, and industry participants. Students at any stage of their PhD can apply to participate. This colloquium is an ideal forum for presenting research in the early stages, as well as work that is nearing completion.

A highlight of the COSIT conference series is its excellent keynote speakers. At COSIT 2009, three keynote speakers were featured. On the first day of full presentations, the speaker was James Pustejovsky, Director of the Brandeis Lab for Linguistics and Computation, Brandeis University. The next speaker on day two of the conference was Neil Burgess, Deputy Director, Institute of Cognitive Neuroscience, at University College London. The third speaker was Bill Hillier, Director of the Space Syntax Laboratory, University College London.

COSIT is held every two years, most recently during the month of September. COSIT 2009 was held in the Centre de la Mer that overlooked the beautiful Baie des Anges, site of the picturesque village of Aber Wrac'h in Brittany, France. In addition to the beautiful location on the coast, conference participants also enjoyed an exhibition by a modern artist, Bruno Kowalski, who kindly organized an exhibition of his recent paintings.

We are very grateful to the numerous generous sponsors of COSIT 2009 that helped us to organize a high-level event, in particular the Region Britanny and city of Brest, Groupement d'Intérêt Scientifique Europôle Mer, Institute Géographique National, Ordnance Survey, UK, Taylor and Francis, and the German Transregional Collaborative Research Center SFB/TR 8 on spatial cognition. The Cognitive Science Society of the United States assisted student participation at the doctoral colloquium. The success of the conference can be also attributed to the members of staff and doctoral students of the French Naval Academy Research Institute that provided continuous support to the organization of the event from the very start. Special thanks to Marie Coz, who supervised much of the administration and day-to-day organizing.

COSIT has both a Steering Committee and a Program Committee. The Steering Committee now has 16 members, while the Program Committee for COSIT 2009 consisted of 63 leading researchers drawn from the fields listed previously. The COSIT 2009 organizers would like to thank both committees for their help with this conference. These committees oversaw all the reviewing and contributed greatly to the shaping of the conference program. We thank also any additional reviewers who contributed reviews on the papers submitted. Sadly, in late May 2009, the COSIT community lost a valued member, Professor Reginald Golledge of the Geography Department at the University of California, Santa Barbara. Professor Golledge's pioneering work in spatial cognition and behavioral geography, for example, developing personal guidance systems for blind travelers, has been immensely important for the field and is widely recognized for its significance. Professor Golledge contributed greatly to the COSIT conference series right from the early years of this conference, serving on both the Steering Committee and the Program Committee, and he will be missed very much by all COSIT participants.

In addition to the Steering Committee and Program Committee, the COSIT organizers would also like to acknowledge Matt Duckham, who maintained the main COSIT website and Jean-Marie le Yaouanc, who set up the local COSIT website. The organizers thank all these individuals for their hard work in making COSIT 2009 a great success.

September 2009

Kathleen Stewart Hornsby
Christophe Claramunt
Michel Denis
Gérard Ligozat

Organization

General Chairs

Christophe Claramunt   Naval Academy Research Institute, France
Michel Denis   University of Paris-Sud, France

Program Chairs

Kathleen Stewart Hornsby   The University of Iowa, USA
Gérard Ligozat   University of Paris-Sud, France

Steering Committee

Anthony Cohn	University of Leeds, UK
Michel Denis	LIMSI-CNRS, Paris, France
Matt Duckham	University of Melbourne, Australia
Max Egenhofer	University of Maine, USA
Andrew Frank	Technical University Vienna, Austria
Christian Freksa	University of Bremen, Germany
Stephen Hirtle	University of Pittsburgh, USA
Werner Kuhn	University of Münster, Germany
Benjamin Kuipers	University of Michigan, USA
David Mark	SUNY Buffalo, USA
Dan Montello	UCSB, USA
Barry Smith	SUNY Buffalo, USA
Sabine Timpf	University of Augsburg, Germany
Barbara Tversky	Stanford University, USA
Stephan Winter	University of Melbourne, Australia
Michael Worboys	University of Maine, USA

Program Committee

Pragya Agarwal, UK
Thomas Barkowsky, Germany
John Bateman, Germany
Brandon Bennett, UK
Michela Bertolotto, Ireland
Thomas Bittner, USA
Mark Blades, UK
Gilberto Camara, Brazil
Roberto Casati, France
Eliseo Clementini, Italy

Table of Contents

4   Spatial Cognition

5   Spatial Knowledge

6   Scene and Visibility Modeling

7   Spatial Modeling

8   Events and Processes

9   Route Planning

A Conceptual Model of the Cognitive Processing of Environmental Distance Information

Daniel R. Montello

Department of Geography
University of California, Santa Barbara
Santa Barbara, CA 93106 USA
montello@geog.ucsb.edu

Abstract. I review theories and research on the cognitive processing of environmental distance information by humans, particularly that acquired via direct experience in the environment. The cognitive processes I consider for acquiring and thinking about environmental distance information include working-memory, nonmediated, hybrid, and simple-retrieval processes. Based on my review of the research literature, and additional considerations about the sources of distance information and the situations in which it is used, I propose an integrative conceptual model to explain the cognitive processing of distance information that takes account of the plurality of possible processes and information sources, and describes conditions under which particular processes and sources are likely to operate. The mechanism of *summing vista distances* is identified as widely important in situations with good visual access to the environment. Heuristics based on time, effort, or other information are likely to play their most important role when sensory access is restricted.

Keywords: Distance information, cognitive processing, spatial cognition.

1   Introduction

Entities and events on Earth are separated by space—this separation is distance. Human activity takes place over distance and involves information about distance. Distance information helps people orient themselves, locate places, and choose routes when traveling. It also helps people evaluate the relative costs of traveling from one place to another and utilize resources efficiently, including food, water, time, and money [1]. Understanding how humans think about and understand distance contributes to predictive and explanatory models of human behavior. For example, it has been axiomatic to geographers, planners, and transportation engineers that humans are effort minimizers and choose routes and destinations partially out of their desire to minimize *functional* distance [2], [3], [4]. Because overcoming the separation between places that are further away generally requires more time, effort, money, or other resources, we expect less interaction between places further away. This generalization has been considered so fundamental to explanation in geography that it has been dubbed the "First Law of Geography" (the repeated use of this phrase in

K. Stewart Hornsby et al. (Eds.): COSIT 2009, LNCS 5756, pp. 1–17, 2009.

textbooks and research literature suggests it is taken quite seriously, e.g., see the Forum [5] in the journal *Annals of the Association of American Geographers*).

Behavioral geographers and others proposed some time ago that is was not objective or actual distance that alone accounted for human activity. Instead, they proposed that models of human spatial activity could be improved by considering *subjective* distance—what people know or believe about distance [4], [6], [7], [8]. For example, when I choose to visit one store rather than another because it is closer, I base this choice on my *belief* that the one is closer, whatever the true distance is. Thus it is evident that understanding human perception and cognition of distances is necessary for understanding human spatial activity and interaction between places.

Although much navigation and spatial planning can occur without precise metric information about distances, or even without distance information at all, I have argued elsewhere that some quantitative information about distances is required to explain human behavior in the environment, for both conceptual and empirical reasons [1], [9]. Neither information about the sequences of landmarks nor information about travel times are sufficient by themselves (of course, travel time could provide the basis for metric information about the separations between places). Conceptually, some quantitative distance information was needed by our evolutionary ancestors in order to navigate creatively; such creativity includes making shortcuts and detours in an efficient manner. Inferring the direction straight back to home after several hours or days of circuitous travel requires distance information, not just information about landmark sequences or travel times. Such creativity clearly is still valuable in present times for many of us, in many situations. Empirically, systematic observation that people can make metrically accurate distance estimates, and can perform shortcut and detour tasks with some accuracy, supports the psychological reality of distance knowledge [1], [10], [11].

Recognizing its importance and pervasive role in human activity, this paper provides a comprehensive and interdisciplinary review of the cognitive processing of distance information by humans. It also proposes a conceptual model of the perception and cognition of environmental distance. As I stated in [1], a complete model of environmental distance knowledge and estimation provides answers to four questions:

1. What is perceived and stored during travel that provides a basis for distance knowledge?
2. What is retrieved from long-term memory (LTM) when distance information is used (e.g., when travel planning is carried out) that determines or influences distance knowledge?
3. What inferential or computational processes, if any, are applied to information retrieved from LTM to produce usable distance knowledge?
4. How does the technique used to measure distance knowledge influence estimates of distance?

Distance knowledge and its expression as measured data in cognitive research result from processes and information sources addressed by the first three questions, in addition to aspects specific to the measurement technique used to collect estimates, addressed by the fourth question. In [12], I reviewed techniques for measuring distance knowledge, comparing techniques based on psychophysical ratio, interval, and ordinal scaling; mapping; reproduction (i.e., retraveling); and route choice. Of course, researchers have uncovered significant new insights about distance estimation since my review. One of the most significant insights about estimating distance (and other spatial properties such as slope) concerns an apparent dissociation between spatial knowledge expressed via direct motoric action, such as retraveling a route as part of distance reproduction, and knowledge expressed via indirect, symbolic techniques, such as verbal estimation in familiar units (a common technique I grouped with ratio scaling methods in [12]), [13], [14].

In [1], I focused on the sources of information for distance knowledge, addressing primarily the first two questions above. To the extent that distance information is acquired via travel through the environment, knowledge of distances must ultimately be based on some kind of environmental information, such as the number of landmarks encountered, or proprioceptive information, such as the bodily sense of travel speed. I organized these sources of information into three classes: (1) number of environmental features, typically but not exclusively visually perceived, (2) travel time, and (3) travel effort or expended energy. I concluded that environmental features enjoys the most empirical support as a source of distance information, although not all types of features are equally likely to influence beliefs about distance. Features noticed by travelers and used by them to organize traveled routes into segments will most impact distance knowledge, e.g., [15], [16]. Two explicit variants of features as a source of distance information are *step counting* and *environmental pattern counting* (e.g., counting blocks).

Travel time is logically compelling as a source of distance information, especially in situations of restricted access to other kinds of information, but it has not been convincingly demonstrated in much research and is often misconceptualized insofar as researchers have failed to consider the role of movement speed. Also, travel effort enjoys very little empirical support but may still function when it provides the only possible basis for judging distances. Since 1997, new research has been reported on the perception of travel speed [17] and its role in distance cognition [18], [19]. Also, research has been reported on the role of effort that suggests it can influence the perception of vista distances when people anticipate they will need to climb a sloped pathway [20], [21]. Nonetheless, showing that experienced effort influences estimates of environmental distances that have actually been traveled remains an elusive phenomenon.

In this paper, I address the remaining question relevant to a complete model of directly experienced environmental distance knowledge and estimation, Question 3. This question asks what inferential or computational processes, if any, are brought to bear on information retrieved from LTM so as to produce usable distance information. To address this question, I describe alternative processes for how humans acquire, store, and retrieve directly experienced distance information. I summarize these processes in the form of a conceptual model that comprehensively presents alternative ways people process distance information and the conditions likely to lead to one alternative or another.

My review and model are organized around a theoretical framework that proposes there are alternative processes accounting for distance knowledge in different situations and multiple, partially redundant information sources that differentially provide information about distances as a function of availability and spatial scale. A few models of environmental distance processing have been proposed in the literature. The model I present below modifies and extends models proposed some time ago by

Briggs [22], Downs and Stea [23], and Thorndyke and Hayes-Roth [24]. These proposals contributed to a comprehensive theory of environmental distance information but have not been significantly updated in over two decades. Furthermore, these older models did not fully express the plurality of plausible distance processes, the idea that a single process can operate on different information sources, nor the idea that a single source might be processed in different ways. Thus, the evidence that researchers have put forth for some aspect of distance cognition is often consistent with multiple specific explanations, making its interpretation ambiguous. What's more, there are partially redundant cognitive systems for processing and estimating traveled distances. More than one system can operate within and between research studies, and even within individual people on different occasions.

2 Environmental Distance, Directly Experienced

As in my earlier review of sources of distance information [1], I am concerned in this manuscript with information about distances in *environmental* spaces [25]. These are physical spaces (typically Earth-surface spaces) that are much larger than the human body and surround it, requiring considerable locomotion for their direct, sensorimotor apprehension. Examples of environmental spaces include buildings, campuses, parks, and urban neighborhoods (it is largely an open research question as to how well spatially talented people can directly apprehend the spaces of large cities and beyond). Their direct apprehension is thus thought to require integrating information over significant time periods, on the order of minutes, hours, days, or more. However, unlike *gigantic* spaces (termed *geographic* spaces in [25]), environmental spaces are small enough to be apprehended through direct travel experience and do not require maps, even though maps may well facilitate their apprehension. Many studies, especially in geography, concern distance information acquired indirectly (symbolically) in naturalistic settings, at least in part, e.g., [26], [27], [28], [29], [30]. The results of theses studies are somewhat ambiguous with respect to how travel-based environmental distance information is processed.

There is a great deal of research on the perception of distance in *vista* spaces, visually perceptible from a single vantage point [31], [32], [33], [34], [35]. This research has often been concerned with evaluating the fit of Stevens's Power Law to vista distance estimates under various conditions. The Power Law states that subjective distance equals physical distance raised to some exponent and multiplied by a scaling constant. Most interest has been in the size of the exponent, which has usually been found to be near 1.0, a linear function (exponents < 1.0, a decelerating function, have been reported more often than exponents > 1.0, but both have been found). This work is relevant to our concern with environmental distance for at least two reasons. First, psychophysical distance scaling has been methodologically important in the study of environmental distance information, as I reviewed above. Second, I propose below that perceived distances in vista spaces provide an important source of information for environmental distance knowledge.

However, it is important to distinguish between "visual" and "spatial." Spatial information expresses properties like size, location, movement, and connectivity along one or more dimensions of space. Most visually-acquired information has a spatial aspect to it, but not all does—color provides perhaps the best example. And although vision provides extremely important spatial information to sighted people, especially spatial information about external reality distant from one's body, other sensory modalities also provide important information about space. These senses include audition, kinesthesis, and haptic and vestibular senses (some evidence even suggests olfaction may play a role for people [36]). There is apparently a spatial mode of cognitive processing that is more abstract than any sensory mode, and it is clear that spatial processing is not limited to or wholly dependent on visual processing [37], [38], [39], [40]. The fact that blind and blindfolded people can accurately estimate distances in the environment shows that vision is not required for the perception and cognition of distance, e.g., [41], [42]. The cognitive processes discussed below differ in their reliance on different sensory modalities, but it is apparent that different modalities provide partially redundant means of picking up distance information.

2.1 Active versus Passive Travel

Even restricting ourselves to distance information acquired directly during travel in the environment, we must consider whether this travel is *active* or *passive* [43], [44], [45], [46]. The terms actually reflect two relevant distinctions. More commonly made is the distinction between voluntarily controlling one's own course and speed versus being led along a given path by another agent—that is, making navigation decisions or not. Active travel in this sense could be called "self-guided." Driving an automobile is typically self-guided; riding as a passenger is not. The distinction is important because distance knowledge depends in part on one's attention to the environment, to one's own locomotion, or to the passage of time. Attention likely varies as a function of the volition of one's locomotory and wayfinding decisions.

A second, less commonly made, distinction is between travel that requires considerable energy output by the body versus travel that does not. Active travel in this sense could be called "self-powered." Walking and running are self-powered; driving an automobile and being carried are not. This distinction is important for distance cognition because of its implications for travel time, speed, and physical effort, all likely influences on distance knowledge. Furthermore, motor feedback resulting from self-powered travel provides input to a psychological system that updates one's location in the environment [47], [48]. These considerations cast doubt on the validity of using desktop virtual environments as environmental simulations in distance cognition research, e.g., [49]. Thus, the two distinctions between active and passive travel are relevant to distance knowledge because of their implications for the relative importance of different information sources and cognitive processes.

3 Cognitive Processes

I turn now to the question of how distance information acquired directly is cognitively processed during its acquisition, storage, and retrieval. In particular, how extensive and elaborate are the mental computations or inferences one must carry out in order to use information about environmental distance? I propose four different classes of processes that answer this question: working-memory, nonmediated, hybrid, and simple-retrieval processes. *Working-memory* processes are those in which relatively effortful (i.e., demanding on limited resources of conscious thought) inferential or computational processes are brought to bear on cognitive representations constructed in working memory (WM) when distance information is used; information about distance per se is not explicitly stored in memory during locomotion. *Nonmediated* processes are those in which distance information is encoded and stored directly during locomotion, without the need for much explicit inference or computation when distance information is used. *Hybrid* processes combine the two: Information about the distances of single segments is directly stored and retrieved, but effortful WM processes are required to combine the segments into knowledge of multi-segment distances. Finally, *simple retrieval* occurs when distance information is well learned and can be retrieved from long-term memory (LTM) as an explicit belief without any inferential processes. For example, one may have stored in LTM that it is about 240 miles from Fargo to Minneapolis, and can directly retrieve (i.e., recall or recognize) that without making an inference or computation. In some cases, a simple-retrieval process results from the explicit storage of distance information originally derived via other processes. Explicit estimates of distance would especially be available for simple retrieval when a person has previously made an explicit estimate based on other processes and then externalized it in words or numbers. In many other cases, it probably results in the first place from knowledge acquired indirectly via maps or language.

Models of spatial working-memory processes typically describe the WM representations as analogue or imagistic, although WM representations may be numeric, verbal, and so on. Two types of analogue representations may be considered. *Travel re-creation* refers to a process in which a temporally-ordered sequence of environmental images is generated that essentially re-creates a sequence of percepts experienced while moving through the environment. *Survey-map scanning* refers to a process in which a unitary, map-like spatial image is generated that represents part of an environment more abstractly, essentially from a vertical or oblique perspective. Foley and Cohen [50] refer to travel re-creation as *scenographic encoding* and survey-map scanning as *abstract encoding*. The distinction between travel re-creation and survey-map scanning is similar to the distinction by Thorndyke and Hayes-Roth [24] between environmental representations learned via navigation and those learned via maps. However, the distinction I make here refers to the nature of the representation and not to its manner of acquisition. Although the nature of one's learning experience almost certainly influences the nature of one's environmental representations (as Thorndyke and Hayes-Roth proposed and empirically supported), the extent to which this is true is still an open question (see review and discussion in [51]).

The generation and use of one or the other type of analogue representation might be empirically distinguishable in several ways. Thorndyke and Hayes-Roth [24] conjectured that patterns of performance on certain distance and angular estimation tasks would differ for the two. For instance, straight-line distance estimates should be less accurate than distance estimates along a route in the case of travel re-creation; the opposite should be true in the case of survey-map scanning. Siegel et al. [52] proposed that when a route is represented and accessed as a linear sequence (travel re-creation), distance estimates in opposite directions would differ in accuracy as a function of the direction in which the route was learned. Palij [53] suggested that what he called *imagined terrains* (re-created travels) should be readily accessible in any alignment that is necessary for the task at hand. *Cognitive maps* (Palij's term for survey maps) should require extra time and effort to access in alignments that differ from a canonical alignment, such as the alignment in which one has viewed the layout. Such alignment effects are robust and well established when involving in-situ navigation maps, e.g., [54], but somewhat inconsistent when involving mental representations acquired from direct experience, e.g., [55]. Either way, however, it is likely that a re-created travel would also be less accessible in non-canonical alignments, such as those not based on the forward direction of travel.

What types of effortful processes might be applied to the representations generated in working memory as part of WM (and hybrid) processes? Thorndyke and Hayes-Roth [24] provided detailed possibilities. In the case of information acquired via navigation (travel re-creation), they proposed that individual straight-line segments are estimated and summed in WM to arrive at an estimate of total route distance (they did not specify how individual segments are estimated). If straight-line estimates were required between points not in the same segment, angular estimation coupled with some "mental trigonometry" would also be required. In the case of information acquired via maps (survey-map scanning), straight-line distance between any two points is estimated from scanning the imaged map, as in image scanning [56]. If route distance is required, individual segments would have to be scanned and the resulting distances summed. Whatever the case, the existence of such WM processes is suggested by introspection, logical analysis of task demands, and scanning-time data, e.g., [57]. Furthermore, research shows that the context created when representations are constructed in WM during estimation can affect the magnitude of estimated distances considerably [58]. Among other things, it can lead to patterns of asymmetries wherein the distance from A to B is estimated to be different than the distance from B to A [59], [60].

Hirtle and Mascolo [61] suggested additional WM processes. They conducted a protocol analysis in which subjects thought aloud while estimating distances between US cities. Although such information would be strongly influenced by maps and other symbolic sources, their work richly suggests many possible processes that could be used to generate estimates from directly-acquired knowledge. Hirtle and Mascolo identified as many as 20 strategies or heuristics claimed to have been used by subjects, including simple retrieval, imagery, translation from retrieval of time, comparisons to other distances, and various forms of mathematical manipulation of segments (e.g., segment addition). They also found that the use of compound strategies (as in a hybrid process) was more likely with longer distances and less familiar places. That is, various indirect heuristics are more likely to be used when people do not have direct travel experience with a particular route.

For the most part, the WM processes described by Thorndyke and Hayes-Roth, and by Hirtle and Mascolo, do not explain what information is used to estimate the lengths of individual segments, nor how it is processed. But it is clear that processes used to access information with WM representations would be demanding of attentional resources—effortful and accessible to consciousness. With both WM and hybrid processes, however, repeated retrieval and inference with some particular distance information could eventually result in its processing by simple retrieval.

The class of nonmediated processes contrasts sharply with the WM and hybrid processes. Nonmediated processes do not rely on effortful inferences operating on WM representations. Instead, nonmediated processes lead to direct storage of distance information. Alternatively, information about time or effort might be acquired via a nonmediated process of some kind. Estimates of distance could then be derived from simple computational processes translating time or effort into distance.

Nonmediated processes essentially offer an alternative to the idea that the generation and manipulation of images in WM is necessary for generating environmental distance knowledge. In the general context of imagery and psychological processing, Gibson [62] wrote that:

> No image can be scrutinized...[a]n imaginary object can undergo an *imaginary* [italics in original] scrutiny...but you are not going to discover a new and surprising feature of the object this way. For it is the very features of the object that your perceptual system has already picked up that constitute your ability to visualize it. (p. 257)

This quote suggests that it would be necessary to "know" how far it is from A to B in order to construct an accurate image of it in WM—that "new" information cannot be extracted from images. If so, the imagery experienced and reported during distance estimation would be epiphenomenal. Pylyshyn [63], whose theoretical orientation otherwise differs radically from Gibson's, offers a related criticism of the functional scanning of images based on a theory of tacit information.

Gibson did not specifically address environmental distance information. However, his framework does suggest one way that nonmediated processes might work to generate distance knowledge. The visual system is attuned to pick up dynamic changes in the *optic array*, called *optic flow*, that specify movement of oneself through the environment (*visual kinesthesis*). When coupled with perceptions of environmental layout, visual kinesthesis might lead to information about traveled distance without the necessity of constructing analogue memory representations. Rieser and his colleagues [48] developed this approach in their theory of *visual-proprioceptive coupling*. According to this, information about distance gained from optic flow is used to calibrate proprioceptive systems. These proprioceptive systems also produce distance information during locomotion, allowing acquisition of environmental distance information by blind or blindfolded subjects (also see [64]). An interesting way in which Rieser and his colleagues demonstrated calibration is to show that reproductions of walked distances can be altered by recalibrating the visual-proprioceptive coupling when research subjects are required to walk on treadmills pulled around on trailers.

Vestibular and kinesthetic sensing would likely play an important role in a nonmediated process for generating distance information [65], [66], although there are apparently situations where these proprioceptive body senses play a restricted role, such as when riding in an automobile [67]. The acceleration picked up by the vestibules and the semicircular canals is integrated over time by the central nervous system, again without the need for effortful scanning or manipulation of images. Information about traveled distance is thus available as a function of relatively automatic *perceptual updating* processes that have evolved to allow humans and other organisms to stay oriented in the environment without great demands on attentional resources [68], [69].

3.1 Processes: Summary and Discussion

I propose four classes of mechanisms by which humans process information about environmental distances: working-memory, nonmediated, hybrid, and simple-retrieval processes. These are primarily distinguished from one another on the basis of the extensiveness of the computations or inferences people carry out in WM in order to use distance information. According to a working-memory process, effortful manipulations are carried out on explicit representations constructed in WM. These representations are frequently analogue representations (i.e., images of path extensions) but need not be. Two major types of relevant analogue representations can be identified—travel re-creation and survey-map scanning; I considered ways the two might be empirically distinguished. I also detailed several ways that WM representations could be manipulated in order to infer explicit estimates of distance (e.g., image scanning).

In stark contrast, a nonmediated process does not require the construction or manipulation of WM representations, analogue or otherwise. Instead, distance information is acquired and stored during locomotion as a result of implicit computational processes that are outside of the conscious awareness of the locomoting person. Hybrid processes combine WM and nonmediated processes. The lengths of single segments are stored and retrieved by a nonmediated process; information about the single segments is manipulated in WM in order to arrive at information about multi-segment distances. Finally, simple retrieval occurs when an explicit distance judgment can be retrieved from LTM without any inferential or computational processes. This would take place with directly experienced extents when an estimate of the length of some particular route has become well learned and stored explicitly in LTM.

Although only one of these processes can operate during a particular occasion in which distance information is used, it is not necessary to conclude that only one of them generally characterizes the processing of distance information. On the contrary, it is likely that all four processes are used in different situations. What determines which process operates? My review and description of the four classes suggests that one of the major factors involved is whether an explicit judgment of distance is required in a given situation, and whether that estimate is already stored as such in LTM. I turn now to a model that proposes some specific conditions that influence when such explicitness is likely to be necessary.

4 A Comprehensive Conceptual Model of the Cognitive Processing of Directly-Acquired Environmental Distance Information

Ideas about processes can be combined with ideas about sources of information in order to formulate a comprehensive conceptual model of the perception and cognition of environmental distance. I propose a model that addresses three questions posed in the introduction: (1) What is perceived and stored during travel that provides a basis for distance knowledge?, (2) what is retrieved from LTM when distance information is used?, and (3) what inferential or computational processes, if any, are brought to bear on the retrieved information so as to produce usable distance knowledge? (The model does not specifically address the influence of the techniques researchers use to

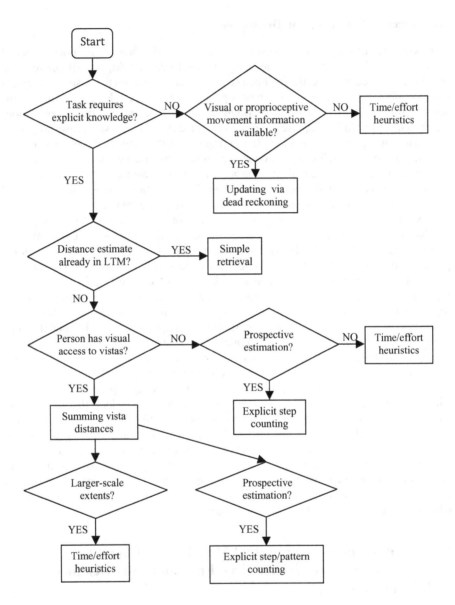

Fig. 1. Proposed model of the multiple processes and information sources for perceiving and cognizing environmental distance. Diamonds are decision nodes, rectangles are end states. All end states make knowledge of distance available in the form needed for the task being performed.

measure distance knowledge.) Figure 1 depicts the model. It is designed to accommodate the availability of alternative processes and multiple, partially redundant information sources. It does this by referring to the demands of the particular task, the availability of particular information sources, the degree of familiarity with the route in question, and its spatio-temporal scale.

At the outset, it is clear that information about environmental distance based on direct travel experience depends on the perception or awareness of body movement or change of position, whether valid or not. This perception generally derives from some combination of vision, kinesthesis, vestibular sensation, audition, and motor efference—any or all of them can contribute in a given situation. As a consequence, information that does not involve any sense or belief of movement, such as a judgment of elapsed time alone, cannot in itself account for distance knowledge.

The model first branches as a function of whether a task requires, or at least tends to activate, explicit knowledge of distance. Some tasks require only implicit information about distance. Locomotion over familiar routes in the environment is an important example; most of us find our way efficiently through the environment on a daily basis without thinking explicitly about our navigational decisions. Nonetheless, our coordinated and efficient travel still requires at least implicit distance knowledge in many situations. The fact that people sometimes have considerable implicit information about distances that guides their behavior in the environment does not, however, ensure that they will be able to externalize that information well using a distance estimation technique. It is therefore possible for subjects to estimate distances explicitly very poorly but do quite well actually navigating, e.g., [70].

In fact, locomotion along familiar routes sometimes does not require much distance information at all, as we observed above, although at least implicit knowledge of distance is often involved. A case in point: Some people can infer the straight-line direction from one place to another rather well, with less than 20° of error, even though they have never traveled directly between the two places [10]. This ability requiresdistance information of some kind. As long as the route is relatively small in scale, so that explicit information is not required, evidence is strong that people can perform this task using only the implicit information about distance provided by the optic flow and/or proprioceptive feedback occurring during locomotion, e.g., [71]. People may have little or no awareness of the operation of this process. As long as visual or proprioceptive information about movement is available, therefore, the model proposes that people (strictly speaking, their cognitive systems) will use a nonmediated process of perceptual updating to "reason" about distances and directions. However, if such perceptual movement information is not available (e.g., a subway ride at constant speed), then the model again suggests that people will need to rely on heuristics about the time and/or effort required to make the trip in order to arrive at knowledge of traveled distance. (As an aside, this situation suggests an important reason why people get lost much more easily when movement information is restricted: Without implicit spatial knowledge, their cognitive system must depend entirely on effortful explicit systems to maintain orientation, which becomes confused without ongoing attention.)

Other tasks require explicit information about distances, i.e., they require conscious awareness of distance quantities to various levels of precision. Notable examples are route planning and giving verbal route directions. In addition, I propose that travelers will require explicit distance information whenever they think about routes of large spatio-temporal scale, no matter what the task (e.g., even when navigating in familiar environments or performing path integration over long distances). Of course, an occasion requiring explicit distance knowledge that is of special interest to behavioral researchers is when a person participates as a research subject in studies of distance cognition. Nearly all such studies require subjects to make explicit (typically numerical, graphical, or verbal) estimates of distances in the environment.

If the task does call for explicit distance information, the model next asks whether an estimate of the length of a given route is already stored in LTM. If it is, the process of simple retrieval operates. This might be the case when a person is very familiar with a particular route and has reasoned about its length in the past.

The model then asks whether the traveler has visual access to vistas; that is, can the person see (rarely, hear) the extents of vistas that end at walls or other visual barriers in the environment? Vistas would be inaccessible to people with severe visual impairment, to people wearing blindfolds, or to people in darkness. If a person does not have access to vistas, then the model asks whether the acquisition of information used to estimate distance occurs under prospective conditions. Prospective conditions exist when a person knows in advance of traveling through the environment that an estimate of distance will be requested. In such cases, step or pattern counting can be used as a way to estimate distance. If prospective conditions do not hold, the person would need to use heuristics about the time and/or effort required to make the trip in order to explicitly estimate distance, after travel is complete. In such cases, subjective distance and time (or effort) will be most strongly related.

If visual access to vistas is available, the model proposes that visually-perceived and retrieved environmental structure will provide the major source of information for distance. Under these conditions, distance knowledge is derived from a hybrid process in which the perceived lengths of route segments that are visible from single vantage points (i.e., vistas) are summed to arrive at estimates for the entire route. This can be termed *summing vista distances*. Any structural features that induce segmentation of routes into vistas, such as opaque barriers, thus tend to elongate estimated environmental distances under the appropriate conditions.

A variety of theoretical and empirical claims motivate my stress on the importance of vista spaces in distance cognition. Gibson [62] emphasized the perception of vistas as integral to the perception of environmental structure under ecologically realistic conditions. A great deal of research on the influence of environmental features (reviewed in [1]), including research on opaque and transparent barriers, points to the import role of discrete pieces of the environment that are visually accessible from particular viewpoints. This stress on vistas also echoes more general theories of human and robotic spatial learning and orientation that posit their central function, [72], [73].

When visual access to vistas is available, and explicit distance information is required, I propose that summing vista distances is the primary mechanism for arriving at distance estimates. In addition, if a prospective estimation situation exists, people can use either step counting or environmental pattern counting if they are aware of such strategies and are not otherwise distracted from using them. The possible moderating influence of various heuristics is also allowed here by the model. These heuristics could be based on travel time or effort, or on such things as route indirectness or the number of features that do not obstruct visibility. The model hypothesizes, however, that heuristic influences will most likely operate with routes of large spatio-temporal scale (i.e., long routes). Under such conditions, the ability to attend to and retrieve relatively continuous information about vistas or elapsed movement is reduced. What's useful for estimating the length of a walk through a building is less useful for estimating the length of a long train trip. For instance, [61] noted that indirect strategies such as time retrieval were more commonly reported with longer distances. Similarly, [74] found an effect of travel effort on estimated distance only for walks that were at least several minutes in duration (as opposed to walks of 45 to 90 seconds).

5 Summary and Conclusions: Future Research Directions

In this paper, I proposed that people process information about environmental distances via one or more of four classes of processes operating on one or more of three sources of information, information acquired during travel through the environment. The four processes include working-memory, nonmediated, hybrid, and simple-retrieval processes. The three sources of information include number of environmental features, travel time, and travel effort. Previous reviews have failed to recognize the plurality of processes and sources that could account for distance knowledge. A comprehensive review of the literature suggests that at different times, people take advantage of alternative processes and multiple, partially redundant sources for acquiring and using information about distances in the environment. The conceptual model presented in Figure 1 attempts to show the conditions that determine which of these multiple processes and information sources will actually operate in a given situation.

It is evident that the perception and cognition of environmental distance is a fruitful research topic for the integration of many aspects of spatial cognition research. The topic involves issues ranging from low-level processes, such as the proprioception of one's movement speed during locomotion, to higher-level processes, such as the representation and manipulation of information via mental imagery. Such research has the potential to help address many interesting theoretical and practical questions related to human behavior in the environment. This review suggests, however, the need for further conceptual refinement and the empirical replication of phenomena that have been previously reported. In particular, we need to understand better the way environmental features of different types will or will not structure mental representations of environments, and the situations in which time and distance heuristics operate. Although I based my proposal that the summing of vista distances is a prominent mechanism for the cognitive processing of environmental distance information, this proposal needs further direct empirical evaluation. Finally, research should address the question of how distance information acquired in various ways, both directly and indirectly (symbolically), is combined or reconciled.

Acknowledgments. I thank several reviewers, especially an exceptionally thoughtful Reviewer 3, for valuable comments. The ideas in this paper were nurtured and modified by conversations with many colleagues, mentors, and students over the years.

References

1. Montello, D.R.: The Perception and Cognition of Environmental Distance: Direct Sources of Information. In: Frank, A.U. (ed.) COSIT 1997. LNCS, vol. 1329, pp. 297–311. Springer, Heidelberg (1997)
2. Bradford, M.G., Kent, W.A.: Human Geography: Theories and Applications. Oxford University Press, Oxford (1977)
3. Deutsch, K.W., Isard, W.: A Note on a Generalized Concept of Effective Distance. Beh. Sci. 6, 308–311 (1961)
4. Golledge, R.G., Stimson, R.J.: Spatial Behavior: A Geographic Perspective. The Guilford Press, New York (1997)
5. Sui, D.Z. (ed.): Forum: On Tobler's First Law of Geography. Ann. Assoc. Amer. Geog. 94, 269–310 (2004)
6. Brimberg, J.: A New Distance Function for Modeling Travel Distances in a Transportation Network. Trans. Sci. 26, 129–137 (1992)
7. Thompson, D.L.: New Concept: "Subjective Distance". J. Ret. 39, 1–6 (1963)
8. Gärling, T., Loukopoulos, P.: Choice of Driving Versus Walking Related to Cognitive Distance. In: Allen, G.L. (ed.) Applied Spatial Cognition: From Research to Cognitive Technology, pp. 3–23. Lawrence Erlbaum, Hillsdale (2007)
9. Montello, D.R.: A New Framework for Understanding the Acquisition of Spatial Knowledge in Large-Scale Environments. In: Egenhofer, M.J., Golledge, R.G. (eds.) Spatial and Temporal Reasoning in Geographic Information Systems, pp. 143–154. Oxford University Press, New York (1998)
10. Ishikawa, T., Montello, D.R.: Spatial Knowledge Acquisition from Direct Experience in the Environment: Individual Differences in the Development of Metric Knowledge and the Integration of Separately Learned Places. Cog. Psych. 52, 93–129 (2006)
11. Schwartz, M.: Haptic Perception of the Distance Walked When Blindfolded. J. Exp. Psych.: Hum. Perc. Perf. 25, 852–865 (1999)
12. Montello, D.R.: The Measurement of Cognitive Distance: Methods and Construct Validity. J. Env. Psych. 11, 101–122 (1991)
13. Creem-Regehr, S.H., Gooch, A.A., Sahm, C.S., Thompson, W.B.: Perceiving Virtual Geographical Slant: Action Influences Perception. J. Exp. Psych.: Hum. Perc. Perf. 30, 811–821 (2004)
14. Wang, R.F.: Action, Verbal Response and Spatial Reasoning. Cog. 94, 185–192 (2004)
15. Berendt, B., Jansen-Osmann, P.: Feature Accumulation and Route Structuring in Distance Estimations—An Interdisciplinary Approach. In: Frank, A.U. (ed.) COSIT 1997. LNCS, vol. 1329, pp. 279–296. Springer, Heidelberg (1997)
16. Jansen-Osmann, P., Berendt, B.: What Makes a Route Appear Longer? An Experimental Perspective on Features, Route Segmentation, and Distance Knowledge. Quart. J. Exp. Psych. 58A, 1390–1414 (2005)
17. Durgin, F.H., Gigone, K., Scott, R.: Perception of Visual Speed While Moving. J. Exp. Psych.: Hum. Perc. Perf. 31, 339–353 (2005)
18. Crompton, A., Brown, F.: Distance Estimation in a Small-Scale Environment. Env. Beh. 38, 656–666 (2006)
19. Hanyu, K., Itsukushima, Y.: Cognitive Distance of Stairways: Distance, Traversal Time, and Mental Walking Time Estimations. Env. Beh. 27, 579–591 (1995)
20. Proffitt, D.R., Stefanucci, J., Banton, T., Epstein, W.: The Role of Effort in Perceiving Distance. Psych. Sci. 14, 106–112 (2003)

21. Witt, J.K., Proffitt, D.R., Epstein, W.: Perceiving Distance: A Role of Effort and Intent. Perc. 33, 577–590 (2004)
22. Briggs, R.: On the Relationship Between Cognitive and Objective Distance. In: Preiser, W.F.E. (ed.) Environmental Design Research, vol. 2, pp. 186–192. Dowden, Hutchinson and Ross (1973)
23. Downs, R.M., Stea, D.: Maps in Minds: Reflections on Cognitive Mapping. Harper & Row, New York (1977)
24. Thorndyke, P.W., Hayes-Roth, B.: Differences in Spatial Information Acquired from Maps and Navigation. Cog. Psych. 14, 560–581 (1982)
25. Montello, D.R.: Scale and Multiple Psychologies of Space. In: Campari, I., Frank, A.U. (eds.) COSIT 1993. LNCS, vol. 716, pp. 312–321. Springer, Heidelberg (1993)
26. Carbon, C.-C., Leder, H.: The Wall Inside the Brain: Overestimation of Distances Crossing the Former Iron Curtain. Psychon. Bull. Rev. 12, 746–750 (2005)
27. Crompton, A.: Perceived Distance in the City as a Function of Time. Env. Beh. 38, 173–182 (2006)
28. Golledge, R.G., Briggs, R., Demko, D.: The Configuration of Distances in Intraurban Space. Proc. Assoc. Amer. Geog. 1, 60–65 (1969)
29. McCormack, G.R., Cerin, E., Leslie, E., Du Toit, L., Owen, N.: Objective Versus Perceived Walking Distances to Destinations: Correspondence and Predictive Validity. Env. Beh. 40, 401–425 (2008)
30. Xiao, D., Liu, Y.: Study of Cultural Impacts on Location Judgments in Eastern China. In: Winter, S., Duckham, M., Kulik, L., Kuipers, B. (eds.) COSIT 2007. LNCS, vol. 4736, pp. 20–31. Springer, Heidelberg (2007)
31. Baird, J.C.: Psychophysical Analysis of Visual Space. Pergamon, New York (1970)
32. Loomis, J.M., Da Silva, J.A., Fujita, N., Fukusima, S.S.: Visual Space Perception and Visually Directed Action. J. Exp. Psych.: Hum. Perc. Perf. 18, 906–921 (1992)
33. Norman, J.F., Crabtree, C.E., Clayton, A.M., Norman, H.F.: The Perception of Distances and Spatial Relationships in Natural Outdoor Environments. Perc. 34, 1315–1324 (2005)
34. Wagner, M.: The Geometries of Visual Space. Lawrence Erlbaum, Mahwah (2006)
35. Wiest, W.M., Bell, B.: Stevens's Exponent for Psychophysical Scaling of Perceived, Remembered, and Inferred Distance. Psych. Bull. 98, 457–470 (1985)
36. Porter, J., Anand, T., Johnson, B., Khan, R.M., Sobel, N.: Brain Mechanisms for Extracting Spatial Information from Smell. Neuron 47, 581–592 (2005)
37. Baddeley, A.D., Lieberman, K.: Spatial Working Memory. In: Nickerson, R.S. (ed.) Attention and Performance VIII, pp. 521–539. Lawrence Erlbaum, Hillsdale (1980)
38. Loomis, J.M., Lippa, Y., Klatzky, R.L., Golledge, R.G.: Spatial Updating of Locations Specified by 3-D Sound and Spatial Language. J. Exp. Psych.: Learn. Mem. Cog. 28, 335–345 (2002)
39. Wickelgren, W.A.: Cognitive Psychology. Prentice-Hall, Englewood Cliffs (1979)
40. Xing, J., Andersen, R.A.: Models of the Posterior Parietal Cortex Which Perform Multimodal Integration and Represent Space in Several Coordinate Frames. J. Cog. Neur. 12, 601–614 (2000)
41. Klatzky, R.L., Loomis, J.M., Golledge, R.G., Cicinelli, J.G., Doherty, S., Pellegrino, J.W.: Acquisition of Route and Survey Information in the Absence of Vision. J. Motor Beh. 22, 19–43 (1990)
42. Rieser, J.J., Lockman, J.L., Pick, H.L.: The Role of Visual Experience in Information of Spatial Layout. Perc. Psychophys. 28, 185–190 (1980)
43. Feldman, A., Acredolo, L.P.: The Effect of Active Versus Passive Exploration on Memory for Spatial Location in Children. Child Dev. 50, 698–704 (1979)

44. Philbeck, J.W., Klatzky, R.L., Behrmann, M., Loomis, J.M., Goodridge, J.: Active Control of Locomotion Facilitates Nonvisual Navigation. J. Exp. Psych.: Hum. Perc. Perf. 27, 141–153 (2001)
45. Sun, H.-J., Campos, J.L., Chan, G.S.W.: Multisensory Integration in the Estimation of Relative Path Length. Exp. Brain Res. 154, 246–254 (2004)
46. Wilson, P.N., Péruch, P.: The Influence of Interactivity and Attention on Spatial Learning in a Desk-Top Virtual Environment. Cur. Psych. Cog. 21, 601–633 (2002)
47. Durgin, F.H., Pelah, A., Fox, L.F., Lewis, J., Kane, R., Walley, K.A.: Self-Motion Perception During Locomotor Recalibration: More than Meets the Eye. J. Exp. Psych.: Hum. Perc. Perf. 31, 398–419 (2005)
48. Rieser, J.J., Pick, H.L., Ashmead, D.H., Garing, A.E.: Calibration of Human Locomotion and Models of Perceptual-Motor Organization. J. Exp. Psych.: Hum. Perc. Perf. 21, 480–497 (1995)
49. Frenz, H., Lappe, M., Kolesnik, M., Buhrmann, T.: Estimation of Travel Distance from Visual Motion in Virtual Environments. ACM Trans. App. Perc. 4, 1–18 (2007)
50. Foley, J.E., Cohen, A.J.: Mental Mapping of a Megastructure. Can. J. Psych. 38, 440–453 (1984)
51. Montello, D.R., Waller, D., Hegarty, M., Richardson, A.E.: Spatial Memory of Real Environments, Virtual Environments, and Maps. In: Allen, G.L. (ed.) Human Spatial Memory: Remembering Where, pp. 251–285. Lawrence Erlbaum, Mahwah (2004)
52. Siegel, A.W., Allen, G.L., Kirasic, K.C.: Children's Ability to Make Bi-Directional Comparisons: The Advantage of Thinking Ahead. Dev. Psych. 15, 656–665 (1979)
53. Palij, M.: On the Varieties of Spatial Information: Cognitive Maps, Imagined Terrains, and Other Representational Forms. Unpublished Manuscript, State University of New York at Stony Brook, Stony Brook, NY (1987)
54. Levine, M.: You-Are-Here Maps: Psychological Considerations. Env. Beh. 14, 221–237 (1982)
55. Waller, D., Montello, D.R., Richardson, A.E., Hegarty, M.: Orientation Specificity and Spatial Updating of Memories for Layouts. J. Exp. Psych.: Learn. Mem. Cog. 28, 1051–1063 (2002)
56. Kosslyn, S.M., Ball, T.M., Reiser, B.J.: Visual Images Preserve Metric Spatial Information: Evidence from Studies of Image Scanning. J. Exp. Psych.: Hum. Perc. Perf. 4, 47–60 (1978)
57. Baum, A.R., Jonides, J.: Cognitive Maps: Analysis of Comparative Judgments of Distance. Mem. Cog. 7, 462–468 (1979)
58. Holyoak, K.J., Mah, W.A.: Cognitive Reference Points in Judgments of Symbolic Magnitudes. Cog. Psych. 14, 328–352 (1982)
59. McNamara, T.P., Diwadkar, V.A.: Symmetry and Asymmetry of Human Spatial Memory. Cog. Psych. 34, 160–190 (1997)
60. Newcombe, N., Huttenlocher, J., Sandberg, E., Lie, E., Johnson, S.: What Do Misestimations and Asymmetries in Spatial Judgment Indicate About Spatial Representation? J. Exp. Psych.: Learn. Mem. Cog. 25, 986–996 (1999)
61. Hirtle, S.C., Mascolo, M.F.: Heuristics in Distance Estimation. In: Proc. 13th Ann. Meet. Cog. Sci. Soc., Lawrence Erlbaum, Hillsdale (1991)
62. Gibson, J.J.: The Ecological Approach to Visual Perception. Houghton Mifflin, Boston (1979)
63. Pylyshyn, Z.W.: The Imagery Debate: Analogue Media Versus Tacit Information. Psych. Rev. 88, 16–45 (1981)

64. Sun, H.-J., Campos, J.L., Young, M., Chan, G.S.W., Ellard, C.: The Contributions of Static Visual Cues, Nonvisual Cues, and Optic Flow in Distance Estimation. Perc. 33, 49–65 (2004)
65. McNaughton, B.L., Chen, L.L., Markus, E.J.: Dead Reckoning, Landmark Learning, and the Sense of Direction: A Neurophysiological and Computational Hypothesis. J. Cog. Neuro. 3, 190–202 (1991)
66. Waller, D., Loomis, J.M., Haun, D.B.M.: Body-Based Senses Enhance Knowledge of Directions in Large-Scale Environments. Psych. Bull. & Rev. 11, 157–163 (2004)
67. Waller, D., Loomis, J.M., Steck, S.D.: Inertial Cues Do Not Enhance Knowledge of Environmental Layout. Psych. Bull. & Rev. 10, 987–993 (2003)
68. Gallistel, C.R.: The Organization of Learning. MIT Press, Cambridge (1990)
69. Loomis, J.M., Klatzky, R.L., Golledge, R.G., Philbeck, J.W.: Human Navigation by Path Integration. In: Golledge, R.G. (ed.) Wayfinding Behavior: Cognitive Mapping and Other Spatial Processes, pp. 125–151. Johns Hopkins University Press, Baltimore (1999)
70. Golledge, R.G., Gale, N., Pellegrino, J.W., Doherty, S.: Spatial Knowledge Acquisition by Children: Route Learning and Relational Distances. Ann. Ass. Amer. Geog. 82, 223–244 (1992)
71. Rieser, J.J.: Access to Knowledge of Spatial Structure at Novel Points of Observation. J. Exp. Psych.: Learn. Mem. Cog. 15, 1157–1165 (1989)
72. Meilinger, T.: The Network of Reference Frames Theory: A Synthesis of Graphs and Cognitive Maps. In: Freksa, C., Newcombe, N.S., Gärdenfors, P., Wölfl, S. (eds.) Spatial Cognition VI. LNCS, vol. 5248, pp. 344–360. Springer, Heidelberg (2008)
73. Yeap, W.K., Jefferies, M.E.: Computing a Representation of the Local Environment. Artif. Intell. 107, 265–301 (1999)
74. Bamford, C.L.: The Effect of Effort on Distance Estimation. Unpublished Master's Thesis, Arizona State University, Tempe, AZ (1988)

Spatial Cognition of Geometric Figures in the Context of Proportional Analogies

Angela Schwering, Kai-Uwe Kühnberger, Ulf Krumnack, and Helmar Gust

University of Osnabrück, Germany
{aschweri,kkuehnbe,krumnack,hgust}@uos.de

Abstract. The cognition of spatial objects differs among people and is highly influenced by the context in which a spatial object is perceived. We investigated experimentally how humans perceive geometric figures in geometric proportional analogies and discovered that subjects perceive structures within the figures which are suitable for solving the analogy. Humans do not perceive the elements within a figure individually or separately, but cognize the figure as a structured whole. Furthermore, the perception of each figure in the series of analogous figures is influenced by the context of the whole analogy. A computational model which shall reflect human cognition of geometric figures must be flexible enough to adapt the representation of a geometric figure and produce a similarly structured representation as humans do while solving the analogy. Furthermore, it must be able to take into account the context, i.e. structures and transformations in other geometric figures in the analogy.

Keywords: computational model for spatial cognition, geometric proportional analogy, re-representation, adaptation, context.

1 Introduction

The cognition of spatial objects involves the construction of a consistent and meaningful overall picture of the environment. Gestalt Psychology (Wertheimer 1912; Köhler 1929; Koffka 1935) argues that human perception is holistic: instead of collecting every single element of a spatial object and afterwards composing all parts to one integrated picture, we experience things as an integral, meaningful whole. The whole contains an internal structure described by relationships between the individual elements.

Perception of the same thing can be different possibly due to differences between humans, due to changes in the context, or due to ambiguity in the figure itself. The following figures show several examples with ambiguous perceptions. The Necker cube shown in Fig. 1 is an example for a multistable perceptual experience where two alternative interpretations tend to pop back and forth unstably. The cube can be seen in two ways, because it is not possible to decide, which one of two crossing lines is in the front or in the back. Fig. 1(b) and (c) show two possible ways to perceive it.

K. Stewart Hornsby et al. (Eds.): COSIT 2009, LNCS 5756, pp. 18–35, 2009.

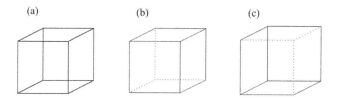

Fig. 1. The Necker Cube (a) is an ambiguous line drawing. Figure (b) and (c) show two possible ways to interpret the Necker Cube.

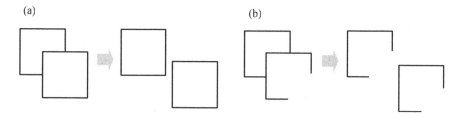

Fig. 2. The perception of a figure is influenced by its context: figure (a) is usually perceived as two complete squares one covering the other, although the covered square is only incompletely visible. In figure (b), the "covered" square is usually perceived as incomplete, because the other square (the context) is incomplete as well.

Fig. 2 is an example where the perception is influenced by the context. Figure (a) shows one complete square and an incomplete square. Most people tend to perceive one square as being covered by the other and therefore complete the non-visible part of the square in their mind to two complete squares. In figure (b), it is more likely that people perceive both squares as incomplete, because the visible square is incomplete as well.

These figures may serve as examples where identical geometric figures are perceived differently and the perception of one element is influenced by its context. A computational model of spatial cognition must be able to compute different perceptions, i.e. different representations for the same spatial object. We will introduce a language for describing geometric figures and show how Heuristic-Driven Theory Projection (HDTP) can adapt representations to reflect different perceptions.

HDTP is a computational approach for analogy making and analogical reasoning. It represents the source and the target stimulus symbolically as two logical theories. In the analogy identification process, HDTP compares both theories for common patterns and establishes a mapping of analogous formulas. The mapping of analogous formulas is captured at an abstract level: The generalized theory formally describes the common patterns of the source and the target stimulus and the analogical relation between them. The symbolic basis of HDTP allows not only the representation of the geometric figures, but also for the representation of general rules which describe how representations can be adapted to reflect different perceptions. The separation of the knowledge about the geometric figure and the abstract knowledge of human perceptions allows HDTP to compute different representations, i.e. compute different conceptualizations of the same geometric figure on-the-fly. HDTP proposes different possible analogies depending on the conceptualization of the figure.

In this paper, we investigate the spatial cognition of simple geometric figures and develop a computational model to compute different perceptions. We conducted experiments with proportional analogies, where subjects have to find a follow-up for a series of geometric figures. Subjects selected different solutions depending on the perception of the geometric figure. In section 2 we describe the experiment, present the results and analyze how subjects perceived the geometric figures in the context of an analogy. Section 3 introduces "Heuristic-Driven Theory Projection" (HDTP), a formal framework to compute analogies. A logical language is used to describe the individual elements in a simple geometric figure in an unstructured manner. From this flat representation it is possible to automatically build up different possible structures and compute "perceptions" which are reasonable to solve the analogy. This mechanism is called re-representation (section 3.3). In section 4, we sketch related work on computational models for solving geometric proportional analogies and discuss the differences to our approach. Section 5 evaluates the applicability of the approach for simple geometric figures and outlines, how HDTP could be used to model human cognition of complex spatial objects.

2   Spatial Cognition of Geometric Figures to Solve Analogies

Here, we give an overview of the experiment focusing only on the results relevant for the computational model. Details about the design and the results of the experiment can be found in (Schwering et al. 2008; Schwering et al. 2009a).

2.1   Setting of the Experiment

The human subject test investigated preferred solutions for proportional analogies of the form (A:B)::(C:D) - read A is to B as C is to D - where A, B and C are a given series of figures and the analogy is completed by inserting a suitable figure for D. All analogies in the test were ambiguous and allowed for different plausible solutions. The analogies were varied in such a way that different perceptive interpretations might be triggered which result in different solutions. For the experiment[1] we used the Analogy Lab, a web-based software platform especially developed for this purpose. Each subject was subsequently shown 20 different analogies randomly chosen from 30 different stimuli: for each analogy they saw the first three objects from an analogy (figure A and B from the source domain and figure C from the target domain) and had to select their preferred solution from three given possible answers (Fig. 3). In every analogy, all three possibilities were reasonable solutions of the analogy; however different solutions required different perceptions of the geometric figures A, B, and C.

---

[1] The experiment consisted of different parts: One part was choosing the preferred solution from three given possible answers. In a second part, participants had to construct themselves via drag&drop their solution. For this analysis we use only data from the choice-part of this experiment.

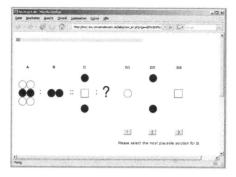

Fig. 3. The analogy lab[2] is a web-based tool to conduct experiments. This screenshot shows one analogy with three possible solutions which can be selected.

The aim of this experiment was to investigate the subjects' perception of geometric figures[3], but also to investigate how the perception changes across different variations of one analogy.

The experiment revealed that subjects applied different strategies to solve the analogies and came up with different solutions. The different solutions can be explained, when the elements in figures A, B, and C are structured differently.

2.2 Different Conceptualization of the Same Stimulus

In the experiment, we investigated 30 different analogies. From this set we selected four analogies to be presented as examples in this paper. We discuss the possible perceptions of the geometric figures, present the preferences of different solutions and discuss how a conceptualization of the figure is related to one solution. We analyze how a computational model could reflect the human perception by reproducing the same groupings and same relations as the subjects did.

Fig. 4 shows the first analogy: the majority of the 161 subjects who solved this analogy selected the geometric figure consisting of one single white square as solution for this analogy. This solution results[4], if the elements in figure A, B and C are grouped into middle elements and outer elements. Figure B can be constructed from figure A by deleting all outer objects. The second preferred solution, the two black circles, results if the subjects group the geometric figures A, B and C according to color and delete all white objects while all black objects remain. The third solution was chosen only two times. It can be explained by keeping the middle elements with

---

[2] http://mvc.ikw.uos.de/labs/cc.php

[3] In a different experiment, we let subjects comment on their solution. From these comments we got evidence that subjects built up different structured representations to solve the analogy in one or the other way. Due to space limitation, we cannot include a detailed comment analysis in this paper.

[4] We would like to point out that these are our interpretations. We base these interpretations on comments that the participants of our experiments gave after solving each analogy. Although in most cases our interpretation seems to be very straight forward, there can be other interpretations that led subjects choose a solution.

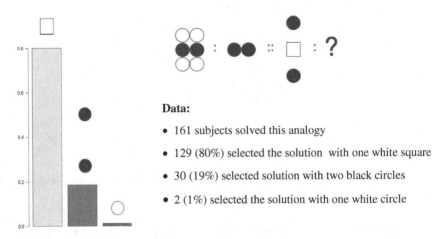

Data:

- 161 subjects solved this analogy
- 129 (80%) selected the solution with one white square
- 30 (19%) selected solution with two black circles
- 2 (1%) selected the solution with one white circle

Fig. 4. The first analogy can be solved by focusing on the position of the elements or on the color. The results show that the majority of subjects preferred to keep the middle object, while several subjects chose to keep the black objects. Only two subjects selected the white circle as solution.

their position and color, but changing the shape to a circular shape. However, this solution is obviously not preferred.

At a more general level, we can reveal different strategies that subjects applied to solve this analogy. The majority of subjects considered the relative position of the elements and grouped elements in middle and outer elements. The second biggest group of participants focused on the color and formed one group with white elements and one group with black elements.

Fig. 5 shows a variation of the first analogy: In figure A, the two top circles are black and all other circles are white and in figure C the colors are flipped compared to figure C in the previous analogy. This variation has a huge effect on the preference distribution and also on the preferred perception. The majority of the subjects chose the figure with one black square as solution for this analogy. Subjects choosing this solution presumably grouped according to colors and deleted all white elements while they kept the black ones. The second preferred solution was one white circle. These subjects focused on the relative position: The top elements form one group and the others form another group. The analogy is solved by keeping the top elements and moving them to the middle of the figure. The third preferred solution keeps the color of the top elements and the shape of the middle elements.

Although both analogies are very similar, the resulting preferences are relatively different. The majority of subjects chose either a grouping strategy based on the position or based on color, but in the first analogy the position-strategy was clearly preferred, while in the second analogy the color was more preferred. The strategy of transferring the color from elements in the source domain but keeping the same shape as in figure C was hardly applied in analogy one (only 1% of the participants), but applied by 20% of the participants in analogy two.

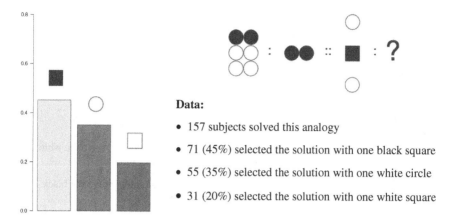

Data:

- 157 subjects solved this analogy
- 71 (45%) selected the solution with one black square
- 55 (35%) selected the solution with one white circle
- 31 (20%) selected the solution with one white square

Fig. 5. The second analogy can be solved by focusing on the color of the elements (preferred solution) or on the position (second preferred solution). It is also possible to treat shape and color differently and transfer only color while the shape remains the same (third preferred solution).

Fig. 6 shows an analogy where the geometric figure B can be perceived as a 180° rotation of figure A. In this case the figure is seen as one whole and is not divided into any subgroups. Subjects who selected the most preferred solution presumably applied this strategy.

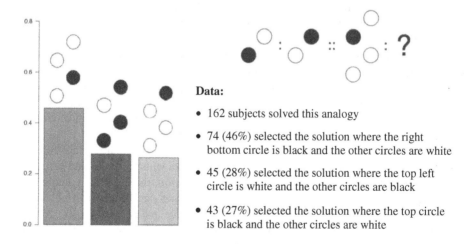

Data:

- 162 subjects solved this analogy
- 74 (46%) selected the solution where the right bottom circle is black and the other circles are white
- 45 (28%) selected the solution where the top left circle is white and the other circles are black
- 43 (27%) selected the solution where the top circle is black and the other circles are white

Fig. 6. The preferred solution of the third analogy is constructed via rotating the whole figure 180°. Participants choosing this solution presumably did not divide the figures into subgroups, but grouped all circles in figure A, all circles in figure B and all circles in figure C in three separate groups independently of their color. The second preferred solution results from a color flip. The third solution can be explained by dividing figure C in two groups: the upper two circles form one group because they repeat figure A and the lower circles from a second group.

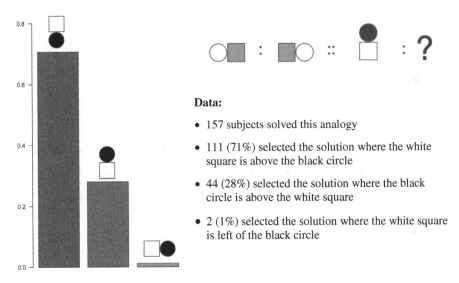

Data:

- 157 subjects solved this analogy

- 111 (71%) selected the solution where the white square is above the black circle

- 44 (28%) selected the solution where the black circle is above the white square

- 2 (1%) selected the solution where the white square is left of the black circle

Fig. 7. The first solution of the fourth analogy has different explanations: the elements can be grouped in circles and squares and switch position. They can be perceived as one whole and rotated. They can be perceived as one whole with a mirroring axis between the circle and the square. If the mirroring axis is defined relative to the figure, the most preferred solution is correct. If the axis is defined absolute, the second preferred solution is the correct one.

The second preferred solution is constructed by flipping the colors, i.e. circles are grouped according to the color and all black circles become white and all white circles become black. In the third preferred solution, figure A is mapped on the two upper circles in figure C. Obviously, figure C is perceived as two groups: one contains the upper two circles and the second one contains the lower two. In this case, one part of figure C is an identical repetition of figure A. The solution is constructed by applying the transformation between A and B to that subgroup of C, that is identical to A. The additional subgroup of C - the two bottom white circles - remain the same.

The fourth analogy is shown in Fig. 7. The most preferred solution has different possible explanations: Each figure consists of two elements: a circle and a square. From figure A to B the circle and square change position, therefore the solution is a white square above the black circle. The same solution can be constructed with a different interpretation: figure A is perceived as a whole and is rotated 180°. A third interpretation is also possible: subjects might have perceived a vertical symmetry axis between the circle and the square. Figure B is mirrored along this axis. If the axis is perceived relative to the elements in the figure, the axis in C runs horizontally between the circle and the square. A very similar explanation exists for the second preferred solution: participants perceived as well a vertical symmetry axis between the circle and the square and mirrored figure C along a vertical axis as well. The third solution was only selected by 2 subjects and is not very preferred.

2.3 Results of the Experiment

The experiment shows that analogies have different solutions depending on how geometric figures are perceived. The preferred perception is influenced by the context, i.e. by the other figures in the analogy (cf. analogy 1 and analogy 2). Proportional analogies are a suitable framework to investigate human perception of geometric figures, because different perceptions can be easily discovered if they lead to different solutions.

Grouping is a common strategy to establish the required structure to solve the analogy. In the examples above, grouping based on similarity (such as grouping of elements with common color or common shape) and grouping based on position play important roles. The position is often defined relative, e.g. middle and outer elements seem to be more prominent than other positions. Spatial proximity or continuous movement are other criteria for structuring geometric figures. In analogy three, figure C is an extended version of figure A. In such cases, the extended figure can be divided into two groups: One group comprising the original figure and the second group comprising the additional elements.

3 A Computational Model for Geometric Analogies

The holistic Gestalt perception contradicts the atomistic way computers process information. A computational model for spatial cognition must be able to compute an overall, holistic representation from a list of single elements. We developed a language to describe geometric figures. The analogy model HDTP[5] computes differently structured representations of a geometric figure based on a flat list of single elements.

3.1 Heuristic-Driven Theory Projection (HDTP)

HDTP is a symbolic analogy model with a mathematically sound basis: The source and the target domain are formalized as theories based on first-order logic. HDTP distinguishes between domain knowledge—facts and laws holding for the source or the target domain—and background knowledge, which is assumed to be generally true. Knowledge about a geometric figure is captured by domain knowledge, while general principles of perception are captured in the background knowledge (Fig. 9).

An analogy is established by aligning elements of the source with analogous elements of the target domain. In the mapping phase, source and target are compared for structural commonalities. HDTP (Gust et al. 2006; Schwering et al. 2009c) uses anti-unification to identify common patterns in the source and target domain. Anti-Unification (Plotkin 1970; Krumnack et al. 2007) is the process of comparing two formulae and identifying the most specific generalization subsuming both formulae.

---

[5] This paper shall present the idea of the computational model and sketch the overall process. A detailed description of the syntactic and semantic properties of HDTP can be found here (Gust et al. 2006; Krumnack et al. 2007; Schwering et al. 2009c).

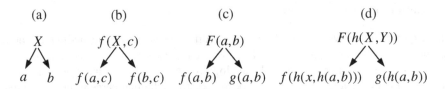

(a)   (b)   (c)   (d)

$$X \qquad f(X,c) \qquad F(a,b) \qquad F(h(X,Y))$$

$$a \quad b \qquad f(a,c) \quad f(b,c) \qquad f(a,b) \quad g(a,b) \qquad f(h(x,h(a,b))) \quad g(h(a,b))$$

Fig. 8. Anti-unification compares two formulae and creates the least general generalization. While (a) and (b) are first-order anti-unification, (c) and (d) require second-order anti-unification to capture the common structure of the formulae.

We use anti-unification to compare the source theory with the target theory and construct a common, general theory which possibly subsumes many common structures of the source and the target domain. Fig. 8 gives several examples for anti-unification. Formulae are generalized to an anti-instance where differing constants are replaced by a variable. In (a) and (b), first-order anti-unification is sufficient. The formulae in (c) and (d) differ also w.r.t. the function symbols. While first-order anti-unification fails to detect commonalities when function symbols differ, higher-order anti-unification generalizes function symbols to a variable and retains the structural commonality. In example (d), F is substituted by f/g, X is substituted by x/a and Y is substituted by $h(a, b)/b$. A detailed description of anti-unification in HDTP can be found in (Krumnack et al. 2007). An example for anti-unification of formulas describing geometric figures is shown below in Fig. 13.

Fig. 9 sketches the HDTP architecture to solve geometric proportional analogies. Figure A and figure B of the analogy are part of the source domain, while figure C

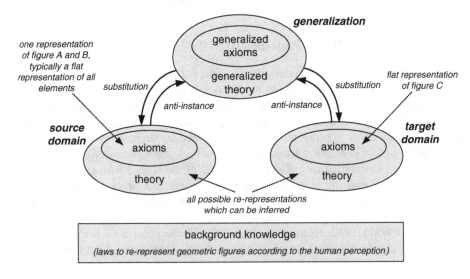

Fig. 9. Overview of the HDTP architecture

(and the still missing figure D) are part of the target domain. All elements in a geometric figure are described by a set of axioms in a formal language (cf. section 3.2). The background knowledge contains laws how to compute structured representations of a geometric figure. Our experiment revealed that one possible strategy is grouping elements with a common color; therefore, the background knowledge contains a law for filtering elements with a common color out of all elements belonging to one figure. Applying these laws to the axiomatic description of a figure leads to a structure (re-)representation of this figure.

To solve the analogy, HDTP compares figure A and figure C for structural commonalities and establishes a mapping between analogous elements in figure A and C. HDTP uses anti-unification for the mapping process and computes a generalization of the commonalities. The generalized theory with its substitutions specifies formally the analogical relation between source and target. Additional information about the source domain - in proportional geometric analogies this is information how to construct figure B from figure A - is transferred to the target domain and applied to figure C to construct figure D (Schwering et al. 2009b).

3.2 Language to Formalize Different Conceptualizations of Geometric Figures

We developed a formal language based on the "Languages of Perception" by (Dastani 1998). Basic elements of a geometric figure can be described by its (absolute) position, shape and color. We can detect groups of elements following the criteria mentioned in section 2.3. For the following example, grouping based on common shape and color is important. The language also supports other structures such as iteration of elements or groups. Since we focus on the basic principle of re-represen-tation and on the changing of flat representations to structured ones, we describe this process exemplary for grouping elements according to their shape and do not elaborate all other possible structures that could be expressed with this language.

The analogy shown in Fig. 10 was solved by grouping all circles in figure A into one group and all remaining elements (in this case a white square) into a second group. Grouping all remaining elements into one group was a common strategy in our experiment. All circles become black, while the remaining elements stay the same. With this strategy the solution to this analogy is keeping the grey square of figure C and changing the color of the circles to black.

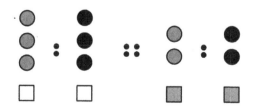

Fig. 10. In this analogy, all circles become black and the squares remain as they are

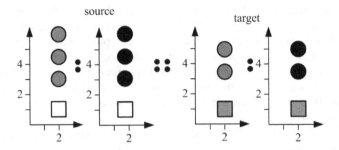

Fig. 11. In HDTP, the analogy is separated into a source and a target domain and a coordinate system determines the absolute position of elements

Fig. 11 shows the same analogy as it would be described in HDTP: figure A and B belong to the source domain and figure C and D belong to the target domain. A coordinate system is used to determine the absolute position of the elements.

HDTP starts with a flat representation of all elements. The elements of figure A are described as follows[6]:

```
% flat representation of figure A
o1 := [shape:square, color:white, position:p(2,1)]
o2 := [shape:circle, color:grey, position:p(2,3)]
o3 := [shape:circle, color:grey, position:p(2,4.5)]
o4 := [shape:circle, color:grey, position:p(2,6)]
```

Based on the flat representation, HDTP has to compute a structured representation which reflects human cognition. First, we show how a structured representation looks like for the running example and how the language supports the re-representation. In the next section, we sketch the process how HDTP automatically detects the correct re-representation steps and computes such structured representations. As we already mentioned, the source domain is perceived as two figures (figure A and figure B) and figure A is divided into a group of circles and the remaining objects (the square):

```
% representation of figure A with structure
group figA := [o1,o2,o3,o4]
group g1 := filter(figA,(shape:circle),+)
group g2 := filter(figA,(shape:circle),-)
```

Groups can be expressed extensionally or intensionally. Extensional groups are defined by listing all members of the group. This is typically the case for the group of elements belonging to one figure such as the group figA. Intensional groups are specified by the defining criteria such as groups g1 and g2. Group g1 is constructed by selecting those elements of group figA which have a circular shape. The plus and the minus sign indicate the polarity: a minus stands for the complement of a group and is used to group the remaining elements in figure A. It is also possible to combine different filters by concatenating different filtering criteria: A group containing all grey circles would be defined as follows:

```
group g1 := filter(figA,(shape:circle, color:grey),+)
```

---

[6] The elements of figure B are constructed from figure A by changing the color of all circles to black and keeping the square. Therefore, they are not described explicitly here.

HDTP uses its background knowledge to transform flat representations into structured ones. The background knowledge contains rules to filter a group for certain elements, i.e. filter group `figA` for all elements which have a circular shape. All circular elements are extracted, added to a list of elements which is used to construct the new group. Analogously, groups can be filtered for a certain color, absolute position, or relative position such as "middle elements".

```
group g1 := filter(figA,(position:top),+)
```

Additional rules are required for groupings based on the relative position. HDTP background knowledge contains rules to compute spatial relations "above", "below", "right", and "left" based on a single cross calculus. For example, top elements are computed by selecting those elements from a group which are not below another element. A single cross calculus is sufficient for the simple geometric analogies used in our experiment. For more complex stimuli one can choose to implement a different calculus to compute spatial relations.

Like figure A, figure C is first represented as a flat list of elements. To establish a mapping, figure C must be regrouped in a way analogous to figure A. If the same subgroups can be constructed, the same transformation can be applied. The following code shows the flat representation and the division into a group of circular elements and a second group of remaining elements.

```
% formalization of figure C as list of flat elements
o5 := [shape:square, color:white, position:p(2,1)]
o6 := [shape:circle, color:grey, position:p(2,3.5)]
o7 := [shape:circle, color:grey, position:p(2,5)]

% representation of figure C in two groups
group figC := [o5,o6,o7]
group g3 := filter(figC,(shape:circle),+)
group g4 := filter(figC,(shape:circle),-)
```

3.3 Solving the Analogy: Re-representation and Anti-unification

The previous section presented the language that is used to describe geometric figures and rules to compute higher structures. Finding the correct conceptualization of a geometric figure within a proportional analogy is an iterative process (Fig. 12): First, HDTP computes different possible conceptualization of figure A using prolog laws in the background knowledge (Schwering et al. 2009b). There are numerous ways in which figure A of the running example could be represented (Schwering et al. 2009a): it could be grouped based on shape, based on color (grey elements, versus white elements), it could be considered as one whole group or any other way of grouping. The re-representation is heuristic-driven:

- It is influenced by Gestalt principles, e.g. according to the law of similarity it makes sense to group grey elements and white elements or circles and squares.
- It is influenced by possible transformations to figure B. If several elements are repeated in B, it is likely that a transformation must exist between the elements in A and the repeated elements in B.

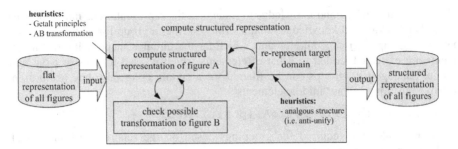

Fig. 12. Iterative process of computing the correct structured representation of the analogy

However, the structure of figure A is not independently created from the overall analogy: Once one (or several) preferred conceptualization of A exist, HDTP tries to re-represent figure C in an analogous way, i.e. it tries to establish the same groupings as in figure A. If this is not possible, the structure of figure A must be revised.

Once figures A and C have a structured representation and the transformation between A and B is known, an analogical mapping can be established via anti-unification and figure D is constructed via analogical transfer.

Fig. 13 shows the anti-unification for an object description and a group definition. The upper part shows an example of a comparison between object o1 and object o5 . Both objects are squares at the position (2,1), but object o1 is white and object o5 is grey. Both formulas differ only with respect to the identifier and with respect to their color. Therefore identifier and color are replaced by a variable X respectively Y in the generalization. The same holds for the group definitions in Fig 13(b): one group is defined on elements in figure A and the other is defined on the elements in figure C of the target domain. The generalization replaces the differing group identifier (g1/g2) with a variable G and figA/figC with the variable F.

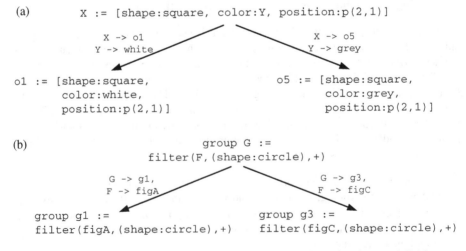

Fig. 13. Anti-unification of two object descriptions and two intentional group definitions

4 Related Work

Proportional analogies were studied in various domains such as the natural-language domain (Indurkhya 1989; Indurkhya 1992), the string domain (Hofstadter and Mitchell 1995), analogical spatial arrangement at a table top scale (French 2002), and in the domain of geometric figures.

In (1962; 1969), Evans developed a heuristic program to solve GPAs. Before the actual mapping process, the program computes meaningful components consisting of several line segments in each figure. Evan's analogy machine determined the relation between A-B, computed a mapping between A-C based on rotation, scaling, or mirroring, and selected an appropriate solution from a list of possible solutions. In contrast to our approach, the representation and the mapping phase are sequentially separated from each other. While we use structural criteria, Evans uses mathematical transformation to detect a suitable mapping between figure A and C.

O'Hara & Indurkhya (1992; 1993) worked on an algebraic analogy model which is able to adapt the representation of line drawing figures during the analogy-making process. Dastani et al. developed a formal language for this algebraic model to describe elements in geometric figures and compute automatically a structural, Gestalt-based representation (Dastani and Scha 2003). This approach accounts also for context effects, i.e. figure C has an effect on the conceptualization of figure A (Dastani and Indurkhya 2001). Both ideas strongly influenced our work. We reuse many ideas developed for this algebraic model and apply them to our logic-based framework.

Mullally, O'Donoghue et al. (2005; 2006) investigated GPAs in the context of maps. They used structural commonalities to detect similar configurations in maps and to automatically classify geographic features. Due to the limitation to maps, they do not support the complex spatial analysis required for our GPAs.

Several other approaches deal with the perception of visual analogies in general. Davies and Goel investigate the role of visual analogies in problem solving (Davies et al. 2008). Forbus et al. (2004) developed an approach to compare sketch drawings. Since GPAs are not the focus of these approaches, we do not discuss them here.

5 Conclusions, Discussion and Future Work

We presented HDTP, a formal framework to automatically compute different conceptualizations of the same figure. We discuss the presented approach and afterwards argue how this approach could be used in a more general context of recognition and classification of spatial objects.

5.1 Summary and Conclusions

Human spatial cognition is a holistic process: we tend to see whole patterns of stimuli when we perceive a spatial object in an environment. According to Gestalt theory, parts of the spatial object derive their meaning from the membership in the entire configuration. Computers, on the other hand, process visual information in an atomistic way. To receive similar patterns as the ones humans perceive, we need a computational model which can generate in a bottom-up manner the structure which is necessary to interpret the stimulus correctly.

Experiments on geometric proportional analogies have shown that subjects perceive the same geometric figure in different ways and that the preferred perception changes, if the context in the analogy is varied. Subjects apply different strategies to solve the analogies: the elements in the geometric figures are often regrouped according to shape, color or position to establish a common structure in source and target domain.

HDTP is a heuristic-driven computational framework for analogy-making and can be used to simulate the human way of solving geometric proportional analogies. We developed a logic-based language to describe geometric figures. HDTP takes such formal descriptions of the figures in the source and the target domain and tries to detect common structures. Usually, source and target are not available in an analogously structured representation at first. HDTP re-represents the descriptions to transform the flat representation into a structured representation of the geometric figure. Different structured representations reflect different conceptualizations of a geometric figure. The process of re-representation is essential to model spatial cognition of geometric figures in the context of proportional analogies: finding the analogous structural patterns in figure A and figure C can be considered as the main task in analogy-making. In the mapping process, HDTP uses the theory of anti-unification to compare a source and a target formula and computes a generalization. The analogical relation between the source and the target is established by creating a generalized theory subsuming all formulae in the source theory and all formulae in the target theory. The proportional analogy is solved by transferring the relation between figure A and B and apply it to figure C to construct figure D.

5.2 Opportunities and Drawbacks of the Approach

In our experiments, we investigated only simple, artificial stimuli so far. The stimuli had different number of elements which varied across three different shapes, three different colors and different positions. The artificial stimuli are simple enough to control variations, to emphasize different aspects and different Gestalt principles and to trigger different perceptions. With systematic variations it is possible to detect how certain variations change the perception. The number of possible re-representations and transformations is limited. The language we are using at the moment supports only simple elements, but it could be extended. Future development shall support complex forms and line drawings like the ones in Fig. 14. It shall become possible to compute an area from a given set of lines and check whether it is a familiar form such as a square or a triangle. The detection of complex structures and forms requires a spatial reasoner which can detect spatial relations.

Analogy making provides a good framework to test spatial cognition, because variations in context may lead to different perceptions which result in different solutions. The different solutions serve as indicator for different conceptualizations.

In section 3.3 we describe a heuristic-driven framework to compute different conceptualizations. We assume the relatively difficult situation where none of the figures is pre-structured. Real-world tasks are often easier. In a recognition task for example, a new stimulus (the target domain) is compared to known stimuli (the source domain). The new stimulus must be restructured to fit to the given structure of the source domain. A pre-defined structure of the source domain reduces drastically the complexity of the underlying framework.

5.3   Spatial Object Recognition as Future Application

In this paper, we discussed the computational framework only in the context of analogies. However, we think that HDTP could serve as a general framework for visual recognition and concept formation. Visual recognition of spatial stimuli is based on matching new stimuli to familiar ones. Often, things are best characterized by their structural (and functional) features, but superficial features do not reveal much about the nature of an object. Therefore, we argue that analogical comparison is very suitable to model the human cognitive process of recognition.

So far, HDTP was tested only with artificial stimuli. The language used at the moment can describe simple elements and express very limited spatial relations. However, the basic principle of the computational model presented in this paper is flexible enough to support complex spatial objects as well. First experiments have shown, that structural commonalities play an important role in object recognition (Stollinski et al. 2009). Future work will investigate HDTP in analogy making between complex stimuli like sketches of real world spatial objects.

Fig. 14 shows different sketches of an oven. Although they differ from each other, they share a lot of structural commonalities: all of them have four hotplates which are inside a polygon representing the top surface of the oven. Five temperature regulators and a spy window (with or without handle to open the door) are inside a polygon representing the front surface of the oven. Similarly to geometric figures, each sketch is represented by its primitive elements (lines and ovals). Background knowledge contains laws how to analyze geometric forms, detect polygons from lines or compute even more complex structures such as a cube.

An effective model for spatial cognition requires a spatial reasoner to compute spatial relations or different 3D perspectives on the same object in space. Already our experiments with simple geometric figures revealed such requirements: Several participants applied three dimensional transformations between figure A and B, which cannot be represented in the two-dimensional model of HDTP. Future work will investigate how existing models for spatial reasoning can be integrated. Also the representation language as well as the re-representation rules must be extended.

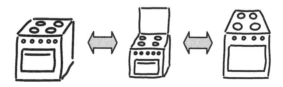

Fig. 14. Analogy-based sketch recognition compares different sketches of spatial objects and detects common structures

5.4   Sketch Map Comparison as Future Application

A second application area for analogical reasoning is the comparison of sketch maps to metric maps. While metric maps such as street maps are constructed from exact measurements, sketch maps are drawn by humans based on their cognitive map. Fig. 15 shows a sketch map and a metric map of the same area. A qualitative comparison of both maps reveals many structural commonalities: the spatial objects lie along streets forming the same street network. The sketch map is a simplified and schematized representation of the metric map.

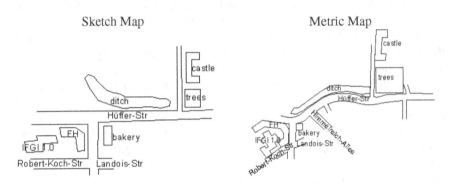

Fig. 15. Analogy-based comparison of a sketch map and a metric map of the same area reveals structural commonalities between spatial objects such as houses, streets, water-bodies and trees

Analogical comparison focuses only on structural commonalities such as the relation of geographic features to streets and streets being connected to other streets. It abstracts from metric details. Therefore, we argue that analogical comparisons are a useful tool for sketch map comparisons.

References

Dastani, M.: Languages of Perception. Institute of Logic, Language, and Computation (ILLC), Amsterdam, Universiteit van Amsterdam (1998)

Dastani, M.M., Indurkhya, B.: Modeling context effect in perceptual domains. In: Akman, V., Bouquet, P., Thomason, R.H., Young, R.A. (eds.) CONTEXT 2001. LNCS, vol. 2116, p. 129. Springer, Heidelberg (2001)

Dastani, M., Scha, R.: Languages for Gestalts of line patterns. Journal of Mathematical Psychology 47, 429–449 (2003)

Davies, C., Goel, A.K., Yaner, P.W.: Proteus: Visual analogy in problem solving. In: Knowledge-Based Systems (2008)

Evans, T.G.: A heuristic program to solve geometric analogy problems, Cambridge, MA, USA, Massachusetts Institute of Technology (1962)

Evans, T.G.: A program for the solution of a class of geometric-analogy intelligence-test questions. In: Minsky, M. (ed.) Semantic information processing, pp. 271–353. MIT Press, Cambridge (1969)

Forbus, K.D., Lockwood, K., Klenk, M., Tomai, E., Usher, J.: Open-domain sketch understanding: The nuSketch approach. In: AAAI Fall Symposium on Making Pen-based Interaction Intelligent and Natural, Washington, DC (2004)

French, R.M.: The computational model of analogy-making. Trends in Cognitive Sciences 6(5), 200–205 (2002)

Gust, H., Kühnberger, K.-U., Schmid, U.: Metaphors and heuristic-driven theory projection (HDTP). Theoretical Computer Science 354(1), 98–117 (2006)

Hofstadter, D.R., Mitchell, J.C.: The copycat project: A model of mental fluidity and analogy-making. In: Hofstadter, D.R., group, F.A.R. (eds.) Fluid Concepts and Creative Analogies, pp. 205–267. Basic Books, New York (1995)

Indurkhya, B.: Modes of analogy. In: Jantke, K.P. (ed.) AII 1989. LNCS, vol. 397. Springer, Heidelberg (1989)

Indurkhya, B.: Metaphor and cognition. Kluwer, Dordrecht (1992)

Koffka, K.: Principles of Gestalt Psychology. Harcourt, New York (1935)

Köhler, W.: Gestalt Psychology. Liveright, New York (1929)

Krumnack, U., Schwering, A., Gust, H., Kühnberger, K.-U.: Restricted higher-order anti-unification for analogy making. In: Orgun, M.A., Thornton, J. (eds.) AI 2007. LNCS, vol. 4830, pp. 273–282. Springer, Heidelberg (2007)

Mullally, E.-C., O'Donoghue, D., Bohan, A.J., Keane, M.T.: Geometric proportional analogies in topographic maps: Theory and application. In: 25th SGAI International Conference on Innovative Techniques and Applications of Artificial Intelligence, Cambridge, UK (2005)

Mullally, E.-C., Donoghue, D.P.: Spatial inference with geometric proportional analogies. Artificial Intelligence Review 26(1-2), 129–140 (2006)

O'Hara, S.: A model of the "redescription" process in the context of geometric proportional analogy problems. In: Jantke, K.P. (ed.) AII 1992. LNCS, vol. 642. Springer, Heidelberg (1992)

O'Hara, S., Indurkhya, B.: Incorporating (re)-interpretation in case-based reasoning. In: Wess, S., Richter, M., Althoff, K.-D. (eds.) EWCBR 1993. LNCS, vol. 837. Springer, Heidelberg (1994)

Plotkin, G.D.: A note on inductive generalization. Machine Intelligence 5, 153–163 (1970)

Schwering, A., Krumnack, U., Kühnberger, K.-U., Gust, H.: Investigating Experimentally Problem Solving Strategies in Geometric Proportional Analogies, Osnabrueck, Germany, University of Osnabrueck, p. 233 (2008)

Schwering, A., Bauer, C., Dorceva, I., Gust, H., Krumnack, U., Kühnberger, K.-U.: The Impact of Gestalt Principles on Solving Geometric Analogies. In: 2nd International Analogy Conference (ANALOGY 2009). New Bulgarian University Press, Sofia (2009a)

Schwering, A., Gust, H., Kühnberger, K.-U., Krumnack, U.: Solving geometric proportional analogies with the analogy model HDTP. In: Annual Meeting of the Cognitive Science Society (CogSci 2009), Amsterdam, The Netherlands (2009b)

Schwering, A., Krumnack, U., Kühnberger, K.-U., Gust, H.: Syntactic principles of Heuristic-Driven Theory Projection. Special Issue on Analogies - Integrating Cognitive Abilities. Journal of Cognitive Systems Research 10(3), 251–269 (2009c)

Stollinski, R., Schwering, A., Kühnberger, K.-U., Krumnack, U.: Structural Alignment of Visual Stimuli Influences Human Object Categorization. In: 2nd International Analogy Conference (ANALOGY 2009). New Bulgarian University Press, Sofia (2009)

Wertheimer, M.: Experimentelle Studien über das Sehen von Bewegung. Zeitschrift für Psychologie 61, 161–265 (1912)

Are Places Concepts? Familarity and Expertise Effects in Neighborhood Cognition

Clare Davies

Research, C530, Ordnance Survey, Romsey Road, Southampton SO16 4GU, U.K.
clare.davies@ordnancesurvey.co.uk

Abstract. Named urban neighborhoods (localities) are often examples of vague place extents. These are compared with current knowledge of vagueness in concepts and categories within semantic memory, implying graded membership and typicality. If places are mentally constructed and used like concepts, this might account for their cognitive variability, and help us choose suitable geospatial (GIS) data models. An initial within-subjects study with expert geographic surveyors tested specific predictions about the role of central tendency, ideals, context specificity, familiarity and expertise in location judgements – theoretically equivalent to categorization. Implications for spatial data models and a further research agenda are suggested.

Keywords: place, neighborhood, vague extents, concepts, expertise, urban spatial cognition.

●

1   Background: Vagueness and Place

Vagueness is a curse and a blessing for spatial information research: a curse pragmatically since GIS (geographic information systems) were not designed to represent it, and a blessing academically since this challenge has inspired many innovative ways to overcome it. As a result, vague extents for adjoining spatial areas can now be handled with a range of modelling techniques [1;2]. However, some types of vagueness seem far more complex and unpredictable than others. The gradient shift between mountain and plain [3] or the thinning of trees at the edge of a forest [4] are measurable, stable gradations. Humans' apparently slippery and changeable concepts of places and their extents may not be [5; 6; 7; 8].

Nevertheless, there is great potential for better data models of (vernacularly recognised) place to enhance geographic information use in many areas of government, infrastructure, research, commerce and health [2]. Thus Mark et al [5], in outlining key themes for geographic spatial cognition research, added almost as an afterthought [p.764] "the issue of place - what are the cognitive models of place and neighborhood, and can these be implemented in computational environments? What would a place-based, rather than coordinate-based GIS look like, and what could it do, and not do?"

K. Stewart Hornsby et al. (Eds.): COSIT 2009, LNCS 5756, pp. 36–50, 2009.

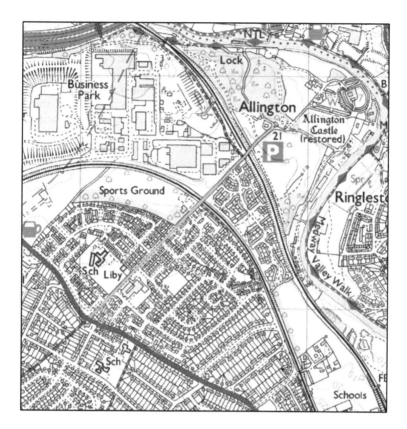

Fig. 1. Example of the problem of defining named localities. To locals 'Allington' means some or all of the central area of this map, ultimately bounded by the river and the old London Road (running from left to centre-bottom). It has no central core or consistent urban style. Its shops and pub are close to London Road, as is the school that is the only key feature of Palace Wood, the named locality to the south. The remains of Allington village, and its castle, are separated from most of the suburb by two rail lines, a trading estate and open space.(Ordnance Survey 1:25000 mapping, © Crown copyright 2009).

Ten years on, the vague extent of many types of place still remains a confusing topic, viewed as marginal by most geographers and environmental psychologists [9].Ideally, empirical work would lead us toward a testable theory of place (extent) cognition, which could help us to evaluate the suitability of computational methods and models for representing place more meaningfully within GIS. However, researchers in the above areas have often eschewed this question, in favour of a stronger focus on residents' affective attachments and social concerns within particular cities or regions, although exceptions do exist [10; 7].

Thus it is still unclear which modelling techniques might best reflect how people mentally store and process knowledge of vaguely defined places in everyday life. To find this out, we first need more basic research into the cognitive phenomenon of vague place extents. This paper will report one initial experiment from an ongoing research programme, designed to examine the question: how do people decide whether a given urban location or feature (e.g. a building or street corner) falls within a named area of a city – loosely termed a 'locality' or 'neighborhood'[1]?

Urban neighborhoods are a well-known example of vague places, both in definition and extent. Therefore, in many cities some locations must lie in indeterminate areas somewhere between two adjacent places. The location may be viewed as falling within either or neither of them, a choice referred to in this paper as a 'location judgment'. A long research tradition has established that the perceived boundaries of urban neighbourhoods can vary greatly between individuals [11] and with increased familiarity [7], and may depend on social (and socially communicated) factors just as much as physical environmental differences [8]. Predicting and modelling neighborhood boundaries has so far seemed an intractable challenge, as individual and contextual differences can seem overwhelming (Figure 1).

This paper will first briefly review some of the literature concerning vagueness of neighbourhoods, then compare it directly to our knowledge of vagueness issues in the more cognitive science domain of concepts and categories (semantic memory). It will then describe a preliminary study to test whether some of the implications of that comparison do in fact apply to people's location judgments. Specifically, this initial experiment used a within-subjects questionnaire method to examine two aspects of potential *intra*-individual differences in location judgments: whether and how place familiarity and professional geographic expertise might cause different choices to be applied by the same person in different contexts. Finally, as well as outlining further ongoing work, I will discuss the implications of the places-as-concepts approach for choosing among existing models of geographic place data.

2   Causal Factors in Perceived Neighborhood Extents

Since the 1960s there has been a regular stream of papers concerning people's conceptions of their own neighborhood, but less on how they view others within a familiar city (with exceptions, e.g. [7]). Within that 'own neighborhood' research area, often the very concept of neighborhood seems to be in danger of confounding named conventional or even administrative districts with people's personal (and usually but not always unnamed) sense of 'home range', so much so that researchers sometimes bend over backwards to avoid using the word 'neighborhood' at all in their instructions to participants, or add 'home range' as well in case it differs (e.g. [12; 10; 13]).

Nevertheless, we could make a tentative assumption that the same criteria people use to define their own neighborhood (particularly where it does coincide with a named locality) may also be used, at least in part, to help them define the other neighborhoods in the same city, since we have long known that cognition of those is not purely spatial either [14]. The relative importance of different criteria seems to differ greatly between studies – often due to limitations in the research method used. They may also be changing over time: perhaps unsurprisingly, more recent studies in the US and UK seem to place less emphasis on use of local within-neighborhood amenities (e.g. shops) and less socialising within neighborhoods than older studies

---

[1] I will use these two words interchangeably in this paper, except when distinguishing named (generally larger) localities from 'neighborhoods' in the more personal sense of 'home range' [10].

from the 1960s and 1970s (see also [15]). Defining factors which have been suggested in the literature include, in no particular order (since relative effect sizes have not been established):

- physical infrastructure boundaries, e.g. major streets or rail lines – although the focus on these may be more likely among people with formal geographic expertise ([16; 13])
- use of local amenities e.g. shops and schools ([11; 17])
- amount and extents of people's walking (and sometimes cycling) from home, for errands or pleasure ([18; 10; 13])
- local social or political activity ([11; 8; 13; 19])
- social cohesion – "people like us" ([11; 17; 13])
- media stories naming specific localities and creating a stronger awareness of their identity ([8; 17])
- environmental aesthetics and similar housing styles ([8; 13])
- familiarity, especially if measured objectively e.g. through landmark identification ([18; 7])

Obviously, not all of these will be used to determine entire named localities as in Figure 1; some will apply only to a smaller and more personal area (Stanton's "home range" concept [10]). Such areas – particularly one's own – are not always viewed by locals as a vaguely-bounded area at all, but sometimes as an experiential network with defined ends for each separate branch ([10; 18]). This can also be true of larger named localities, and seems to remain true even for some long-term residents, contradicting the "route-to-survey knowledge" school of thought on cognitive mapping, but not more recent studies of it [20]. Meanwhile one study of newcomers learning a city from scratch found that their knowledge of its localities from one test to the next was strangely uncorrelated, rather than showing a smooth learning curve [7], perhaps suggesting a dynamic rather than static assessment of those localities' identities and extents.

In fact, many authors have expressed the concern that there may not be a stable, strong, consistent or even *any* concept of neighborhoods with definable (even vague) extents, in many people's minds. Schnell et al [17] found that for many residents of Tel Aviv-Jaffa, there was no sense of specific locality at all in their part of the city. Beguin & Romero [7] bemoaned the unknown "black box" of people's place cognition which "forces us" to assume that "there is a unique cognition of an urban item (either a neighborhood or any element of it) in an individual's mind at a given time" [p.688]. Yet Bardo [12], discussing people's own home neighbourhood concept, recognised that "a particular individual may define more than one neighborhood depending upon the frame of reference he or she is given" [p.348]. Martin [8] went further to claim [p.362] that "we do not know neighborhoods when we see them; we construct them, for purposes of our research or social lives, based on common ideals of what we expect an urban neighborhood to be. The neighborhoods that we define through research or social exchange are always subject to redefinition and contention; they are not self-evident."

3 Neighborhoods as Concepts?

So cognitively speaking, neighborhoods – and perhaps other types of vernacular place such as larger regions – are potentially vague (sometimes), and may be context-specific constructions rather than stable spatial entities. This may also be true of another well studied cognitive phenomenon: concepts. Concepts used to be viewed as stable yet often vague stored mental entities, until Barsalou [21] showed that providing different contexts could change the way that people categorized the same instances of people or objects. Although his data did not actually demonstrate within-subject variability, Barsalou claimed in his conclusion that human conceptual ability was "extremely dynamic" [p.648] in that "people may not retrieve the same concept from long-term memory every time they deal with a particular category" [p.646]. To Barsalou, concepts are constructed in working memory, not merely retrieved from a stable store. This has the important implication that the same instance may be classified differently – as one category or another – depending on context. Yet participants will also happily, when asked, pile cards depicting the same objects into sharply delineated piles, and the literature rarely reports any reluctance to do so.

Barsalou also made another distinction which may prove valuable if we were to decide to treat neighborhoods and other places as concepts. Until then the notion of *central tendency* had dominated theorists' thinking on (largely taxonomic types of) concepts for around a decade – the idea that in any category of instances (even if formally defined), some are seen as better examples of it than others. The most popular theory [22] described this in terms of resemblance to a central core or prototype. Barsalou showed that many categories are instead goal-derived – existing to fulfil a function or need – and that people's ratings of goodness-of-example in such categories tend towards an extreme *'ideal'* that would be the ultimate fulfilment of that need, rather than a central 'typical' average. Additionally, he argued for personal *familiarity* and also *frequency of instantiation* (how often you come across the instance classified within the category) as potential extra influences on goodness-of-example judgements.

Work since Barsalou's paper has shown that, if anything, we seem to use ideals for more types of categories than he predicted – sometimes even for natural-kind categories that are obviously taxonomic such as trees or birds [23]. Furthermore, experts seem to rely on ideals much more than the novice student participants of most lab studies. Yet different types of expert (e.g. professional landscapers versus taxonomists) will apply different ideals, and at the highest levels of expertise personal familiarity may have also a greater impact than central tendency [24]. However, we may assume that expertise is applied primarily within the professional context in which it is usually used, and that experts are able to think like laypeople when dealing with more everyday contexts. Thus we might expect to see within- as well as between-subjects differences in the effects of expertise, where contexts are changed.

More recently, Hampton [25] hazarded three potential ways in which categorization of an instance (e.g. an object) might vary with context within the same individual: changes occurring in either (a) the representation of the instance, (b) the representation of the category, or (c) the threshold of similarity required to categorize the instance into it. Hence the same person could classify the same entity differently due to circumstance – sometimes into one category, and sometimes into another – creating the effect of vagueness. Hampton also argued that if people consider vagueness acceptable, the same person may be aware that they could classify a given instance as both "X and not X" – a result which he claimed is difficult to represent in the fuzzy logic often used to model vaguely bounded entities.

Hampton also asked "How do we live with vagueness?" [p.378] His suggested answer to this was that social constraints, and the need to communicate effectively with others, mean that we are bound to build and use consensus, rather than having no sense of boundaries at all between related categories.

3.1 Location Judgments as Category Membership

Why apply concept theories to place? After all, place is inherently spatial and hence, even if we were to imagine a place as a category and a single location as an instance of it, we would usually assume that central tendency in the form of distance from a core would be a more realistic model of fuzzy location judgements than Barsalou's goal-derived categories and their indefinable 'ideals'. Yet the brief literature review above seems to suggest otherwise. Furthermore, a recent review and study of cognitive mapping of urban environments [20] has suggested that while people may store their route knowledge through a city as a network of individual locations or 'vista spaces' (as defined by [26]), those locations must also still be categorised into places or regions if we are to explain the well-known hierarchical biases in spatial reasoning (e.g. [14]). This would help to explain how studies of cognitive mapping seem to emphasise a mental route network model, even though people can readily draw polygons on a map to show separate districts or neighborhoods in the same city.

If this is reasonable, then even the few papers cited above on concepts (out of a literature mountain) have clear implications for those neighborhoods, and for location judgements as a form of categorization task:

1. Under different circumstances the same people may produce different representations of the same neighborhood, even though most will also happily draw crisp boundaries around it on a map when requested.
2. Neighborhoods may prove to be more of a goal-derived category for many residents in many situations, implying that locations seen as 'good examples' might tend towards a goal-fulfilling ideal (e.g. historic buildings) rather than a core central prototype (spatial or otherwise typical). However, again this may vary with context.
3. Expertise – particularly formal knowledge – may increase and change the use of ideals in defining a place.
4. As people communicate more about a place, social consensus will create increased similarity between and within people's judgements of it.
5. People may be willing to accept that a location can be seen simultaneously as in and not in a place; if this is true then any computational model of the resulting vagueness needs to accommodate this.

Note that points 3 and 4 could contradict each other. Greater familiarity may cause a more personal and goal-derived 'expert' understanding of a place, or it may create greater consensus with others which might suggest increasing reliance on central tendency. This obviously requires more explicit empirical testing than it has had so far in the neighborhood literature, where some authors have noted greater consistency among longer-term residents ([7], while others have found them to have very personal familiarity-based and non-consensus definitions [10].

4 Expertise, Familiarity and Location Judgements

To begin to test the above predictions, and thus start to evaluate the plausibility of places as concepts, the present study focused on the criteria people use for making location judgements under different (hypothetical) circumstances, particularly varying the roles of expertise and familiarity. Focusing directly on the criteria used for these choices may allow us to start teasing out the cognitive factors underpinning the vague vernacular geography of neighborhoods.

Expert-novice studies in cognition are often problematic, due to the many non-expertise-related differences that tend to exist between two different groups of people: e.g. often the expert group is older and has more general experience of both general and professional life, as well as the specific task domain under study. Therefore the present study adopted a within-subjects approach – comparing a group of geographic place experts with themselves under different circumstances.

4.1 Participants

22 professional field surveyors working for Ordnance Survey, the national mapping agency of Great Britain, took part in the experiment. These surveyors (19 male, 1 female and 2 undisclosed, median age group 45-54 years) work mostly from home and are spread geographically across the country. Their main roles are to survey changes in their local area mainly at a highly detailed level, intended for 1:1250-scale urban and 1:2500-scale rural mapping (effectively to an accuracy of <1 metre on the ground), and to gather information for generalised smaller-scale products such as maps and gazetteers.

Although formally authorising and recording place names and extents is no longer a part of this role, all but one of the surveyors were experienced enough (mean=28.5 years, s.d.=9, min=8, max=38) to remember having had to do it formally, and/or to have more recently done it informally as part of their general updates of mapping data. Most of them also used the questionnaire to express their views and ideas on the organisation's ongoing issues with, and potential means of collecting, place information.

4.2 Method

The experiment was administered via a paper questionnaire, which participants completed in their own time and mailed back to the researcher. Between initial briefing and instructions, and final debriefing with space for comments, the questionnaire consisted of three independent sections whose order was counterbalanced between participants. Each section consisted of two sets of Likert rating-scale items, preceded by a page of context-setting questions. Participants were instructed not to look back to previous sections while completing each one.

The first ('Work') page of one of the three sections focused on the surveyor's work context, asking questions about their role, length of experience, current and previous geographical regions, and experience and views concerning collecting place names and extents. The participant was then asked to name a town where they had done some surveying, but which they did not know very well, and asked whether they had ever had to collect placename information there. On turning over the page, they were asked to imagine that they were trying to decide whether a specific location in the town fell within a particular named neighborhood. 13 potential factors were then listed, with Likert rating scales[2] from 0-5; participants circled a number for how much each factor would help them decide that the location fell within the neighborhood. (Participants could also write in and rate their own factors, although few did so[3].) The next page asked them which of the factors (now in a different randomised order) would help them to decide that another location was *not* in the neighborhood.

Another ('Home') section related to the surveyor's home, with the first page asking them to name their home town, say how long they had lived there, and name a neighborhood within it that they knew very well either from living or frequently visiting there (and if the latter, how often and why they did so). They were then asked to think of and name two locations near the edge of that neighborhood (but not necessarily near each other). For each location, the participant indicated his confidence (from 0-10) that the location was within the named neighborhood, and also indicated the percentage of local people whom he thought would agree that it was. The next two pages asked him to take one of these locations and imagine he was arguing for its being within the named neighborhood, rating the same 13 (differently randomised) factors as before for helpfulness, then doing the same while imagining themselves arguing that the other location was *not* in the neighborhood.

The first page of the remaining section ('New Area') asked the participant to imagine moving home to a new area, and to consider the problem of trying to choose a new home while getting to know the town. They were asked to say whether they had ever moved home, and how long ago this was. The next two pages asked him to imagine trying to decide that a given house was within a certain desirable neighborhood, and then trying to prove that an estate agent was stretching the truth by such a claim, once again using the same 13 (again re-randomised) factors.

Thus each section of the questionnaire took the same basic form – setting a context and then asking participants to rate the helpfulness of 13 potential factors in arguing for, and then against, a location classification within a real or imaginary neighborhood.

4.3 Criteria and Hypotheses

The 13 factors were described in full sentences, but are presented in shortened form below for ease of reference:

---

[2] Piloting showed that a longer scale would not be meaningful, and would be too fatiguing for this many questions.

[3] Only 9/22 participants added (at the most) one or two extra criteria per scenario – too few and too disparate to analyse. They included (with number of participants): local government officials' views (2); local residents' views (2); neighborhood names appearing on streetname signs, as happens in some UK towns particularly post-war 'new towns' (2); the location having visible dominance or centrality within the neighborhood (3); thematic groupings of related street names (1); being on a different bus route from the rest of the neighborhood (1).

1. Right/wrong side of a physical barrier e.g. road or embankment
2. Same/different administrative, electoral or school district
3. Neighborhood name does (not) appear in Royal Mail official address
4. Named this way (or not) by real estate agents or property developers
5. Always/never referred to this way by local people
6. Mentioned (or not) using this name in the local media, tourist or other local information
7. Name (dis)similar to that of the neighborhood
8. Close to/far from a key feature e.g. shops, church, park, main street
9. Close to/far from the placement of the name on maps
10. (Dis)similar visual appearance to the rest of the neighborhood
11. People in this neighborhood like (or don't like) to be associated with it
12. Same/different function/use as the rest of the neighborhood
13. Age or other sense of belonging to/differing from it

These factors partly reflect the above literature, and partly known (and somewhat British-specific) issues such as the Royal Mail's use of some locality names in their 'official' address designations, and people's frequently stated reliance on Ordnance Survey (OS) maps to define an assumed 'correct' geography.

It will be noted that factors 1-3 imply crisp and definitive 'in/out' factors, and thus preclude any notion of graded membership as found with most concepts. As suggested earlier, we might expect formal geography experts to prefer these factors, particularly when imagining acting within their professional context. Factors 4-7 are dependent on potentially variable use of the neighborhood name by other people in non-definitive contexts, and thus evoke Barsalou's frequency of instantiation. Items 8 and 9 concern central tendency: potential distance from an imagined spatial core or prototype. Items 10-13 involve non-spatial factors that could evoke either central tendency or, perhaps more probably, an 'ideal'. For example, a beautiful historic house may be seen as the ultimate ideal of an older neighborhood, while not at all typical of it.

Based on Barsalou's ideas, we could tentatively hypothesise significant differences between the imagined scenarios in their mean score across criteria, with the more formal work context treating the fewest criteria as strongly relevant, and with these being more ideals-related (for ideals relevant to the experts' job). Nevertheless we might also expect some criteria to be preferred to others by this participant group across scenarios (since, for instance, the experts in Lynch et al's study still applied their expertise to an out-of-context experimental task, with perhaps some differential treatment of certain criteria between different scenarios (an interaction effect).

If Hampton was correct that people are able to appreciate vagueness and "X and not X" situations when either the instance, category or threshold of acceptance changes in some way, then within each scenario (but not necessarily across them) we might expect to also see differences in criteria relevance between the 'in' and the 'not in' decisions (i.e. an interaction of scenario and decision). Finally, we could see an interaction of all three factors together (criterion, scenario and decision), if each individual set of ratings was treated as a separate situation by the participants despite their formal geographic background, which encourages and rewards more consistent rule application. This would be the strongest test of both Barsalou's and Hampton's theories' applicability to location judgements.

4.4 Results

A within-subjects (repeated-measures) ANOVA[4] was run on the mean rating scores for Scenario x Criterion x Decision (3x13x2). Strongly significant main effects were found for Scenario (Greenhouse-Geisser-corrected $F_{1.39,1113.33}=8.84$, p=0.0009) and for Criterion ($F_{4.26,567.29}=14.08$, p<0.0001), but only weakly for Decision ($F_{1,1598}=3.60$, p=0.06).

Post hoc tests by Scenario showed the nature of the main effect: the mean score for the New Area scenario was significantly lower across criteria and decision than either Work ($t_{1126}=-3.71$, p=0.0002) or Home ($t_{1123}=-2.91$, p=0.004). This suggests that the participants found more of the location judgement criteria to be more relevant (on average) in the unfamiliar New Area scenario than in the course of their familiar work or home lives.

The weak main effect for Decision suggested that overall, participants gave a slightly higher mean relevance rating across criteria when they were considering *including* a location in a neighbourhood, than when considering *excluding* it. However, there was a very strong interaction effect between Scenario and Decision ($F_{1.30,1041.26}=10.69$, p=0.0003), which is illustrated in Figure 2 (left), and a weaker interaction between Criterion and Decision ($F_{5.93,790.21}=2.05$, p=0.06) also shown in Figure 2 (right). The interaction between Scenario and Criterion was weaker still (p=0.2), with no three-way interaction (p=0.4).

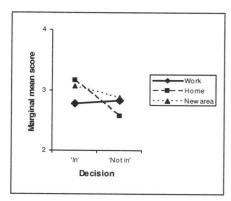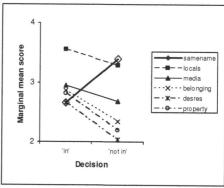

Fig. 2. Left: interaction between scenario and decision, showing the most consistency in the Work scenario, and the least in the Home one. Right: interaction between criterion and decision (showing only criteria differing by >0.3). Note that the relevance of comparing the location's and neighborhood's names was deemed much stronger when judging exclusion than inclusion, whereas other strongly-varying criteria changed in the opposite direction.

---

[4] The ANOVA was also rerun including a main effect of questionnaire order to check for any effect of this, but none was found (p>0.8).

Post hoc tests did not reveal clear levels of separation among the criteria. The three highest-scorers were criteria 1 (physical barrier, mean=3.56), 3 (Royal Mail address, mean=3.48) and 5 (locals' naming, mean=3.40), followed by 8 (distance from key features, mean=3.20), 2 (named administrative area, mean=3.20) and 9 (nearness to the name on a map, mean=3.17). The four lowest scorers were 12 (shared function, mean=2.28), 11 (desirability of association, mean=2.28), 10 (similarity of appearance, mean=2.30) and 4 (property industry naming, mean=2.47). It will be noted that these four means falling below 2.5 implies that these four factors were less than averagely popular with the surveyors across the six contexts. Also note that the top two criteria imply crisp, definitive geography of a neighbourhood as discussed earlier, whereas the lowest three are the most likely to invoke the notion of 'ideals'.

The ANOVA was also replicated while grouping the criteria by those types, i.e. reflecting either crisp definitions, frequency of instantiation, central tendency or ideals. This found the same pattern of results for the other main and interaction effects, and a very strong main effect for criteria type ($F_{3,1652}=42.76$ $p<0.0001$), but no significant interactions of it with the other factors. The raw means and confidence intervals for the four criterion types are shown in Figure 3. It confirms that ideals were used the least as criteria, followed by frequency of instantiation, spatial central tendency and, most importantly, crisp definitive factors. Post hoc tests showed that the strongest locus of the effect was the distinction between ideals (the least used type) and the other three types (e.g. comparing ideals to frequency of instantiation, $t_{20}=-2.91$, $p=0.009$).

Finally, the participants' own confidence ratings out of 10 that the locations specified in the Home section of the questionnaire did belong within the neighborhood (mean=9.77, sd=9.59), and their estimate of the percentage of the population who

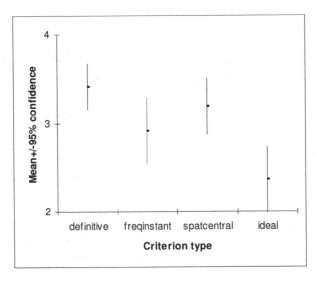

Fig. 3. Means and 95% confidence intervals for the four criterion types, across scenarios and decisions

would agree with this (mean=77%, sd=26%), were compared and correlated. Hampton [25] predicted that as people's concepts are socially influenced, social consensus on category membership should correlate with individuals' strength of conviction about it. Both measures were strongly skewed towards 100%; Spearman's non-parametric correlation showed $r_{22}=0.56$, $p=0.006$. (However, since both measures were estimated by the same participant, confirmatory bias is likely to have been involved, so further tests with independent samples are needed.)

5 Discussion

Field surveyors are obliged to optimise precision, accuracy and certainty in their work. Crisp delineation of geographic features is often required for topographic map surveying, and even when Ordnance Survey surveyors formally collected placenames these were always verified as 'authoritatively' as possible. Surveyors were once instructed to consult local administrators and documents or, in their absence, such supposed authority figures as school heads, clergy or doctors [27], or in more recent egalitarian times an opportunity sample of local residents. These days certain OS data products include Royal Mail address definitions, which may or may not include a named locality as delineated by that organisation (although their definitions are often ignored by or unknown to many locals). Surveyors' professional objectivity also requires downplaying any aesthetic or subjective issues in recording information. Overall, for this population, these factors would predict exactly the pattern of biases that were found in the data. Thus, contradicting studies such as that by Lynch et al [23], in this domain experts may be *less* likely to rely on ideals.

Nevertheless, if neighborhoods can be treated as concepts to some extent, these participants demonstrated the flexibility of concept definitions predicted by both Barsalou [21] and more recently by Hampton [25], differentiating both among scenarios (at least in strength of criteria ratings, although not clearly in the actual choice of criteria) and between decisions of inclusion versus exclusion. This seems to imply that, as Hampton suggested, the 'threshold' for category membership can shift between scenarios when applied to locations as potential members of places – a quantitative but perhaps not qualitative effect of context. However, since people tend towards consistency when answering similar questions repeated within the same questionnaire, this response bias may have masked a stronger distinction between scenarios, by encouraging repeated selection of the same criteria. Further research is needed to overcome this, either with a between-subjects design or by gathering responses on separate occasions or in different forms.

It should be noted that when treated as a 13-level factor, differences among criteria interacted with scenario, but this was not replicated when the criteria were collapsed into the four types discussed above. Thus it would seem that this classification did not account for all of the variance among individual criteria. Indeed, the third highest-scoring criterion was the (potentially vague) neighborhood extent implied by locals' use of its name – more reminiscent of Barsalou's frequency of instantiation than of crisp delineation – while the fourth was distance from key features, implying Rosch's older notion of central tendency [22]. Yet one of the four below-chance-scoring criteria, naming by property industry agents such as in real estate, was arguably also an example of frequency of instantiation. The difference, may lie in the level of authenticity and hence authority that may be assigned to these two local sources.

This study obviously raises further questions, which ongoing research is currently attempting to answer. Do non-professional laypeople (i.e. formal 'novices', perhaps using a between-subjects experiment design) differ from the present sample in the criteria they apply to location judgements? If so, does this difference tend further toward the use of ideals, or is it purely based on central tendency (either in terms of distance from core features or typicality of appearance/function?) Does it differ more or less than for experts between familiar and less familiar places?

Validation is also required: do people's *actual* location judgements (categorisations) draw on the same criteria that they would identify in a rated list, and vary in a similar way between contexts as the present study implies? A more controlled (but less ecologically valid) experiment could also artificially induce familiarity and expertise to some degree, allowing all participants to consider the same locations and places. This would remove any chance of any unknown potential confounding factors, given that participants in the present study were all thinking about different places while rating the criteria relevance in the 'Home' and (to a lesser extent) the 'Work' scenarios.

Even for this professional group apparently varies in location judgement criteria between deciding inclusion and exclusion. This seems to indicate potential problems for some proposed methods of modelling vague place extents. Just as people are asymmetric in their distance judgements between two points [28], so their criteria for (and hence potential values of) place extents may not be consistent even for the same place when considered under different circumstances. According to Hampton fuzzy logic-based models cannot easily accommodate such "X and not X" outcomes. This could shed doubt on such models' applicability to place in the GIS context [29].

At the same time, however, the recognition even by this sample of the role of factors such as frequency of instantiation means that the eggyolk model of place (e.g. [30; 6]) may oversimplify variations among locations by trying to include all "maybe" or "disputed" locations in a single 'eggwhite' polygon, and all "definitely agreed" locations in a neat 'eggyolk' one. The messy familiarity surfaces noted by authors such as Aitken and Prosser [18] imply that even when aggregated across local residents, the 'yolk' and 'white' may not form neat contiguous surfaces. Obviously, this hypothesis requires further testing. Hampton [25] talked of conceptual vagueness as a problem of 'psychological supervaluation', which suggests that the supervaluations we may require for place might have to cope with intrapersonal as well as cross-population differences in implied extents.

An alternative way forward might be to evaluate the computational models which have been proposed within cognitive science for modelling category formation and use (e.g. [31; 32]). However, it is not clear how the inputs and outputs of such models could help us to handle place in the GIS context, nor whether they can be scaled up practically to cope with the vastness of topographic data. Ideally within that data, or linked to it, place membership information would be available or calculable for every location and/or object in the urban and rural landscape. This is not trivial. For example, the OS MasterMap® topographic database of Great Britain contains over 450 million features[5]. There is a real danger that any computational models which have not been tested beyond a few dozen categories with a handful of categories and exemplars (dog is an animal, flower and tree are plants, etc.) might not be quite up to the job of a national dataset reflecting everyday place cognition.

Acknowledgments and Disclaimer. Many thanks to the field surveyors for generously giving their time and very conscientious attention to this study; Nicky Green for help with data coding; Dave Capstick, Jenny Harding and Jenny Green for help with piloting; Robin Kent and various Ordnance Survey colleagues for helping to distribute the questionnaire.

This article has been prepared for information purposes only. It is not designed to constitute definitive advice on the topics covered and any reliance placed on the contents of this article is at the sole risk of the reader.

References

1. Jones, C.B., Purves, R.S., Clough, P.D., Joho, H.: Modelling vague places with knowledge from the Web. International Journal Of Geographical Information Science 22, 1045–1065 (2008)
2. Davies, C., Holt, I., Green, J., Harding, J., Diamond, L.: User needs and the implications for modelling place. In: International Workshop on Computational Models of Place (PLACE 2008), Park City, Utah, USA, September 23 (2008),
   http://www.ordnancesurvey.co.uk/oswebsite/
   partnerships/research/publications/docs/2008/
   PLACE08Davies_etalfinal_geo.pdf
3. Smith, B., Mark, D.M.: Do mountains exist? Towards an ontology of landforms. Environment And Planning B-Planning & Design 30, 411–427 (2003)
4. Bennett, B.: What is a forest? On the vagueness of certain geographic concepts. Topoi 20, 189–201 (2001)
5. Mark, D.M., Freksa, C., Hirtle, S.C., Lloyd, R., Tversky, B.: Cognitive models of geographical space. International Journal Of Geographical Information Science 13, 747–774 (1999)
6. Montello, D.R., Goodchild, M.F., Gottsegen, J., Fohl, P.: Where's Downtown?: Behavioral methods for determining referents of vague spatial queries. Spatial Cognition and Computation 3, 185–204 (2003)
7. Beguin, H., Leiva Romero, V.: Individual cognition of urban neighbourhoods over space and time: a case study. Environment and Planning A 28, 687–708 (1996)
8. Martin, D.G.: Enacting Neighborhood. Urban Geography 24, 361–385 (2003)
9. Cresswell, T.: Place: a Short Introduction. Blackwell, Oxford (2004)
10. Stanton, B.H.: The Incidence of Home Grounds and Experiential Networks: Some Implications. Environment and Behavior 18, 299–329 (1986)
11. Lee, T.: Urban neighborhood as a socio-spatial schema. Human Relations 21, 241–267 (1968)
12. Bardo, J.W.: A Reexamination of the Neighborhood as a Socio-Spatial Schema. Sociological Inquiry 54, 346–358 (1984)

---

[5] http://www.ordnancesurvey.co.uk/oswebsite/products/osmastermap/whatisosmm.html

13. Talen, E., Shah, S.: Neighborhood evaluation using GIS - An exploratory study. Environment and Behavior 39, 583–615 (2007)
14. Hirtle, S.C., Jonides, J.: Evidence of hierarchies in cognitive maps. Memory and Cognition 13, 208–217 (1985)
15. Talen, E.: Sense of community and neighbourhood form: An assessment of the social doctrine of new urbanism. Urban Studies 36, 1361–1379 (1999)
16. Pacione, M.: The temporal stability of perceived neighborhood areas in Glasgow. Professional Geographer 35, 66–73 (1983)
17. Schnell, I., Benjamini, Y., Pash, D.: Research note: Neighborhoods as territorial units: The case of Tel Aviv-Jaffa. Urban Geography 26, 84–95 (2005)
18. Aitken, S.C., Prosser, R.: Residents' spatial knowledge of neighborhood continuity and form. Geographical Analysis 22, 301–325 (1990)
19. Cope, M.: Patchwork neighborhood: children's urban geographies in Buffalo, New York. Environment and Planning A 40, 2845–2863 (2008)
20. Meilinger, T.: The network of reference frames theory: a synthesis of graphs and cognitive maps. In: Freksa, C., Newcombe, N.S., Gärdenfors, P., Wölfl, S. (eds.) Spatial Cognition VI. LNCS, vol. 5248, pp. 344–360. Springer, Heidelberg (2008)
21. Barsalou, L.W.: Ideals, central tendency, and frequency of instantiation as determinants of graded structure in categories. Journal of Experimental Psychology: Learning, Memory and Cognition 11, 629–654 (1985)
22. Rosch, E., Mervis, C.B.: Family resemblances: Studies in the internal structure of categories. Cognitive Psychology 7, 573–605 (1975)
23. Lynch, E.B., Coley, J.B., Medin, D.L.: Tall is typical: Central tendency, ideal dimensions, and graded category structure among tree experts and novices. Memory & Cognition 28, 41–50 (2000)
24. Johnson, K.E.: Impact of varying levels of expertise on decisions of category typicality. Memory & Cognition 29, 1036–1050 (2001)
25. Hampton, J.A.: Typicality, graded membership, and vagueness. Cognitive Science 31, 355–384 (2007)
26. Montello, D.: Scale and Multiple Psychologies of Space. In: Campari, I., Frank, A.U. (eds.) COSIT 1993. LNCS, vol. 716, pp. 312–321. Springer, Heidelberg (1993)
27. Harley, J.B.: Place-Names on the Early Ordnance Survey Maps of England and Wales. The Cartographic Journal 8, 91–104 (1971)
28. McDonald, T., Pellegrino, J.: Psychological Perspectives on Spatial Cognition. In: Gaerling, T., Golledge, R.G. (eds.) Behavior and Environment: Psychological and Geographical Approaches, pp. 47–82. Elsevier, Amsterdam (1993)
29. Fisher, P.: Boolean and Fuzzy Regions. In: Burrough, P.A., Frank, A.U. (eds.) Geographic Objects With Indeterminate Boundaries, pp. 87–94. Taylor & Francis, London (1996)
30. Cohn, A.G., Gotts, N.M.: The "Egg-Yolk" Representation of Regions With Indeterminate Boundaries. In: Burrough, P.A., Frank, A.U. (eds.) Geographic Objects With Indeterminate Boundaries, pp. 171–187. Taylor and Francis, London (1996)
31. Rogers, T.T., McClelland, J.L.: Semantic Cognition: a Parallel Distributed Processing Approach. MIT Press, Cambridge MA (2004)
32. Love, B.C., Medin, D.L., Gureckis, T.: SUSTAIN: A Network model of category learning. Psychological Review 111, 309–332 (2004)

A Metric Conceptual Space Algebra

Benjamin Adams and Martin Raubal

Departments of Computer Science and Geography
University of California, Santa Barbara
badams@cs.ucsb.edu, raubal@geog.ucsb.edu

Abstract. The modeling of concepts from a cognitive perspective is important for designing spatial information systems that interoperate with human users. Concept representations that are built using geometric and topological conceptual space structures are well suited for semantic similarity and concept combination operations. In addition, concepts that are more closely grounded in the physical world, such as many spatial concepts, have a natural fit with the geometric structure of conceptual spaces. Despite these apparent advantages, conceptual spaces are underutilized because existing formalizations of conceptual space theory have focused on individual aspects of the theory rather than the creation of a comprehensive algebra. In this paper we present a metric conceptual space algebra that is designed to facilitate the creation of conceptual space knowledge bases and inferencing systems. Conceptual regions are represented as convex polytopes and context is built in as a fundamental element. We demonstrate the applicability of the algebra to spatial information systems with a proof-of-concept application.

1 Introduction

In recent years there has been an increasing demand for research on the representation and modeling of cognitive phenomena for spatial information systems [21,22]. Semantic similarity measurement in particular has been an active area of research for spatial applications [17,29]. Since human users interface and interoperate with these systems, they must have a means for representing the conceptual structures that exist in the users' minds, especially those concepts that are related to spatial cognition. Although geometric modes of concept representation have not been as widely adopted as other representational frameworks for cognitive modeling, they have garnered interest from researchers in the spatial sciences, because many spatial concepts are intrinsically thought of in terms of their geometric and topological features.

Models of human cognitive processes require a formal representation that a computer system can interpret. The two prevailing frameworks for representing cognitive processes are the symbolic and connectionist methods [9]. The symbolic method aims at modeling high-level abstract concepts using symbol manipulation schemes. Inferences are often the result of first-order logical operations on the symbols in the model. The connectionist method attempts to model cognition in a way that more closely compares to the biological neural structure of the brain, mathematically represented as nodes and their weighted connections.

K. Stewart Hornsby et al. (Eds.): COSIT 2009, LNCS 5756, pp. 51–68, 2009.
© Springer-Verlag Berlin Heidelberg 2009

While useful for many cognitive computational tasks, both of these representational frameworks do not perform well at modeling certain aspects of cognition, specifically semantic similarity and concept combination. Semantic similarity measurement is fundamental to the task of cognitive categorization, because conceptual units are classified with other conceptual units with which they are most similar [28]. Using symbolic representations, the combination of concepts is most often measured as the set-theoretic intersection of the properties of two classes, but this method fails when combining concepts without shared properties [10]. For example, *stuffed gorilla* combines the concepts *stuffed* and *gorilla*. However, the *gorilla* concept has no intersecting properties with the *stuffed* concept because it is a living thing. In addition, ad hoc concepts, which by definition are concepts that combine from different domains, are similarly difficult to represent using symbolic representations [16]. In the case of connectionist representations, even small connectionist models can be highly complex and become unwieldy for representing semantics on the level needed for operating with concepts.

As a complement to the two representational frameworks listed above, Gärdenfors has introduced conceptual space theory [9]. Conceptual spaces are geometric and topological structures that represent concepts as convex regions in multi-dimensional domains. This theory constitutes a mid-level spatialization approach to concept representation and is particularly suited as a framework for spatial information systems. Conceptual spaces can model semantic similarity naturally as a function of distance within a geometric space, and conceptual regions are subject to geometric operations such as projections and transformations that result in new concept formations.

Critics of conceptual space theory have contended that its usefulness has only been demonstrated for simplistic cases with little abstraction and using formalizations that are designed for specific contexts [33]. It is our position that rather than being due to theoretical limitations, the difficulty in assessing the experimental worth of conceptual spaces has been in part that no conceptual space algebra exists with well-defined operations that allow one to build and reason with complex conceptual space structures. To help rectify that situation, in this paper we present a metric conceptual space algebra, consisting of formal definitions of its components and operations that can be applied to them. The work builds upon previous formalizations of conceptual spaces but aims to be more comprehensive both as a mathematical model and as a launching pad for computational implementation. Our key contributions are the formalizations of query operations for semantic similarity measurement and concept combination.

Section 2 introduces conceptual space theory and previous formalization approaches. In section 3, we define a conceptual space algebra with its components. Concepts are thereby represented as convex polytopes. In addition, contrast classes and context are formally defined. Section 4 presents the algebraic operations, i.e., core metric operations, and query operations for similarity and concept combination. Section 5 applies the conceptual space algebra to the problem of comparing countries and regions of the world with different contexts. The final section presents conclusions and directions for future research.

2 Related Work

The theory of conceptual spaces was introduced as a framework for representing information at the conceptual level [9]. Conceptual spaces are based on the paradigm of cognitive semantics, which emphasizes that meanings are mental entities, i.e., mappings from symbols to conceptual structures, which refer to the real world [20]. They can be utilized for knowledge representation and sharing, and account for the fact that concepts are dynamic and change over time [4,26].

A conceptual space is a set of quality dimensions with a geometric or topological structure for one or more domains. Domains are represented by sets of integral dimensions, which are distinguishable from all other dimensions, e.g., the color domain. Concepts cover multiple domains and are modeled as n-dimensional regions. Every instance of the corresponding category is represented as a point in the conceptual space. This allows for expressing the similarity between two instances as a function of the spatial distance between their points. Recent work has investigated the representation of actions and functional properties in conceptual spaces [12].

Vector algebra offers a natural framework for representing conceptual spaces. A conceptual vector space can be formally defined as $C^n = \{(c_1, c_2, \ldots, c_n) | c_i \in C\}$ where the c_i are the quality dimensions [25]. Vector spaces have a metric and therefore allow for the calculation of distances between points in the space. This can also be utilized for measuring distances between concepts, either based on their approximation by prototypical points or regions [30]. Calculating semantic distances between instances of concepts requires that all quality dimensions of the space must be represented in the same relative unit of measurement. Given a normal distribution, this can be achieved by applying the z-transformation for these values [7]. Different contexts can be represented by assigning weights to the quality dimensions of a conceptual vector space. C^n is then defined as $\{(w_1 c_1, w_2 c_2, \ldots, w_n c_n) | c_i \in C, w_j \in W\}$ where W is the set of real numbers. The use of convex hull and Voronoi tessellation algorithms can be used to learn conceptual space regions from a set of data points [13].

Work has been done to link conceptual space theory to established representational frameworks. Conceptual spaces are mid-level representations and they have been bridged to higher-level symbol representations [3]. The geometric representation of concepts has been extended to a fuzzy graph representation as well [27]. However, the work done so far has not provided an integrated framework that encompasses the full suite of conceptual space principles within a mathematically defined geometric and topological structure, which is the aim of the algebra presented here.

3 Formal Definitions

In this section we present a formal definition of a metric conceptual space and its components. The conceptual space definition is mathematical and designed for the practical goal of facilitating the construction of conceptual space knowledge bases. For this reason, the convex regions used to represent concepts are specified with more explicitness than in previous formalizations of conceptual spaces. Convex regions are defined as a convex polytopes [14], which are generalizations of polygons to n dimensions. The $n-1$ dimensional faces of a convex polytope are called facets. This definition of a concept region was chosen because operations on polytopes are computationally tractable for domains composed of more than two dimensions. Curved regions can be approximated using polytopes in much the same way that polygon primitives are used to describe more complex structures in geographic information systems (GIS) and computer graphics applications. In addition, there are mathematical representations for convex polytopes that are generalizable over any number of dimensions. This is important, because unlike in GIS and graphics applications, the number of dimensions in a domain can be arbitrarily large.

A designation of the context is required for many conceptual space algebra operations. Methodologies for representing context for similarity measurement has been an active research area [19]. We extend the notion of context for similarity as weights on domains as well as quality dimensions. Take, for example, a conceptual space with a *color* domain that is composed of three quality dimensions: *hue*, *value*, and *saturation*. It is conceivable that one may want to weight the entire *color* domain lower in a *night* context [36], while also weighting *value* higher than *hue* and *saturation*. This secondary weighting has the effect of making a dark red color more similar to a dark blue color than to a light red color.

The role of context is not confined to similarity measurement. When combining concepts the salience of the domains for each concept helps to determine which regions override other regions. Given the ubiquity of context, we define a context as a set of salience weights that can be applied to components of any type in the conceptual space. This definition leaves open the option of applying salience to objects in the conceptual space in a manner beyond what is discussed in this paper, which will facilitate extending the operation set of the algebra. For example, one can create a context for a set of instances where each instance in the set is given its own salience weight for prototype learning.

3.1   Metric Conceptual Space Structure

A metric conceptual space is a multi-leveled structure. A distinction is made between the representation of the geometric elements (regions and points) and the conceptual elements (concepts, properties, and instances). In contrast to other formalizations of conceptual spaces, regions and points are associated with only one domain each, and not with the conceptual space as a whole. Concepts and instances, on the other hand, span across one or more domains. This structure facilitates semantic similarity measurements for concepts and instances that take into account different distance measurements for within and between domains as well as concept combination operations that operate domain-by-domain.

The following definitions are organized in a top-down way beginning with the definition of a conceptual space and defining each component of this space in turn. We refer to concepts, properties, and instances as *objects* in the conceptual space.

Definition 1. *A* metric conceptual space *is defined as a 6-tuple,* $S = \langle \Delta, \Gamma, \breve{I},$ $\blacklozenge, K, c \rangle$.

- Δ *is a finite set of* domains, *where a* domain $\delta \in \Delta$.
- Γ *is a finite set of* concepts, *where a* concept $\gamma \in \Gamma$.
- $\breve{I}$ *is a finite set of* instances, *where an* instance $\breve{i} \in \breve{I}$.
- $\blacklozenge$ *is a finite set of* contrast classes, *where a* contrast class $\blacklozenge \in \blacklozenge$.
- K *is a finite set of* contexts, *where a* context $k \in K$.
- c *is a* constant similarity sensitivity *parameter.*

The set components of a conceptual space are dynamic and can be modified by applying algebraic operations. The conceptual space algebra defines a number of operations that take elements from one or more of the components of the conceptual space and produce values. In some cases query operations will produce numeric or Boolean values, but in other cases they will produce higher-level structures such as new concepts and modify the existing set. In the latter case, the products are inserted into the appropriate set component. For example, an operation to learn a new concept will add the new concept into the Γ component of the conceptual space.

3.2 Domains and Quality Dimensions

Definition 2. *A* domain *is defined as a set of quality dimensions,* $\delta = Q$. *Q is the finite set of integral quality dimensions that form the domain, where a quality dimension* $q \in Q$. $\forall q, q \in \delta \wedge \delta \neq \delta' \Rightarrow q \notin \delta'$.

Definition 3. *A* quality dimension *is defined as a triple,* $q = \langle \hat{\mu}, \hat{r}, \hat{o} \rangle$.

- $\hat{\mu}$ *indicates the* measurement level *or scale of the dimension, where* $\hat{\mu} \in \{ratio, interval, ordinal\}$.
- $\hat{r}$ *indicates the* range *of the dimension, where* $\hat{r}$ *is a pair* $\hat{r} = \langle min, max \rangle$.
- $\hat{o}$ *indicates whether the dimension is* circular, *where* $\hat{o} \in \{true, false\}$.

The quality dimensions in a conceptual space represent a means for measuring and ordering different quality values of objects in the space (in the case of concepts these values might be a range of values). There are four widely-recognized scales of measurement – nominal, ordinal, interval, and ratio – that can be used to assign values to data, and each of these measurement levels has associated with it different mathematical properties [32]. The $\hat{\mu}$ component of a quality dimension can specify the quality dimension scale as ordinal, interval, or ratio. Interval and ratio scales both work naturally for quality dimensions because differences in measurements can be easily compared due to the fact that the units for these scales are equalized. An ordinal scale's values are rank ordered and are consistent with the ordering operations of a conceptual space algebra. However, conclusions of semantic similarity for ordinal quality dimensions should be made with care, taking into account the fact that ordinal scales do not have equalized scales. Psychometric analysis techniques for imposing a distance measurement on an ordinal scale (e.g., Rasch models) can be used to convert an ordinal scale to interval scale [24]. Since, we are primarily interested in quality dimensions as ways of specifying measurement and for use in ordering operations, nominally scaled quality dimensions are not directly supported by this definition. However, it is possible to represent the values on a nominal scale as properties in a domain. One approach for modeling geographic nominal data values as regions in a conceptual space has been shown in [2].

Quality dimensions can be phenomenal or scientific, which means they represent subjective psychological dimensions or are defined by positivist scientific theories, respectively [11]. Conceptual spaces are equally capable of representing both types of dimensions and there is no distinction made between scientific and phenomenal dimensions in the formal definition. A circular dimension is one that wraps around, so the maximum distance value is $\frac{\hat{r}_{max} - \hat{r}_{min}}{2}$. For example, the hue dimension in the color domain is a circular dimension with value range of $[0, 2\pi]$, and any measured distance will be $< \pi$.

3.3   Concepts, Properties, and Instances

Definition 4. *A concept is defined as a pair, $\gamma = \langle \Diamond, P \rangle$.*

- *$\Diamond$ is a finite set of convex regions, where there is an injective relation between $\Diamond$ and Δ. That is, there is a one-to-one relationship from regions in the set to domains and there can only be one region per domain.*
- *P is a prototypical instance.*

Definition 5. *A property is defined as a concept with $|\Diamond| = 1$.*

A concept is a collection of convex regions across one or more domains and an associated prototypical instance. $\forall p, p \in P \Rightarrow \exists \Diamond, \Diamond \in \Diamond \wedge p \in \Diamond$. The prototypes or representative members of a concept play an important role in categorization [28]. There is experimental evidence that the perceived similarity of an object to a prototypical exemplar is used by humans during classification [15]. Given the prototypical instance(s) of one or more concepts, one can derive the regions that compose it using a Voronoi tessellation technique [13]. Conversely, a prototypical instance can be identified by finding the point of central tendency for a set of exemplar points in each domain. The measurement level of the quality dimensions determines how the central tendency is calculated: the geometric mean (or the arithmetic mean of the natural logarithm scale) for ratio scaled, the arithmetic mean for interval scaled, and the median for ordinal scaled dimensions.

Definition 6. *A convex region $\Diamond$ is defined as a convex polytope in the n-dimensional space corresponding to a given domain, δ.*

As defined, a convex region in a conceptual space can be represented as either 1) a set of vertices that constitute the convex hull of the region or 2) a bounded intersection of half-spaces that can be written as a set of linear inequalities [14].

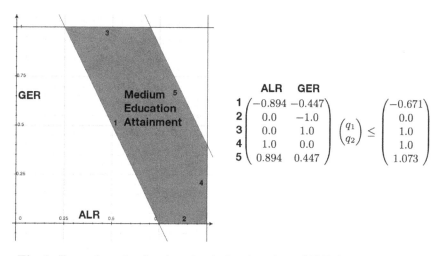

Fig. 1. Example region in education index domain and H-Polytope representation

The first representation is called a V-polytope and the second representation a H-polytope. The H-polytope and V-polytope representations are equivalent, but some important core operations, such as point inclusion, have much better combinatorial complexity when starting with a H-polytope representation.

A region within an n-dimensional domain can be written in the H-polytope matrix form $Aq \leq b$, where q is a variable transpose vector and each q_i corresponds with an $\in Q$:

$$\begin{pmatrix} a_{11} & a_{12} & \cdots & a_{n1} \\ a_{21} & a_{22} & \cdots & a_{n2} \\ \vdots & \vdots & \ddots & \vdots \\ a_{m1} & a_{m2} & \cdots & a_{nm} \end{pmatrix} \begin{pmatrix} q_1 \\ q_2 \\ \vdots \\ q_n \end{pmatrix} \leq \begin{pmatrix} b_1 \\ b_2 \\ \vdots \\ b_m \end{pmatrix}$$

The values in any row i of the A matrix and b transpose vector correspond to the coefficients of the linear inequality that defines the ith half-space boundary of the polytope.

Figure 1 shows a region with five facets in a domain built from two dimensions: adult literacy rate (ALR) and gross enrollment ratio (GER). The domain is based on the United Nations' measure of educational attainment [35]. A country's education index is equal to $\frac{2}{3} \times ALR + \frac{1}{3} \times GER$. Here we define a concept of medium education attainment as the region where $0.5 \leq$ education index ≤ 0.8.

Definition 7. *An instance $\breve{\imath}$ is defined as a finite set of points with an injective relation to Δ. That is, there is a one-to-one relationship from points in the set to domains and there can only be one point per domain.*

Instances, which can be thought of as real-world objects or data points for training sets, are represented by a set of points (or vectors) in one or more domains.

Definition 8. *A point p is defined as a vector of quality dimension values in the n-dimensional space corresponding to a given domain, δ.*

3.4   Contrast Class

Definition 9. *A contrast class* ◆ *is defined as a region in a unit hypercube that corresponds to a domain in the conceptual space. Each dimension of the hypercube corresponds to one quality dimension in the domain. The region is specified by one or two parallel hyperplanes that intersect the unit hypercube:*

$$min : -a_1x_1 - a_2x_2 - \cdots - a_nx_n \leq -b_1$$
$$max : a_1x_1 + a_2x_2 + \cdots + a_nx_n \leq b_2$$

Contrast classes are described as a special type of property (or class of properties) by Gärdenfors. We define it as a unique type of element for the algebra, because it is used differently from a normal property by algebraic operations. It is more appropriately thought of as a function that describes sub-regions relative to regions in a domain. In a one-dimensional domain the contrast class region is bounded by two points on a unit line segment; in two dimensions the region is bounded by two parallel lines intersecting a unit square; and so on. The operation for combining a contrast class with a concept is detailed in section 4.

Some common contrast classes are *large, small, old, young, northern, southern*. The example shown in figure 1 is the result of combining a *medium* contrast class to the *education attainment* property. *Education attainment* covers the square area [0,1] on both dimensions, and the *medium* contrast class is projected on that region resulting in the *medium education attainment* sub-region.

3.5   Context

Definition 10. *A context k is defined as a finite set of salience weight, conceptual space component pairs k = ⟨ω, component⟩. The conceptual space components in the context must all be of the same type (e.g., quality dimensions), which is referred to as the* context type. *In addition,* $\sum_{i=1}^{n} \omega_i = 1$ *where each weight* ω_i *in a context has a value* $0 \leq \omega_i \leq 1$.

If a context is used in an operation that is applied to a context typed conceptual space component that is not included in the context then the salience weight of that component is 0. For example, if a similarity operation is applied to two instances that span the same three domains and the context only contains weights for two of the three domains then the third dimension's salience weight will be 0 for the operation.

4   Algebraic Operations

In this section we introduce the algebraic operations that can be placed on the components of a conceptual space. We organize the operations into core metric operations on points and regions followed by similarity and concept combination query operations. We use shorthand notations for the components (Table 1).

4.1   Core Operations

Regions and points are the most primitive objects in a conceptual space, each existing within a single $\delta \in S$. Since a domain is a metric and topological space, all of the operations that can be applied to regions and points in a topological vector space are applicable. The computational implementation of these operations essentially reduces to problems of numeric linear algebra, which are well-established. With the understanding that there are many more operations possible on these primitive types, we present the particulars of how the H-polytope representation of conceptual regions can be used to implement intersection and inclusion operations algorithmically. In addition, the within domain distance metric for points and regions is defined.

$\diamond_1 \bigcap \diamond_2 \rightarrow \diamond_{new}$. This function calculates the intersection of two convex regions. The result is a new convex region or an empty set. The first step of an algorithm to calculate the intersection of two H-polytopes is to get the union of the two systems of linear inequalities. It is possible that this union will result in redundant inequalities, so a linear programming problem is constructed to remove these redundancies: Given two regions $\diamond_1$ and $\diamond_2$ represented as $Aq \leq b$ and $Sq \leq t$, respectively, maximize each inequality $s^T q$ in $\diamond_2$ subject to $Aq \leq b$ and only add the inequality to the union if the optimal value is less than or equal to t [8]. There are several algorithmic techniques used for solving linear programming problems, the most popular being the Simplex method [6]. The Simplex method performs very well in most cases, averaging a number of iterations that is less than three times the number of inequalities in the set [23]. The total computational complexity of the intersection operation is therefore linear with respect to the number of inequalities in most cases.

$p \in \diamond \rightarrow$ **Boolean.** The inclusion operation given a point and a region represented as a H-polytope is equivalent to testing whether the point satisfies the entire system of linear inequalities. The computational complexity is linear with respect to the number of facets in the polytope.

Table 1. Notation for named elements

Notation	Meaning	Example
$\gamma^{concept}$	named concept	$\gamma^{european\ state}$
$\diamond^{concept}_{domain}$	region of concept in domain	$\gamma^{european\ state}_{coordinates}$
δ^{domain}	named domain	$\delta^{coordinates}$
$\delta^{domain}_{quality\ dim}$	quality dimension in domain	$\delta^{coordinates}_{latitude}$
$\blacklozenge^{domain}_{contrast\ class}$	contrast class in domain	$\blacklozenge^{coordinates}_{southern}$
$\delta\left(\gamma_P\right)$	domain of property	$\delta\left(\gamma^{temperate\ zone}\right) = \delta^{coordinates}$
$Q\left(\delta^{domain}\right)$	quality dimensions of domain	$Q\left(\delta^{coordinates}\right) = \{longitude, latitude\}$

$dist(p_1, p_2, k) \rightarrow \mathbb{R}$. The distance metric for two points that exist within the same domain is defined as a weighted Euclidean metric, where the weights are determined by a quality dimension-typed context, k.

$$dist(p_1, p_2, k) = \sqrt{\sum_{i=1}^{n} \omega_i (p_{1_i} - p_{2_i})^2}$$

where n is $|Q|$, i is an index to an ordering of Q, q_i is the ith element in that ordered set, and $(\omega_i, q_i) \in k$. If a dimension is circular then the difference $p_{1_i} - p_{2_i}$ in the above equation will be modulo the range of the quality dimensions divided by two. Euclidean distance measure, as opposed to another instance of the Minkowski metric, was chosen because of experimental results that show its cognitive plausibility for measuring the similarity of concepts composed of integral qualities [9,18]. When the qualities are separable, the city-block distance was found to be more appropriate, which is captured by the similarity operation.

Normalization of dimensions is important to ensure that a change in units does not result in a different distance measurement and subsequently a different similarity measurement. However, given the variety of options depending on the structure of domains, we reserve normalization as a preprocessing operation rather than an integral component of the distance measure.

4.2   Query Operations

Similarity. Experiments on similarity cognition have shown that the similarity of two objects can be measured as an exponentially decaying function of the distance between the two objects: $sim(d) = e^{-cd}$ [31]. The following similarity operation utilizes a compound distance function that takes into account the structural distinction of separable and integral dimensions.

$sim(\breve{\imath}_A, \breve{\imath}_B, k, K) \rightarrow \mathbb{R}$. Given two instances, $\breve{\imath}_A$ and $\breve{\imath}_B$, a domain-type context, k, and a set of quality dimension-type contexts, K, this function calculates a distance between $\breve{\imath}_A$ and $\breve{\imath}_B$. Let $\Delta_{\breve{\imath}} = \Delta(\breve{\imath}_A) \cap \Delta(\breve{\imath}_B)$. The distance between two instances is a weighted sum of all of the within domain Euclidean distance measures for each $p \in \breve{\imath}$:

$$d(\breve{\imath}_A, \breve{\imath}_B, k, K) = \sum_{j=1}^{|\Delta_{\breve{\imath}}|} k_j \times \sqrt{|Q(\delta_j)|} \times dist(p_j(\breve{\imath}_A), p_j(\breve{\imath}_B), K_j)$$

where j is an index to an ordering of $\Delta_{\breve{\imath}}$. Ignoring context weights, the result of this distance function is a composite value that is $\geq$ the Euclidean distance and $\leq$ the city-block distance, if all the quality dimensions were in one multi-dimensional space. The context and context set parameters allow one to apply saliences on both domain and quality dimension levels. Because Euclidean and city-block metrics are being mixed, each weighted within-domain Euclidean distance measure is also normalized by the square root of the cardinality of the domain to prevent low dimensional domains from having more salience than high dimensional domains. Following this distance function we get the following similarity function for two instances:

$$sim(\breve{\imath}_A, \breve{\imath}_B, k, K) = e^{-cd(\breve{\imath}_A, \breve{\imath}_B, k, K)}$$

$sim(\gamma_A, \gamma_B, k, K) \rightarrow \mathbb{R}$. There are many possible methods for measuring the distance between two regions within a domain. The simplest method is just to measure it as the distance between prototypical points within each region. The analogue to that method for a similarity measure between concepts would be to measure the distance between prototypical *instances* for the two concepts using the similarity function just described. Other methods have been proposed to measure distance between spatial conceptual regions using the vertices of the convex hull of the region [30]. The advantage to these methods is that they allow for asymmetrical distance measurements, the lack of which is a common criticism of geometric models of similarity [34]. Using the V-polytope form of the regions, the methods can be used to calculate the within-domain distance for each domain, which can then be summed in a weighted form as above. The distance from an instance to a concept can also be calculated.

Concept combination. Gärdenfors describes techniques for combining concepts in conceptual spaces, but his methodology has not been formalized yet [9]. Here we describe three concept combination operations using the components of a conceptual space as defined above. The operations are property-concept, concept-concept, and contrast class-concept combinations. For these operations it is important to note that one concept is the *modifier* concept and the other is the *modified* concept. This distinction is linked with the importance of the ordering of concepts in linguistic expressions. For example, the concept combination *green village* is distinct from the concept combination *village green*. We follow a convention that the *modifier* concept is the first parameter and the *modified* concept is the second parameter of any combination operation.

The following algorithms describe how the regions of concept combinations are formed. With all three concept combination operations not only new regions but also new prototypical points need to be learned. The process is the same for all three. For any newly created region $\diamond_{new}$, the new prototypical point is set equal to the centroid of $\diamond_{new}$. Alternately, in the case that an associated instance set is available, the prototypical point can be learned from the set of instances $\in \diamond_{new}$.

$combine\,(\gamma_P, \gamma_C) \rightarrow \gamma_{new}$. Algorithm 1. The combination of a property and a concept is the simplest case. There is no need to specify the salience of the domains, because it is understood that $\delta(\gamma_P)$ is of higher salience. In the case that γ_C does not have a region specified for the domain of γ_P, the property's region is added to the concept. When the property region is part of a specified domain for γ_C then there are two possible outcomes. The property region and the concept region for that domain can overlap, in which case the new concept region is the intersection of the regions. If they do not overlap, the property region will override the concept region in that domain. For example, the property-concept combination *purple mountain* will override the *mountain* region in the *color* domain to the *purple* region assuming that the *color* region for *mountain* is in another area of the *color* domain. The other domains are unaffected.

Algorithm 1. Property-Concept Combination

operation combine$(\gamma_P, \gamma_C) \rightarrow \gamma_{new}$
if $\Delta(\gamma_C) \ni \delta(\gamma_P)$ then
   if $\diamond_{\delta(\gamma_P)}^{\gamma_P} \cap \diamond_{\delta(\gamma_P)}^{\gamma_C} = \emptyset$ then
      $\gamma_{new} \Leftarrow \left(\diamond(\gamma_C) - \left\{ \diamond_{\delta(\gamma_P)}^{\gamma_C} \right\} \right) \cup \left\{ \diamond_{\delta(\gamma_P)}^{\gamma_P} \right\}$
   else
      $\gamma_{new} \Leftarrow \left(\diamond(\gamma_C) - \left\{ \diamond_{\delta(\gamma_P)}^{\gamma_C} \right\} \right) \cup \left\{ \diamond_{\delta(\gamma_P)}^{\gamma_P} \cap \diamond_{\delta(\gamma_P)}^{\gamma_C} \right\}$
   end if
else
   $\gamma_{new} \Leftarrow \diamond(\gamma_C) \cup \left\{ \diamond_{\delta(\gamma_P)}^{\gamma_P} \right\}$
end if
return γ_{new}

---

combine $(\gamma_A, \gamma_B, K_A, K_B) \rightarrow \gamma_{new}$. Algorithm 2. As with similarity, context plays an important role in concept combinations. When combining two concepts that both span more than one domain, only the regions in a subset of the domains will be affected depending on the context. For both concepts a salience weight is given for each domain (i.e., a domain-type context). If the domains are shared by the two concepts then the context will determine which concept has precedence. Otherwise, the new concept will adopt the region from the concept for which the domain is specified. Currently, the weights are only for comparison, therefore values of 0 and 1 are sufficient. However, room is left for more complex combination operations that take into account the differences in weight values.

combine$(\blacklozenge, \gamma) \rightarrow \gamma_{new}$. The operation to apply a contrast class to a concept only affects the domain for which the contrast class is defined, which we refer to as δ_{CC}. For the sake of brevity, the following is a high-level description of the operation, which at a low-level relies on standard geometric operations. Let $\diamond$ be the region of γ in δ_{CC} and p the prototypical point $\in \diamond$. Find the minimum bounding box around $\diamond$, which gives a range magnitude for each dimension of $\diamond$. Stretch the contrast class unit hypercube (and min, max hyperplanes accordingly) to the size of the minimum bounding box. Center this stretched version over p and intersect the hyperplane(s) with $\diamond$ to get $\diamond_{new}$. γ_{new} is equal to γ in all other domains with $\diamond_{new}$. Figure 2 illustrates these steps with a contrast class *tall* and concept *mountain* in a *size* domain with two dimensions: *height* and *width*. This operation can be applied recursively. For example, *tall* can be applied again to *tall mountain* to obtain a region for *very tall mountain*.

Algorithm 2. Concept-Concept Combination

---

operation combine$(\gamma_A, \gamma_B, K_A, K_B) \rightarrow \gamma_{new}$
Let $\Delta_{new} = \Delta(\gamma_A) \bigcup \Delta(\gamma_B)$
Let $\gamma_{new} = \emptyset$
for all $\delta \in \Delta_{new}$ **do**
  if $\delta \notin \Delta(\gamma_B)$ **then**
    insert $\diamond_\delta^{\gamma_A}$ into γ_{new}
  else if $\delta \notin \Delta(\gamma_A)$ **then**
    insert $\diamond_\delta^{\gamma_B}$ into γ_{new}
  else { // check context}
    if $K_{A\delta} > K_{B\delta}$ **then**
      $\gamma_{new} = \gamma_{new} \bigcup combine(\gamma_{A\delta}, \gamma_{B\delta})$
    else
      insert $\diamond_\delta^{\gamma_B}$ into γ_{new}
    end if
  end if
end for
return γ_{new}

---

5 Application: Country Concept Comparison

The developed metric conceptual space algebra provides a means for creating complex conceptual space structures and applying similarity and concept combination operations to concepts represented in the space. In order to demonstrate the functionality of these algebraic operations and its use for spatial problems with high dimensional data, we present a case study where the algebra is used for the comparison of countries and regions of the world.

The countries of the world and the groups to which they are classified are complex concepts. The United Nations Development Program (UNDP), for example, divides the countries of the world into eight mutually exclusive classes: Arab States, East Asia and Pacific, Latin America and the Caribbean, South Asia, Southern Europe, Sub-Saharan Africa, Central and Eastern Europe and the CIS, and High-Income OECD [35]. This classification scheme is based on a combination of cultural, geographic, political, historical, and economic factors, but as such it does not afford the ability to make more nuanced comparisons between countries and regions. With a conceptual space representation these factors can be organized into separable domains. The similarity of countries can be compared based on context, and concept combination can generate new classes.

5.1 Data Collection

Data for 155 countries were aggregated from the CIA World Factbook, UNDP, and the World Resources Institute [5,35,37] and were used to represent each country as an instance in a conceptual space with 16 quality dimensions organized into six domains (Table 2). The UNDP classes were represented as concepts

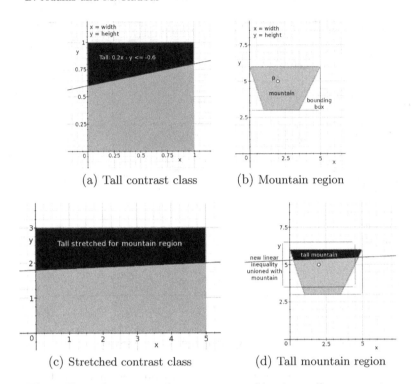

Fig. 2. Example contrast class-concept combination: tall + mountain

formed by taking the convex hulls of country instances in that class for each domain. We defined the prototypes of these classes as the set of points formed by mean values of all the instances in the class. For the population and size dimensions the values were scaled logarithmically. All data values and dimensions were then normalized so that each quality dimension had a range [0,1].

5.2 Results

For demonstrating the similarity operation, three contexts were created, which we refer to as "natural resources", "geographic", and "human issues"[1]. Table 2 presents the weights for each context. Using the instance similarity operation we found the similarity between every pair of countries for a given context. Table 3 shows a sample of these similarity results for Turkey.

To demonstrate the use of contrast classes we defined a *northern* contrast class for the *coordinates* domain as $\delta_{latitude}^{coordinates} \geq 0.5$ and combined it with the eight UNDP classes. This contrast class corresponds roughly to the top half – along the latitude dimension – of any region (re-centered over the prototype) in

---

[1] These context weights and their associated labels were chosen merely to illustrate the similarity operation, without making a claim for cognitive plausibility.

Table 2. Country conceptual space domains and contexts

Domains & Quality Dimensions		Context N.R.		Context Geo.		Context H.I.	
Size	*Land area*	0.4	0.8	0.3	0.5	0.1	0.5
	Water area		0.2		0.5		0.5
Coordinates	*Latitude*	0.0	–	0.4	0.7	0.0	–
	Longitude		–		0.3		–
Land type %	*Forests*	0.5	0.5	0.3	0.2	0.1	0.0
	Grasslands		0.0		0.2		0.0
	Wetlands		0.1		0.2		0.0
	Croplands		0.4		0.2		0.5
	Barren		0.0		0.2		0.0
	Urban		0.0		0.0		0.5
Education	*Adult lit. rate*	0.0	–	0.0	–	0.3	0.7
	Gross enrol. ratio		–		–		0.3
Economic	*GDP index*	0.1	1.0	0.0	–	0.2	1.0
Demographic	*Population*	0.0	–	0.0	–	0.3	0.4
	Pop. growth rate		–		–		0.4
	Urban pop. (%)		–		–		0.2

Table 3. Top 5 most similar countries to Turkey by context

Natural Res.	Geography	Human Issues
Thailand 0.877	Kyrgyzstan 0.833	Spain 0.890
Malaysia 0.875	Spain 0.832	Zimbabwe 0.872
Colombia 0.872	Italy 0.831	Uruguay 0.840
Mexico 0.867	Nepal 0.817	Greece 0.827
Viet Nam 0.855	Uzbekistan 0.815	Italy 0.815

the *coordinates* domain. Figure 3a shows the results of combining *northern* with the coordinates region of *sub-Saharan Africa*. Also shown are all the instances with points in the coordinates domain that lie within that region.

Next we created a property region in the land type domain to describe the property *desert-like*. This region is defined as the area where *barren* ≥ 0.6 and *wetlands* ≤ 0.1. The property was combined with the *arab state* concept to create a *desert-like arab state*. Figure 3b shows a two-dimensional projection of the land type domain with the result of this combination.

Finally, to test complex concept combination we created an ad hoc category for arid, highly educated countries with a large urban population. In the land type domain it was represented by a region where *barren* ≥ 0.5, *forests* ≤ 0.1, and *wetlands* ≤ 0.05; in the education domain it was the region where $\frac{2}{3} \times ALR + \frac{1}{3} \times GER \geq 0.8$; and in the demographic domain it was the region where urban population > 0.8. The result was that one could represent, for example, the combination of this ad hoc category with the concept of

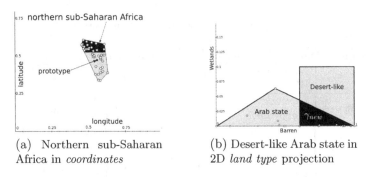

(a) Northern sub-Saharan Africa in *coordinates*

(b) Desert-like Arab state in 2D *land type* projection

Fig. 3. Concept combinations

Latin America to posit a scenario of an arid, highly developed version of Latin America.

6   Conclusions and Future Work

In this paper we presented a metric conceptual space algebra that defines conceptual spaces as multi-leveled representational structures with operations that are well-suited for computational implementation. A clearer distinction is made between domains and quality dimensions than in previous formalizations of conceptual spaces, which allows for more expressivity for similarity and complex concept combination operations. The conceptual regions that exist in domains are defined as convex polytopes, so that primitive geometric and topological operations can be implemented using algorithms that are tractable. We demonstrated a practical application of conceptual space algebraic operations for a spatial information system, though the algebra presented in this paper is designed for general use [11].

This paper presented a theoretical framework for a metric conceptual space algebra, and there exist many avenues for extending it. The described operations assume no correlation between the different regions that compose a concept, but in reality the quality values of concepts are very often correlated. One possible solution is to integrate multivariate analysis with existing operations. In addition, concepts are often uncertain and dynamic. The algebra described in this paper should be extended to accommodate convex regions that are fuzzy. Parameterized rough fuzzy sets may be one component to such an extension [1]. Introducing operations that allow one to query about the shape of conceptual regions over time would also be a valuable addition to the algebra [26].

Acknowledgments

This work was supported by NSF IGERT Grant #DGE-0221713. The suggestions from four anonymous reviewers helped improve the content of the paper.

References

1. Ahlqvist, O.: A Parameterized Representation of Uncertain Conceptual Spaces. Transactions in GIS 8(4), 492–514 (2004)
2. Ahlqvist, O., Ban, H.: Categorical Measurement Semantics: A New Second Space for Geography. Geography Compass 1(3), 536–555 (2007)
3. Aisbett, J., Gibbon, G.: A general formulation of conceptual spaces as a meso level representation. Artificial Intelligence 133, 189–232 (2001)
4. Barsalou, L.: Situated simulation in the human conceptual system. Language and Cognitive Processes 5(6), 513–562 (2003)
5. Central Intelligence Agency - The World Factbook (2009), https://www.cia.gov/library/publications/the-world-factbook/
6. Dantzig, G.B.: Linear Programming and Extensions. Princeton University Press, Princeton (1963)
7. Devore, J., Peck, R.: Statistics - The Exploration and Analysis of Data, 4th edn. Duxbury, Pacific Grove (2001)
8. Fukuda, K.: Frequently Asked Questions in Polyhedral Computation (2004), http://www.ifor.math.ethz.ch/~fukuda/polyfaq/polyfaq.html
9. Gärdenfors, P.: Conceptual Spaces: The Geometry of Thought. MIT Press, Cambridge (2000)
10. Gärdenfors, P.: How to Make the Semantic Web More Semantic. In: Varzi, A., Vieu, L. (eds.) Formal Ontology in Information Systems, Proceedings of the Third International Conference (FOIS 2004). Frontiers in Artificial Intelligence and Applications, vol. 114, pp. 153–164. IOS Press, Amsterdam (2004)
11. Gärdenfors, P.: Conceptual Spaces as a Framework for Knowledge Representation. Mind and Matter 2, 9–27 (2004)
12. Gärdenfors, P.: Representing actions and functional properties in conceptual spaces. In: Ziemke, T., Zlatev, J., Frank, R. (eds.) Body, Language and Mind, pp. 167–195. Mouton de Gruyter, Berlin (2007)
13. Gärdenfors, P., Williams, M.: Reasoning About Categories in Conceptual Spaces. In: Proceedings of the Fourteenth International Joint Conference on Artificial Intelligence, pp. 385–392. Morgan Kaufmann, San Francisco (2001)
14. Grünbaum, P.: Convex Polytopes. In: Kaibel, V., Klee, V., Ziegler, G. (eds.), 2nd edn., Springer, New York (2003)
15. Harnad, S.: Categorical Perception: The Groundwork of Cognition. Cambridge University Press, Cambridge (1987)
16. Janowicz, K., Raubal, M.: Affordance-Based Similarity Measurement for Entity Types. In: Winter, S., Duckham, M., Kulik, L., Kuipers, B. (eds.) COSIT 2007. LNCS, vol. 4736, pp. 133–151. Springer, Heidelberg (2007)
17. Janowicz, K., Raubal, M., Schwering, A., Kuhn, W.: Semantic Similarity Measurement and Geospatial Applications. Transactions in GIS 12(6), 651–659 (2008)
18. Johannesson, M.: The Problem of Combining Integral and Separable Dimensions, Lund, Sweden. Lund University Cognitive Studies, p. 16 (2002)
19. Keßler, C.: Similarity measurement in context. In: Kokinov, B., Richardson, D.C., Roth-Berghofer, T.R., Vieu, L. (eds.) CONTEXT 2007. LNCS, vol. 4635, pp. 277–290. Springer, Heidelberg (2007)
20. Lakoff, G.: Cognitive Semantics. In: Eco, U., Santambrogio, M., Violi, P. (eds.) Meaning and Mental Representations, pp. 119–154. Indiana University Press, Bloomington (1988)

21. Mark, D., Freksa, C., Hirtle, S., Lloyd, R., Tversky, B.: Cognitive models of geographical space. International Journal of Geographical Information Science 13(8), 747–774 (1999)
22. Montello, D., Freundschuh, S.: Cognition of Geographic Information. In: McMaster, R., Usery, E. (eds.) A research agenda for geographic information science, pp. 61–91. CRC Press, Boca Raton (2005)
23. Nocedal, J., Wright, S.J.: Numerical Optimization. Springer, New York (1999)
24. Rasch, G.: On General Laws and the Meaning of Measurement in Psychology. In: Proceedings of the Fourth Berkeley Symposium on Mathematical Statistics and Probability, IV, pp. 321–334. University of California Press, Berkeley (1961)
25. Raubal, M.: Formalizing Conceptual Spaces. In: Varzi, A., Vieu, L. (eds.) Formal Ontology in Information Systems. Proceedings of the Third International Conference (FOIS 2004), pp. 153–164. IOS Press, Amsterdam (2004)
26. Raubal, M.: Representing Concepts in Time. In: Freksa, C., Newcombe, N.S., Gärdenfors, P., Wölfl, S. (eds.) Spatial Cognition VI. LNCS, vol. 5248, pp. 328–343. Springer, Heidelberg (2008)
27. Rickard, J.T.: A concept geometry for conceptual spaces. Fuzzy Optimal Decision Making 5, 311–329 (2006)
28. Rosch, E.: Cognitive Representations of Semantic Categories. Journal of Experimental Psychology: General 104, 192–232 (1975)
29. Schwering, A.: Approaches to Semantic Similarity Measurement for Geo-Spatial Data: A Survey. Transactions in GIS 12(1), 5–29 (2008)
30. Schwering, A., Raubal, M.: Measuring Semantic Similarity between Geospatial Conceptual Regions. In: Rodríguez, M.A., Cruz, I., Levashkin, S., Egenhofer, M.J. (eds.) GeoS 2005. LNCS, vol. 3799, pp. 90–106. Springer, Heidelberg (2005)
31. Shepard, R.N.: Toward a universal law of generalization for psychological science. Science 237, 1317–1323 (1987)
32. Stevens, S.: On the Theory of Scales of Measurement. Science 103, 677–680 (1946)
33. Tanasescu, V.: Spatial Semantics in Difference Spaces. In: Winter, S., Duckham, M., Kulik, L., Kuipers, B. (eds.) COSIT 2007. LNCS, vol. 4736, pp. 96–115. Springer, Heidelberg (2007)
34. Tversky, A.: Features of Similarity. Psychological Review 84(4), 327–352 (1977)
35. UNDP (United Nations Development Program): Human Development Report 2007/2008. Palgrave Macmillan, New York (2007)
36. Winter, S., Raubal, M., Nothegger, C.: Focalizing Measures of Salience for Wayfinding. In: Meng, L., Zipf, A., Reichenbacher, T. (eds.) Map-based Mobile Services – Theories, Methods and Implementations, pp. 127–142. Springer, Berlin (2005)
37. World Resources Institute: EarthTrends: Environmental Information. World Resources Institute, Washington, D.C (2006), http://earthtrends.wri.org

Grounding Geographic Categories in the Meaningful Environment

Simon Scheider, Krzysztof Janowicz, and Werner Kuhn

Institute for Geoinformatics, Westfälische Wilhelms-Universität Münster,
Weseler Straße 253 D-48151 Münster, Germany
{simon.scheider,janowicz,kuhn}@uni-muenster.de

Abstract. Ontologies are a common approach to improve semantic interoperability by explicitly specifying the vocabulary used by a particular information community. Complex expressions are defined in terms of primitive ones. This shifts the problem of semantic interoperability to the problem of how to ground primitive symbols. One approach are *semantic datums*, which determine reproducible mappings (measurement scales) from observable structures to symbols. Measurement theory offers a formal basis for such mappings. From an ontological point of view, this leaves two important questions unanswered. Which qualities provide semantic datums? How are these qualities related to the primitive entities in our ontology? Based on a scenario from hydrology, we first argue that human or technical *sensors implement semantic datums*, and secondly that primitive symbols are *definable* from the *meaningful environment*, a formalized quality space established through such sensors.

Keywords: Semantic Heterogeneity; Symbol Grounding Problem; Semantic Datum; Meaningful Environment.

1 Introduction

The symbol grounding problem [13] remains largely unsolved for ontologies: ultimately, the semantics of the primitive terms in an ontology has to be specified outside a symbol system. Tying domain concepts like river and lake to data about their instances (as proposed, for example, in [5]) constrains these in potentially useful ways, but defers the grounding problem to the symbol system of the instance data. While these data may have shared semantics in a local geographic context, they do not at higher levels, such as in an INSPIRE scenario of integrating data and ontologies across Europe (http://inspire.jrc.ec.europa.eu/). It does not seem practical for, say, Romanian hydrologists, to ground their lake and river concepts in British geography, or vice versa. Furthermore, grounding domain concepts in a one-by-one manner is an open-ended task. One would prefer a method for grounding ontological primitives in *observation procedures* in order to support more general ontology mappings.

In this paper, we propose such a method and demonstrate its applicability by the category *water depth*. We provide an ontological account of Gibson's meaningful environment [12] and use Quine's notion of observation sentences [20] as a

K. Stewart Hornsby et al. (Eds.): COSIT 2009, LNCS 5756, pp. 69–87, 2009.

basis for grounding ontological concepts in reproducible observation procedures. At first glance it seems improbable that highly elaborated scientific concepts, like those of INSPIRE, could be reconstructed from meaningful primitives. Although such a wider applicability of the method remains to be shown, we suggest however - like Quine - that even the elaborated language of natural science must eventually be grounded in observational primitives. After discussing basic issues from measurement theory, philosophy and cognition (section 2), we review the core ideas of Gibson's meaningful environment and formalize them (section 3). Using the example of water depth, we apply the theory in section 4, before drawing some conclusions on what has been achieved and what remains to be done (section 5).

2   Measurement and the Problem of Human Sensors

In this section we introduce the notion *semantic datum* and claim that successful grounding ultimately rests on the existence of human sensors for body primitives.

2.1   Semantic Datums for Languages about Qualities

Measurement scales are maps from some observable structure to a set of symbols [26]. Measurement theory merely provides us with formal constraints for such mappings, namely scale types. It does not disambiguate scales themselves. For example, we can distinguish ratio scales from interval scales, because ratio scales can be transformed into each other by a similarity transformation, while for interval scales we need a linear transformation [26]. But individual scales are never uniquely determined by their formal structure. This is called the *uniqueness problem* of measurement [26]. Therefore the symbol grounding problem [13] remains unsolved: In order to disambiguate scales, we need to know about the conventions of *measurement standards*, like unit lengths or unit masses.

One approach to this problem are semantic datums [14][18]. A semantic datum interprets free parameters. It provides a particular interpretation for the *primitive symbols* of a formal system. An interpretation is a function from all symbols (terms, attributes, and relations) in a formal symbol system to some particular other structure which preserves the truth of its sentences. Typically, formal systems allow for more than one interpretation satisfying their sentences, and therefore they have an ambiguous meaning. As non-primitive symbols in a formal system are definable from the primitive ones, a semantic datum can fix one particular interpretation. The structure in which the symbols are interpreted can be either *other formal systems* (*reference frames* in the sense of Kuhn et al. [15]) or *observable real world structures* (*qualities*). We say that a formal system is *grounded*, if there exist semantic datums for an interpretation into qualities. Examples for such semantic datums are *measurement standards*. A formal system may also be *indirectly grounded* by chaining several semantic datums. Simple examples of grounded formal systems are *calibrated measurement devices*, like a thermometer. A semantic datum is given by the observable freezing and boiling events of water at a standard air pressure. More complex examples are *datums for geodetic positions*: A directly observable semantic datum for the positions on a Bessel ellipsoid consists of a named spot on the earth's surface like "Rauenberg" near Berlin (*Potsdam Datum*) and a standard position and orientation for the ellipsoid.

How can we expand these ideas to arbitrary languages about qualities? Better: For which language primitives do such semantic datums exist?

2.2 Sensors as Implemented Scales

In this paper, we consider a sensor to be a device to *reproducibly* transform observable structures into symbols. A sensor implements a measurement scale including a semantic datum, and therefore establishes a source for grounding. The main requirement for such sensors is reproducibility, i.e. to make sure that multiple applications of the sensor assign symbols in a uniform way.

We already said that *calibrated measurement devices* are grounded formal systems, and therefore we call them *technical sensors*. Following Boumans [8], any reliable calibration of a technical sensor is based ultimately on *human sensation,* because it needs *reproducible gauging* by human observers [8].

So any grounding solution based on measurement ultimately rests on the existence of *human sensors*. But what is a human sensor supposed to be? For human sensors, reproducibility means something similar to *inter-subjective word meaning* in linguistic semantics. More specifically, we mean with a human sensor that an information community shares symbols describing a *certain commonly observable situation*.

2.3 Are There Human Sensors for Body Primitives?

Embodiment and Virtuality of Language Concepts. A commonly observable situation is exactly what Quine [20] described with *occasion sentences* and more specifically with *observation sentences*. Quine's argument is that natural language sentences vary in their semantic indeterminacies. There are certain occasion sentences, utterable only on the occasion, with relatively low indeterminacy and high *observationality*. These sentences are called *observation sentences*. Symbols of a language in general inherit their meaning from such sentences, but the further away they are from such observation sentences inside of a language, the more abstract and indeterminate they get. Thus a symbol like "Lake Constance" (a name for an individual) is less virtual than "Lake" (a general term), which is again less virtual than a social construction like "Wetland". Whether a symbol is less virtual than another is primarily dependent on its reference to *bodies and their parts*. This is what Quine called *divided reference* ([20], chapter 3): Humans individuate bodies like "Mama" by reference to (pointing at) their observable parts and using a criterion of individuation. And they quantify over general terms (categories) like "Mother" by reference to a yet undetermined number of similar but virtual bodies.

The empirical arguments from *cognitive linguists*, e.g. [16], that embodiment is the semantic anchor for more virtual language concepts via metaphors, seem to underpin these early ideas about body based primitives. Langacker [16] suggests that *imagined bodies and fictive entities* are a semantic basis for formal logic and quantifier scopes in the sense of Quine.

Following these lines of thought, we take the view that language semantics, especially the semantics of formal ontologies, is anchored in the individuation of bodies as unified wholes of parts of the environment. Individuation rests on perceivable qualities, e.g. their shape. But how can we imagine humans to perceive such properties of bodies and their parts?

Scanning. A certain kind of virtuality in perception is especially interesting for us. Drawing on ideas of Talmy, Langacker [16] also suggests that *mental scanning*, a fictive motion of a virtual body in the perceived or imagined environment, is central to language semantics. The sentence "The balloon rose quickly" thus denotes an actual body movement, whereas "This path rises quickly near the top of the mountain" can only be understood by imagining a virtual body movement in an actual environment. This view is also supported by recent work on grounded cognition [1] which claims that human cognition is grounded through situated simulation. The motion oriented notion of scanning proposed here is a special case of such situated simulation. We assume that *some perception* is *scanning*: a *series of virtual steps in an environment*, with each step leading from one *locus of attention* to the next.

Ostension and Agreement on Names. In order to assume a human sensor for an observable language symbol, it is necessary that different actors (as well as the same actor on different occasions) will reliably agree on the truth of its observation sentences in every observable situation. This effectively means that there must be consensus about names for bodies and body parts. Quine [20] suggested that *observable names*, like "Mama", can be agreed upon in a language community by *pointing* at a body. According to Quine, the agreement on names for bodies can be based on an observable action such as ostension, given the situation is simultaneously observed and the viewpoints of a language teacher and a learner are enough alike. In the same manner, the correct word usage is inculcated in the individual child of a language community by social training on the occasion, that is by the child's disposition to respond observably to socially observable situations, and the adults disposition to reward or punish its utterances[1] ([20], chapters 1 and 3). In this way, agreement on names for bodies actually spreads far beyond the concretely observable situation, involving a whole language community. *Ostension* is nothing else than a communicative act including a virtual movement, because an observer has to scan a pointing body part, e.g. a finger, and extend it fictively into space. So this fits our assumptions. Further on, we will just assume that body parts and body primitives can always be given unambiguous names by using ostension.

---

[1] According to Quine, observation sentences are the *entrance gate* to language, because they can be easily learned directly by ostension without reference to memory.

3   Gibson's Meaningful Environment

In this section we discuss a formal grounding method based on Gibson's meaningful environment. We begin with two examples illustrating perceptual primitives for bodies and then proceed with a discussion of Gibson's ideas.

3.1   The Blind Person in a Closed Room

How does a blind man perceive the geometric qualities of a closed room? Standing inside the room, he can rely on his tactile and hearing sensors to detect its surface qualities. Because he knows his body takes some of the space of the room, the room must be higher than his body. If he can turn around where he stands, he knows that a roughly cylindrical space is free and part of the room. By taking a step forward, he concludes that an elongated "corridor" is free and part of the room. Because he can repeat steps of the same length into the same direction, he can step diametrically through the room and even measure one of its diameters. His last step may be shortened, because his foot bumps into the wall. He thus detected the inner surface of the room. If he continues summing up paths through the room, he can individuate the whole room by its horizontal extent.

3.2   The Child and the Depth of a Well

Imagine a child in front of a water well trying to assess its depth. It cannot see the ground in the well as no light reaches it. Nevertheless, the child can perform a simple experiment. It drops a brick from the top of the well and waits until it hits the water surface. The child cannot see this happening but can hear the sound. It can repeat the experiment and count the seconds from dropping to hearing the sound, and hence it can measure the depth of the well. The child assesses the depth by simulating a motion that it cannot do by its own using a brick.

3.3   A Short Synopsis and Extension of Gibson's Ideas

Gibson [12] sketches an informal ontology of elements of the environment that are accessible to basic human perception, called *the meaningful environment*. In the following, we try to condense his ideas about what in this environment is actually directly perceivable and complement them with the already discussed ideas about fictive motion.

   The environment is mereologically structured at all levels from atoms to galaxies. Gibson claims that at the ecological scale, so called *nested units* are basic for perception: Canyons are nested within mountains, trees are nested within canyons, leaves are nested within trees. The structure of ecological units depends on the environment as well as the perceiving actor in it. But the perception of individual units, of their composition and of certain aspects of their form and texture are common to humans in a certain ecological context. Thus we can assume a *human sensor* for them. Although there are no a priori atomic units, perceptual limits do exist for geometric properties (we will call this perceivable granularity). Biological cells are beyond these limits, and therefore not directly perceivable, whereas leaves are.

The composition of ecological units determines the surface qualities of meaningful things, i.e. their shape, called layout, and their surface texture (including colors). Furthermore, these surface qualities individuate the meaningful things in the environment by affordances. Due to ecological reasons, things with surfaces are able to afford actions: for example to support movements, to enclose something as hollow objects, to afford throwing as detached objects. Surface perception is therefore considered by Gibson to be a reliable mechanism for object individuation: Surfaces are the boundaries of all meaningful things humans can distinguish by perception. Beyond each surface lies another meaningful thing (exclusiveness), and the meaningful environment is exhaustively covered by such meaningful things (there is no part of it that is not covered by them). Furthermore, all major categories of meaningful things can be individuated by some affordance characteristic based on surface qualities.

The most important top level categories of such things are substances, media and surfaces. A part of a medium is a unit of the environment that affords locomotion through it, is filled with illumination (affords seeing) and odor (affords smelling) and bears the perceivable vertical axis of gravity (affords vertical orientation). In this paper, we will restrict our understanding of a medium to locomotion affordances. It is clear that the classification of a medium is stable only in a certain locomotion context: water is a medium for fish or divers, but not for pedestrians. So there will be different media for different locomotion contexts, but for most cases, including this paper, it will be enough to consider two of them: water and air. Substances simply denote the rigid things in a meaningful environment that do not afford locomotion through them. Surfaces are a thin layer of medium or substance parts located exactly where any motion must stop.

We complement Gibson's ideas by drawing on the concept of virtual places and fictive motion outlined in 2.3. In doing so, we reaffirm the idea of affordances as central to the perception of the environment, because we assume that substances, media and surfaces can be perfectly conceived through the imagination of virtual bodies moving through them. So if people say that the branches of a biological tree are thicker than their forearm, we consider them conceiving parts of the tree and the forearm as places for a virtual body. We call such a place locus of attention. Furthermore, we also consider humans being able to refer to parts of places that are beyond perception and therefore even more virtual: Humans can refer for example to the cells of a leaf without perceiving them. We closely stick to the idea that this perceivable environment is the source for human conceptualization, and all other categories are refinements or abstractions from them. In particular, our notion of place is e.g. much less abstract than Casati and Varzi's notion [9]: Entities and their places are not distinguishable, and therefore two things occupying the same place are the same. This also distinguishes our approach from that of an ecological niche [22].

3.4   Nested Places

The central methodological question is: for which structures in a Gibson environment can we assume sensors, and what are their formal properties? In this section, we will introduce and discuss the domain G (denoting the domain of virtual places) and the part-of relation $P[partof]$ on G. In Sect. 3.5, 3.6 and 3.7, we discuss geometrical properties of virtual steps, $Step$, $=_L$ $[equallength]$ and $OnL[equaldirection]$ in G, and in sections 3.8 and 3.9 we introduce medium connectedness $AirC$ and $WaterC$ and verticality $VertAln$ on G, respectively. In the formal part of these sections, we introduce a first order theory and assume for convenience that all quantifiers range over G unless indicated otherwise, and that all free variables are implicitly all-quantified.

Following our discussion above, we take the view that our *domain G* consists of *nested places for the actual and virtual things that can be perceived* in the environment. Mathematical artifacts like infinitely thin planes, lines and points are not in this domain, because they cannot contain extended things and are not perceivable. So we must construct the whole environment from something equivalent to regular regions in Euclidean space. Places have well-behaved mereological and geometric structures, which will be discussed in the subsequent sections. Our ideas about this structure were influenced by [2], [6].

As discussed in Sect. 3.3, our domain of places has a part-whole structure humans can refer to by pointing, and this structure is assumed to be *atomless* (even though we assume a granularity for perceiving their geometrical or topological properties):

Axiom1. We assume the axioms of a *closed extensional mereology* (CEM) [9] for a primitive part-of relation $P : G \times G$ on places G (meaning the first place is a part of the second), so that the mereological sum of every collection of places is another existing place and two places having the same parts are mereologically equal. The usual mereological symbols PP (proper-part-of), O (overlap), PO (proper overlap) and $+$ (sum) are definable.

3.5   Steps and Their Length and Direction

We suppose that humans experience the geometrical and topological structures of the meaningful environment by a primitive binary relation $Step(a, b)$, meaning the *virtual or actual movement of a locus of attention* from place a to b. Humans perceive *length* and *direction* of steps, because (in a literal sense) they are able to repeat steps of equal length and of equal direction. And thereby, we assume, they are able to observe and measure lengths of arbitrary things in this environment. The ratio scale properties of these lengths, as described e.g. by extensional systems in [26], must then be formally derivable from the structural properties of steps. What formal properties can we assume for such perceivable steps?

The visual perception of *geometrical properties*, like equality of distances and straightness of lines, is a source of puzzles in the psychological literature, because human judgment tests revealed e.g. that perceived straight lines are not

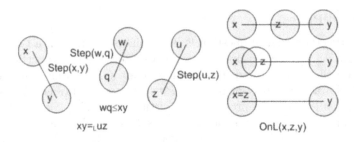

Fig. 1. Equal length and equidirection for steps

equivalent to the usual Euclidean straight lines [23]. Like Roberts et al. [21] suggested, it is nevertheless plausible to assume that the usual *Euclidean properties*, like congruence of shapes under rotation, are reconstructed by humans through learning. We adopt this view because Euclidean properties seem indispensable for human orientation and the recognition of body concepts.

Perceived geometry should have a *finite* and *discrete* structure, because of the *resolution properties* of sensors [10], and human perception in particular [17]. There are *finite approaches to geometry* available that seem to fit well to our problem (see Suppes [24]), but would also require finite and approximate accounts of a length scale (an example for such a scale can be found in [25]).

As a first step, we confined ourselves in this paper to an *infinite and dense version of a theory of steps with granularity*, based on the pointless axiomatization of *Euclidean geometry* given originally by Tarski [27]. We write $xy =_L uz$ for two steps from locus x to y and from u to z having *equal length*, and $OnL(x, z, y)$, if locus z is *on a line* between x and y or equal to any of them (compare Fig. 1). Note that *OnL* implies collinearity and betweenness. For the rest of the paper, we assume that the primitives *OnL* and $=_L$ are *only defined for loci of attention*.

Our equidistance $=_L$ and equidirection *OnL* primitives for steps satisfy a 3-D version of Tarki's *equidistance* and *betweenness* axioms for *Elementary Geometry* [27], similar to the approach in [2]. The quantifiers on points in Tarski's version or on spheres in Bennett's version can be replaced by quantifiers over the domain and range of steps: Our quantifier $\forall^{Locus}x.F(x)$ for example, meaning $\forall x.Step(x, x) \rightarrow F(x)$, restricts the domain of places to the *loci of attention*. Unlike Tarski and Bennett [2], we do not assume sphericity but the step relation as a primitive. Furthermore, *identity of points* in Tarski's or *concentricity of spheres* in Bennett's version just means to *"step on the spot"* in our theory. We therefore assume a version of Tarski's axioms with identity of points replaced by mereological equality of loci of attentions:

Axiom2. We assume the following axioms for *equidistance* $=_L$: Symmetry, identity, transitivity (compare the three axioms for *equidistance* in [2]), and for *equidirection OnL*: Identity, transitivity and connectivity (compare the three axioms for *betweenness* in [2]). We also assume the axioms of Pasch, Euclid, the Five-Segment axiom, the axiom of Segment Construction, the Weak

Continuity axiom, and a 3-D version of the Upper and Lower Dimension axiom, as described in [2].

Tarski's axioms ensure that there are loci of attention centered at all the "points" of a Euclidean space. It follows that steps and their lengths and direction have the expected Euclidean properties, and in particular that *each pair of loci of attention* forms a step with these properties.

3.6   Loci of Attention

A locus of attention is a smallest perceivable place. It can be thought of as a granular sphere with congruent shape giving rise to a minimal resolution for geometric properties in general. This is because we assume that humans perceive the geometrical qualities of arbitrary places by covering them with loci of attention. In fact, loci of attention are our simplified version of *just noticeable differences* in psycho-physics [17], but being independent of the stimulus.

We can define a notion of *shorter than* $\leq$, holding iff a step from y to x is shorter than a step from q to z:

Definition 1. *(shorter than)* $\forall^{Locus} y, x, q, z.(yx \leq qz) \leftrightarrow \exists x'.OnL(y, x, x') \land yx' =_L qz$

We now can define a topological notion of *touching* or *weakly connected*, which applies for a smallest step with non-overlapping loci:

Definition 2. *(touching)* $\forall^{Locus} x, y.Touching(x, y) \leftrightarrow \neg O(x, y) \land (\forall z.Step (x, z) \land xz \leq xy \rightarrow O(x, z))$

All loci of attention are required to be *congruent* to each other. This can be expressed by requiring congruent lengths for all pairs of touching loci:

Axiom3. *(locus of attention)* $\forall^{Locus} x, y, z, u.Touching(x, y) \land Touching (z, u) \rightarrow xy =_L zu$

It follows that loci are spheres of a fixed size. Similar to [3], *significant places* are the ones that are big enough to contain a granular locus of attention:

Definition 3. *(significant place)* $Significant(r) \leftrightarrow \exists^{Locus} x.P(x, r)$

We hence assume that insignificant places fall beyond the perceivable resolution for geometric properties. A *discrete* theory of steps would allow for an even stronger notion of resolution based on a minimal step length (see [24]).

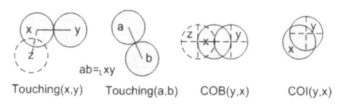

Touching(x,y)   Touching(a,b)   COB(y,x)   COI(y,x)

Fig. 2. Touching steps, steps centered on boundary (COB) and interior (COI)

3.7   The Environment Is Wholly Covered by Steps

A step is the central perceivable relation giving rise to affordances, because if we *virtually step through an environment*, it affords the *continuous transfer* of a virtual body from one place to another. This is only possible if those places are *strongly connected*. A *strongly self-connected* place always contains a *sphere* which, when we split the place at any point, overlaps both halves of the split (compare definitions in [4], [7]). So there is a 2-D surface in the middle corresponding to any cut (like "cutting in wood"), and not a line or a point. Strong connectedness can be expressed based on our primitives by introducing *paths*:

We call the minimal elongated place one can step in a *path*. We assume that if there is *a step from locus x to y on a path p*, then x and y are part of p, and there is always exactly one other locus z on p with equal distance from x and y :

Definition 4. (path) $EndsOfPath(x, y, p) \leftrightarrow Step(x, y) \land P(x + y, p) \land$
$(\exists! z.Step(x, z) \land P(z, p) \land xz =_L zy) \land (\neg \exists p'.EndsOfPath(x, y, p') \land PP(p', p))$

The idea is that *paths* are the smallest elongated places of minimal thickness enclosing a step and all the closer steps in between them. Because there is exactly one locus z in the middle of x and y on the path p, we assure that the path has minimal thickness and is elongated (see Fig. 3). Because there is no smaller path with that property, we make sure that x is the beginning and y is the end of the path. The definition implies strong connectedness of a path, because all pairs of loci have a continuous collection of loci in between them, and loci are spherical.

A general definition of strong connectedness for significant places is then straightforward:

Definition 5. (strong connectedness) $SC(x) \leftrightarrow Significant(x) \land$
$(\forall^{Locus} u, z.P(u + z, x) \rightarrow \exists p.EndsOfPath(u, z, p) \land P(p, x))$

We take the view that our domain of places, the meaningful environment, is *wholly covered by steps and paths*, because we assume that this is the common way how people experience their environment. It turns out that our environment of places is quite similar to the ideas outlined in [4], [2], [6], especially Bennett's region based geometry RBG. In order to establish the link between steps on

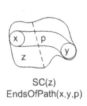

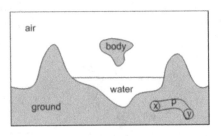

Fig. 3. Strongly connected places are connected by paths. The meaningful environment is wholly covered by strongly connected places.

one side and arbitrary significant places on the other, we will introduce 4 axioms along the lines of thought in [4], which ensure that each significant place coincides with a set of centers of loci of attention.

To this end we need a topological notion called *centered on the boundary* $COB(y,x)$ (compare Fig. 2), meaning y is just on the boundary of x,

Definition 6. *(centered on boundary)* $\forall^{Locus} y, x.COB(y,x) \leftrightarrow$
$(\exists z.Step(x,z) \wedge OnL(z,x,y) \wedge (zx =_L xy) \wedge Touching(z,y))$

from which a further notion, *centered on the interior* (Fig. 2), is definable.

Definition 7. *(centered on interior)* $\forall^{Locus} y, x.COI(y,x) \leftrightarrow$
$(\exists z.Step(x,z) \wedge COB(z,x) \wedge (\neg zx =_L xy) \wedge (xy \leq xz))$

For *arbitrary places of a significant size* x we can now define a predicate meaning that a step is centered on its interior.

Definition 8. *(centered on interior)* $\forall^{Locus} y.COI(y,x) \leftrightarrow$
$(\exists z.Step(y,z) \wedge COI(y,z) \wedge P(z,x))$

We first have to make sure that there is a sum of loci of attention corresponding to every (significant) open 3-ball in our Euclidean environment. We therefore assume that for each step xy of at least half a locus length, there is a perceivable ball z centered on x and topologically bounded by y. This is actually a granular variant of Bennett's Axiom 6 in [4]:

Axiom4. *(steps give rise to significant 3-balls)* $\forall^{Locus} x, y.\neg COI(x,y) \rightarrow$
$(\exists z.(\forall w.COI(w,z) \leftrightarrow (xw \leq xy \wedge \neg xw =_L xy)))$

We also assume that the domain of steps is *extensible*, so there are always larger balls constructible (compare Axiom 7 in [4]):

Axiom5. *(steps are extensible)* $\forall^{Locus} x, y.\exists z.Step(x,z) \wedge (x \neq z) \wedge (\forall z'.xz \leq$
$xz' \leftrightarrow (xy \leq xz' \wedge \neg xz' =_L xy))$

Secondly, we have to make sure that center points on *arbitrary significant places* behave in correspondence with their mereological structures. So parts of places always imply interior steps (compare Axiom 8 in [4]):

Axiom6. *(parts imply interior steps)* $P(x,y) \leftrightarrow (\forall^{Locus} u.COI(u,x) \rightarrow$
$COI(u,y))$

And third, we must assure that all places are actually covered by steps (compare Axiom 9 in [4]):

Axiom7. *(places overlap with loci of attention)* $\forall r \exists^{Locus} x.O(x,r)$

As was shown by Bennett [4], our Axioms 1-7 provide an axiom system for 3-dimensional regular open sets of Euclidean space: It can be proven that the sets of interior loci ("points") of arbitrary sums of loci are *regular open*. Because of granularity of perception, places are not in general constructible from steps in our theory. It can nevertheless be proven that they are *coverable* by loci of attention, and these covering sums of loci must have the expected geometrical properties:

Proposition 1. *Every place is part of a sum of loci of attention.*

To see this, be aware that from Axiom 7 it follows that every place has a part that is part of a locus of attention. With Axiom 1 we assume that every place is a sum of such parts. Thus every place is coverable by steps.

3.8   Media and Substances Are Wholes under Simple Affordance

Closely following Gibson, we assume that humans can directly perceive whether the environment *affords a certain type of movement*, and are thereby able to *individuate bodies, media and their surfaces*. Like in the previous section, we can think of such movements as fictive motions, and therefore the affordance primitives in this section are just refinements of our already introduced *step* primitive. In general, we assume that humans can perceive a multitude of *such simple affordances*. For our purpose, we will describe two of them, $AirC$ and $WaterC$, for movement in *air* and *water*, respectively. Because they have the power of individuation, we assume that those relations are *mutually exclusive*. Then we can *define media* as *unified wholes* under the respective affordance primitive.

Media. Connected by the same medium implies a step on a medium-connected *path* and is *symmetric* and *transitive*:

Axiom8. *(connected by the same medium)* $MediumC(x,y) \rightarrow$
$\exists p.EndsOfPath(x,y,p) \wedge (\forall^{Locus} z.P(z,p) \rightarrow MediumC(z,x)$
$\wedge MediumC(z,y)) \wedge (\forall u.MediumC(y,u) \rightarrow MediumC(x,u))$

The actually observable primitives are its two mutually exclusive sub-relations $AirC$ and $WaterC$,

Definition 9. *(a medium is either air or water)* $MediumC(x,y) \leftrightarrow$
$AirC(x,y) \vee WaterC(x,y)$

Axiom9. *(mutual exclusiveness)* $MediumC(x,y) \wedge MediumC(u,z) \wedge (z + u)O(x+y) \rightarrow ((WaterC(z,u) \wedge WaterC(x,y))Xor(AirC(z,u) \wedge AirC(x,y)))$

which give rise to media *water* and *air* by using them as *unity criterion*: A *(water/air) medium* is any maximal medium-connected whole:

Definition 10. *(media)* $Air(x) \leftrightarrow Whole(x,AirC) \wedge$
$Water(x) \leftrightarrow Whole(x,WaterC) \wedge Medium(x) \leftrightarrow Whole(x,MediumC)$

For a definition of *whole* as a *maximal sum* of parts *connected* by a partial equivalence relation, we refer to [11]. Informally, a medium is just any place which has all places of an *equivalence class* of *AirC* or *WaterC* as parts. Due to Axiom 8 and by definition, media must be composed of paths and therefore must have a significant size.

Substances and Bodies. According to Gibson, substances are not directly perceivable, only via the perception of a certain medium and its surface: just like walls are only perceivable as obstacles for the locomotion of light and other bodies through media. In this view, substances can be defined from media and define all other forms of places:

Definition 11. (substance) $Substance(x) \leftrightarrow \neg \exists z.(Medium(z) \wedge O(xz))$

Nevertheless, humans can obviously recognize the shape of certain significant strongly connected substances, called *bodies*. Therefore we must conceive bodies, similar to media, as individual wholes made up of *substance paths*. This is possible because the whole environment is covered with virtual paths by proposition 1. Bodies are maximal strongly connected substance wholes:

Definition 12. (connected by a body) $BodyC(x, y) \leftrightarrow Substance(x + y) \wedge SC(x + y)$,

meaning two places are connected by the same body if they are substances and their sum is strongly connected, and

Definition 13. (body) $Body(x) \leftrightarrow Whole(x, BodyC)$,

meaning a body is any maximal strongly-connected whole of substances.

Surfaces. Surfaces must also be definable using steps, and therefore have a *minimal thickness*. Thereby we avoid the philosophical question whether an infinitely thin topological boundary belongs to a region or its complement (see Casati and Varzi [9]). We take Borgo's view advocated in [7] and assume that every individuated body or medium has its own surface, which is simply a thin layer of paths making up the surface part of the body or medium (compare Fig. 4), such that every step's locus is touching a locus from the outside of the body or medium:

Definition 14. (connected by a surface) $SurfaceC(x, y, r) \leftrightarrow$
$(Medium(r) \vee Body(r)) \wedge (\exists p.EndsOfPath(x, y, p) \wedge P(p, r) \wedge (\forall^{Locus} z.P(z, p) \rightarrow$
$(\exists^{Locus} u.\neg O(r, u) \wedge Touching(z, u))))$

A surface is then a maximal surface connected part of a body or a medium:

Definition 15. (surface) $SurfaceOf(x, r) \leftrightarrow$
$Whole(x, (\lambda u, v.SurfaceC(u, v, r)))$

From these definitions and Axioms 8 and 9 it is provable that substances and media in fact make up the whole meaningful environment:

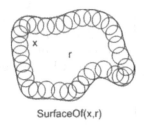

Fig. 4. Surfaces of bodies consist of a thin layer of virtual paths touching the outside of the body

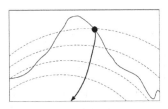

Fig. 5. A curved gravity line and equipotential surfaces

Proposition 2. *Substances and Water and Air "partition" the meaningful environment: Any place that is part of one category cannot be part of another one (mutual exclusion), and any place is part of a sum of places of these categories (exhaustiveness).*

3.9   Verticality and Absolute Orientation

We assume that there is a human sensor asserting that a step is aligned with gravity, as illustrated by the *well* example in Sect. 3.2. This primitive is called *VertAln*. Assuming collinearity and parallelism for this primitive would be an oversimplification, because *gravity lines through the earth's body are not straight lines* (compare Fig. 5).

   We assume that *VertAln* describes a symmetric and transitive step, and that it always exists for arbitrary loci of attention:

Axiom10. *(ubiquity)* $\forall^{Locus} x . \exists^{Locus} z . VertAln(x, z)$
Axiom11. *(symmetry and transitivity)* $VertAln(x, y) \rightarrow Step(x, y) \wedge$
   $VertAln(y, x) \wedge \forall^{Locus} z . (VertAln(x, z) \rightarrow VertAln(y, z))$

4   Defining Water Depth from Direct Perception of a Water Body

It remains to show that the Gibson environment is applicable to our water depth scenario, such that all geographic categories, including the water depth quality, are definable in it.

4.1   Deriving Lengths from the Meaningful Environment

We claimed that all involved categories can be defined from the meaningful environment. What is still missing is a definition for a symbol space of lengths. The domain of lengths L is not part of the environment and is assumed to be an *abstract extensional system* in the sense of [26]. It has two binary relations "smaller than or equal" and "sum" that satisfy Suppes' extensional axioms, and is therefore on a *ratio scale* with one degree of freedom.

   A length function *Length* can be derived as a homomorphism from the set of steps of the meaningful environment into the length space L. It is clear that this mapping is conventional and must itself rely on a semantic datum. We therefore first have to fix a *unit step Step(0,u)*, for example the two ends of the platinum bar called *"Mètre des Archives"*. We call one end of this bar 0 and the other one u:

Axiom12. *(non-trivial unit step)* $\forall x, y.(x = 0 \wedge y = u) \rightarrow (Step(x,y) \wedge x \neq y)$

Second, we map the quaternary symbol $\leq$ on loci to the "smaller than or equal to" symbol on the length space. Third, we map the following definable summation symbol on loci to the summation symbol of the length quality space:

Definition 16. *(sum of lengths)* $cx \oplus ez =_L ky \leftrightarrow \exists x', y'.(0x' =_L cx) \wedge (0y' =_L ky) \wedge (x'y' =_L ez) \wedge (OnL(0, x', y') \vee OnL(x', 0, y') \vee OnL(0, y', x'))$

A sum of the lengths of two steps is a third step having the expected length. We fourth map all steps with the same length as *Step(0,u)* to the symbol "1", and all other steps *homomorphically* to a number symbol: All steps with equal length are mapped to one and only one number symbol such that the truth of all sentences about $\leq$ and $\oplus$ is preserved.

   For convenience, we write $\forall^L$ for a quantifier over lengths in L and use the symbols $=_L, \oplus, \leq$, defined for steps, analogously on lengths. Once a length space for steps is established, we can define a *chain length of a path* recursively as the sum of lengths of any chain of steps on it:

Definition 17. *(chain length)* $\forall^L k.ChainLength(p) =_L k \leftrightarrow \exists^{Locus} x, y.EndsOfPath(x, y, p) \wedge (k =_L Length(x, y) \vee (\exists z, p', p''.(p = p' + p'') \wedge (k =_L ChainLength(x, z, p') \oplus ChainLength(z, y, p''))))$

The *length of an arbitrarily shaped path* is then its maximal chain length. Because - in our current infinite theory - steps are assumed to be dense, this must be an infinitesimal approximation:

Definition 18. *(length of a path)* $\forall^L k.Length(p) =_L k \leftrightarrow \forall k'.ChainLength(p) = k' \wedge k \leq k' \rightarrow k =_L k'$

4.2   Water Depth

The ratio scaled meter water depth space needs a semantic datum to fix its interpretation, because there is no direct sensor available for it. It is a quality that

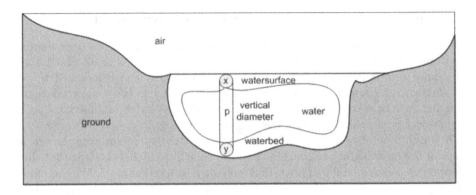

Fig. 6. Water depth is the length of a vertical diameter path in a water body

has to be constructed from others. Informally, in the meta-data of a database, we could say that a water depth of a river is the vertical distance measured between a point on the water surface and the river bed. We now can define these notions from the meaningful environment:

Let us start by defining a diameter of a medium in the meaningful environment: A *diameter* of a medium r is any medium path connecting two parts of its surface, such that *the path is contained in r*:

Definition 19. (diameter) $Diameter(p, r) \leftrightarrow Medium(r) \wedge P(p, r) \wedge$
$SurfaceOf(s, r) \wedge (\exists^{Locus} x, y, u.EndsOfPath(x, y, p) \wedge P(x + y, s) \wedge x \neq y \wedge$
$P(u, p) \wedge \neg P(u, s))$

Note that what normally would be considered as the *water surface* is a part of our medium surface. We define a *depth of a medium* as the length of a diameter path whose steps are vertically aligned:

Definition 20. (depth) $\forall^L k.Depth(k, p, r) \leftrightarrow Diameter(p, r) \wedge$
$k =_L Length(p) \wedge (\forall^{Locus} z, u.P(z, p) \wedge P(u, p) \rightarrow VertAln(z, u))$

We can now state that a water depth is a depth of a water body connecting the medium air with the ground. There is an infinite number of water depths for a water body (compare [19]):

Definition 21. (water depth) $\forall^L k.Waterdepth(k, p, r) \leftrightarrow Depth(k, p, r) \wedge$
$Water(r) \wedge (\exists^{Locus} x, y.EndsOfPath(x, y, p) \wedge (\exists^{Locus} a, b.Touching(x, a) \wedge$
$Touching(y, b) \wedge AirC(a, a) \wedge BodyC(b, b)))$

Sticks and sonars both can be considered as realizations of such a virtual water depth path.

5   Conclusion and Future Work

Our approach resolves semantic heterogeneity of basic symbols in an ontology by introducing semantic datums in the form of an observation procedure. Water depth is an example of a basic symbol in a navigation ontology in need of grounding. We show that Gibson's meaningful environment is a sufficient basis to establish such semantic datums. For this, we introduced a formal theory of Gibson's *meaningful environment*, supplying observable primitives for our theory. We show how to ground water depth in our theory by asserting formal characteristics of observable primitives and defining water depth in terms of these.

We take the view that sensors (human or technical) are implementations of semantic datums that reproducibly interpret observable primitives into observable real world structures. Technical sensors are based on human sensors, and human sensors detect bodies and their movements. Natural language semantics is grounded in *bodily experience* and *scanning,* that is the imaginative movement of virtual bodies in a perceivable environment. This is plausible because humans can always unambiguously refer to bodies and their parts by ostension. We claim that Gibson's meaningful environment is fully equipped with body related sensors, including sensors for steps, equal length, equidirection, verticality and various sensors for simple affordance primitives that can be used to individuate media, surfaces and substances.

The presented theory of the meaningful environment is a first sketch of our ideas. As outlined in Sect. 3.5, the theory could be made more appropriate to sensors by making it *finite* and *discrete*, so that we admit a resolution for lengths, along the lines of thought in [24] and [25]. Furthermore, it is still an open question which ontologies are amenable to our method. The general applicability to geospatial concepts needs to be demonstrated with additional case studies. We are currently working on flow velocity and street network categories.

Even though the proposed observation primitives are bound to have cognitive interpretations, it is important to note that they have low indeterminacy. Observable properties on this basic body level do not seem to have a graded structure, too. For example, there is no graded truth value for the sentence "this is the wall of this room" for blind men being in that room, as there will be perfect agreement on this sentence. Any theory about semantic grounding must primarily be able to explain how humans actually accomplish inter-subjective measurement and observation of surface qualities, despite all the cognitive and linguistic ambiguities involved.

Acknowledgments

This work is funded by the *Semantic Reference Systems II* project granted by the German Research Foundation (DFG KU 1368/4-2), as well as the International Research Training Group on *Semantic Integration of Geospatial Information* (DFG GRK 1498).

References

1. Barsalou, L.W.: Grounded cognition. Annual Review of Psychology 59, 617–645 (2008)
2. Bennett, B.: A Categorical Axiomatisation of Region-Based Geometry. Fundamenta Informaticae 46, 145–158 (2001)
3. Bennett, B.: Physical Objects, Identity and Vagueness. In: Fensel, D., McGuinness, D., Williams, M.A. (eds.) Proc. of the 8th International Conference on Principles of Knowledge Repr. and Reasoning. Morgan Kaufmann, San Francisco (2002)
4. Bennett, B., Cohn, A.G., Torrini, P., Hazarika, S.M.: A Foundation for Region-Based Qualitative Geometry. In: Horn, W. (ed.) Proc. 14th European Conf. on Artificial Intelligence, pp. 204–208. IOS Press, Amsterdam (2000)
5. Bennett, B., Mallenby, D., Third, A.: An Ontology for Grounding Vague Geographic Terms. In: Eschenbach, C., Gruninger, M. (eds.) Proc. 5th Intern. Conf. on Formal Ontology in Information Systems. IOS Press, Saarbrücken (2008)
6. Borgo, S., Guarino, N., Masolo, C.: A Pointless Theory of Space Based on Strong Connection and Congruence. In: Proc. Principles of Knowledge Repr. and Reasoning (KR 1996), Boston (1996)
7. Borgo, S., Guarino, N., Masolo, C.: An Ontological Theory of Pysical Objects. In: Ironi, L. (ed.) Proc. Qualitative Reasoning 11th Intern. Workshop, Paris (1997)
8. Boumans, M.: Measurement Outside the Laboratory. Pilosophy of Science 72, 850–863 (2005)
9. Casati, R., Varzi, A.: Parts and Places: The Structures of Spatial Representation. MIT Press, Cambridge (1999)
10. Frank, A.: Scale is Introduced by Observation Processes. In: Proc. 6th Intern. Symp. on Spatial Data Quality, ISSDQ 2009 (2009) (forthcoming)
11. Gangemi, A., Guarino, N., Masolo, C., Oltramari, A.: Understanding top-level ontological distinctions. In: Stuckenschmidt, H. (ed.) Proc. IJCAI 2001 workshop on Ontologies and Information Sharing (2001)
12. Gibson, J.J.: The Ecological Approach to Visual Perception. LEA Publishers, Hillsdale (1986)
13. Harnad, S.: The Symbol Grounding Problem. Physica D 42, 335–346 (1990)
14. Kuhn, W.: Semantic Reference Systems. Int. J. Geogr. Inf. Science 17(5), 405–409 (2003)
15. Kuhn, W., Raubal, M.: Implementing Semantic Reference Systems. In: Gould, M., Laurini, R., Coulondre, S. (eds.) 6th AGILE Conference on Geographic Information Science, pp. 63–72. Presses Polytechniques et Universitaires Romandes, Lyon (2003)
16. Langacker, R.W.: Dynamicity, Fictivity, and Scanning. The Imaginative Basis of Logic and Linguistic Meaning. In: Pecher, D., Zwaan, A. (eds.) Grounding Cognition. The Role of Perception and Action in Memory, Language and Thinking, Cambridge University Press, Cambridge (2005)
17. Michell, J.: Measurement in Psychology: Critical History of a Methodological Concept. Cambridge Univ. Press, Cambridge (1999)
18. Probst, F.: Semantic Reference Systems for Observations and Measurements. PhD Thesis, University of Münster (2007)
19. Probst, F., Espeter, M.: Spatial Dimensionality as Classification Criterion for Qualities. In: Proc. Intern. Conf. on Formal Ontology in Information Systems (FOIS 2006). IOS Press, Baltimore (2006)

20. Quine, W.V.O.: Word and Object. The MIT Press, Cambridge (1960)
21. Roberts, F.S., Suppes, P.: Some Problems in the Geometry of Visual Perception. Synthese 17, 173–201 (1967)
22. Smith, B., Varzi, A.C.: The Niche. Noûs 33, 214–238 (1999)
23. Suppes, P.: Is Visual Space Euclidian? Synthese 35, 397–421 (1977)
24. Suppes, P.: Finitism in Geometry. Erkenntnis 54, 133–144 (2001)
25. Suppes, P.: Transitive Indistinguishability and Approximate Measurement with Standard Finite Ratio-scale Representations. J. Math. Psych. 50, 329–336 (2006)
26. Suppes, P., Zinnes, J.L.: Basic Measurement Theory. In: Luce, R.D., Bush, R.R., Galanter, E. (eds.) Handbook of Mathematical Psychology, ch. 1, vol. 1 (1963)
27. Tarski, A.: What is Elementary Geometry? In: Henkin, L., Suppes, P., Tarski, A. (eds.) The Axiomatic Method with Special Reference to Geometry and Physics, Amsterdam, pp. 16–29 (1959)

Terabytes of Tobler: Evaluating the First Law in a Massive, Domain-Neutral Representation of World Knowledge

Brent Hecht[1] and Emily Moxley[2]

[1] Electrical Engineering and Computer Science, Northwestern University, Frances Searle Building #2-419, 2240 Campus Drive, Evanston, IL 60208
brent@u.northwestern.edu
[2] Electrical and Computer Engineering, UC Santa Barbara, Mailbox 217, Harold Frank Hall, Santa Barbara, CA 93106
emoxley@ece.ucsb.edu

Abstract. The First Law of Geography states, "everything is related to every-thing else, but near things are more related than distant things." Despite the fact that it is to a large degree what makes "spatial special," the law has never been empirically evaluated on a large, domain-neutral representation of world knowledge. We address the gap in the literature about this critical idea by sta-tistically examining the multitude of entities and relations between entities pre-sent across 22 different language editions of Wikipedia. We find that, at least according to the myriad authors of Wikipedia, the First Law is true to an over-whelming extent regardless of language-defined cultural domain.

Keywords: Tobler's Law, First Law of Geography, Spatial Autocorrelation, Spatial Dependence, Wikipedia.

1 Introduction and Related Work

When he first posited the statement that "everything is related to everything else, but near things are more related than distant things" [1] almost 40 years ago[1], Waldo Tobler had no idea it would have such staying power. Widely accepted as the First Law of Geography and also frequently known as simply Tobler's Law or Tobler's First Law (TFL), this assertion appears in nearly every geography and Geographic Information Systems (GIS) textbook printed today. Moreover, many social and physi-cal sciences have adopted as existentially essential the ideas of spatial dependence or spatial autocorrelation, both of which are accessibly and succinctly defined by Tobler [2]. TFL has even spawned a First Law of Cognitive Geography, which states that in the context of information visualization "people *believe* closer things are more simi-lar" [3, 4]. Obviously, all of these ideas have proven to be of major applied use in a

---

[1] Although his paper was published in 1970, he first presented his work at a 1969 meeting of the International Geographic Union's Commission on Quantitative Methods, making this year arguably the law's true 40th anniversary.

K. Stewart Hornsby et al. (Eds.): COSIT 2009, LNCS 5756, pp. 88–105, 2009.
© Springer-Verlag Berlin Heidelberg 2009

vast array of well-known work solving a vast array of specific problems. However, Tobler's statement is high-level, domain-neutral, and problem-independent in scope and it has never been empirically evaluated in these terms. Many authors [5-9] have opined on the topic at a philosophical level, but no experiments have been done. A data-based investigation of such a broad statement has enormous challenges associated with it, and at least part of the reason for this gap in the literature has been the lack of available data to examine.

However, that hurdle was overcome with the development and rise to immense popularity of Wikipedia, the collaboratively authored encyclopedic corpus of unprecedented scale. While it is by no means perfect as a representation of the sum of world knowledge, it is by far the closest humanity has come to having such a data set. As of this writing, Wikipedia, consistently ranked as one of the top ten most-visited websites on the Internet, contains 2.76 million articles in its English edition, and has a total of 25 language editions with over 100,000 articles (see Table 1 for descriptive statistics of the data used in our studies). Each of these articles describes a unique entity.

All of these facts are relatively well known by the Internet-using population. What is less understood is the scope of the quantity of *relations* between these entities present in Wikipedia. The relations, encoded by contributors ("Wikipedians"), and viewed as links to other Wikipedia pages by visitors, number well into the hundreds of millions. Although these unidirectional relations are not typed (except in some demonstration versions of Wikipedia such as "Semantic Wikipedia" [10]), they can still tell us which of the millions of entities are related in some way, and which are not.

We seek to leverage the entities and relations in this enormous data set to examine the validity of Tobler's Law in the very general context described above. While our experiment is, to our knowledge, the most broad empirical investigation of Tobler's Law done to date, it does have its limitations. Critically, we of course do not claim to evaluate the First Law on a representation of *all* spatial data in existence. Rather, due to our data source, our results will only confirm or deny the validity of TFL in the world *as humans see it*. We do assert that Wikipedia data is a reasonable, although flawed, proxy for the world as it is understood by humans. Ignoring this proxy, our experiments will at least determine the validity of TFL in the context of the world knowledge that has been represented by the millions of people who have contributed to Wikipedia (although most of it has been authored by a smaller number of people [11]), is accessed by countless millions more, and is used by dozens of systems in AI and NLP (e.g. [12, 13])

Along the same lines, it is important to at least briefly address the question of accuracy. While it has been found that Wikipedia's reputation for questionable intellectual reliability has been somewhat unfairly earned [14], the nature of our study almost entirely sidesteps the accuracy concern. Because we examine entities and relations in aggregate and rely far more on their existence than their specific details, we can to a large degree ignore accuracy risks. An Internet user would have to very purposely manipulate massive amounts of specific data across many languages of Wikipedia to be able to change the results of our experiments. Non-malicious systemic characteristics of Wikipedia do create their problems, but we describe in detail how we address these, point out when we are unable to, and discuss the problems therein. In summary, like all science, this study is subject to the rule of "garbage in, garbage out." However, judging by the number of papers published in the past few years on Wikipedia or using Wikipedia data in the fields of computer science, psychology, geography, communication, and more [7, 11, 15-17], Wikipedia has been assumed to be far better than garbage by *large* numbers of our peers. More specific to this project, many of the exact same structures leveraged in this study have been used, for instance, to calculate semantic relatedness values between words with much greater accuracy than WordNet [18].

Very rare even in Wikipedia research is our intensely multilingual approach. Less than a handful of papers attempt to validate conclusions in more than one language's Wikipedia. Almost always, the English Wikipedia is considered to be a proxy for all others, a problematic assumption at best. As a way to distinguish the opportunities and challenges of including a double-digit number of languages in our study, we describe our work as using not a multilingual data set, but a "hyperlingual" data set [19] (in analogy to multispectral and hyperspectral imagery). In the end, we consider our results to be much more valid because they are similar across the entire hyperlingual Wikipedia data set rather than being restricted to a single or small number of Wikipedias.

2 Data Preparation

2.1 WikAPIdia

The foundation of our work is WikAPIdia, a Java API to hyperlingual Wikipedia data that we have developed. Available upon request for academic research[2], WikAPIdia provides spatiotemporally-enabled access to any number of Wikipedias (a language edition of Wikipedia is often referred to as simply "a Wikipedia"). WikAPIdia also has a large number of graph analysis and natural language processing features built-in.

WikAPIdia initially takes as input the XML database dumps provided by the Wikimedia Foundation, the non-profit that manages Wikipedia. These XML files contain a "snapshot" in time of the state of the Wikipedia from which they are derived. The user must input the XML dump of each language she wishes her instance of WikAPIdia to support. For this study, we used XML dumps for the 22 different Wikipedias that had around 100,000 articles or more at the time of the dump (see Table 1). Since maximum temporal consistency is desirable in order to minimize external effects, snapshots from as close as possible to the most recent English dump, that of 8 October, 2008, were used.

The parsing of these 22 dump files takes a relatively new, moderately powered desktop PC approximately several days to complete. During this process, a large number of structures are extracted from the very semi-structured Wikipedia dataset, some of which are not used in this study. Those that are of importance to our experiment are described in detail below.

---

[2] Contact the first author for more details.

Table 1. Descriptive statistics of the Wikipedia Article Graph (WAG) and the number of spatial articles for each of the Wikipedias included in this study

Language	Vertices (Articles) = $\lvert V \rvert$	No. Edges (Links) = $\lvert E \rvert$	Spatial Articles = $\lvert V_{spatial} \rvert$
Catalan	141,277	3,478,676	13,474
Chinese	203,824	5,566,490	14,177
Czech	112,057	3,089,517	8,599
Danish	97,825	1,714,025	7,118
Dutch	497,902	9,679,755	103,977
English	2,515,908	76,779,588	174,906
Finnish	208,817	3,782,563	11,559
French	716,557	20,578,831	67,042
German	827,318	21,456,176	85,906
Hungarian	120,850	3,009,814	6,939
Italian	516,120	14,968,632	67,433
Japanese	532,496	20,946,112	21,621
Norwegian	193,298	3,774,509	16,607
Polish	555,563	12,678,608	58,367
Portuguese	437,640	8,577,435	79,844
Romanian	118,345	1,434,939	20,349
Russian	341,197	7,762,322	30,346
Slovakian	102,089	1,931,138	7,708
Spanish	443,563	12,576,477	42,431
Swedish	295,605	5,555,219	18,816
Turkish	120,689	2,260,241	5,431
Ukrainian	131,297	1,743,304	4,692
TOTAL	**9,230,237**	**243,344,371**	**867,342**

2.1.1 The Wikipedia Article Graph

From each Wikipedia one of the key structures extracted by WikAPIdia is the Wikipedia Article Graph (WAG). The WAG is a graph (or network) structure that has as its vertices (nodes) the articles of the Wikipedia and the links between the articles as its edges. In graph theory terminology, the WAG is a directed sparse multigraph because its edges have direction (a link from one article to another), each node is connected to a relatively small number of other nodes, and it can contain more than one link from an article to another article. This last characteristic is a problem for our study, and our workaround is described in section three. Even the "smallest" Wikipedias have relatively enormous WAGs. An overview of the basic size characteristic of each WAG is found in Table 1.

2.1.2 Interlanguage Links

Encoded by a dedicated and large band of Wikipedians aided by bots that propagate their work across the various language editions of Wikipedia, interlanguage links are essentially multilingual dictionary entries placed in each Wikipedia article. By parsing out these links, WikAPIdia is able to recognize that the English article "Psychology", the German article "Psychologie", and the Chinese article "心理学" all refer to the same concept. Although interlanguage links provide much of the raw material for creating WikAPIdia's multilingual concept dictionary, a significant amount of detailed post-processing is necessary to mediate conflicts contained within these links.

2.1.3 Spatial Articles

The spatial data used by WikAPIdia all derives from the latitude/longitude tags included in tens of thousands of articles by contributors to several different Wikipedias[3]. The motivation to contribute a tag is to allow readers of a tagged article to click a link to see the location of the subject of the article in any Internet mapping application such as Google Maps. However, WikAPIdia uses these tags for a modified purpose. In addition to providing their specific location in a geographic reference system[4], the tags inform WikAPIdia that the subjects of tagged articles are *geographic* entities. In other words, articles tagged with latitude and longitude coordinates can be called *spatial articles*. For example, the English article "Country Music" is very unlikely to contain a latitude/longitude tag, whereas the article "Country Music Hall of Fame and Museum" does include a tag and can be included in the class of spatial articles.

Although contributors to any of the Wikipedias included in this study are theoretically capable of tagging articles in their language of choice with spatial coordinates, in practice, this is not done in many of the "smaller" Wikipedias with any significant degree of coverage. As such, WikAPIdia uses interlanguage links to copy a tag across all languages' articles that refer to the same concept. For instance, although the German article "Country Music Hall of Fame" does not possess a geotag, it does include an interlanguage link to the English article mentioned above. WikAPIdia copies the tag from the English article to the German article, the Spanish article "Museo y Salón de la Fama del Country", etc. The number of spatial articles found for each Wikipedia is listed in Table 1.

2.2 The Scale Problem

Once our instance of WikAPIdia had successfully parsed and processed all 22 language dump files, an additional stage of data preparation was necessary due to what we call the "Geoweb Scale Problem" (GSP). Stated simply, GSP arises because most Web 2.0 spatial data representation schemas only support point vector features. The "blame" for this limited representational expressiveness can probably be split between designers' lack of education about geographic information as well as a dearth of popular tools that support vector features of greater than zero dimensions. For many

---

[3] WikAPIdia has its own spatial data parsers, but also supports the tags collected by the Wikipedia-World Project
(http://de.wikipedia.org/wiki/Wikipedia:WikiProjekt_Georeferenzierung/Wikipedia-World/en).

[4] WikAPIdia assumes all latitude and longitude tags are derived from World Geodetic System 1984 (WGS1984) due to its popularity amongst the general public.

geoweb applications, GSP does not restrict functionality a great deal, but in some cases, the points-only paradigm borders on ridiculous.

We are able to sidestep the smaller inaccuracies introduced by Wikipedia's point-based geotagging system by choosing the appropriate scale for analysis. In our experiments, we adopt a 50-kilometer precision for this purpose. We thus consider, for example, all points between 0 and 50km from each other to be of equal distance from each other. The problem that both all cities and all articles about places within those cities are encoded as points, for instance, can be almost entirely solved as long as we "flatten" our distance function in this way. For example, the spatial article pair "Chicago" and "O'Hare International Airport", which lies within Chicago, would be correctly placed in the 0-50km distance class instead of being assigned the false precision of 25km.

However, certain egregious point spatial representations cannot be reasonably handled using the above methodology. The U.S. state of Alaska, for instance, is encoded as a point (64°N, 153°W) in Wikipedia. Cartographically speaking, the only scale at which this representation would be valid is perhaps if one were examining Earth from Mars. Unfortunately, the relatively small number of cases that suffer from representation issues of this degree play a disproportionately large role in our study because these articles tend to have the most inlinks (or relations directed at them). Without further steps, the link from "Anchorage, Alaska" to "Alaska" and thousands like it would be falsely considered in a very high distance class (if the point representations are used, the "Anchorage" / "Alaska" pair would be in the 300-350km distance class, for example).

The only way to tackle this issue is to associate more appropriate representations for this set of particularly problematic point geotags using external data sources. To accomplish this task, we used the polygonal data and toponyms (place names) for countries and first-order administrative districts (like Alaska) included with ESRI's ArcGIS product[5]. We matched this data to Wikipedia articles in all languages by using Wikipedia's "redirect" structures and interlanguage links. Redirects are intended to forward users who search for "USA", for example, to the article with the title "United States." However, they also represent an immense synonymy corpus. The combination of these synonyms and ESRI's toponym-matched polygon data resulted in a relatively effective polygonal gazetteer for countries and first-order administrative districts.

We tried two different methods of leveraging this polygonal information. First, we used a point-in-polygon algorithm to map all point/polygon pairs to a distance of 0, as discussed in section three. Second, we simply stripped all the countries and first-order administrative districts found by our georeferencer from our data set in order to maintain a consistent scale of analysis. Interestingly, the main and secondary characteristics of our results – as described below – were very similar or identical between both methods (Figure 3.3, for example, is nearly identical trend-wise in both cases). As such, since the point-in-polygon algorithm was sufficiently computational complex to limit our sample size extensively (see section 3) and since the details of the

---

[5] Specifically, we used data in the "admin" and "countries" shapefiles.

choices we made crossing scales are too numerous to fit in this limited space, we discuss our second method in the remainder of this paper and leave analysis of multiple scales to future work. We also note that one of the several proposed "second laws" of geography states "everything is related to everything else, but things observed at a coarse spatial resolution are more related than things observed at a finer resolution." [20] This would suggest that by choosing a more local scale of analysis, we are selecting the option at which less spatial dependence in relations would occur, so TFL's validity should be most tested at this scale.

3 The Main Experiment

3.1 Hypothesis

Let us consider three spatial articles A, B, and C where $distance(A,B) < distance(A,C)$ (assuming the function $distance()$ calculates the straight-line distance from the entities described in the articles in the function's parameters)[6]. If Tobler's statement that "near things are more related than distant things" is indeed true, it is expected that spatial article A would be more likely to contain a relation/link to B than to C. In other words, we hypothesize that given three spatial articles A, B, and C, if TFL is valid, $P_{relation}(A,B)$ will be generally greater than $P_{relation}(A,C)$ if $distance(A,B) < distance(A,C)$, where $P_{relation}$ is the probability of the first spatial article containing a link/relation to the second. We also hypothesize, however, given TFL's first clause "everything is related to everything else," that $P_{relation}(A,C) > 0$, even if A and C are separated by a very large distance.

3.2 Methods

We test this hypothesis on the hyperlingual data set parsed and prepared by WikAPIdia. Stated simply, our basic methodology is to examine all pairs of spatial articles [A, B] (excluding the identity case, or [A, A]) and record for each pair the straight-line distance between them and whether or not A contains a link to B or B contains a link to A. We perform this analysis *separately* for each Wikipedia in order to compare the results for each language-defined data set.

Taking a page from the field of geostatistics, once we have all the relation existence / distance tuples (examples of tuples include (1) A and B are 242km apart and A links to B, and (2) A and C are 151km apart and A does not link to C), we group these tuples into distance *lag* classes of 50 kilometers[7] and evaluate the overall probability, P_d, of a link existing between two spatial articles that are separated by each distance class d. We measure the overall probability by calculating the number of existing

---

[6] As has been noted by many authors writing about Tobler's Law, straight-line distance is only the simplest of the many possible distance metrics that could be used. We leave experiments with more complex measures to future work.

[7] We identified 50km as the minimum precision possible due to the reasons presented above in our discussion of scale.

relations in each distance class and dividing that by the total number possible links, as shown below:

$$\frac{\sum_{i=1}^{|PAIR_d|} relation(PAIR_{di})}{|PAIR_d|} = P_d$$

where $PAIR_d$ is the set of all the spatial article pairs in distance class d, $relation()$ evaluates to 1 if there exists an outlink from the first spatial article in the pair to the second and 0 otherwise, and $|PAIR_d|$ is the number of spatial article pairs. It is important to note that $[A, B]$ and $[B, A]$ are considered to be different pairs because A can contain a relation to B, but also vice versa, and both must be considered in evaluating the number of possible links. From these lag-based statistics, we are able to derive empirical functions that form the basis of our results and analysis presented in the following subsections. These functions are effectively "pseudo-covariograms", with probabilities of relation/link existence in the place of covariance. We call these functions representing the relatedness between points spatial "relatograms" ("relate-oh-grams").

Even for the "smaller" Wikipedias, executing the above algorithm is a very computationally complex task. Consider the Catalan Wikipedia, for example. Since it contains 13,474 spatial articles, it was necessary to query WikAPIdia for the distance and existence of links between $|V_{spatial}|^2 - |V_{spatial}| = 13,474^2 - 13,474 = 181,535,202$ pairs. Since we must consider the pairs $[A, B]$ and $[B, A]$ to be different, we are unable – at least in the link existence portion of the measurements – to take advantage of the computational benefits of symmetry.

Since WikAPIdia makes extensive use of MySQL tables to store data, a great number of iterations require disk operations. Even with the large number of optimizations we wrote, this makes the algorithm even more time intensive. As such, for the English Wikipedia, for instance, doing the requisite $|V_{spatial}|^2 - |V_{spatial}| = \sim 30.5$ billion iterations was simply impractical. Thus, for all Wikipedias in which $|V_{spatial}| > 50,000$, we used a random subset (without replacement) of size 50,000. Since 50,000 spatial articles represents 28.5 percent at minimum of a Wikipedia's dataset, we are able to attain tractability without risking statistical insignificance.

In order to compare our results across all the Wikipedias, it was necessary to normalize by a measure of each Wikipedia's overall "linkiness." The measure we use is the probability of a link occurring from any random article (not necessarily spatial) X to any other random article Y. We calculated this measure by using the following formula:

$$P_{relation}(X,Y) = \frac{|E_{adjusted}|}{|V|^2 - |V|} = P_{random}$$

where $|V|$ = the number of total articles in the Wikipedia (not just spatial articles), and $|E_{adjusted}|$ is the number of non-duplicate links in the Wikipedia. Duplicate links occur when editors add two or more outlinks in any X to any Y. We evaluated $|E_{adjusted}|$ by calculating the average link duplication over a random sample of 10,000 links in each Wikipedia and dividing $|E|$ by this number. The average link duplication ranged from about 1.08 (or, each link from an X to a Y appears on average 1.08 times) to 1.26, so this was an important normalization.

Finally, in our results and analysis below, we frequently make use of a ratio that allows us to complete the sentence, *"If spatial article A is separated from spatial article B by distance class d, it is ___ times as likely as random to contain a link to B."* Given distance class *d*, this ratio is simply:

$$\frac{P_d}{P_{random}}$$

3.3 Results and Basic Analyses

As can be seen in Figure 1, if spatial articles *A* and *B* are within 50km of each other, they are around *245 times* as likely to have a relation connecting them than if they were any two random Wikipedia articles on average[8]. This spatial bias drops off rapidly, however, and by 650km or so, all significant positive spatial dependence goes away. In other words, Figure 1 clearly shows that our hypothesis that $P_{relation}(A,B)$ is

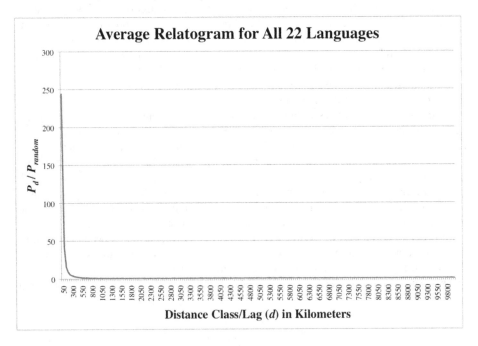

Fig. 1. A "relatogram" of the unweighted average of P_d / P_{random} across all 22 Wikipedias included in our study. The *y*-axis thus describes the average multiple of random probability that a link will occur from *A* to *B* given *d*. Along the *x*-axis are the distance classes/lags considered, or all the *d*s (0-50km, 50-100km, etc). The graph looks like that of any variable showing a great degree of spatial dependence: a large amount of relatedness at small distance classes, and a very large drop-off as larger distances are considered.

---

[8] By "average", we mean the average of all 22 languages' relatograms without weighting by number of articles considered.

generally greater than $P_{relation}(A,C)$ if $distance(A,B) < distance(A,C)$ is true[9]. Despite the fact that Wikipedians by no means attempted to create a resource that displayed relation spatial dependence, they nonetheless did so in dramatic fashion, and did so regardless of their language-defined cultural domain. Without exploring further, we can state firmly that Tobler's Law has been validated empirically on a massive repository of world knowledge. In Tobler's words but our emphasis, "near things *are* more related than distant things."

We also see in figures 1, 2, and 3 that no matter the distance class, the probability of A and B having a relation is never consistently zero, affirming the accepted meaning of the non-spatial dependence clause of TFL, that "everything is related to everything else." In some of the smaller Wikipedias, P_d occasionally drops to zero, but never does so consistently. This can be seen in the intermittent gaps in the series in figure 3 ($log(0)$ is undefined, so is displayed as a gap).

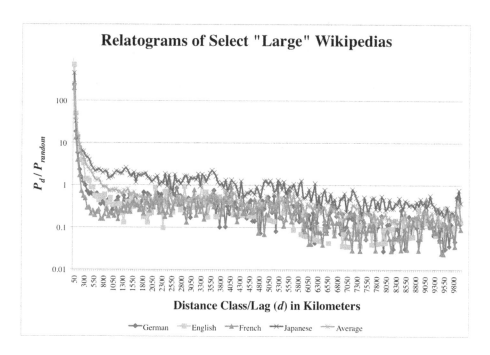

Fig. 2. "Relatograms" of selected large Wikipedias and the unweighted average of P_d / P_{random} across all 22 Wikipedias included in our study. As opposed to Figure 1, the *y*-axis is displayed on a logarithmic scale, allowing easier discrimination of the variation occurring at higher distance classes and lower probability ratios. A value of $y = 1.0$ means that P_d, or the probability of a link occurring between articles pairs in distance class d, is equal to P_{random}, or the probability of any two articles having a link to one another in the Wikipedia being examined.

---

[9] Since for most language we consider the whole dataset and for those we do not our n is in the millions, we do not show error bars as they would be undefined or microscopically small.

Beyond confirming our hypothesis, however, this experiment also produced several second-order observations and, as is usual, raised more questions than it answered. One of the most important secondary patterns we noted is that, in all Wikipedias, beyond a certain threshold $d = \varphi$, P_d drops below P_{random} and thus the ratio displayed in the charts becomes less than 1.0. Once this occurs, it nearly always stays that way, as shown in Figures 2 and 3. In many cases, P_d is consistently less than 25 percent of P_{random}, meaning a spatial article A is at least four times more likely to have a link to any *random* article Y than it is to a spatial article B after *distance*(A,B) reaches φ. In other words, while short distances have a remarkably strong positive effect on the probability that two spatial articles will be related, larger distances actually have a noticeable *negative* effect on these probabilities.

Additionally, although the primary signal of spatial dependence is obvious, the variation among the different languages is fascinating and somewhat of an enigma. Why does the French Wikipedia demonstrate such an immediate drop to a near-asymptote while the Japanese Wikipedia displays a much more gradual decent? What results in the highly varying initial probability ratios for the 0-50km distance class? No correlation with any common network measures (i.e. number of nodes, etc.) explains these or other notable differences. As such, we have to assume Wikipedia-centric, cultural or linguistic variation to be the cause. We discuss our ideas with regard to this beguiling phenomenon and our plan for further research in this area in the future work section.

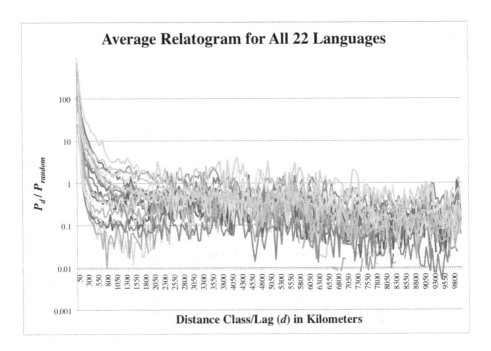

Fig. 3. The variation between and similarities in the relatograms in all 22 languages. The purpose of this chart is not to be able to follow individual languages, but to see overall trends.

3.4 Advanced Analyses

In the previous section, we qualitatively discussed our results. In this section, we seek to analyze them with more mathematical rigor so as to better understand the empirical meaning of TFL as suggested by Wikipedia. Our primary aim is to show that the relatograms can be reasonably described as power law distributions, at least as compared to the "distance doesn't matter" model of the uniform distribution.

Power laws are observed frequently in both the manmade and natural world. An observation of a phenomenon, g, that follows a power law with scaling exponent k over varying x, is governed by the equation:

$$g(x) = ax^k + o(k)$$

where $o(k)$ is asymptotically small. The scaling invariance of the distribution, that is $g(cx) \propto g(x)$, becomes clear when we examine the distribution in log-log space:

$$\log(g(x)) = k\log(x) + \log(a)$$

From this we can see that a necessary condition for a power law is a straight line in log-log space. The slope of the line provides scaling exponent k.

Examining the probability of a link, P_d / P_{random}, at varying lags for *some* Wikipedias gives the appearance of a power law. The straight lines seen in log-log plots of the relatograms of these Wikipedias (Figure 3.4) reveal that indeed they appear to follow a power law over a selected distance range. We fit a power law distribution to the data, limiting the lag to 1000km, using Bayesian probability theory to find the best-fit parameters [21]. This amounts to finding the parameters a,k in the power law equation above, that given the observed distribution $f(x)$ and an assumption H_{pl} of power-law distributed data, give the function that is most in accordance with the data. A maximum likelihood estimation is equivalent to the maximum a posteriori (MAP) optimization in the event of a uniform prior probability $P(a)$ and $P(k)$. We assume a Gaussian error model at each point, $f_i(x)$, and use a uniform prior distribution for a in log space and a uniform prior distribution of k in angular space. Using Bayesian estimation to maximize the posterior probability $P(a,k|f(x),H_{pl})$ as formulated in the equation below, we find the most probable parameters. The best-fit scaling exponents are given in Table 2, and, expectedly, all scaling exponents are negative since the link rate decays as we consider links at further distances.

$$P(a,k \mid f(x), H_{pl}) = \frac{\dfrac{1}{(2\pi)^{N/2}\sigma^N}\exp(\dfrac{-\sum_i(\log f_i(x)-\log(a)-k\log(x))^2}{\sigma^2})P(a)P(k)}{P(f(x)\mid H_{pl})}$$

While finding a "best-fit" power law is possible no matter what the underlying distribution of the data, we would like to be able to compare such a fit with other plausible

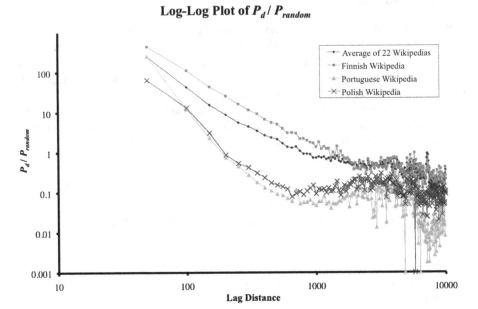

Fig. 4. Log-log plots for P_d / P_{random} over lag distance for select Wikipedias. Some Wikipedias fit a power law at small lags better than others. For example, Polish does not very closely follow a power law even at small distances, while Finnish does. Differences in power-law scaling parameter k is evident in the varying slopes.

distributions. In particular, we would like to *reject* a uniform distribution, call this hypothesis H_{unif}. A uniform distribution would indicate that the $P_{relation}$ of two points does not change with spatial distance. We can accomplish this easily by finding the odds ratio between the two hypotheses, power-law distributed versus uniformly distributed [22]. This allows us to compare the likelihood that the distribution is drawn from a power law distribution, following the formulation above, versus a uniform distribution, following a formulation $g(x)=c$ for some constant c. We can specifically compare the two using an odds ratio of the evidence, $P(f(x)|H)$, for each model, formulated as:

$$\frac{P(f(x)|H_{pl})}{P(f(x)|H_{unif})} = \frac{\int \int P(f(x)|a,k,H_{pl})\, da\, dk}{\int P(f(x)|c,H_{unif})\, dc}$$

This odds ratio is also given in Table 2, and shows that a power law fits the distribution far better than a uniform assumption. The link rate as we move farther away in geographic space may be reasonably characterized as "failing", or decreasing, as a power law.

Table 2. The scale exponent for the best-fit power law of each Wikipedia over lag classes up to 1000km. Comparison with evidence of a uniform distribution gives an odds ratio that rejects a uniform distribution as compared to a power law.

Language	Scaling Exponent	Odds Ratio $P(f(x)|H_{pl})/P(f(x)|H_{unif})$
Catalan	-1.6187	2.6971e+055
Czech	-1.3887	6.3573e+057
Danish	-1.7257	1.4333e+067
German	-1.5687	9.8629e+061
English	-2.0467	4.3767e+076
Spanish	-1.6187	6.1007e+058
Finnish	-1.7837	1.6816e+085
French	-1.4747	7.5283e+055
Hungarian	-1.4307	5.2172e+056
Italian	-1.7837	1.0075e+054
Japanese	-1.3487	2.4968e+071
Dutch	-2.2017	5.4446e+058
Norwegian	-1.9077	4.7396e+076
Polish	-1.7257	3.635e+044
Portuguese	-2.1217	2.0573e+059
Romanian	-1.5207	4.6351e+034
Russian	-1.4307	3.0116e+051
Slovakian	-1.1667	5.9496e+041
Swedish	-1.7837	1.1618e+076
Turkish	-1.5207	2.0533e+070
Ukrainian	-1.6187	9.1704e+085
Chinese	-1.3887	1.6234e+075
Average	-1.6187	7.0424e+067

3.5 Network Analogy

Some readers may have recognized the similarity of the unnormalized statistic, P_d, with the *clustering coefficient* statistic frequently used to analyze networks. We have measured the clustering coefficient of a network neighborhood consisting of only edges that exist between binned geographical distances. That is, each data point in Figure 2 represents the local clustering coefficient of the neighborhood created by activating only the possible edges that connect nodes at distance d=50km, 100km, etc. The clustering coefficient is an important indicator in special types of networks, such as scale-free and small-world. A small-world network is characterized by the necessary conditions of a large average local clustering coefficient and a low path length (number of hops to get from one node to another). While our current analysis lacks analysis of path length, Figure 2 shows that, compared to a random network, networks consisting of nodes at small geographical distances have a much higher clustering coefficient than networks representing larger geographical distances. Our work here shows that the literal interpretation of the term *small-world network* may ultimately be provable in the strict geographic sense. After adding an analysis of path length, this dataset may show that networks consisting of nodes at short geographical distances are indeed small-world.

The current formulations lack crucial information to analyze whether they are also *scale-free*, a quality characterized by a power-law distribution in node degree (the number of connections out of a node). Future analysis will incorporate this concept into analysis of spatial Wikipedia.

4 Future Work

The results discussed in this paper have generated numerous further research questions. Most notably, as mentioned, we are actively seeking an explanation for the variation amongst the different languages' relatograms. Our current preliminary hypotheses can be split into two separate but overlapping categories: cultural causes and linguistic causes. On the cultural side, could it be that since the standard activity space of individuals is much smaller than 50km in many of the cultures examined, this causes the 0-50km distance class to have a much lower relation probability in these cultures' Wikipedias? Do certain cultures describe spatial entities in more relational terms, resulting in a higher average probability over large numbers of distance classes? Could we also be seeing differential culturally defined regional cognition effects, as is suggested by [23]? As for the linguistic causes, do certain languages' norms and/or grammatical structures make it more or less difficult to express relations to locations that are closer or further away? Since Wikipedia is a written language corpus and links must occur inline in the corpus, even a slight linguistic proclivity in this area could have a somewhat large effect, relatively speaking. Similarly, given the nature of Wikipedia, does the reference frame used by each language have an effect? Languages that default to a relative reference frame in formal writing will have at least more *opportunities* to encode spatial relations as links than those that use absolute frames. This is simply because contributors to Wikipedia writing in relative frame languages must mention more spatial entities (as opposed to cardinal directions), allowing them the chance to add a links to these entities while they are at it.

We are also working with the hyperlingual Wikipedia dataset to examine another vital and unique aspect of spatial information: scale. For instance, do the WAGs hierarchical structures' mimic urban spatial hierarchies? In other words, can we evaluate central place theory using the hyperlingual Wikipedia? How does this work in "home" countries of languages versus in foreign countries? We are preparing a manuscript repeating this study from a multiple scales perspective, as is discussed in section 2.2.

Also important is to consider more advanced models of relatedness. We have used here a straightforward binary "link existence" approach to avoid the many complications involved with using recently published Wikipedia-based semantic relatedness (SR) measures [7, 12, 18]. However, we are currently working to compare the present results with those from these SR methods. We hope to elucidate how well spatial relations are captured, a very important consideration given that Wikipedia-based semantic relatedness has quickly become a popular tool in the artificial intelligence community.

Our experiments were focused at a general, theoretical level, but the results do have applied value as a crude quantitative description of spatial relatedness in the absence of more specialized knowledge. While we are by *absolutely* no means recommending that scientists use our "pseudo-covariograms" in place of real co-variograms developed on data relevant to their specific research project, at times no such data is available. Take as an example the work of Gillespie, Agnew and colleagues [24], who used a model from biogeography to predict terrorist movement. We assert that our general model of spatial correlation might be more valid in that context than Gillespie et al.'s approach, especially if/when the Arabic and/or Pashto Wikipedias become large enough for our analyses[10]. Future work will involve improving the applied functionality of our methodology even further by including a more sophisticated and/or localized distance function than universal straight-line distance and providing crude relation type information through Wikipedia's category structures.

5 Conclusion

In this paper, we have shown empirically that in the largest attempt to describe world knowledge in human history, the First Law of Geography proves true: nearby spatial entities in this knowledge repository have a much higher probability of having relations than entities that are farther apart, although even entities very far apart still have relations to each other. In other words, we have seen that the very medium that was supposed to oversee the "death of distance" – the Internet – has instead facilitated the reaffirmation of a theory about the importance of distance that is almost 40 years old and that has roots dating back centuries.

Finally, we would also like to reiterate the significance of the fact that TFL proved true in the knowledge repositories constructed by people who speak *twenty-two* different languages. The discussion of what are the universal truths about humanity that span cultural boundaries is a prickly one, but here we have seen at least some evidence that the tendency to see spatial entities as more related to nearer entities than ones that are further away at least deserves mention in that debate.

Acknowledgements

The authors would like to thank Phil Marshall, Ken Shen, Nada Petrovic, Darren Gergle, and Martin Raubal for consulting on the methods and/or idea generation behind this study.

---

[10] Because of this potential utility, we have made the data from all of our results publicly available at http://www.engr.ucsb.edu/~emoxley/HechtAndMoxleyData.zip. We have also included the algorithms used to perform our experiments in the general WikAPIdia software, which as noted above, is available upon request for academic purposes.

References

1. Tobler, W.R.: A computer movie simulating urban growth in the Detroit region. Economic Geography 46, 234–240 (1970)
2. Longley, P., Goodchild, M., Maguire, D., Rhind, D.: The nature of geographic data. Geographic information systems and science (2005)
3. Fabrikant, S.I., Ruocco, M., Middleton, R., Montello, D.R., Jörgensen, C.: The First Law of Cognitive Geography: Distance and Similarity in Semantic Spaces. In: GIScience (2002)
4. Montello, D.R., Fabrikant, S.I., Ruocco, M., Middleton, R.S.: Testing the first law of cognitive geography on point-display spatializations. In: Kuhn, W., Worboys, M.F., Timpf, S. (eds.) COSIT 2003. LNCS, vol. 2825, pp. 316–331. Springer, Heidelberg (2003)
5. Miller, H.J.: Tobler's First Law and Spatial Analysis. Annals of the Association of American Geographers 94, 284–289 (2004)
6. Sui, D.Z.: Tobler's First Law of Geography: A Big Idea for a Small World. Annals of the Association of American Geographers 94, 269–277 (2004)
7. Hecht, B., Raubal, M.: GeoSR: Geographically explore semantic relations in world knowledge. In: Bernard, L., Friis-Christensen, A., Pundt, H. (eds.) AGILE 2008: Eleventh AGILE International Conference on Geographic Information Science, The European Information Society: Taking Geoinformation Science One Step Further, pp. 95–114. Springer, Heidelberg (2008)
8. Tobler, W.R.: On the First Law of Geography: A Reply. Annals of the Association of American Geographers 94, 304–310 (2004)
9. Goodchild, M.: The Validity and Usefulness of Laws in Geographic Information Science and Geography. Annals of the Association of American Geographers 94, 300–303 (2004)
10. Völkel, M., Krötzsch, M., Vrandecic, D., Haller, H., Studer, R.: Semantic Wikipedia. In: 15th International Conference on World Wide Web WWW 2006, Edinburgh, Scotland, pp. 585–594 (2006)
11. Ortega, F., Gonzalez-Barahona, J.M., Robles, G.: The Top-Ten Wikipedias: A Quantative Analysis Using WikiXRay. In: International Conference on Software and Data Technology ICSOFT 2007, pp. 46–53 (2007)
12. Gabrilovich, E., Markovitch, S.: Computing Semantic Relatedness using Wikipedia-based Explicit Semantic Analysis. In: IJCAI 2007: Twentieth Joint Conference for Artificial Intelligence, Hyberabad, India, pp. 1606–1611 (2007)
13. Milne, D., Witten, I.H.: Learning to Link with Wikipedia. In: ACM 17th Conference on Information and Knowledge Management, CIKM 2008, Napa Valley, California, United States, pp. 1046–1055 (2008)
14. Giles, J.: Special Report: Internet encyclopaedias go head to head. Nature 438, 900–901 (2005)
15. Halavais, A., Lackaff, D.: An Analysis of Topical Coverage of Wikipedia. Journal of Computer-Mediated Communication 13, 429–440 (2008)
16. Pfeil, U., Panayiotis, Z., Ang, C.S.: Cultural Differences in Collaborating Authoring of Wikipedia. Journal of Computer-Mediated Communication 12 (2006)
17. Kittur, A., Chi, E., Pendleton, B.A., Suh, B., Mytkowicz, T.: Power of the Few vs. Wisdom of the Crowd: Wikipedia and the Rise of the Bourgeoisie. In: 25th International Conference on Human Factors in Computing Systems CHI 2007, pp. 1–9 (2007)
18. Milne, D., Witten, I.H.: An effective, low-cost measure of semantic relatedness obtained from Wikipedia. In: AAAI 2008 Workshop on Wikipedia and Artificial Intelligence WIKI-AI 2008, Chicago, IL (2008)

19. Hecht, B., Gergle, D.: Measuring Self-Focus Bias in Community-Maintained Knowledge Repositories. In: Communities and Technologies 2009: Fourth International Conference on Communities and Technologies, University Park, PA, USA, p. 10 (2009)

20. Arbia, G., Benedetti, R., Espa, G.: Effects of the MUAP on image classification. Geographical Systems 1, 123–141 (1996)

21. Clauset, A., Shalizi, C.R., Newman, M.E.J.: Power law distributions in empirical data. SIAM Review (2009)

22. Goldstein, M.L., Morris, S.A., Yen, G.G.: Fitting to the power-law distribution. The European Physical Journal B - Condensed Matter and Complex Systems 41, 255–258 (2004)

23. Xiao, D., Liu, Y.: Study of cultural impacts on location judgments in eastern china. In: Winter, S., Duckham, M., Kulik, L., Kuipers, B. (eds.) COSIT 2007. LNCS, vol. 4736, pp. 20–31. Springer, Heidelberg (2007)

24. Gillespie, T.W., Agnew, J.A., Mariano, E., Mossler, S., Jones, N., Braughton, M., Gonzalez, J.: Finding Osama bin Laden: An Application of Biogeographic Theories and Satellite Imagery. MIT International Review (2009)

Merging Qualitative Constraint Networks Defined on Different Qualitative Formalisms

Jean-François Condotta, Souhila Kaci,
Pierre Marquis, and Nicolas Schwind

Université d'Artois
CRIL CNRS UMR 8188, F-62307 Lens
{condotta,kaci,marquis,schwind}@cril.univ-artois.fr

Abstract. This paper addresses the problem of merging qualitative constraint networks (QCNs) defined on different qualitative formalisms. Our model is restricted to formalisms where the entities and the relationships between these entities are defined on the same domain. The method is an upstream step to a previous framework dealing with a set of QCNs defined on the same formalism. It consists of translating the input QCNs into a well-chosen common formalism. Two approaches are investigated: in the first one, each input QCN is translated to an equivalent QCN; in the second one, the QCNs are translated to approximations. These approaches take advantage of two dual notions that we introduce, the ones of *refinement* and *abstraction* between qualitative formalisms.

1 Introduction

Using a qualitative representation of information in spatial applications is needed when information is incomplete or comes from the natural language, or when quantitative information is unavailable or useless. In many spatial applications such as Geographic Information Systems (GIS), information often comes from sentences like "parcels A and B are connected" or "Paris is at north or at northeast of Toulouse", and one has to deal with such qualitative descriptions. Starting from Allen's work in the particular field of temporal reasoning [1], many approaches to deal with qualitative representation and reasoning have been proposed in the last two decades [6,18,20]. Different aspects of representation are dealt with topological relations [17] or precedence relations when orientation between entities is required [12]. These qualitative formalisms allow us to represent a set of spatial entities and their relative positions using qualitative constraint networks (QCNs).

In some applications (e.g. distributed knowledge systems) information can be gathered from several sources, and the multiplicity of sources means that combining information often leads to conflicts. Merging information has attracted much attention in the literature these past ten years. When information provided by sources is expressed by means of multi-sets of propositional formulae [14,21,10], dealing with inconsistency consists in computing a consistent propositional formula representing a global view of all input formulae.

K. Stewart Hornsby et al. (Eds.): COSIT 2009, LNCS 5756, pp. 106–123, 2009.
© Springer-Verlag Berlin Heidelberg 2009

In Spatial Databases, knowledge about a set of spatial entities is represented by QCNs, which are provided by different sources, making them often conflicting. Consider for instance the following example: given three parcels A, B and C, one of the sources states that Parcel A is included in Parcel B, and Parcel B is disconnected from Parcel C, while another source declares that Parcels A and C are at the same location. Clearly there is a conflict and this calls for merging.

A first framework for merging QCNs has been proposed in [4], when the input QCNs are defined on the same qualitative formalism. The method computes a non-empty set of consistent scenarios which are the closest ones to all QCNs. The present work generalizes this method with QCNs defined on different qualitative formalisms. In general the sources provide QCNs based on different formalisms, although they are based on the same domain and consider similar relationships between entities. For instance, even if both RCC-5 [9] and RCC-8 [17] theories consider topological relationships between regions, the existing approach cannot deal with two QCNs defined respectively on each one of them.

In the particular context of qualitative spatial reasoning, an ultimate goal for the merging problem would be to deal with QCNs defined on heterogeneous formalisms without any restriction. Specific studies of calculi combinations have been investigated in the literature [8,11,16].

In this paper, we present an extension of the framework proposed in [4] by taking into account QCNs defined on different formalisms. We restrict our model to qualitative formalisms where the entities and the relationships between these entities are defined on the same domain. For instance, Allen's interval algebra [1] and $\mathcal{INDU}$ calculus [15] fulfill this requirement, since they both consider qualitative relationships between temporal intervals of the rational line. Our method consists of adding an upstream step to the method described in [4] by translating the input QCNs into a common formalism.

The rest of this paper is organised as follows. In Section 2 we recall some necessary background about qualitative algebras and QCNs. In Section 3 we describe the problem and the merging procedure given in [4]. Then we present in Section 4 how we can translate the input QCNs into a common *refinement* of the related qualitative algebras, using *bridges* between them. In Section 5 we give an alternative of the method proposed in the previous section by approximating the QCNs into a common *abstraction* of the qualitative algebras. Lastly we conclude and present some perspectives for further work.

2 Background

This section introduces necessary notions of qualitative algebras and definitions around qualitative constraint networks. A qualitative calculus (or qualitative algebra) considers $\mathcal{B}$, a finite set of binary relations over a domain $\mathcal{D}$, the universe of all considered entities. Each relation of $\mathcal{B}$ (called a basic relation) represents a particular qualitative position between two elements of $\mathcal{D}$. We make some initial assumptions on such a set $\mathcal{B}$. Let us first introduce the notion of *partition scheme* [13].

Definition 1 (Partition scheme). *Let $\mathcal{D}$ be a non-empty set and $\mathcal{B}$ be a set of binary relations on $\mathcal{D}$. $\mathcal{B}$ is called a* partition scheme *on $\mathcal{D}$ iff the following conditions are satisfied:*

- *The basic relations of $\mathcal{B}$ are* jointly exhaustive and pairwise disjoint, *namely any couple of $\mathcal{D}$ satisfies one and only one basic relation of $\mathcal{B}$.*
- *The identity relation $eq = \{(a,b) \in \mathcal{D} \times \mathcal{D} \mid a = b\}$ is one of the basic relations of $\mathcal{B}$.*
- *$\mathcal{B}$ is closed under converse, namely if r is a basic relation of $\mathcal{B}$, then so is its converse $r^{\smile} = \{(a,b) \mid (b,a) \in r\}$.*

In the rest of this paper, we will require that any considered set $\mathcal{B}$ of binary relations on $\mathcal{D}$ is a partition scheme on $\mathcal{D}$.

The set $2^{\mathcal{B}}$, the set of all subsets of $\mathcal{B}$, with the usual set-theoretic operators union ($\cup$), intersection ($\cap$), complementation ($\sim$), and weak composition ($\diamond$) [19] is called a *qualitative algebra*. Any element R of $2^{\mathcal{B}}$ is called a *relation* and represents a relation $rel(R)$ defined as $rel(R) = \bigcup\{r \mid r \in R\}$. This means that a pair of elements $(X,Y) \in \mathcal{D} \times \mathcal{D}$ satisfies a relation $R \in 2^{\mathcal{B}}$ if and only if $(X,Y) \in rel(R)$. The converse $R^{\smile}$ of a relation $R \in 2^{\mathcal{B}}$ is defined as $R^{\smile} = \{r \in \mathcal{B} \mid r^{\smile} \in R\}$.

For illustration, we consider the Cardinal Directions Algebra $2^{\mathcal{B}_{card}}$ [12], generated by the partition scheme $\mathcal{B}_{card}$ on $\mathbb{R}^2$ illustrated on Figure 1.a. The Cardinal Directions Algebra allows us to represent relative positions of points of the Cartesian plane, provided a global reference direction defined by two orthogonal lines. For each point $B = (b_1, b_2)$, the plane is divided in nine disjoint zones forming the set of basic relations $\mathcal{B}_{card}$. For example, the basic relation n corresponds to the area of points $A = (a_1, a_2)$, with $a_1 = b_1$ and $a_2 > b_2$.

Given any qualitative algebra $2^{\mathcal{B}}$, pieces of knowledge about a set of spatial or temporal entities can be represented by means of qualitative constraint networks (QCNs for short). This structure allows us to represent incomplete information

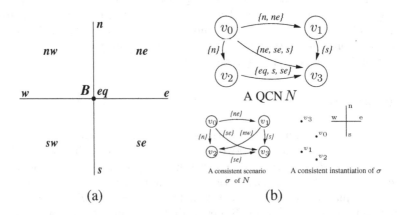

(a) (b)

Fig. 1. The nine basic relations of $\mathcal{B}_{card}$ (a) and a QCN on $2^{\mathcal{B}_{card}}$ and one of its consistent scenarios (b)

about the relations between entities. Formally, a QCN N defined on $2^{\mathcal{B}}$ is a pair (V, C), where $V = \{v_0, \cdots, v_{n-1}\}$ is a finite set of variables representing the spatial or temporal entities and C is a mapping which associates to each pair of variables (v_i, v_j) an element R of $2^{\mathcal{B}}$. R represents the set of all possible basic relations between v_i and v_j. We write C_{ij} instead of $C(v_i, v_j)$ for short. For all $v_i, v_j \in V$, we suppose that $C_{ji} = C_{ij}^{\smallsmile}$ and $C_{ii} = \{eq\}$.

A QCN can be represented by a graph, using some conventions: for all $v_i, v_j \in V$, we do not represent the constraint C_{ji} if C_{ij} is represented since $C_{ji} = C_{ij}^{\smallsmile}$; we do not represent either the constraint C_{ii} since $C_{ii} = \{eq\}$; lastly when $C_{ij} = \mathcal{B}$ (i.e. no information is provided between the variables v_i and v_j), we do not represent it.

With regard to a QCN $N = (V, C)$ we have the following definitions:

Definition 2. *A consistent instantiation of N over $V' \subseteq V$ is a mapping α from V' to $\mathcal{D}$ such that $\alpha(v_i)\, C_{ij}\, \alpha(v_j)$, for all $v_i, v_j \in V'$. A solution of N is a consistent instantiation of N over V. N is a consistent QCN iff it admits a solution. A sub-network of N is a QCN $N' = (V, C')$ where $C'_{ij} \subseteq C_{ij}$ for all $v_i, v_j \in V$. A consistent scenario of N is a consistent sub-network of N in which each constraint is composed of exactly one basic relation of $\mathcal{B}$.*

$[N]$ denotes the set of all consistent scenarios of N. A QCN defined on $2^{\mathcal{B}_{card}}$ over 4 variables and one of its consistent scenarios are depicted in Figure 1.b.

3   Related Work: Merging QCNs Defined on Same Qualitative Algebras

Before summarizing a merging method for QCNs which has been proposed in the literature [4], let us introduce the merging process through an example, which will be our running example for this section.

Example 1. We consider three agents A_1, A_2 and A_3 having incomplete knowledge about the configurations of a common set $V = \{v_0, v_1, v_2, v_3\}$ of four variables. Each agent A_i provides a QCN $N_i = (V, C_i)$ defined on $2^{\mathcal{B}_{card}}$ representing the qualitative relations between pairs of V. Figure 2 depicts the three related QCNs. All consistent scenarios of each QCN are also depicted below through a qualitative representation on the plane.

The merging process takes as input a set of QCNs $\mathcal{N} = \{N_1, \ldots, N_m\}$ defined on the same qualitative algebra $2^{\mathcal{B}}$ and on the same set of variables V. Given such a set of QCNs describing different points of view about the configurations of the same entities, we would like to derive a global view of the system, taking into account each input QCN. A natural way to deal with this problem is to return as the result of merging the information on which all sources agree. For example, we can consider the set $\bigcap_{N_i \in \mathcal{N}} [N_i]$, that is the set of all consistent scenarios which belongs to each QCN N_i. However this set can be empty. It is the case in our example, since for instance the two variables v_1 and v_3 only satisfy the relation $\{w\}$ in N_1 while they only satisfy the relation $\{eq\}$ in N_3.

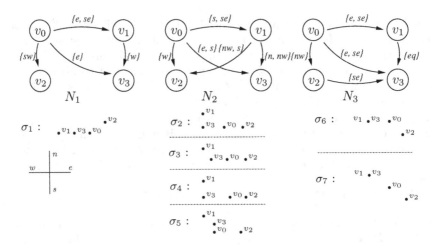

Fig. 2. Three $QCNs$ defined on V and their respective consistent scenarios

Condotta et al. [4] have proposed a parsimonious method for the merging problem. Their method is inspired from propositional merging [21,10,2]. It consists in computing a non-empty set of consistent scenarios that are the closest ones to each input QCN, with respect to a *distance*. The result is mainly based on the definition of the distance which represents the degree of closeness between scenarios and the set of QCNs. A QCN merging operator Θ is typically a mapping which associates to a set of QCNs $\mathcal{N}$ defined on $2^{\mathcal{B}}$ and V a set of consistent scenarios on V. More precisely, the merging process follows three steps:

First, we need to compute a local distance d between a scenario σ and a QCN N, which is the smallest distance between σ and each consistent scenario of N. Formally, this distance is defined as follows:

$$d(\sigma, N) = \begin{cases} \min\{d(\sigma, \sigma') \mid \sigma' \in [N]\} & \text{if } N \text{ is consistent,} \\ 0 & \text{otherwise.} \end{cases}$$

Here we need a distance d between two scenarios σ, σ'. This distance is a mapping which associates a positive number to every pair of scenarios, and satisfies the following two conditions:

$$\forall \sigma, \sigma' \text{ scenarios on } V, \quad \begin{cases} d(\sigma, \sigma') = d(\sigma', \sigma) \\ d(\sigma, \sigma') = 0 \text{ iff } \sigma = \sigma'. \end{cases}$$

A particular distance between scenarios has been proposed in [4]. It takes advantage of a notion of conceptual neighbourhood specific to the considered set of basic relations $\mathcal{B}$. Figure 3.a depicts the conceptual neighbourhood graph of the Cardinal Directions Algebra. This graph corresponds to the Hasse diagram of the corresponding lattice defined in [12]. There is an intuitive meaning behind this graph. For example, assume that two points satisfy the relation ne. Then continuously moving one of the two points can lead them to directly satisfy the relation n. Thus these two relations are considered as "close". Let us denote by $\sigma(v_i, v_j)$ the basic relation satisfied between v_i and v_j in the scenario σ. The neighbourhood distance between two scenarios is defined as follows, $\forall \sigma, \sigma'$ scenarios on V:

$$d_{NB}(\sigma, \sigma') = \sum_{i<j} d_{nb}(\sigma(v_i, v_j), \sigma'(v_i, v_j)),$$

with $d_{nb}(r_1, r_2)$ the length of the smallest path between the basic relations r_1 and r_2 in the related neighbourhood graph. We use this particular distance in our running example.

Example 1 (continued). Consider the scenarios σ_1 and σ_2 depicted in Figure 2. The neighbourhood distance between σ_1 and σ_2 is computed as follows:

$$\begin{aligned}
d_{NB}(\sigma_1, \sigma_2) = {}& d_{nb}(\sigma_1(v_0, v_1), \sigma_2(v_0, v_1)) + d_{nb}(\sigma_1(v_0, v_2), \sigma_2(v_0, v_2)) \\
& + d_{nb}(\sigma_1(v_0, v_3), \sigma_2(v_0, v_3)) + d_{nb}(\sigma_1(v_1, v_2), \sigma_2(v_1, v_2)) \\
& + d_{nb}(\sigma_1(v_1, v_3), \sigma_2(v_1, v_3)) + d_{nb}(\sigma_1(v_2, v_3), \sigma_2(v_2, v_3)) \\
= {}& d_{nb}(e, se) + d_{nb}(sw, w) + d_{nb}(e, e) \\
& + d_{nb}(sw, nw) + d_{nb}(w, n) + d_{nb}(ne, e) \\
= {}& 1 + 1 + 0 + 2 + 2 + 1 = 7.
\end{aligned}$$

The second step of the merging process consists in aggregating local distances computed in the first step to get a global distance between a scenario and the set of QCNs $\mathcal{N}$. For example, the sum operator $\sum$ is appropriate when the result of merging has to represent the point of view of the majority of the agents. $\sum$ is so called a *majority operator* [14].

Example 1 (continued). Consider the scenario σ depicted in Figure 3.b. The global distance $d_\sum$ (using the majority operator $\sum$) between σ and the set of three QCNs (cf Figure 2) is computed as follows (we do not detail computations for the sake of conciseness) :

$$\begin{aligned}
d_\sum(\sigma, \{N_1, N_2, N_3\}) &= d(\sigma, N_1) + d(\sigma, N_2) + d(\sigma, N_3) \\
&= 7 + 3 + 2 = 12.
\end{aligned}$$

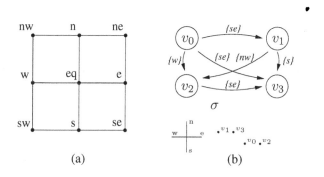

(a)   (b)

Fig. 3. The conceptual neighbourhood graph of the Cardinal Directions Algebra (a) and a consistent scenario σ (b)

The last step consists in computing the result of the merging. It is the set of all consistent scenarios which are the "closest" ones to the set $\mathcal{N}$. Formally,

$$\Theta(\mathcal{N}) = \{\sigma \mid \sigma \text{ is consistent and } d_{\Sigma}(\sigma, \mathcal{N}) \text{ is minimal}\}.$$

In our running example, $\Theta(\{N_1, N_2, N_3\})$ is composed of one consistent scenario, namely σ, depicted in Figure 3.b. Condotta et al. [4] have pointed out that this merging operator has a "good" logical behavior in the sense of [10,3].

Note however that the merging method described above requires the enumeration of all possible consistent scenarios on V, which makes this process hardly practicable.

4 Dealing with Different Qualitative Algebras: Toward a Common Refinement

4.1 The Model

The approach presented in the previous section permits us to deal with a set of possibly conflicting QCNs defined on the same qualitative algebra. We aim in this section at extending it by taking into account different qualitative algebras, provided that all the related partition schemes are on the same domain $\mathcal{D}$.

Formally, let $\mathcal{N} = \{N_1, \ldots, N_m\}$ be the input QCNs and $\mathcal{N} = \{\mathcal{N}_1, \ldots, \mathcal{N}_p\}$ be a partition of $\mathcal{N}$ such that $\forall k \in \{1, \ldots, p\}$, any QCN of $\mathcal{N}_k$ is defined on the qualitative algebra $2^{\mathcal{B}_k}$. Let $\mathcal{A} = \{2^{\mathcal{B}_1}, \ldots, 2^{\mathcal{B}_p}\}$ be the set of all related qualitative algebras. We require that each partition scheme $\mathcal{B}_k$ is on $\mathcal{D}$.

We first need to introduce the notion of *equivalence* between two QCNs. Given a QCN N defined on V, we denote by $Sol(N)$ the set of solutions of N, i.e., the possibly infinite set of consistent instantiations of N over V, different from the set of its consistent scenarios $[N]$.

Definition 3. *Let $2^{\mathcal{B}}$, $2^{\mathcal{B}'}$ be two qualitative algebras on $\mathcal{D}$, and N, N' two QCNs on V respectively defined on $2^{\mathcal{B}}$ and $2^{\mathcal{B}'}$. Then N and N' are equivalent iff $Sol(N) = Sol(N')$.*

The method can be summarized in two main steps:

(1) We suppose that there exists a qualitative algebra $Ref(\mathcal{A})$ such that any QCN of $\mathcal{N}$ can be translated in an equivalent QCN on $Ref(\mathcal{A})$. The set of all translated QCNs is denoted by $\mathcal{N}'$.
(2) We use a merging process of QCNs defined on the same qualitative algebra for merging the set $\mathcal{N}'$ (see Section 3).

This section is devoted to deal with the first step of this process. We show that such a qualitative algebra $Ref(\mathcal{A})$ always exists and how to define it. This algebra will be called a *refinement* of all input qualitative algebras. Let us first introduce the notion of refinement between qualitative algebras.

Definition 4. *Let $\mathcal{B}$, $\mathcal{B}'$ be two partition schemes on $\mathcal{D}$. The set $2^{\mathcal{B}'}$ is called a refinement of $2^{\mathcal{B}}$ iff there exists a mapping $Ref_{\mathcal{B}\rightarrow\mathcal{B}'}$ which associates to each relation of $2^{\mathcal{B}}$ a relation of $2^{\mathcal{B}'}$ such that $\forall R \in 2^{\mathcal{B}}$, $rel(Ref_{\mathcal{B}\rightarrow\mathcal{B}'}(R)) = rel(R)$. Such a mapping $Ref_{\mathcal{B}\rightarrow\mathcal{B}'}$ is called an r-bridge from $2^{\mathcal{B}}$ to $2^{\mathcal{B}'}$.*

As a typical example, consider the RCC-8 algebra [17] and the RCC-5 algebra [9] which are both used to express topological relationships between regions. Then following Definition 4 the RCC-8 algebra is a refinement of the RCC-5 algebra. Let us make some remarks around this definition. First, an r-bridge $Ref_{\mathcal{B}\rightarrow\mathcal{B}'}$ from $2^{\mathcal{B}}$ to $2^{\mathcal{B}'}$ is fully characterized by its restriction to the set of singleton relations of $2^{\mathcal{B}}$ (namely the set $\{\{r\} \mid r \in \mathcal{B}\}$) to $2^{\mathcal{B}'}$. Indeed, we have for all relation $R \in 2^{\mathcal{B}}$,

$$Ref_{\mathcal{B}\rightarrow\mathcal{B}'}(R) = \bigcup\{Ref_{\mathcal{B}\rightarrow\mathcal{B}'}(\{r\}) \mid r \in R\}.$$

Moreover, let us notice that $Ref_{\mathcal{B}\rightarrow\mathcal{B}'}(\{eq\}) = \{eq\}$ and that for any relation $R \in 2^{\mathcal{B}}$, $Ref_{\mathcal{B}\rightarrow\mathcal{B}'}(R^\smile) = Ref_{\mathcal{B}\rightarrow\mathcal{B}'}(R)^\smile$. One can also state that if $2^{\mathcal{B}'}$ is a refinement of $2^{\mathcal{B}}$, there exists only one r-bridge $Ref_{\mathcal{B}\rightarrow\mathcal{B}'}$ from $2^{\mathcal{B}}$ to $2^{\mathcal{B}'}$. Lastly, $Ref_{\mathcal{B}\rightarrow\mathcal{B}'}$ is an injective function.

The definition of such an r-bridge $Ref_{\mathcal{B}\rightarrow\mathcal{B}'}$ comes from the following proposition:

Proposition 1. *$2^{\mathcal{B}'}$ is a refinement of $2^{\mathcal{B}}$ iff $\forall r' \in \mathcal{B}'$, $\forall r \in \mathcal{B}$, either $r' \subseteq r$ or $r' \cap r = \emptyset$.*

Proof

$(\Rightarrow)$ Let $2^{\mathcal{B}'}$ be a refinement of $2^{\mathcal{B}}$. There exists an r-bridge $Ref_{\mathcal{B}\rightarrow\mathcal{B}'}$ from $2^{\mathcal{B}}$ to $2^{\mathcal{B}'}$. Let $r \in \mathcal{B}$. Then there exists $R \in 2^{\mathcal{B}'}$ such that $rel(Ref_{\mathcal{B}\rightarrow\mathcal{B}'}(\{r\})) = r = rel(R)$. Let $r' \in \mathcal{B}'$. If $r' \in R$, then $r' \subseteq r$ since $rel(R) = r$; if $r' \notin R$, then $r' \cap rel(R) = \emptyset$ (since the basic relations of $\mathcal{B}'$ are pairwise disjoint). Thus $r' \cap r = \emptyset$.

$(\Leftarrow)$ Suppose that $\forall r' \in \mathcal{B}'$, $\forall r \in \mathcal{B}$, either $r' \subseteq r$ or $r' \cap r = \emptyset$. Then $\forall r \in \mathcal{B}$ we get $r = \bigcup\{r' \in \mathcal{B}' \mid r' \subseteq r\}$ since basic relations of $\mathcal{B}'$ are jointly exhaustive relations on $\mathcal{D} \times \mathcal{D}$. Define the mapping $Ref_{\mathcal{B}\rightarrow\mathcal{B}'}$ from $2^{\mathcal{B}}$ to $2^{\mathcal{B}'}$ such that $\forall r \in \mathcal{B}$, $Ref_{\mathcal{B}\rightarrow\mathcal{B}'}(\{r\}) = \{r' \in \mathcal{B}' \mid r' \subseteq r\}$ and $\forall R \in 2^{\mathcal{B}}$, $Ref_{\mathcal{B}\rightarrow\mathcal{B}'}(R) = \bigcup\{Ref_{\mathcal{B}\rightarrow\mathcal{B}'}(\{r\}) \mid r \in R\}$. We can assert that $\forall R \in 2^{\mathcal{B}}$, $rel(Ref_{\mathcal{B}\rightarrow\mathcal{B}'}(R)) = rel(R)$. □

The previous proposition can be also written as follows:

$$2^{\mathcal{B}'} \text{ is a refinement of } 2^{\mathcal{B}} \text{ iff } \forall r' \in \mathcal{B}', \exists! r \in \mathcal{B}, r' \subseteq r.$$

Thus, if $2^{\mathcal{B}'}$ is a refinement of $2^{\mathcal{B}}$, we can define the restriction of the r-bridge $Ref_{\mathcal{B}\rightarrow\mathcal{B}'}$ from the singleton relations of $2^{\mathcal{B}}$ to $2^{\mathcal{B}'}$ as follows:

$$\forall r \in \mathcal{B}, Ref_{\mathcal{B}\rightarrow\mathcal{B}'}(\{r\}) = \{r' \in \mathcal{B}' \mid r' \subseteq r\}.$$

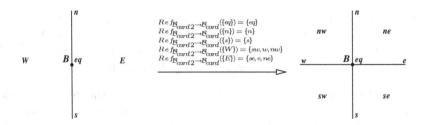

The five basic relations of $\mathcal{B}_{card2}$ The nine basic relations of $\mathcal{B}_{card}$

Fig. 4. $2^{\mathcal{B}_{card}}$ is a refinement of $2^{\mathcal{B}_{card2}}$

Example 2. Let $2^{\mathcal{B}_{card2}}$ be the qualitative algebra generated by the partition scheme $\mathcal{B}_{card2}$ on $\mathbb{R}^2$, with $\mathcal{B}_{card2} = \{eq, n, E, s, W\}$. The five basic relations of $\mathcal{B}_{card2}$ are depicted in the left-hand side of Figure 4. We point out on the same figure the r-bridge $Ref_{\mathcal{B} \to \mathcal{B}'}$ from singleton relations of $2^{\mathcal{B}_{card2}}$ to $2^{\mathcal{B}_{card}}$. Therefore $2^{\mathcal{B}_{card}}$ is a refinement of $2^{\mathcal{B}_{card2}}$.

It is now time to extend the notion of refinement to the $QCNs$.

Definition 5. *Let $N = (V, C)$ be a QCN defined on $2^{\mathcal{B}}$, and $2^{\mathcal{B}'}$ be a refinement of $2^{\mathcal{B}}$. We define the QCN $Ref_{\mathcal{B} \to \mathcal{B}'}(N)$ on $2^{\mathcal{B}'}$ as the QCN (V, C') with $C'_{ij} = Ref_{\mathcal{B} \to \mathcal{B}'}(C_{ij})$ $\forall v_i, v_j \in V$.*

Proposition 2. *Given a QCN N on $2^{\mathcal{B}}$, N and $Ref_{\mathcal{B} \to \mathcal{B}'}(N)$ are equivalent QCNs.*

Proof. Given a QCN N on $2^{\mathcal{B}}$ and $Ref_{\mathcal{B} \to \mathcal{B}'}(N) = (V, C')$, for all $v_i, v_j \in V$, we have $C'_{ij} = Ref_{\mathcal{B} \to \mathcal{B}'}(C_{ij})$. Moreover, we know that $rel(C_{ij}) = rel(Ref_{\mathcal{B} \to \mathcal{B}'}(C_{ij}))$. It follows that $rel(C_{ij}) = rel(C'_{ij})$. We can conclude that for all $a, b \in \mathcal{D}$, a C_{ij} b iff a C'_{ij} b. Hence, α is a solution of N iff α is a solution of $Ref_{\mathcal{B} \to \mathcal{B}'}(N)$. □

Example 3. Figure 5 depicts a QCN defined on $2^{\mathcal{B}_{card2}}$ and its equivalent QCN $Ref_{\mathcal{B} \to \mathcal{B}'}(N)$ defined on $2^{\mathcal{B}_{card}}$.

We now define an ordering over the qualitative algebras onto a domain $\mathcal{D}$ based on the notion of refinement.

A QCN N defined on $2^{\mathcal{B}_{card2}}$ The QCN $Ref_{\mathcal{B}_{card2} \to \mathcal{B}_{card}}(N)$
defined on $2^{\mathcal{B}_{card}}$ equivalent to N

Fig. 5. Two equivalent QCNs

Definition 6. *Given* $2^{\mathcal{B}}, 2^{\mathcal{B}'}$ *two qualitative algebras on* $\mathcal{D}$, $2^{\mathcal{B}} \leq_{ref}^{\mathcal{D}} 2^{\mathcal{B}'}$ *iff* $2^{\mathcal{B}}$ *is a refinement of* $2^{\mathcal{B}'}$.

Proposition 3. $\leq_{ref}^{\mathcal{D}}$ *is a weak partial ordering over qualitative algebras on* $\mathcal{D}$, *i.e., a reflexive, antisymmetric and transitive relation.*

Proof. Let $\mathcal{B}$, $\mathcal{B}'$ and $\mathcal{B}''$ be three partition schemes on $\mathcal{D}$.

- By defining the mapping $Ref_{\mathcal{B}\to\mathcal{B}}$ from $2^{\mathcal{B}}$ to $2^{\mathcal{B}}$ by $Ref_{\mathcal{B}\to\mathcal{B}}(R) = R$ for all $R \in 2^{\mathcal{B}}$ we obtain an r-bridge from $2^{\mathcal{B}}$ to $2^{\mathcal{B}}$. Hence, $2^{\mathcal{B}} \leq_{ref}^{\mathcal{D}} 2^{\mathcal{B}}$.
- Let $\mathcal{B}$ and $\mathcal{B}'$ be two partition schemes such that $2^{\mathcal{B}}$ and $2^{\mathcal{B}'}$ are refinements of respectively $2^{\mathcal{B}'}$ and $2^{\mathcal{B}}$. Let r be a basic relation belonging to $\mathcal{B}$. Since $2^{\mathcal{B}}$ is a refinement of $2^{\mathcal{B}'}$ there exists a unique basic relation $r' \in \mathcal{B}'$ such that $r \subseteq r'$. Moreover, since $2^{\mathcal{B}'}$ is a refinement of $2^{\mathcal{B}}$ there exists a unique basic relation $r'' \in \mathcal{B}$ such that $r' \subseteq r''$. Hence, $r \cap r'' \neq \emptyset$. We can conclude that $r = r'' = r'$. Consequently, $2^{\mathcal{B}} = 2^{\mathcal{B}'}$.
- Let us suppose that $2^{\mathcal{B}''} \leq_{ref}^{\mathcal{D}} 2^{\mathcal{B}'} \leq_{ref}^{\mathcal{D}} 2^{\mathcal{B}}$ and consider the two r-bridges $Ref_{\mathcal{B}\to\mathcal{B}'}$ and $Ref_{\mathcal{B}'\to\mathcal{B}''}$ from $2^{\mathcal{B}}$ to $2^{\mathcal{B}'}$ and from $2^{\mathcal{B}'}$ to $2^{\mathcal{B}''}$ respectively. Let us define the mapping $Ref_{\mathcal{B}\to\mathcal{B}''}$ from $2^{\mathcal{B}}$ to $2^{\mathcal{B}''}$ by $Ref_{\mathcal{B}\to\mathcal{B}''}(R) = Ref_{\mathcal{B}'\to\mathcal{B}''}(\ Ref_{\mathcal{B}\to\mathcal{B}'}(R))$ for all $R \in 2^{\mathcal{B}}$. We have $rel(Ref_{\mathcal{B}\to\mathcal{B}''}(R)) = rel(Ref_{\mathcal{B}'\to\mathcal{B}''}(Ref_{\mathcal{B}\to\mathcal{B}'}(R))) = rel(Ref_{\mathcal{B}\to\mathcal{B}'}(R)) = rel(R)$ for all $R \in 2^{\mathcal{B}}$. Hence, $Ref_{\mathcal{B}\to\mathcal{B}''}$ is an r-bridge from $2^{\mathcal{B}}$ to $2^{\mathcal{B}''}$. Consequently, $2^{\mathcal{B}''} \leq_{ref}^{\mathcal{D}} 2^{\mathcal{B}}$. $\square$

Definition 7. *Let* $\mathcal{A} = \{2^{\mathcal{B}_1}, \ldots, 2^{\mathcal{B}_p}\}$, *with* $p \geq 1$, *a set of qualitative algebras on* $\mathcal{D}$. *Then* $2^{\mathcal{B}}$ *is called a* common refinement *of* $\mathcal{A}$ *iff* $2^{\mathcal{B}} \leq_{ref}^{\mathcal{D}} 2^{\mathcal{B}_k}$ *for all* $k \in \{1, \ldots, p\}$.

Consider a set $\mathcal{A} = \{2^{\mathcal{B}_1}, \ldots, 2^{\mathcal{B}_p}\}$ of qualitative algebras on $\mathcal{D}$, we will define a common refinement denoted by $Ref(\mathcal{A})$.

In the sequel, we will denote by $v(k)$, with $k \in \{1, \ldots, p\}$ and $p \geq 1$, the kth component of a p-tuple $v \in \mathcal{B}_1 \times \cdots \times \mathcal{B}_p$, that is, the basic relation of v corresponding to $\mathcal{B}_k$.

Definition 8. *Let* $\mathcal{A} = \{2^{\mathcal{B}_1}, \ldots, 2^{\mathcal{B}_p}\}$ *be a set of qualitative algebras on* $\mathcal{D}$. *The qualitative algebra* $Ref(\mathcal{A}) = 2^{\mathcal{B}_{Ref}(\mathcal{A})}$ *is defined as follows :*

$$\mathcal{B}_{Ref}(\mathcal{A}) = \{ \bigcap_{1 \leq k \leq p} v(k) \mid v \in \mathcal{B}_1 \times \cdots \times \mathcal{B}_p\} \setminus \{\emptyset\}.$$

Firstly, let us prove that $Ref(\mathcal{A})$ is well-defined.

Proposition 4. *Let* $\mathcal{A} = \{2^{\mathcal{B}_1}, \ldots, 2^{\mathcal{B}_p}\}$ *be a set of qualitative algebras on* $\mathcal{D}$. $\mathcal{B}_{Ref}(\mathcal{A})$ *is a partition scheme on* $\mathcal{D}$.

Proof

- We first prove that $\mathcal{B}_{Ref}(\mathcal{A})$ is a partition of $\mathcal{D} \times \mathcal{D}$. For all $(a, b) \in \mathcal{D} \times \mathcal{D}$, since each $\mathcal{B}_k$ is a partition of $\mathcal{D} \times \mathcal{D}$, there exists a unique basic relation $r_k \in \mathcal{B}_k$ such that $(a, b) \in r_k$. Thus there exists a unique p-tuple $v \in \mathcal{B}_1 \times \cdots \times \mathcal{B}_p$ such that for each $k \in \{1, \ldots, p\}$, $(a, b) \in v(k)$ (v is defined by $v(k) = r_k$). Hence, there exists a unique relation $r \in \mathcal{B}_{Ref}(\mathcal{A})$ such that $(a, b) \in r$.

- For all $2^{\mathcal{B}_k} \in \mathcal{A}$, $eq \in \mathcal{B}_k$. Consequently, we can assert that the identity relation eq onto $\mathcal{D}$ is an element of $\mathcal{B}_{Ref}(\mathcal{A})$.
- Let $r \in \mathcal{B}_{Ref}(\mathcal{A})$ and v the p-tuple of $\mathcal{B}_1 \times \cdots \times \mathcal{B}_p$ such that $r = \bigcap_k v(k)$. By defining the p-tuple $v' \in \mathcal{B}_1 \times \cdots \times \mathcal{B}_p$ by $v'(k) = (v(k))^{\smile}$ for all $k \in \{1, \ldots, p\}$. We have $\bigcap_k v'(k)$ which is a relation belonging to $\mathcal{B}_{Ref}(\mathcal{A})$ and which is the converse of r. □

Let us now prove that the built refinement is the greatest common refinement of $Ref(\mathcal{A})$ w.r.t. $\leq_{ref}^{\mathcal{D}}$.

Proposition 5. *Let* $\mathcal{A} = \{2^{\mathcal{B}_1}, \ldots, 2^{\mathcal{B}_p}\}$ *be a set of qualitative algebras on* $\mathcal{D}$. $Ref(\mathcal{A})$ *is the greatest common refinement of* $\mathcal{A}$ *w.r.t.* $\leq_{ref}^{\mathcal{D}}$.

Proof

- First we show that $\forall 2^{\mathcal{B}_i} \in \mathcal{A}$, $Ref(\mathcal{A}) = 2^{\mathcal{B}_{Ref}(\mathcal{A})}$ is a refinement of $2^{\mathcal{B}_i}$. Let $2^{\mathcal{B}_i} \in \mathcal{A}$, and r be a basic relation of $\mathcal{B}_i$. Consider an element $(a, b) \in r$. Denote by r_k the basic relation of $\mathcal{B}_k$ containing (a, b) for each $k \in \{1, \ldots, p\}$. Note that $r_i = r$. Let $r' = \bigcap_k r_k$. We have r' which is a basic relation belonging to $\mathcal{B}_{Ref}(\mathcal{A})$, moreover $r \subseteq r'$. From Proposition 1, we can assert that $Ref(\mathcal{A}) \leq_{ref}^{\mathcal{D}} 2^{\mathcal{B}_i}$.
- Let $2^{\mathcal{B}}$ be a common refinement of $\mathcal{A}$. We now show that $Ref(\mathcal{A}) \leq_{ref}^{\mathcal{D}} 2^{\mathcal{B}}$. Let $r \in \mathcal{B}$. Since $2^{\mathcal{B}}$ is a common refinement of $\mathcal{A}$, for all $i \in \{1, \ldots, p\}$ there exists $r_i \in \mathcal{B}_i$ such that $r \subseteq r_i$. Let $r' = \bigcap_k r_k$. We have r' which is a basic relation belonging to $\mathcal{B}_{Ref}(\mathcal{A})$, moreover $r \subseteq r'$. We can conclude that $2^{\mathcal{B}}$ is a refinement of $Ref(\mathcal{A})$, that is, $2^{\mathcal{B}} \leq_{ref}^{\mathcal{D}} Ref(\mathcal{A})$. □

Thus, for each $2^{\mathcal{B}_i} \in \mathcal{A}$ there exists an r-bridge $Ref_{\mathcal{B}_i \to \mathcal{B}_{Ref}(\mathcal{A})}$ from $2^{\mathcal{B}_i}$ to $Ref(\mathcal{A})$ defined for each basic relation $r \in \mathcal{B}_i$ as $Ref_{\mathcal{B}_i \to \mathcal{B}_{Ref}(\mathcal{A})}(\{r\}) = \{r' \in \mathcal{B}_{Ref}(\mathcal{A}) \mid r' \subseteq r\}$, and for each relation $R \in 2^{\mathcal{B}_i}$ as $Ref_{\mathcal{B}_i \to \mathcal{B}_{Ref}(\mathcal{A})}(R) = \bigcup \{Ref_{\mathcal{B}_i \to \mathcal{B}_{Ref}(\mathcal{A})}(\{r\}) \mid r \in R\}$. Then, for each QCN $N_i \in \mathcal{N}$ defined on $2^{\mathcal{B}_j} \in \mathcal{A}$, we translate N_i into an equivalent QCN $N_i' = Ref_{\mathcal{B}_i \to \mathcal{B}_{Ref}(\mathcal{A})}(N_i)$ on $Ref(\mathcal{A})$ (cf Definition 5). We then get a set of QCNs $\mathcal{N}' = \{N_1', \ldots, N_m'\}$ defined on the same qualitative algebra $Ref(\mathcal{A})$, and we can use a merging process for QCNs defined on the same qualitative algebra, as described in Section 3.

4.2   Instantiating our Framework on the Star Algebra

The Star algebra [20] is a generalization of two kinds of algebras around cardinal directions distinguished by Frank [7], the coned-shaped directions and the projection-based directions [12]. It considers relations of the domain $\mathcal{D} = \mathbb{R}^2$ and it is set by a level of granularity m. The parameter m specifies the number of lines which intersect in a reference point B. Each line j, $1 \leq j \leq m$ forms an angle δ_j with the reference direction. The plane is split into $4m + 1$ zones ($2m$ half-lines, $2m$ two-dimensional zones, and the relation eq), forming the partition scheme $\mathcal{STAR}_m[\delta_1, \ldots, \delta_m](\delta_1)$ on $\mathbb{R}^2$. We call such a partition scheme a *Star partition scheme*. When all two-dimensional zones are of equal size, we rather write $\mathcal{STAR}_m(\delta_1)$, with δ_1 the angle formed by the first line w.r.t. the reference

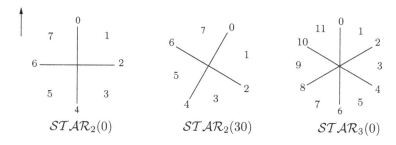

Fig. 6. Three Star partition schemes

direction. The basic relations of a Star partition scheme are numbers from 0 to $4m - 1$ identifying a zone, and the relation eq.

Figure 6 depicts three Star partition schemes $\mathcal{STAR}_2(0)$, $\mathcal{STAR}_2(30)$ and $\mathcal{STAR}_3(0)$. Notice that $\mathcal{STAR}_2(0)$ corresponds to the set $\mathcal{B}_{card}$.

Let $\mathcal{A} = \{2^{\mathcal{B}_1}, 2^{\mathcal{B}_2}, 2^{\mathcal{B}_3}\}$ with $\mathcal{B}_1 = \mathcal{STAR}_2(0)$, $\mathcal{B}_2 = \mathcal{STAR}_2(30)$ and $\mathcal{B}_3 = \mathcal{STAR}_3(0)$.

Let us now suppose that we have to merge a set of QCNs $\mathcal{N} = \{N_1, N_2, N_3\}$. The QCNs N_1, N_2, N_3, respectively defined on $2^{\mathcal{B}_1}$, $2^{\mathcal{B}_2}$ and $2^{\mathcal{B}_3}$, are depicted in Figure 7.

We aim to get a set $\mathcal{N}'$ of QCNs equivalent to those of $\mathcal{N}$ and defined on the same qualitative algebra. First, we define the qualitative algebra $Ref(\mathcal{A})$ which is the greatest common refinement of $\mathcal{A}$. This qualitative algebra is generated by the partition scheme $\mathcal{B}_{Ref}(\mathcal{A})$ defined as $\mathcal{B}_{Ref}(\mathcal{A}) = \{v(1) \cap v(2) \cap v(3) \mid v \in \mathcal{B}_1 \times \mathcal{B}_2 \times \mathcal{B}_3, (v(1) \cap v(2) \cap v(3)) \neq \emptyset\}$. We obtain from this definition the partition scheme depicted in Figure 8.

For each $2^{\mathcal{B}_i} \in \mathcal{A}$, recall that the bridge from $2^{\mathcal{B}_i}$ to $Ref(\mathcal{A})$ is defined for each basic relation $r \in \mathcal{B}_i$ as $Ref_{\mathcal{B}_i \to \mathcal{B}_{Ref}(\mathcal{A})}(\{r\}) = \{r' \in \mathcal{B}_{Ref}(\mathcal{A}) \mid r' \subseteq r\}$, and for each relation $R \in 2^{\mathcal{B}_i}$ by $Ref_{\mathcal{B}_i \to \mathcal{B}_{Ref}(\mathcal{A})}(R) = \bigcup\{Ref_{\mathcal{B}_i \to \mathcal{B}_{Ref}(\mathcal{A})}(\{r\}) \mid r \in R\}$. Then for each $\mathcal{B}_i \in \mathcal{A}$, using the bridge $Ref_{\mathcal{B}_i \to \mathcal{B}_{Ref}(\mathcal{A})}$ from $2^{\mathcal{B}_i}$ to $Ref(\mathcal{A})$ we translate each QCN $N_i \in \mathcal{N}$ on $2^{\mathcal{B}_i}$ into an equivalent QCN $N_i' = Ref_{\mathcal{B}_i \to \mathcal{B}_{Ref}(\mathcal{A})}(N_i)$ on $Ref(\mathcal{A})$ (cf Definition 5). Figure 9 depicts the three QCNs N_1', N_2' and N_3' on $Ref(\mathcal{A})$.

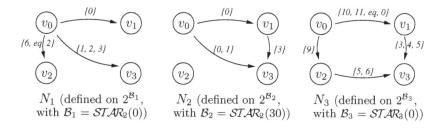

Fig. 7. The set $\mathcal{N}$ of three QCNs defined on different Star partition schemes

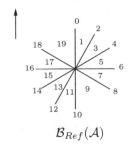

$$\mathcal{B}_{Ref}(\mathcal{A})$$

Fig. 8. The partition scheme $\mathcal{B}_{Ref}(\mathcal{A})$

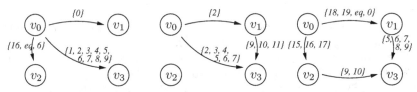

N'_1 (defined on $Ref(\mathcal{A})$) N'_2 (defined on $Ref(\mathcal{A})$) N'_3 (defined on $Ref(\mathcal{A})$)

Fig. 9. The three QCNs N'_1, N'_2 and N'_3 defined on $Ref(\mathcal{A})$

We get a set $\mathcal{N}' = \{N'_1, N'_2, N'_3\}$ of QCNs defined on the same qualitative algebra, with each N'_i equivalent to $N_i \in \mathcal{N}$. Thus we can use a merging process for QCNs defined on the same qualitative algebra as the one described in Section 3.

Let us now point out an interesting property when elements of $\mathcal{A}$ are all Star algebras. Consider again the partition scheme $Ref(\mathcal{A})$ on Figure 8, and recall that $Ref(\mathcal{A})$ is a common refinement of three Star algebras. We can notice that $\mathcal{B}_{Ref}(\mathcal{A})$ is the Star partition scheme $\mathcal{STAR}_5[0, 30, 60, 90, 120](0)$. More generally, the following proposition holds:

Proposition 6. *Let $\mathcal{A}$ be a set of Star algebras. Then $Ref(\mathcal{A})$ is a Star algebra.*

For lack of space, we will omit the proofs of the propositions in the sequel.

5   Another Alternative: Toward a Common Abstraction

5.1   The Model

In the previous section we explained how to associate to each input QCN on a qualitative algebra $2^{B_i} \in \mathcal{A}$ a QCN which is defined on the greatest common refinement $Ref(\mathcal{A})$. The merging method described in [4] can now be used since each QCN is translated into an equivalent QCN on the same qualitative algebra. However the number of basic relations of $\mathcal{B}_{Ref}(\mathcal{A})$ can be large and this number has an important role in the next step of the merging process (see Section 3). In this section, instead of translating the QCNs into a common refinement of $\mathcal{A}$, we look at the consequences of using a common abstraction of $\mathcal{A}$, a notion we first have to define. The notion of abstraction is dual to the notion of refinement and is given by the following definition.

Definition 9. *Let $2^{\mathcal{B}}$ and $2^{\mathcal{B}'}$ be two qualitative algebras on $\mathcal{D}$. $2^{\mathcal{B}'}$ is an abstraction of $2^{\mathcal{B}}$ iff $2^{\mathcal{B}}$ is a refinement of $2^{\mathcal{B}'}$.*

Assume $2^{\mathcal{B}'}$ to be an abstraction of $2^{\mathcal{B}}$. Recall that there exists a unique r-bridge $Ref_{\mathcal{B}'\to\mathcal{B}}$ from $2^{\mathcal{B}'}$ to $2^{\mathcal{B}}$ which is an injective, not necessary bijective function (in case of a bijective function, we have $2^{\mathcal{B}'} \leq^{\mathcal{D}}_{ref} 2^{\mathcal{B}}$ and $2^{\mathcal{B}} \leq^{\mathcal{D}}_{ref} 2^{\mathcal{B}'}$, namely $2^{\mathcal{B}} = 2^{\mathcal{B}'}$ since $\leq^{\mathcal{D}}_{ref}$ is an ordering relation). Thus in the general case we cannot find an r-bridge $Ref_{\mathcal{B}\to\mathcal{B}'}$ from $2^{\mathcal{B}}$ to $2^{\mathcal{B}'}$. This means that while it is always possible to translate a QCN on $2^{\mathcal{B}'}$ to an equivalent QCN on $2^{\mathcal{B}}$ (cf Proposition 2), in general the inverse is not possible.

However we can define a "weaker" bridge from $2^{\mathcal{B}}$ to $2^{\mathcal{B}'}$: we consider the mapping $Abs_{\mathcal{B}\to\mathcal{B}'}$ from $2^{\mathcal{B}}$ to $2^{\mathcal{B}'}$ such that for each basic relation $r \in \mathcal{B}$, $Abs_{\mathcal{B}\to\mathcal{B}'}(\{r\})$ is the relation $\{r'\}$ of $2^{\mathcal{B}'}$ such that $r' \supseteq r$, and for each relation R of $2^{\mathcal{B}}$, $Abs_{\mathcal{B}\to\mathcal{B}'}(R) = \bigcup\{Abs_{\mathcal{B}\to\mathcal{B}'}(\{r\}) \mid r \in R\}$. We will call such a mapping $Abs_{\mathcal{B}\to\mathcal{B}'}$ an a-bridge from $2^{\mathcal{B}}$ to $2^{\mathcal{B}'}$.

Example 4. Consider again the qualitative algebras $2^{\mathcal{B}_{card}}$ and $2^{\mathcal{B}_{card2}}$ (cf Example 2) and recall that $2^{\mathcal{B}_{card}} \leq_{ref} 2^{\mathcal{B}_{card2}}$. For example, we have $Abs_{\mathcal{B}_{card}\to\mathcal{B}_{card2}}(\{nw\}) = \{W\}$ and $Abs_{\mathcal{B}_{card}\to\mathcal{B}_{card2}}(\{sw, s\}) = \{W, s\}$.

Definition 10. *Let $N = (V, C')$ be a QCN defined on $2^{\mathcal{B}}$, and $2^{\mathcal{B}'}$ be an abstraction of $2^{\mathcal{B}}$. We define the QCN $Abs_{\mathcal{B}\to\mathcal{B}'}(N) = (V, C)$ on $2^{\mathcal{B}'}$ as $\forall v_i, v_j \in V$, $C_{ij} = Abs_{\mathcal{B}\to\mathcal{B}'}(C'_{ij})$.*

For any QCN N on $2^{\mathcal{B}}$, its translation $Abs_{\mathcal{B}\to\mathcal{B}'}(N)$ on $2^{\mathcal{B}'}$ is not an equivalent QCN in the general case. Nevertheless, we have the following weaker property.

Proposition 7. *Let N be a QCN defined on $2^{\mathcal{B}}$, and $2^{\mathcal{B}'}$ be an abstraction of $2^{\mathcal{B}}$. We have $Sol(N) \subseteq Sol(Abs_{\mathcal{B}\to\mathcal{B}'}(N))$.*

This means that the translation $Abs_{\mathcal{B}\to\mathcal{B}'}(N)$ of a QCN N on $2^{\mathcal{B}}$ can be viewed as an approximation of N.

Now similarly to Definition 5, we define the notion of *common abstraction*.

Definition 11. *Let $\mathcal{A} = \{2^{\mathcal{B}_1}, \ldots, 2^{\mathcal{B}_p}\}$ be a set of qualitative algebras on $\mathcal{D}$. Then $2^{\mathcal{B}}$ is called a common abstraction of $\mathcal{A}$ iff $\forall 2^{\mathcal{B}_k} \in \mathcal{A}$, $2^{\mathcal{B}_k} \leq_{ref} 2^{\mathcal{B}}$.*

We claim that a common abstraction of a set $\mathcal{A}$ of qualitative algebras on $\mathcal{D}$ always exists. Indeed consider the partition scheme $\mathcal{B}_{\neq}$ on $\mathcal{D}$ with $\mathcal{B}_{\neq} = \{eq, (\mathcal{D} \times \mathcal{D}) \setminus eq\}$. Then it is easy to see that $2^{\mathcal{B}_{\neq}}$ is an abstraction of any qualitative algebra on $\mathcal{D}$.

Definition 12. *Let $\mathcal{A} = \{2^{\mathcal{B}_1}, \ldots, 2^{\mathcal{B}_p}\}$ be a set of qualitative algebras on $\mathcal{D}$. The qualitative algebra $Abs(\mathcal{A}) = 2^{\mathcal{B}_{Abs}(\mathcal{A})}$ is defined as follows:*

$$\mathcal{B}_{Abs}(\mathcal{A}) = \{r \mid \forall \mathcal{B}_k \in \mathcal{A}, \exists R_k \in 2^{\mathcal{B}_k} : Rel(R_k) = r$$
$$\text{and } \nexists r' \subset r : \forall \mathcal{B}_k \in \mathcal{A}, \exists R'_k \in 2^{\mathcal{B}_k} : Rel(R'_k) = r'\}.$$

We claim that $Abs(\mathcal{A})$ is well defined and is the least common abstraction of $\mathcal{A}$. Similarly to Propositions 4 and 5, the following proposition holds:

Proposition 8. *Let* $\mathcal{A} = \{2^{\mathcal{B}_1}, \ldots, 2^{\mathcal{B}_P}\}$ *be a set of qualitative algebras on* $\mathcal{D}$.

- $\mathcal{B}_{Abs}(\mathcal{A})$ *is a partition scheme on* $\mathcal{D}$.
- $Abs(\mathcal{A})$ *is the least common abstraction of* $\mathcal{A}$.

Therefore, the process consists in translating each QCN N_i defined on $2^{\mathcal{B}_j} \in \mathcal{A}$ in a QCN $N_i' = Abs_{\mathcal{B}_j \to \mathcal{B}_{Abs}(\mathcal{A})}(N_i)$ on $Abs(\mathcal{A})$, using the a-bridge $Abs_{\mathcal{B}_j \to \mathcal{B}_{Abs}(\mathcal{A})}$ from $\mathcal{B}_j$ to $Abs(\mathcal{A})$. We get a new set $\mathcal{N}'$ of QCNs defined on $Abs(\mathcal{A})$ which are approximations of the QCNs of $\mathcal{N}$. Although we do not get equivalent QCNs, the counterpart is that since the size of the partition scheme $\mathcal{B}_{Abs}(\mathcal{A})$ is much smaller than the size of $\mathcal{B}_{Ref}(\mathcal{A})$, we reduce the complexity of the following step of the merging process, namely, merging the set of the translated QCNs defined on the same qualitative algebra. Furthermore, since this method provides a set of QCNs $\mathcal{N}'$ which are approximations of the initial set of QCNs $\mathcal{N}$, each translated QCN $N_i' \in \mathcal{N}'$ admits a larger set of solutions than its corresponding QCN $N_i \in \mathcal{N}$. Then in some cases, even if the QCNs of the initial set $\mathcal{N}$ are conflicting (namely, if they do not admit any common solution), the translated QCNs of $\mathcal{N}'$ can be not conflicting. Thus, this method can be viewed as a first step to deal with contradictions.

5.2   An Example on the Star Algebra

Similarly to Proposition 6, notice that the following proposition holds:

Proposition 9. *Let* $\mathcal{A}$ *be a set of Star algebras. Then* $Abs(\mathcal{A})$ *is a Star algebra.*

Now consider the three Star partition schemes $\mathcal{STAR}_3[0, 45, 90](0)$, $\mathcal{STAR}_4[0, 60, 90, 150](0)$ and $\mathcal{STAR}_4[0, 30, 90, 120](0)$ depicted in Figure 10 and forming the set $\mathcal{A}$. The partition scheme of the least common abstraction $Abs(\mathcal{A})$ is depicted in this same figure.

Let $\mathcal{N} = \{N_1, N_2, N_3\}$ be the three QCNs respectively defined on $2^{\mathcal{B}_1}$, $2^{\mathcal{B}_2}$ and $2^{\mathcal{B}_3}$, with $\mathcal{B}_1 = \mathcal{STAR}_3[0, 45, 90](0)$, $\mathcal{B}_2 = \mathcal{STAR}_4[0, 60, 90, 150](0)$ and $\mathcal{B}_3 = \mathcal{STAR}_4[0, 30, 90, 120](0)$ (these three partition schemes are depicted in Figure 10). Figure 11 depicts the three QCNs of $\mathcal{N}$ and their translation to their least common refinement $Abs(\mathcal{A})$ (with $\mathcal{B}_{Abs}(\mathcal{A}) = \mathcal{STAR}_2(0)$), forming the set $\mathcal{N}' = \{N_1', N_2', N_3'\}$.

Let R_1 be the relation of $2^{\mathcal{B}_1}$ with $R_1 = \{1, 2\}$, and R_2 be the relation of $2^{\mathcal{B}_2}$ with $R_2 = \{2, 3\}$. We can notice that the constraint between v_0 and v_1 is assigned to R_1 in N_1 and this same constraint is assigned to R_2 in N_2. Since $rel(R_1) \cap rel(R_2) = \emptyset$ (see Figure 10), there does not exist any solution of both N_1 and N_2. Thus the QCNs of $\mathcal{N}$ are conflicting. Nevertheless, the QCNs of $\mathcal{N}'$ are not conflicting, since there exists a consistent scenario of all QCNs of $\mathcal{N}'$. This consistent scenario is depicted in Figure 12. In this example there is no need to apply a merging process of the set $\mathcal{N}'$ a posteriori.

Let $QA_{\mathcal{D}}$ be the set of all qualitative algebras on $\mathcal{D}$. From Propositions 5 and 11, there is a particular structure induced by the ordering $\leq_{ref}^{\mathcal{D}}$ on $QA_{\mathcal{D}}$:

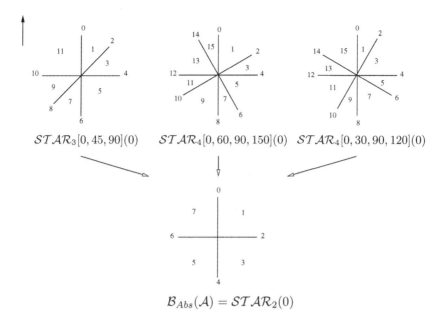

$$\mathcal{B}_{Abs}(\mathcal{A}) = \mathcal{STAR}_2(0)$$

Fig. 10. Three Star Algebras and the partition scheme $\mathcal{B}_{Abs}(\mathcal{A})$ of their least common abstraction $Abs(\mathcal{A})$

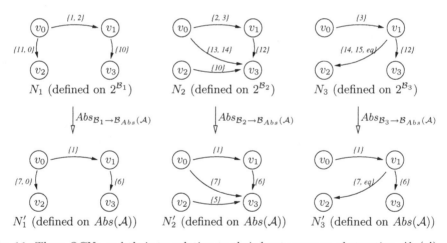

Fig. 11. Three QCNs and their translation to their least common abstraction $Abs(\mathcal{A})$

Lemma 1. $(QA_{\mathcal{D}}, \leq^{\mathcal{D}}_{ref})$ *is a lattice, namely,* $\forall 2^{\mathcal{B}}, 2^{\mathcal{B}'} \in QA_{\mathcal{D}}$, $2^{\mathcal{B}}$ *and* $2^{\mathcal{B}'}$ *have a least upper bound* $Abs(2^{\mathcal{B}}, 2^{\mathcal{B}'})$ *and a greatest lower bound* $Ref(2^{\mathcal{B}}, 2^{\mathcal{B}'})$.
Notice that the lattice is not bounded, since for any qualitative algebra $2^{\mathcal{B}}$ of $QA_{\mathcal{D}}$, we can find a qualitative algebra $2^{\mathcal{B}'}$ of $QA_{\mathcal{D}}$ such that $2^{\mathcal{B}'} \leq^{\mathcal{D}}_{ref} 2^{\mathcal{B}}$ and not $2^{\mathcal{B}} \leq^{\mathcal{D}}_{ref} 2^{\mathcal{B}'}$. However, the lattice has an upper bound, which corresponds to the qualitative algebra $2^{\mathcal{B}_{\neq}}$, with $\mathcal{B}_{\neq} = \{eq, (\mathcal{D} \times \mathcal{D}) \setminus eq\}$.

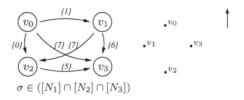

$$\sigma \in ([N_1] \cap [N_2] \cap [N_3])$$

Fig. 12. A consistent scenario of all QCNs $N_i' \in \mathcal{N}'$

6   Conclusion and Perspectives

In this paper, we addressed the problem of merging qualitative constraint networks (QCNs) when defined on different qualitative formalisms. The extension proposed here consists of translating the input QCNs into a common formalism. Therefore it is an upstream process to existing methods dealing with QCNs which are defined on the same formalism. We established a structure for all qualitative algebras which consider relations defined on the same domain. This structure is given by a weak partial ordering relation and forms a lattice over the qualitative algebras. The relation is based on the dual notions of *refinement* and *abstraction*, allowing us to define bridges between these qualitative algebras.

We are currently implementing the structure into QAT (Qualitative Algebra Toolkit) [5], a JAVA library allowing us to handle QCNs. An experimental study has to be made to measure the interest of using common abstractions instead of common refinements for the QCNs merging problem.

A future work will pursue study about the properties of this structure and about the links between qualitative algebras. This could allow us to increase the range of this work. For example, we will investigate the correspondences (in terms of consistency) between a QCN and its "translation" using the bridges involved by the structure. Indeed, translating an inconsistent QCN into another one using a well-chosen bridge could lead to restore the consistency of the QCN. We also intend to include subalgebras in the structure in order to define some bridges (in terms of refinements/abstractions) between qualitative algebras and their subalgebras (in particular their tractable subclasses), and study the consequences on the QCNs. We will also study how to define a bridge between heterogeneous formalisms, depending on a degree of compatibility between the related relations. In order to deal with this very general setting, we will have to think about new and possibly weaker structures.

References

1. Allen, J.-F.: An interval-based representation of temporal knowledge. In: Proc. of 7th the International Joint Conference on Artificial Intelligence (IJCAI), pp. 221–226 (1981)
2. Cholvy, L.: Reasoning about merging information. Handbook of Defeasible Reasoning and Uncertainty Management Systems 3, 233–263 (1998)

3. Condotta, J.-F., Kaci, S., Marquis, P., Schwind, N.: Merging qualitative constraints networks using propositional logic. In: Proc. of the 10th European Conference on Symbolic and Quantitative Approaches to Reasoning with Uncertainty (EC-SQARU), pp. 347–358 (2009)
4. Condotta, J.-F., Kaci, S., Schwind, N.: A Framework for Merging Qualitative Constraints Networks. In: Proc. of the 21th FLAIRS Conference, pp. 586–591 (2008)
5. Condotta, J.-F., Ligozat, G., Saade, M.: A qualitative algebra toolkit. In: Proc. of the 2nd IEEE International Conference on Information Technologies: from Theory to Applications, ICTTA (2006)
6. Egenhofer, M.-J.: Reasoning about binary topological relations. In: Günther, O., Schek, H.-J. (eds.) SSD 1991. LNCS, vol. 525, pp. 143–160. Springer, Heidelberg (1991)
7. Frank, A.U.: Qualitative spatial reasoning about cardinal directions. In: Proc. of the 7th Austrian Conference on Artificial Intelligence, pp. 157–167 (1991)
8. Gerevini, A., Renz, J.: Combining topological and size information for spatial reasoning. Artificial Intelligence 137(1-2), 1–42 (2002)
9. Jonsson, P., Drakengren, T.: A complete classification of tractability in RCC-5. Journal of Artificial Intelligence Research 6, 211–221 (1997)
10. Konieczny, S., Lang, J., Marquis, P.: DA^2 merging operators. Artificial Intelligence 157(1-2), 49–79 (2004)
11. Li, S.: Combining topological and directional information for spatial reasoning. In: Proc. of the 20th International Joint Conference on Artificial Intelligence (IJCAI), pp. 435–440 (2007)
12. Ligozat, G.: Reasoning about cardinal directions. Journal of Visual Languages and Computing 9(1), 23–44 (1998)
13. Ligozat, G., Renz, J.: What Is a Qualitative Calculus? A General Framework. In: Zhang, C., Guesgen, H.W., Yeap, W.-K. (eds.) PRICAI 2004. LNCS, vol. 3157, pp. 53–64. Springer, Heidelberg (2004)
14. Lin, J.: Integration of weighted knowledge bases. Artificial Intelligence 83, 363–378 (1996)
15. Pujari, A.K., Kumari, G.V., Sattar, A.: $\mathcal{INDU}$: An interval duration network. In: Proc. of the 16th Australian Joint Conference on Artificial Intelligence (AI), pp. 291–303 (2000)
16. Ragni, M., Wölfl, S.: Reasoning about topological and positional information in dynamic settings. In: Proc. of the 21th FLAIRS Conference, pp. 606–611 (2008)
17. Randell, D.-A., Cui, Z., Cohn, A.: A spatial logic based on regions and connection. In: Proc. of the 3rd International Conference on Principles of Knowledge Representation and Reasoning (KR), pp. 165–176 (1992)
18. Renz, J.: Qualitative spatial reasoning with topological information. Springer, Heidelberg (2002)
19. Renz, J., Ligozat, G.: Weak Composition for Qualitative Spatial and Temporal Reasoning. In: van Beek, P. (ed.) CP 2005. LNCS, vol. 3709, pp. 534–548. Springer, Heidelberg (2005)
20. Renz, J., Mitra, D.: Qualitative direction calculi with arbitrary granularity. In: Zhang, C., Guesgen, H.W., Yeap, W.-K. (eds.) PRICAI 2004. LNCS, vol. 3157, pp. 65–74. Springer, Heidelberg (2004)
21. Revesz, P.Z.: On the semantics of arbitration. Journal of Algebra and Computation 7(2), 133–160 (1997)

Semi-automated Derivation of Conceptual Neighborhood Graphs of Topological Relations

Yohei Kurata

SFB/TR 8 Spatial Cognition, Universität Bremen
Postfach 330 440, 28334 Bremen, Germany
ykurata@informatik.uni-bremen.de

Abstract. Conceptual neighborhood graphs are similarity-based schemata of spatial/temporal relations. This paper proposes a semi-automated method for deriving a conceptual neighborhood graph of topological relations, which shows all pairs of relations between which a smooth transformation can be performed. The method is applicable to various sets of topological relations distinguished by the 9+-intersection. The method first identifies possible primitive-level transitions, combines those primitive-level transitions, and removes invalid combinations that do not satisfy some necessary conditions. As a demonstration, we develop conceptual neighborhood graphs of topological region-region relations in $\mathbb{R}^2$, $\mathbb{S}^2$, and $\mathbb{R}^3$, topological relations between a directed line and a region in $\mathbb{R}^2$, and Allen's interval relations.

Keywords: conceptual neighborhood graphs, conceptual neighbors, topological relations, 9+-intersection, smooth transformation.

1 Introduction

Conceptual neighborhood graphs [1] (in short, *CN-graphs*) are similarity-based schemata of spatial/temporal relations, in which pairs of relations called *conceptual neighbors* are linked. Conceptual neighbors are, in the original sense, a pair of relations between which a smooth transformation can be performed [1]. CN-graphs have been developed for various relation sets (for instance, [1-11]) in order to schematize the relations and analyze their properties. CN-graphs are also used in qualitative spatio-temporal reasoning to infer possible transitions of spatial configurations [2, 7, 12] or to relax constraints in constraint networks [13].

In previous studies, conceptual neighbors are sometimes determined by a specific type of smooth transformations [4, 8, 9, 11] or even another similarity measures [4, 14]. As a result, some CN-graphs show only a small portion of relation pairs between which a smooth transformation can be performed and, therefore, they are insufficient for inferring possible transitions of spatial configurations. In addition, the diversity of conceptual neighbors is confusing for the users of CN-graphs. Nevertheless, various types of conceptual neighbors have been developed, because for each relation set a specific type of conceptual neighbor allows quick development of a schematic

K. Stewart Hornsby et al. (Eds.): COSIT 2009, LNCS 5756, pp. 124–140, 2009.
© Springer-Verlag Berlin Heidelberg 2009

CN-graph (e.g., [2, 4, 8, 9, 11, 14]), whereas identifying all possible smooth transformations between relations is usually a time-consuming and error-prone process.

To tackle these problems in terms of topological relations, we propose a semi-automated method for deriving a CN-graph of a given set of topological relations. The derived CN-graph shows all pairs of topological relations between which a smooth transformation can be performed. This method is powerful, because (i) the process of detecting potential neighbors is *fully* automated and (ii) the method is applicable to a variety of topological relations distinguished by the 9^+-*intersection* [9, 15]. With the 9^+-intersection and its universal constraints [15], we can easily identify a set of topological relations between arbitrary combination of objects. Once a set of topological relations is identified, it is able to schematize these relations quickly and analyze their properties based on the CN-graph derived by our method.

In this paper, we focus on the topological relations in 1D, 2D and 3D Euclidian spaces ($\mathbb{R}^1$, $\mathbb{R}^2$, and $\mathbb{R}^3$,), 1-sphere $\mathbb{S}^1$ (i.e., a linear loop), and 2-sphere $\mathbb{S}^2$ (i.e., a spherical surface). *Region* refers to a surface embedded in a 2D or 3D space. *Simple regions* are regions without holes, spikes, cuts, and disconnected interior parts.

The remainder of this paper is structured as follows: Section 2 reviews the related studies on CN-graphs. Section 3 summarizes the major concepts of the 9^+-intersection. Section 4 redefines conceptual neighbors and introduces the families of conceptual neighbors. Section 5 describes our method for deriving CN-graphs. As a demonstration, Section 6 derives several CN-graphs. Finally, Section 7 concludes with a discussion of future problems.

2 Conceptual Neighbors: Definitions and Applications

The idea of CN-graphs was proposed by Freksa [1] in order to analyze Allen's [16] interval relations. He linked two interval relations as conceptual neighbors if a *smooth transformation* can be performed between them. He distinguished three types of smooth transformations; moving (dragging) an endpoint of one interval while keeping the location of another endpoint, sliding one interval entirely, and stretching/shortening one interval. These three types of smooth transformations yield three types of conceptual neighbors, namely *A-*, *B-*, and *C-neighbors*. Fig. 1a shows the CN-graph formed by these three types of conceptual neighbors.

After Freksa's proposal, conceptual neighbors of relations have been discussed not only for interval relations [1, 5], but also for topological relations [2, 4, 8-10, 17] and other qualitative spatial relations (for instance, [3, 6, 7, 11, 18]). An early example is [2], in which Egenhofer and Al-Taha developed a CN-graph of topological region-region relations in $\mathbb{R}^2$ (Fig. 1b). They first developed an approximated graph, in which each relation was connected to the relations with the smallest number of different elements in their 9-intersection matrices [19]. This approach, later called the *snapshot model*, enables us to identify the conceptual neighbors of topological relations computationally, even though the meaning of conceptual neighbors is not the same as that based on the possibility of smooth transformations.

Egenhofer and Mark [4] developed two CN-graphs of topological line-region relations in $\mathbb{R}^2$; one was based on the snapshot model and another was based on the possibility of smooth transformations (only those established by dragging either endpoint
or interior of the line). Fig. 1c shows the latter CN-graph. The developed two graphs
had similar structures, but the former graph was planar while the latter one was more
complicated (Fig. 1c).

Kurata and Egenhofer [8] developed a CN-graph of topological relations between
two directed lines (*DLines*) in $\mathbb{R}^2$ based on the snapshot model. Interestingly, the
same graph can be derived based on the possibility of smooth transformations by
dragging either the starting point, interior, or ending point of the DLine while maintaining the intersection state of non-dragged parts of the DLine. The same type of
smooth transformations were used in [9] as a foundation for developing a CN-graph
of topological DLine-region relations in $\mathbb{R}^2$ (Fig. 1d).

In many studies on qualitative spatial relations, CN-graphs are used to schematize
the sets of spatial relations of concern. If the relations are arranged appropriately in a
diagrammatic space, the CN-graph highlights several properties of the relations, such
as pairs of converse relations and symmetric relations. Making use of this schematic

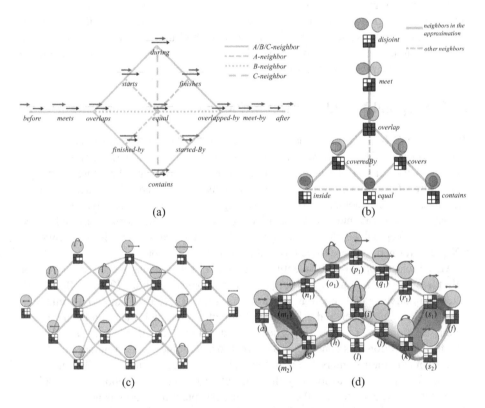

Fig. 1. CN-graphs of (a) Allen's interval relations [1], (b) topological region-region relations in
$\mathbb{R}^2$ [19], (c) topological line-region relations in $\mathbb{R}^2$ [4], and (d) topological DLine-region relations in $\mathbb{R}^2$ [9]

Fig. 2. A CN-graph-based icon that represents a subset of topological line-line relations in $\mathbb{R}^2$

appearance, some studies introduced the CN-graph-based icons that represent a subset of spatial relations by black nodes [6, 9, 14] (Fig. 2). Such icons are useful, because (i) computational operations on the spatial relations often result in the sets of relations that form a cluster in a CN-graph [8, 20] and similarly (ii) linguistic expressions that describe spatial arrangements often correspond to the clusters of relations in a CN-graph [9, 21, 22] (Fig. 2).

In qualitative spatio-temporal reasoning, CN-graphs have more roles. One is to list every possible transitions of spatial relations [2, 7, 12]. Such information is used, for instance, to infer a possible sequence of spatial configurations between two snapshots of spatial scenes. Another role of CN-graphs emerges when we have to relax constraints in constraint networks. Qualitative spatial calculi use constraint networks in which each constraint represents possible relations between objects. If a constraint network turns out to be inconsistent but we still want to find a certain constraint scenario, the constraints are relaxed by adding neighboring relations [13].

3   The 9^+-intersection

The 9^+-intersection [9] is a model of topological relations, which extends the 9-intersection [19]. Both models presume the distinction of three *topological parts* of spatial objects, namely *interior*, *boundary*, and *exterior*. Based on point-set topology [23], the interior of a spatial object X, denoted $X°$, is defined as the union of all open sets contained in X, X's exterior X^- is defined as the union of all open sets that do not contain X, and the boundary ∂X is defined as the difference between X^-'s complement and $X°$. In the 9-intersection, the topological relations between two objects are distinguished typically by the presence or absence of pairwise intersections of their topological parts. On the other hand, the 9^+-intersection considers the pairwise intersections of the objects' *topological primitives*. Topological primitives are self-connected and mutually-disjoint subsets of the objects' topological parts. The concept of topological primitives is useful when a certain topological part of the objects consists of multiple disjoint subparts. For instance, the boundary of a simple line consists of two distinctive points. By distinguishing these two points as different topological primitives, the 9^+-intersection can capture the topological relations between a directed line and another object.

In the 9-intersection, the presence/absence of pairwise intersections of topological parts is represented by an icon with 3×3 black-and-white cells (Fig. 3a) [21]. Following this convention, in the 9^+-intersection, the presence/absence of pairwise intersections of topological primitives is represented by a nested icon like Fig. 3b [9].

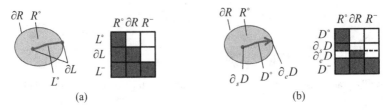

Fig. 3. Iconic representations of topological relations in (a) the 9-intersection and (b) the 9⁺-intersection

In the 9⁺-intersection, topological primitives are classified according to their spatial dimensions (*0D-3D*) and boundedness (*bounded, looped,* and *unbounded,* represented by prefixes *B-, L-,* and *U-,* respectively) [15]. For instance, the interior, boundary, and exterior of a region in $\mathbb{R}^2$ belong to *B-2D, L-1D,* and *U-2D,* respectively. With these notations, the structure of each object can be represented by a *structure graph,* which shows the class and connectivity of all primitives of the object (Fig. 4) [15].

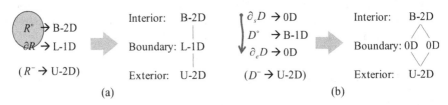

Fig. 4. Representations of the topological structures of (a) a simple region and (b) a directed line, both embedded in $\mathbb{R}^2$

4 Conceptual Neighbors and Their Families

In this section, we redefine the conceptual neighbors of topological relations, and then introduce two families of the conceptual neighbors. First, we redefine the conceptual neighbors of (generic) spatial/temporal relations, following Freksa's [1] original notion of conceptual neighbors. Let $\mathcal{R}$ be a set of spatial/temporal relations between two objects A and B.

Definition (*conceptual neighbor*): A relation $r_i \in \mathcal{R}$ is called a *conceptual neighbor* of a relation $r_j \in \mathcal{R} \setminus \{r_i\}$ if at least one instance of r_j can switch *directly* to r_i by a *smooth transformation* of the configuration (i.e., switch to r_i without passing through any third relation $r_k \in \mathcal{R} \setminus \{r_i, r_j\}$).).

This definition leaves the interpretation of smooth transformations open. For determining conceptual neighbors of topological relations, we follow the following definition of smooth transformations:

Definition (*smooth transformation*): A smooth transformation of a configuration of two objects in a space S is to deform the shape of either or both objects continuously in S, without changing the topological structure of each object.

Recall that the topological structure of each object is represented by the structure graph (Fig. 4). This definition is generic, covering various types of smooth transformations such as rotation, translation, expansion/contraction, and deformation by dragging a part of an object.

When judging the conceptual neighborhood of two topological relations, we do not lose generality if we consider the deformation of only one object in a relativistic view. Thus, to simplify discussion, let us consider that an object A is deformed while an object B is fixed. Then, B's primitives are regarded as the **partitions** of the space, denoted $\mathcal{P} = \{\mathcal{P}_i\}$, because B's primitives are jointly exhaustive, pairwise disjoint, and now fixed in the space.

By A's deformation, A's primitive a_x may experience one or more of the following *primitive-level events*:

- *propagation*: a_x, which initially intersects with a partition $\mathcal{P}_i$, gains a sequence of intersections with $\mathcal{P}_i$ and its adjacent partition $\mathcal{P}_j$ (Figs. 5a$_1$-a$_2$);
- *inverse-propagation*: a_x, which initially has a sequence of intersections with two adjacent primitives P_i and $\mathcal{P}_j$, loses the intersection with $\mathcal{P}_j$ while keeping the intersection with P_i (Figs. 5a$_1$-a$_2$);
- *penetration*: a_x, which initially intersects with a partition $\mathcal{P}_i$, gains a sequence of intersections with $\mathcal{P}_i$, $\mathcal{P}_i$'s adjacent primitive P_j, and $\mathcal{P}_j$'s adjacent primitive P_k (Figs. 5b$_1$-b$_2$);
- *inverse-penetration*: a_x, which initially has a sequence of intersections with three adjacent primitives $\mathcal{P}_i$, $\mathcal{P}_i$, and P_k (where $\mathcal{P}_j$ is adjacent to both $\mathcal{P}_i$ and P_k), loses the intersections with $\mathcal{P}_j$ and P_k while keeping the intersection with $\mathcal{P}_i$ (Figs. 5b$_1$-b$_2$);
- *transfer$^+$*: a_x becomes intersecting with a partition $\mathcal{P}_i$ and not intersecting with one or more of $\mathcal{P}_i$'s lower-dimensional adjacent partitions (Figs. 5c$_1$-c$_4$); and
- *transfer$^-$*: a_x becomes not intersecting with a partition $\mathcal{P}_i$ and intersecting with one or more of $\mathcal{P}_i$'s lower-dimensional adjacent partitions (Figs. 5c$_1$-c$_4$);

Otherwise, a_x does not gain or lose any intersection. Note the difference of the expressions *to become intersecting with X* and *to gain an intersection with X*. The former expression presumes the absence of intersections with X before the smooth transformation, while the latter one does not. This implies that *transfer$^+$* and *transfer$^-$* always yield a change in the intersection state of the primitive, but *propagation* and *penetration* do not necessarily (e.g., Fig. 5b$_2$). Similarly, *to become not intersecting with X* presumes the absence of intersections with X after the smooth transformation, while *to lose an intersection with X* does not.

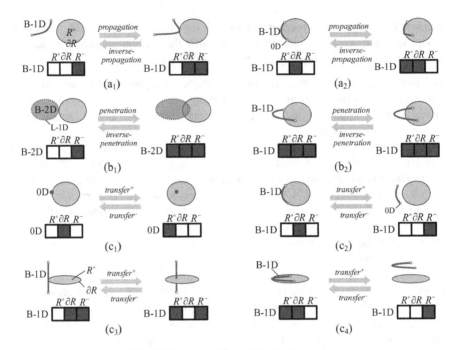

Fig. 5. Examples of primitive-level events

There are certain dependences between the primitive-level events that the primitives of one object may experience at the same time. For instance:

- in Fig. 5a$_2$, the *propagation* of the B-1D is triggered by the *transfer* of the 0D;
- in Fig. 5b$_1$, the *penetration* of the B-2D is triggered by the *propagation* of the L-1D; and
- in Fig. 5c$_2$ the *transfer⁻* of the B-1D triggers the *transfer⁻* of the 0D.

When a primitive-level event triggers other primitive-level events, the set of these events is called an *event sequence*. We consider that a primitive-level event, which does not trigger other primitive-level events, also forms an event sequence by itself.

Now we are ready to introduce two families of conceptual neighbors, called *SE-neighbor* (*Single-Event-based* neighbor) and *SES-neighbors* (*Single-Event-Sequence-Based neighbors*):

Definition (*SE-neighbor*): A relation $r_i \in \mathcal{R}$ is called a *SE-neighbor* of a relation $r_j \in \mathcal{R} \setminus \{r_i\}$ if an instance of r_j can switch directly to r_i by a smooth transformation where only one of A's primitives experiences only one primitive-level event (Fig. 6a)

Definition (*SES-neighbor*): A relation $r_i \in \mathcal{R}$ is called a *SES-neighbor* of a relation $r_j \in \mathcal{R} \setminus \{r_i\}$ if an instance of r_j can switch directly to r_i by a smooth transformation where all primitive-level events that A's primitives experience jointly form one event sequence (Fig. 6b)

Fig. 6. Smooth transformations, which establish (a) SE-neighbors, (a-b) SES-neighbors, and (a-c) conceptual neighbors, respectively

The definition of SE-neighbors generalizes the definition of concept neighbors of DLine-related relations in [9], while that of SES-neighbors generalizes the definition of conceptual neighbors of line-region relations in [4]. All SE-neighbors of a relation r_i are also SES-neighbors of r_i, and all SES-neighbors of r_i are also conceptual neighbors of r_i (Figs. 6a-c). As a result, given a set of topological relations, the set of SE-neighbors are a subset of SES-neighbors, while the set of SES-neighbors are a subset of conceptual neighbors. Neighborhood graphs based on SE- and SES-neighbors are, therefore, potentially useful when the CN-graphs need to be simplified for visualization.

5 A Method for Deriving CN-graphs

Given a set of relations, the process of deriving its CN-graph is divided into two steps; first, all pairs of conceptual neighbors are identified. By linking these pairs, we already obtain a CN-graph in a mathematical sense. However, we often go one further step, in which the relations are arranged in a diagrammatic space, such that the CN-graph looks visually schematic. This study focuses on the first step, while the second step is left for other studies (e.g., [17]).

Our method is summarized as follows: for each relation r_i in a given set of topological relations $\mathcal{R}$, the potential conceptual neighbors of r_i, called r_i's *neighbor candidates*, are derived computationally. Then, people manually check the validity of each neighbor candidate; i.e., whether r_i has a geometric instance that switches directly to the relation of the neighbor candidate by a smooth transformation. Fig. 7 illustrates the process of deriving the neighbor candidates of *covers* relation. Since our method is based on the 9$^+$-intersection, relations are represented by the 9$^+$-intersection icons. First, the 9$^+$-intersection icon for r_i is decomposed into rows. For each row, all possible patterns resulting from a smooth transformation are derived, using precomputed lists of possible primitive-level transitions (Section 5.1). These possible patterns of each row are combined to rebuild the 9$^+$-intersection icons. From these icons, some are removed if they do not represent any relation in $\mathcal{R}$. In addition, some are removed if they do not satisfy the necessary conditions in Section 5.2. We also conduct a similar process, starting from the decomposition of the 9$^+$-intersection icon into columns. Finally, we pick up the icons derived from both processes, which represent r_i's neighbor candidates. For instance, the process in Fig. 7 results in three icons, which indicate that *contains*, *equal*, and *overlap* relations are the potential conceptual neighbors of *covers* relation.

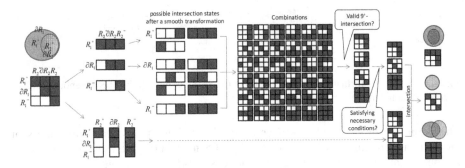

Fig. 7. A process of deriving the potential conceptual neighbors of *covers* relation

5.1 Listing Possible Primitive-Level Transitions

In order to list all possible primitive-level transitions, we first have to identify all intersection states (i.e., the patterns of rows/columns) that each primitive may take. The set of intersection states that a primitive p may take depends on p's class (0D, B-1D, etc.) and the structure of the partner object (Fig. 8). Practically, this set is identified simply by listing all patterns of the 9^+-intersection icon's row/column that corresponds to the primitive p. For instance, from the eight 9^+-intersection icons in Fig. 1b that represent the topological region-region relations in $\mathbb{R}^2$, we can find that the region's interior primitive (B-2D), boundary primitive (L-2D), and exterior primitive (U-2D) may take three, six, and two intersection states when the partner object is a simple region in $\mathbb{R}^2$ (Figs. 8b-d). Precisely speaking, this solution does not exclude the possibility that B-2D, L-2D, and U-2D may theoretically take other intersection states, but these additional intersection states, if they exist, are not relevant for deriving conceptual neighbors.

Next, for each of p's possible intersection states, we judge the possibility of transitions to other intersection states by a smooth transformation. During a smooth transition, p may experience primitive-level events (*propagation*, *penetration*, *inverse-propagation*, *inverse-penetration*, *transfer*$^+$, and *transfer*$^-$) that may yield a

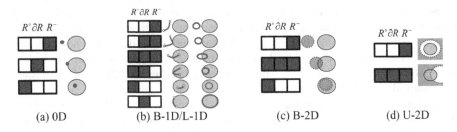

(a) 0D   (b) B-1D/L-1D   (c) B-2D   (d) U-2D

Fig. 8. Possible intersection states of 0D, B-1D, L-1D, B-2D, and U-2D primitives when the partner object is a simple region in $\mathbb{R}^2$

change of p's intersection state (Section 4). What primitive-level events may occur is restricted by the following constraints (suppose the primitives of the partner object form the partitions $\mathcal{P} = \{\mathcal{P}_i\}$ and p initially intersects with all partitions in $\mathcal{P}^{\text{init}} \subseteq \mathcal{P}$ but not others):

- p may experience a *propagation* in which p becomes intersecting with $\mathcal{P}_i$ ($\in \mathcal{P} \setminus \mathcal{P}^{\text{init}}$) only if p is not 0D and $\mathcal{P}_i$ has an adjacent partition in $\mathcal{P}^{\text{init}}$ with which p intersects before and after the smooth transformation (Figs. 5a$_1$-5a$_2$);
- p may experience an *inverse-propagation* in which p becomes not intersecting with $\mathcal{P}_i$ ($\in \mathcal{P}^{\text{init}}$) only if $\mathcal{P}_i$ has an adjacent partition in $\mathcal{P}^{\text{init}}$ with which p intersects before and after the smooth transformation (Figs. 5a$_1$-5a$_2$);
- p may experience a *penetration* in which p becomes intersecting with $\mathcal{P}_i$ and $\mathcal{P}_j$ ($\in \mathcal{P} \setminus \mathcal{P}^{\text{init}}$, $\dim(\mathcal{P}_i) < dim(\mathcal{P}_j)$) only if p is not 0D, $\mathcal{P}_i$ and $\mathcal{P}_j$ are adjacent, and $\mathcal{P}^{\text{init}}$ contains a $\mathcal{P}_i$'s adjacent higher-dimensional partition with which p intersects before and after the smooth transformation (Figs. 5b$_1$-5b$_2$);
- p may experience an *inverse-penetration* in which p becomes not intersecting with $\mathcal{P}_i$ and $\mathcal{P}_j$ ($\in \mathcal{P}^{\text{init}}$, $\dim(\mathcal{P}_i) < dim(\mathcal{P}_j)$) only if $\mathcal{P}_i$ and $\mathcal{P}_j$ are adjacent, $\mathcal{P}^{\text{init}} \setminus \{\mathcal{P}_i, \mathcal{P}_j\}$ contains a $\mathcal{P}_i$'s adjacent higher-dimensional partition with which p intersects before and after the smooth transformation (Figs. 5b$_1$-5b$_2$);
- p may experience *transfer$^+$* in which a_x becomes intersecting with $\mathcal{P}_i$ ($\in \mathcal{P} \setminus \mathcal{P}^{\text{init}}$) and not intersecting with a set of partitions $\mathcal{P}^{\text{lose}}$ ($\subseteq \mathcal{P}^{\text{init}}$) only if $\mathcal{P}^{\text{lose}}$ contains a $\mathcal{P}_i$'s adjacent lower-dimensional partition; and
- p may experience *transfer$^-$* in which a_x becomes not intersecting with $\mathcal{P}_i$ ($\in \mathcal{P}^{\text{init}}$) and intersecting with a set of partitions $\mathcal{P}^{\text{gain}}$ ($\subseteq \mathcal{P} \setminus \mathcal{P}^{\text{init}}$) only if $\mathcal{P}^{\text{gain}}$ contains a $\mathcal{P}_i$'s adjacent lower-dimensional partition.

With these constraints we can identify all possible intersection states of p resulting from a smooth transformation. By repeating this process for every intersection state

Table 1. Possibility of primitive-level transitions of a B-1D primitive when the partner object is a simple region in $\mathbb{R}^2$. The primitive-level events that establish the transitions are also indicated (pr: propagation, pe: penetration, ipr: inverse-propagation, ipe: inverse-penetration, tr$^+$: transfer$^+$, and tr$^-$: transfer$^-$).

		After					
		√	√ (tr⁻)	–	√ (pr)	–	√ (pe)
		√ (tr⁺)	√	√ (tr⁺)	√ (pr)	√ (pr)	√ (pr+pr)
Before		–	√ (tr⁻)	√	–	√ (pr)	√ (pe)
		√ (ipr)	√ (ipr)	–	√	√ (ipr+pr)	√ (pr)
		–	√ (ipr)	√ (ipr)	√ (pr+ipr)	√	√ (pr)
		√ (ipe)	√ (ipr⁻+ipr)	√ (ipe)	√ (ipr)	√ (ipr)	√

that p may take, we obtain a table of possible transitions of p's intersection state. For instance, Table 1 shows the possible transitions of the intersection state of a B-1D primitive (e.g., a DLine's interior) when the partner object is a simple region in $\mathbb{R}^2$. We can see that some transitions presume multiple primitive-level events.

5.2 Necessary Conditions

Given the neighbor candidates that are derived as the combination of primitives' possible intersection states resulting from a smooth transformation, the following five conditions are used to remove invalid candidates. The last four conditions were relevant to the dependency between primitive-level events.

- If a_x is entirely contained by $\mathcal{P}_i$ before the smooth transformation, then a_x cannot contain $\mathcal{P}_i$ entirely after the smooth transformation, and vice versa (Fig. 9a) because there must be a moment that a_x is equal to $\mathcal{P}_i$;
- If A's primitive a_x experiences a *penetration* in which A becomes intersecting with two adjacent partitions $\mathcal{P}_i$ and $\mathcal{P}_j$ $(\dim(\mathcal{P}_i) < \dim(\mathcal{P}_j))$, then A must have at least one primitive that is a_x's lower-dimensional neighbor and intersects with $\mathcal{P}_i$ before the smooth transformation and with $\mathcal{P}_j$ after it (Fig. 9b);
- Conversely, if a_x experiences an *inverse-penetration* in which A becomes not intersecting with $\mathcal{P}_i$ and P_j $(\dim(\mathcal{P}_i) < \dim(\mathcal{P}_j))$, then A must have at least one primitive that is a_x's lower-dimensional neighbor and intersects with $\mathcal{P}_j$ before the smooth transformation and with $\mathcal{P}_i$ after it;

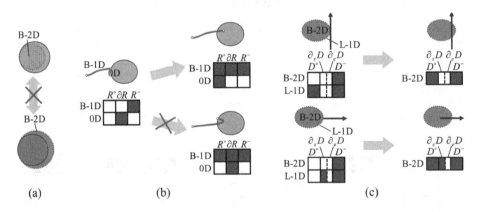

(a)   (b)   (c)

Fig. 9. Illustration of necessary conditions: (a) if B-2D is entirely contained by R°, it cannot contain R° entirely after the smooth transformation, (b) in order for B-1D to experience a *penetration*, 0D (B-1D's neighbor) should intersect with ∂R before the smooth transformation and R° after it, and (c) in order for B-2D to gain an intersection with D°, L-1D (B-2D's neighbor) should intersects with D°, $\partial_s D$, or $\partial_e D$ before the smooth transformation

- If a_x, which initially intersects only with an equal-dimensional partition $\mathcal{P}_i$, becomes intersecting with $\mathcal{P}_i$'s adjacent lower-dimensional partition $\mathcal{P}_j$, then A must have at least one primitive that is a_x's lower-dimensional neighbor and intersects with either $\mathcal{P}_j$ or one of $\mathcal{P}_j$'s adjacent lower-dimensional partitions before the smooth transformation (Fig. 9c); and

- Conversely, if a_x, which initially intersects with a lower-dimensional partition $\mathcal{P}_i$, becomes intersecting only with an equal-dimensional partition $\mathcal{P}_j$ that is adjacent to $\mathcal{P}_i$, then A must have at least one primitive that is a_x's lower-dimensional neighbor and intersects with either $\mathcal{P}_i$ or one of $\mathcal{P}_i$'s adjacent lower-dimensional partitions after the smooth transformation.

6 Case Studies

To demonstrate the potential of the method proposed in Section 5, this section derives CN-graphs of five relation sets—topological region-region relations in $\mathbb{R}^2$, $\mathbb{S}^2$, and $\mathbb{R}^3$, Allen's [16] interval relations, and topological DLine-region relations in $\mathbb{R}^2$—and compares them with the CN-graphs reported in previous studies. Recall that, given a set of n relations, our method computationally derives the potential conceptual neighbors of each relation. After checking the validity of these potential neighbors, we obtain a $n \times n$ Boolean matrix that shows the neighborhoods of the relations. How to represent these neighborhoods in a diagrammatic space is not supported by the current method, but in this section we visualized the CN-graphs such that they look schematic and structurally similar to existing CN-graphs (Figs. 10-13).

Case 1: Topological relations between two regions in $\mathbb{R}^2$

The 9-intersection (and naturally the 9^+-intersection as well) distinguishes eight region-region relations in $\mathbb{R}^2$ [19]. We computed the potential conceptual neighbors of these eight relations. We found that these potential neighbors are all valid. In addition, we found they are symmetric (i.e., if r_1 is a conceptual neighbor of r_2, then r_2 is also a conceptual neighbor of r_1). Fig. 10a shows the CN-graph we obtained. This CN-graph is exactly identical to the CN-graph developed by Egenhofer and Al-Taha [2] based on the possibility of smooth transformations (Fig. 1b).

Case 2: Topological relations between two regions in $\mathbb{S}^2$

The 9-intersection (and the 9^+-intersection as well) distinguishes eleven region-region relations in a spherical surface [14]. Again, we found that all computationally-derived potential conceptual neighbors of these relations are valid and symmetric. The obtained CN-graph (Fig. 10b) contains the CN-graph in Fig. 10a as a sub-graph and three more relations specific to $\mathbb{S}^2$ (*embraces*, *attaches*, and *entwined*). This CN-graph is similar to Egenhofer's [14] CN-graph, but ours shows six more neighbors: *attaches–embraces*, *attaches–overlap*, *attaches–disjoint*, *equal–overlap*, *equal–inside*, and *equal–contains*. This is because Egenhofer's CN-graph is based on the snapshot model instead of the possibility of smooth transformations.

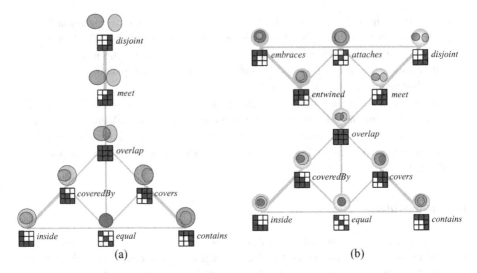

Fig. 10. CN-graph of topological region-region relations in (a) $\mathbb{R}^2$ and (b) $\mathbb{S}^2$, derived by the method in Section 5

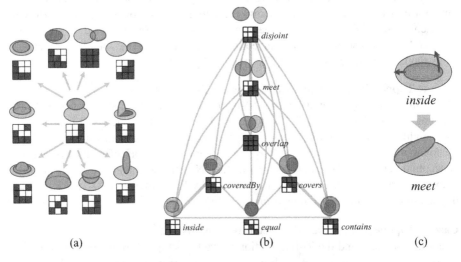

Fig. 11. (a) Various possibility of smooth transformations in $\mathbb{R}^3$. (b) A CN-graph of a subset of eight topological region-region relations in $\mathbb{R}^3$ [14], derived by the method in Section 5. (c) A smooth transformation from *inside* to *meet*, which is possible in $\mathbb{R}^3$, but not in $\mathbb{R}^2$.

Case 3: Topological relations between two regions in $\mathbb{R}^3$

The 9^+-intersection (essentially the 9-intersection) distinguishes 43 region-region relations in $\mathbb{R}^3$ [15]. No CN-graph of these relations was reported before. One critical problem is that in $\mathbb{R}^3$ each relation has many conceptual neighbors (e.g., Fig. 11a). Indeed, our method derives on average 40.7 potential conceptual neighbors for each relation. For simplification, we consider only eight region-region relations that are common in $\mathbb{R}^2$, $\mathbb{S}^2$, and $\mathbb{R}^3$. All computationally-derived potential conceptual neighbors of these eight relations were found to be valid and symmetric. The obtained CN-graph (Fig. 11b) is structurally similar to its two-dimensional counterpart in Fig. 10a, containing it as a sub-graph. This is reasonable because all smooth transformations possible in $\mathbb{R}^2$ are also possible in $\mathbb{R}^3$. Additional links represent the smooth transformations possible only in $\mathbb{R}^3$. We found that *disjoint*, *meet*, and *equal* relations are conceptual neighbors of all other relations. The reader might feel strange that *meet* is a conceptual neighbor of *inside* or *contains*, but Fig. 11c shows the possibility of a smooth transformation between them established by two simultaneous primitive-level events. This implies that *meet* is not a SES-neighbor of *inside* or *contains*.

Case 4: Topological relations between two uni-directed lines in $\mathbb{R}^1$

The 9^+-intersection distinguishes 26 DLine-DLine relations in $\mathbb{R}^1$ [15]. Half of these relations, in which two DLines have the same direction, correspond to Allen's [16] interval relations. We computationally derived the potential conceptual neighbors of these 13 relations, which were found to be valid and symmetric. The obtained CN-graph (Fig. 12) looks similar to Freksa's [1] CN-graph (Fig. 1a), but we found two more neighbors: *starts–finishes* and *finished-by–started-by*. These two neighbors presume a smooth transformation by dragging two endpoints of one DLine simultaneously. This transformation does not belong to the three types of smooth transformations discussed in [1]. On the other hand, *starts* and *finished-by* are not conceptual neighbors, because during a smooth transformation between *starts* and *finished-by* there is a moment when two DLines have the same length (i.e., *equal* holds).

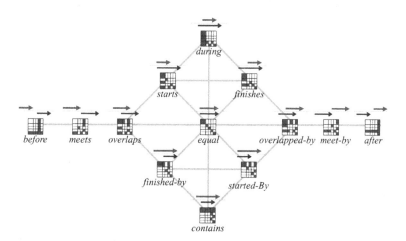

Fig. 12. A CN-graph of topological relations between two uni-directed lines in $\mathbb{R}^1$ (essentially a CN-graph of Allen's [16] interval relations), derived by the method in Section 5

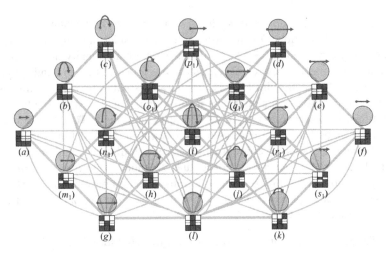

Fig. 13. A CN-graph of a subset of topological DLine-region relations in $\mathbb{R}^2$ [9], derived by the method in Section 5

Case 5: Topological relations between a DLine and a region in $\mathbb{R}^2$

The 9^+-intersection distinguishes 26 DLine-region relations in $\mathbb{R}^2$ [9]. Again, all computationally-derived potential conceptual neighbors of these relations were found to be valid and symmetric. Fig. 13 shows a sub-graph of the obtained CN-graph, which contains 19 DLine-region relations (the omitted seven relations are derived from the relations (m_1)-(s_1) by reversing the DLine's direction). This graph looks complicated, but it is remarkably systematic. First, we can see a lattice that has five queues of relations from left-top to right-bottom and another five queues from right-top to left-bottom. The upper-half of Kurata and Egenhofer's [9] CN-graph in Fig. 1d is homeomorphic to this lattice. Most members of each queue are mutually conceptual neighbors. In addition, each member of one queue is a conceptual neighbor of most members in the next queue. Only two irregular neighbors that jump a queue are found between (p_1) and (l) and between (n_1) and (r_1).

7   Conclusions and Future Work

CN-graphs are important for both schematization of spatial/temporal relations and qualitative spatio-temporal reasoning. Previous studies have developed a variety of conceptual neighborhoods using various concepts of conceptual neighbors. In this paper, we proposed a semi-automated method for deriving CN-graphs of topological relations, where conceptual neighbors are determined by the possibility of smooth transformations. The reliability of this method is indicated in our cases studies, where all computationally-derived candidates for the conceptual neighbors were found valid. Since our method is based on the 9^+-intersection, CN-graphs can be derived for various sets of topological relations. For instance, 28 sets of topological relations between simple points, directed lines, regions, and bodies embedded in $\mathbb{R}^1$, $\mathbb{R}^2$, $\mathbb{R}^3$, $\mathbb{S}^1$, and $\mathbb{S}^2$ are identified in [15] based on the 9^+-intersection. It is left for future work to examine for each relation set whether computationally-derived potential conceptual neighbors are always valid or not. In case this is not true, we have to find out additional constraints to remove invalid neighbor candidates.

In this paper we did not discuss the methods for deriving SE- and SES-neighbors. Actually, SE-neighbors can be derived by a similar method. In this method, we use alternative lists of primitive-level transitions, which exclude all transitions that presume multiple primitive-level events with respect to A's primitives. With these alternative lists, the neighbor candidates of a relation r_i are derived by the same algorithm. Then, some candidates are removed if their 9^+-intersection icon is different from the 9^+-intersection icon of r_i in multiple rows. The thick links in Figs. 10-13 already show the SE-neighbors derived by this method. The method for deriving SES-neighbors is now under development. For this method, we have to clarify all dependencies between primitive-level events. We expect that the simplified version of CN-graphs based on SE- or SES-neighbors will be useful for visual schematization of spatial relations when the CN-graphs are complicated (e.g., Fig. 13).

This paper has not discussed the design issues of CN-graphs; i.e., how to arrange the relations in a diagrammatic space such that the CN-graph looks visually schematic. Some design heuristics are discussed in [17], although the validity of these heuristics should be examined carefully with more examples. It is then an interesting future problem to integrate the work in this paper and the work on the design aspect to establish a consecutive method for deriving visually schematic CN-graphs.

Acknowledgement

This work is supported by DFG (Deutsche Forschungsgemeinschaft) through the Collaborative Research Center SFB/TR 8 Spatial Cognition.

References

1. Freksa, C.: Temporal Reasoning Based on Semi-Intervals. Artificial Intelligence 54(1-2), 199–227 (1992)
2. Egenhofer, M., Al-Taha, K.: Reasoning about Gradual Changes of Topological Relationships. In: Frank, A.U., Formentini, U., Campari, I. (eds.) GIS 1992. LNCS, vol. 639, pp. 196–219. Springer, Heidelberg (1992)
3. Galton, A.: Lines of Sight. In: Keane, M., Cunningham, P., Brady, M., Byrne, R. (eds.) AI and Cognitive Science 1994, pp. 103–113. Dublin University Press, Dublin (1994)
4. Egenhofer, M., Mark, D.: Modeling Conceptual Neighborhoods of Topological Line-Region Relations. International Journal of Geographical Information Systems 9(5), 555–565 (1995)
5. Hornsby, K., Egenhofer, M., Hayes, P.: Modeling Cyclic Change. In: Kouloumdjian, J., Roddick, J., Chen, P.P., Embley, D.W., Liddle, S.W. (eds.) ER Workshops 1999. LNCS, vol. 1727, pp. 98–109. Springer, Heidelberg (1999)
6. Gottfried, B.: Reasoning about Intervals in Two Dimensions. In: Thissen, W., Wieringa, P., Pantic, M., Ludema, M. (eds.) IEEE International Conference on Systems, Man and Cybernetics, pp. 5324–5332. IEEE Press, Los Alamitos (2004)
7. Van de Weghe, N., De Maeyer, P.: Conceptual Neighbourhood Diagrams for Representing Moving Objects. In: Akoka, J., Liddle, S.W., Song, I.-Y., Bertolotto, M., Comyn-Wattiau, I., van den Heuvel, W.-J., Kolp, M., Trujillo, J., Kop, C., Mayr, H.C. (eds.) ER Workshops 2005. LNCS, vol. 3770, pp. 228–238. Springer, Heidelberg (2005)

8. Kurata, Y., Egenhofer, M.J.: The Head-Body-Tail Intersection for Spatial Relations Between Directed Line Segments. In: Raubal, M., Miller, H.J., Frank, A.U., Goodchild, M.F. (eds.) GIScience 2006. LNCS, vol. 4197, pp. 269–286. Springer, Heidelberg (2006)

9. Kurata, Y., Egenhofer, M.: The 9⁺-Intersection for Topological Relations between a Directed Line Segment and a Region. In: Gottfried, B. (ed.) 1st Workshop on Behavioral Monitoring and Interpretation, TZI-Bericht, vol. 42, pp. 62–76. Technogie-Zentrum Informatik, Universität Bremen, Germany (2007)

10. Reis, R., Egenhofer, M., Matos, J.: Conceptual Neighborhoods of Topological Relations between Lines. In: Ruas, A., Gold, C. (eds.) 13th International Symposium on Spatial Data Handling, Headway in Spatial Data Handling. Lecture Notes in Geoinformation and Cartography, pp. 557–574. Springer, Berlin (2008)

11. Billen, R., Kurata, Y.: Refining Topological Relations between Regions Considering Their Shapes. In: Cova, T.J., Miller, H.J., Beard, K., Frank, A.U., Goodchild, M.F. (eds.) GIScience 2008. LNCS, vol. 5266, pp. 20–37. Springer, Heidelberg (2008)

12. Dylla, F., Moratz, R.: Exploiting Qualitative Spatial Neighborhoods in the Situation Calculus. In: Freksa, C., Knauff, M., Krieg-Brückner, B., Nebel, B., Barkowsky, T. (eds.) Spatial Cognition IV. LNCS, vol. 3343, pp. 304–322. Springer, Heidelberg (2005)

13. Hernández, D., Zimmermann, K.: Default Reasoning and the Qualitative Representation of Spatial Knowledge. Technical report (FKI-175-93), Institute fir Informatik, Technischen Universität München (1993)

14. Egenhofer, M.: Spherical Topological Relations. Journal on Data Semantics III, 25–49 (2005)

15. Kurata, Y.: The 9⁺-Intersection: A Universal Framework for Modeling Topological Relations. In: Cova, T.J., Miller, H.J., Beard, K., Frank, A.U., Goodchild, M.F. (eds.) GIScience 2008. LNCS, vol. 5266, pp. 181–198. Springer, Heidelberg (2008)

16. Allen, J.: Maintaining Knowledge about Temporal Intervals. Communications of the ACM 26(11), 832–843 (1983)

17. Kurata, Y.: A Strategy for Drawing a Conceptual Neighborhood Diagram Schematically. In: Stapleton, G., Howse, J., Lee, J. (eds.) Diagrams 2008. LNCS, vol. 5223, pp. 388–390. Springer, Heidelberg (2008)

18. Schlieder, C.: Reasoning about Ordering. In: Kuhn, W., Frank, A.U. (eds.) COSIT 1995. LNCS, vol. 988, pp. 341–349. Springer, Heidelberg (1995)

19. Egenhofer, M., Herring, J.: Categorizing Binary Topological Relationships between Regions, Lines and Points in Geographic Databases. In: Egenhofer, M., Herring, J., Smith, T., Park, K. (eds.) NCGIA Technical Reports 91-7. National Center for Geographic Information and Analysis, Santa Barbara, CA, USA (1991)

20. Egenhofer, M.: Deriving the Composition of Binary Topological Relations. Journal of Visual Languages and Computing 5(2), 133–149 (1994)

21. Mark, D., Egenhofer, M.: Modeling Spatial Relations between Lines and Regions: Combining Formal Mathematical Models and Human Subjects Testing. Cartography and Geographical Information Systems 21(3), 195–212 (1994)

22. Mark, D., Comas, D., Egenhofer, M., Freundschuh, S., Gould, M., Nunes, J.: Evaluating and Refining Computational Models of Spatial Relations through Cross-Linguistic Human-Subjects Testing. In: Kuhn, W., Frank, A.U. (eds.) COSIT 1995. LNCS, vol. 988, pp. 553–568. Springer, Heidelberg (1995)

23. Alexandroff, P.: Elementary Concepts of Topology. Dover Publications, Mineola (1961)

Exploiting Qualitative Spatial Constraints for Multi-hypothesis Topological Map Learning

Jan Oliver Wallgrün

SFB/TR 8 Spatial Cognition, University of Bremen
Enrique-Schmidt-Str. 5, 28359 Bremen, Germany
`wallgruen@informatik.uni-bremen.de`

Abstract. Topological maps are graph-based representations of space and have been considered as an alternative to metric representations in the context of robot navigation. In this work, we seek to improve on the lack of robustness of current topological mapping systems against ambiguity in the available information about the environment. For this purpose, we develop a topological mapping system that tracks multiple graph hypotheses simultaneously. The feasibility of the overall approach depends on a reduction of the search space by exploiting spatial constraints. We here consider qualitative direction information and the assumption that the map has to be planar. Qualitative spatial reasoning techniques are used to check the satisfiability of individual hypotheses. We evaluate the effects of absolute and relative direction information using relations from two different qualitative spatial calculi and combine the approach with a topological mapping system based on Voronoi graphs realized on a real robot.

1 Introduction

The problem of learning and maintaining a spatial model of an initially unknown environment is generally regarded as a fundamental problem of mobile robot research. During the last decades, work on this problem has been concentrated on coordinate-based spatial representations like occupancy grids and feature-based representations (see [1] for an overview). An alternative to these representation approaches are graph-based representations, often referred to as *topological maps* [2,3]. In these approaches the environment is typically conceptualized as a *route graph* [4] consisting of nodes that stand for distinctive places or navigational decision points and edges that stand for the distinctive paths connecting these places (cmp. Fig. 1(b)) [2,3]. The problem of computing the correct graph model from a history of local observations has been investigated theoretically for graph environments without any geometric information. For instance, [5] showed that without further information successful map learning cannot be guaranteed without the help of at least one movable marker.

In this text, we are concerned with the problem of making topological mapping robust in the presence of uncertainty and ambiguity in the available spatial information. In the majority of topological mapping approaches only a single

---

K. Stewart Hornsby et al. (Eds.): COSIT 2009, LNCS 5756, pp. 141–158, 2009.

map hypothesis is maintained with the consequence that the map construction process tends to fail as soon as a wrong decision is made. A more promising approach, for instance suggested in [6] and [7], is to keep track of all possible map hypotheses simultaneously. However, this approach can increase the computational costs dramatically as the number of possible topological map hypotheses can grow exponentially with the number of exploration steps. Hence, additional available information needs to be exploited in order to eliminate as many hypotheses as possible and make the multi-hypothesis approach feasible.

In this work, we extend multiple-hypothesis topological mapping and earlier work on qualitative spatial reasoning in route graphs [8,9]. We adopt Kuipers abductive learning approach [7] and prefer among all valid graph hypotheses one that has a minimal number of nodes. The resulting mapping approach incrementally incorporates observations performing a best-first search through the tree of possible graph hypotheses. In addition, we incorporate qualitative information about the directions of leaving hallways and the assumption that the environment is planar (a constraint which has already been individually investigated in [10]). Qualitative spatial reasoning and incremental planarity testing are used to discard invalid hypotheses and thereby prune the search space. We employ and compare information from two different qualitative spatial constraint calculi, the absolute cardinal direction calculus [11] and the relative $\mathcal{OPRA}_2$ calculus [12]. Furthermore, we combine the described approach with a topological mapping approach based on generalized Voronoi graphs [13] and extensively evaluate our approach using simulation experiments as well as real exploration data from a mobile robot. The experiments show that direction information and the planarity constraint lead to a huge increase in solution quality and decrease in search space. However, the application also reveals shortcomings of existing spatial calculi and reasoning methods.

We start by presenting our general multi-hypothesis topological mapping framework in Sect. 2. We then explain the incorporation of spatial consistency and planarity checking (Sect. 3). Sect. 4 is about combining our approach with the Voronoi graph-based representation and Sect. 5 describes the different experiments performed for evaluation.

2   Multi-hypothesis Topological Mapping

Let us consider the following scenario: A robot is roaming through a graph-like environment like the one shown in Fig. 1(a). The environment consists of junctions and straight hallways connecting the junctions. For every passed junction, the robot stores a *junction observation* J_i consisting of a cyclically ordered set of leaving hallways $\langle l_1^{[J_i]}, l_2^{[J_i]}, ..., l_n^{[J_i]} \rangle$ and a spatial description consisting of spatial relations over the set of observed leaving hallways. For instance, for junction observation J_1 in Fig. 1(a) the spatial description could be $\{\text{southwest}(l_1^{[J_1]}), \text{south}(l_2^{[J_1]})\}$ using cardinal directions or, alternatively, it could be $\{\text{obtuse}(l_1^{[J_1]}, l_2^{[J_1]})\}$ when using some qualitative categories for the angles formed by pairs of hallways.

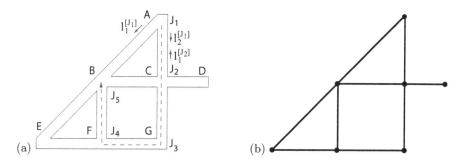

Fig. 1. (a) Walk of a robot through a graph-like environment, (b) a topological map representing the environment

Junction observations are connected by *hallway traversal actions* consisting of leaving the current junction via one of the observed leaving hallways and arriving at the next junction via one of the leaving hallways belonging to the next junction observation, e.g., $l_2^{[J_1]} \rightarrow l_1^{[J_2]}$ for traversing the hallway connecting A and C where $l_1^{[J_2]}$ would be the observed leaving hallway leading north in J_2. A list $\langle J_1, T_1, J_2, T_2, ..., T_{n-1}, J_n \rangle$ of alternating junction observations J_i and hallway traversals T_j forms the *history* of one particular exploration run through the graph environment.

The goal of a topological mapping algorithm now is to incrementally process the history of observations and actions and for each step determine one or all route graph hypotheses that can be considered valid explanations of the information processed so far. Each route graph hypothesis consists of an undirected graph with a combinatorial embedding into the plane (e.g., represented by specifying the cyclic orders of leaving edges for each node in the graph) and the position and orientation of the robot at the beginning of the exploration run (e.g., given by a node and a leaving edge).

During exploration, a currently valid hypothesis may turn out to be invalid when the next junction observation is processed. Hence, instead of committing to a single hypothesis, we track all valid hypotheses simultaneously. Fig. 2 shows in the top row three possible hypotheses assuming that the robot has just arrived at junction G in the example observing J_1 to J_3 (and assuming that all hallways are straight and that the junction observations are given in terms of qualitative cardinal direction relations from Ligozat's cardinal direction calculus [11], north, northwest, west, etc.). Black nodes here stand for junctions that have been observed, while white nodes are introduced for the end points of hallways that have not been traversed so far. When moving on to F and processing the new observation J_4, the first hypothesis can be complemented in two different ways leading to two successors in the search tree. Similarly, the third hypothesis has five successors. For the second hypothesis, however, the new observation leads to a contradiction: no hallway leading northeast is observed and, hence, this hypothesis can be discarded completely based on the direction information.

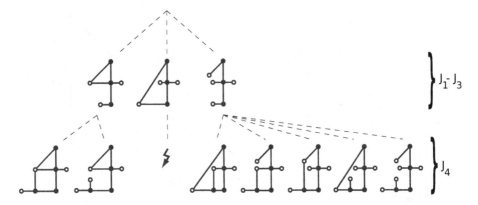

Fig. 2. Part of the search space of valid route graph hypotheses for the example from Fig. 1(a)

2.1 Minimal Route Graph Model Finding

The approach sketched above performs an exhaustive search through the tree of possible hypotheses. A modification of this approach proposed by Kuipers [7] is to prefer among all valid hypotheses the one that offers the simplest explanation. In this text, we interpret *simplest* as meaning a hypothesis that contains a minimal number of nodes which we will call a *minimal route graph model*. However, other criteria for minimality are conceivable.

The number of nodes in the graph hypotheses grows monotonically with increasing depth in the search tree because new nodes and edges will be added but never removed when new observations are incorporated and successor hypotheses are formed. As a result, we can search for the minimal route graph model in a best-first manner, always expanding the currently minimal hypothesis. This means that in the example the right hypothesis in the top row would not have been expanded because it already has the same number of nodes as the previously generated hypothesis at the bottom left. We will refer to the resulting mapping algorithm as the *minimal (route graph) model finding* algorithm in the following.

2.2 Valid Route Graph Models

The search tree as depicted in Fig. 2 only contains *valid* route graph hypotheses in the sense that we can assign coordinates to the junction nodes so that the the available spatial information is reproduced correctly. In addition, we might demand that additional constraints stemming from background knowledge need to be satisfied. Exploiting these kind of constraints is crucial for the minimal model finding algorithm to counteract the exponential growth of the search tree with the length of the exploration history and the otherwise high degree of ambiguity.

In this work, we assume that the available spatial information consists of qualitative direction information (absolute or relative) about the leaving hallways.

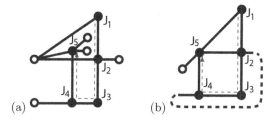

Fig. 3. Two invalid hypotheses for the history from Fig. 1(a)

As background knowledge we make the assumption that the environment is planar. Other kind of spatial or general constraints could be incorporated in the overall framework.

In our case, a hypothesis has to satisfy three conditions to be considered valid:

1. repeating the sequence of actions specified in the history within the hypothetical route graph yields a sequence of node degrees identical to the original sequence of leaving hallway numbers (*structural constraint*),
2. there must exist a way to draw the hypothetical route graph into the plane without crossing edges that is in accordance with the specified combinatorial embedding (*planarity constraint*), and
3. given this drawing, repeating the actions also reproduces the direction relations provided by the original junction observations (*direction constraints*).

When generating the successor hypotheses in the search tree, we only generate hypotheses which satisfy the structural constraint. In addition, we take into account that two junction observations can only correspond to the same node in a hypothesis if the perceived directions match. As a result, we can simply store the direction information as constraints to the edges in the graphs.

In Fig. 3 we see two examples of invalid hypotheses, this time for the complete walk depicted in Fig. 1(a). Both are depicted by one particular drawing of the route graph into the plane and both satisfy the structural constraint. The drawing of the first hypothesis would also reproduce the observed direction relations. However, it has crossing edges and, more importantly, no drawing without crossing edges exists that is in accordance with the underlying combinatorial embedding because the combinatorial embedding itself is not planar.

The drawing of the second hypothesis is planar but the positions assigned to the nodes do not reproduce the direction information correctly as the hallway that directly connects the junctions labeled J_2 and J_4 is supposed to lead east from J_2 and arrive at J_4 from the west. Hence, J_4 would have to be to the east of J_2. However, from the knowledge that the hallway connecting J_2 with J_3 leads south and the hallway connecting J_3 with J_4 leads to the west it can be concluded that J_4 has to be somewhere to the west of J_2. As a result of this reasoning, we know that no drawing satisfying the direction constraints for this hypothesis can exist at all as the contained direction information is inconsistent.

The minimal route graph model finding problem we have described here is a combinatorial optimization problem. For deciding whether a given hypothesis is valid or not, we need to determine whether a drawing exists that satisfies the planarity as well as the direction constraints. This is a constraint satisfaction problem over infinite domains (points in the plane) which makes it computationally challenging. However, as the two examples demonstrate that many structurally valid hypotheses generated during the search process can be ruled out by testing planarity of the combinatorial embedding and the global consistency of annotated direction constraints individually. This is the approach we will take in this work and it allows us to employ the efficient techniques for deciding consistency of spatial constraints developed in the area of qualitative spatial reasoning (see [14] for an overview). Nevertheless, the approach is incomplete in the sense that it may not filter out all invalid map hypotheses: There may exist a drawing for a given map hypothesis that is planar and one that is compliant with the direction constraints but none that is both. Results on how well this approach works in practice will be given in the section on experimental evaluation. The details of incorporating the constraints into the minimal model finding algorithm will be explained in the next section on rejection based on spatial constraints.

2.3   Two Mapping Variants

Up to now, we have described a version of the minimal model finding problem in which each model is a complete closed environment that might contain unvisited junctions which form the end points of perceived but never traversed hallways. A less complex version of the problem can be obtained by restricting the models to visited places and allowing hallways with open endings without stating how these are connected (cmp. Fig. 4).

We will investigate both variants and refer to them as CompEnv (complete environment) and VisOnly (visited only), respectively. For the exploration run from Fig. 1(a) only a single valid hypothesis exists in the VisOnly case (the one shown in the right of Fig. 4). However, for longer exploration runs we would also have to track multiple valid hypotheses.

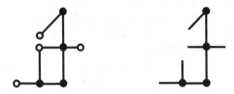

Fig. 4. A graph hypothesis in the CompEnv version (left) containing visited and unvisited nodes and for the VisOnly version (right) containing hallways for which the end points are not specified

3   Rejection Based on Spatial Constraints

In the following, we describe how checking of planarity and, in particular, of consistency of the direction constraints using the absolute cardinal direction calculus [11] and the relative $\mathcal{OPRA}_2$ calculus [12] are realized in our mapping approach.

3.1   Planarity Constraint

Each graph hypothesis for which the cyclic order information derived from the cyclic orders of perceived hallways does not describe a planar embedding can be immediately discarded. Checking whether a combinatorial embedding is planar takes $O(n)$ time [15]. The criterion used is whether the genus of the graph given by Euler's formula is zero.

We employ an incremental approach of planarity checking which is similar to the one described in [10]. Planarity checking is integrated into our search algorithm by representing the route graph hypotheses as bidirected graphs and by updating the information about faces of the embedding whenever we modify the graph structure. When the genus becomes non-zero, the hypothesis at hand can be discarded as the planarity constraint is violated.

3.2   Qualitative Direction Information

To incorporate direction constraints, we formulate observed directions by using the relations from a qualitative constraint calculus. Fig. 5 illustrates the base relations of the employed cardinal direction calculus relating two point objects and the relative $\mathcal{OPRA}_2$ calculus which describes the relative orientation of two oriented points.

In our approach absolute direction information is exploited to enforce three conditions:

1. Valid direction orderings: When adding a new edge to a node, it can only be inserted into the cyclic edge order at a position so that the edges preserve the cyclic order of cardinal directions. For instance, a resulting cyclic order of edges with directions north, south, west is not valid as west would have to appear between north and south.
2. Valid junction matchings: When trying to unify two nodes, corresponding edges need to have the same directions.
3. Global consistency: As discussed, there needs to be a way of assigning coordinates to the nodes such that all direction constraints are satisfied.

For a relative calculus like $\mathcal{OPRA}_2$, valid direction matching cannot be employed because there is no inherent cyclic order between the base relations.

Enforcing the first two conditions when constructing new hypotheses is trivial. For the global consistency check, we first extract a constraint network from the route graph hypothesis and then employ the qualitative spatial reasoning toolbox SparQ [16] for the consistency check using the standard algebraic closure

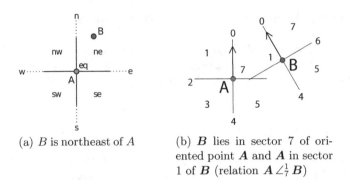

(a) B is northeast of A

(b) B lies in sector 7 of oriented point A and A in sector 1 of B (relation $A \angle_7^1 B$)

Fig. 5. The employed spatial calculi: the cardinal direction calculus and the $\mathcal{OPRA}_2$ calculus

algorithm [17,18]. The algorithm is based on operations on the relations of the calculus at hand, in particular the binary composition operation which yields the relation holding between objects A and C when the relations holding between A and B and between B and C are given. The algebraic closure algorithm runs in $O(n^3)$ time where n is the number of objects in the constraint network.

Employing both mentioned calculi allows us to compare the effects of absolute and relative direction information. However, both calculi have their individual shortcomings. The cardinal direction calculus, on the one hand, while being rather efficient because a large tractable subset exists for which the algebraic closure algorithm decides consistency, does not allow for expressing the cyclic order information about the leaving edges in the route graph. As a result, it can happen that a constraint network deemed consistent by the consistency check, only has solutions for which the cyclic order information is not preserved.

$\mathcal{OPRA}_2$, on the other hand, can express the cyclic order information. However, for $\mathcal{OPRA}_2$ algebraic closure does not decide consistency even for atomic constraint networks which also means that inconsistent hypotheses may not be discovered. We still chose these two calculi as to our knowledge no suitable direction or orientation calculi currently exist which at the same time have good computational properties, are expressive enough so that they will rule out many hypotheses, can express the cyclic edge ordering, and defines relations which are easily and reliably identifiable by an autonomous agent.

When using the absolute cardinal direction calculus, the extracted constraint network contains one variable for each node in the route graph hypothesis and the constraints holding between them are directly derived from the direction relations annotated to the edges.

In contrast, the $\mathcal{OPRA}_2$ calculus is a relative calculus describing the relative orientation of two oriented points (points in the plane with an additional direction parameter) towards each other. In order to determine the right $\mathcal{OPRA}_2$ relation, a robot would only need to be able to assess the angles formed by leaving hallways instead of needing a compass. In the constraint network, one oriented point variable is introduced for each pair of node and incident edge (meaning one for every leaving hallway). Hence, we end up with $2 \times n$ variables where n is the number of edges in the hypothesis.

4   Application to Voronoi Graph Representations

Besides evaluating the minimal route graph finding approach described in the previous sections using randomly created graph environments (see next section), we combined the approach with a topological map representation developed for indoor environments perceived via range sensors described in [13]. This representation approach is based on the idea of employing the generalized Voronoi diagram (GVD) to derive a route graph representation from sensor data (see for instance [19]).

As shown in Fig. 6(a), the GVD is a retraction of free space to a network of one-dimensional curve segments (the *Voronoi curves*) which meet at so-called *meet points*. The GVD can be abstracted into an undirected graph called the *generalized Voronoi graph* (GVG) as depicted in Fig. 6(b). To increase its suitability as a spatial representation, the GVG is annotated with additional information, e.g. a combinatorial embedding into the plane, local node descriptions, and relative geometric information. When applied for topological mapping, the problem is to incrementally build up the global GVG from small locally observed subgraphs, while the robot moves along the Voronoi curves.

Our overall mapping system extracts local Voronoi graphs from local grid maps of the robot's immediate surrounding and incrementally generates the history information about the observed Voronoi nodes and traversed Voronoi edges. The minimal model finding module updates the search tree based on new history information and computes a new minimal route graph model.

To apply the minimal model approach to the Voronoi graph representation several adaptations were made:

1. Multiple connections between two nodes are allowed.
2. Observed local Voronoi graphs can contain multiple nodes and edges which are translated into history information without actually traversing the edges.

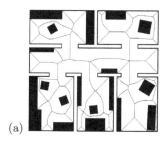   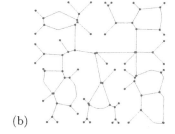

(a)   (b)

Fig. 6. (a) The generalized Voronoi diagram (fine lines) derived from a polygonal 2D environment, (b) the corresponding generalized Voronoi graph

3. In practice, it may not be possible to reliably determine the exact direction relations. Therefore, we utilize disjunctions of base relations when the perceived direction is a linear relation or lies close to the boundary of a relation sector (e.g., $\{ne, n, nw\}$ for observed relation $\{n\}$).
4. Voronoi curves are typically not straight line connections. Hence, we only employ direction constraints in the global consistency check if both connected Voronoi nodes have been perceived simultaneously. Otherwise, the direction information for this edge is only used for matching junction observations.

The results of applying this overall mapping system on the data set from a real world exploration run can be found in Sect. 5.2.

5   Experimental Evaluation

To evaluate the application of qualitative spatial reasoning approaches to the topological mapping problem, we performed several simulated exploration experiments in randomly generated graph environments of varying size and using random walks of varying length through the graphs. In addition, we evaluated the combination with the mapping system based on Voronoi graphs applying it to a data set from a real-world exploration run.

5.1   Simulation Experiments

In the simulation experiments, we investigated several aspects of our approach. The main results are summarized below.

Solution quality. We first investigated how much the planarity constraint and qualitative direction information helps in order to improve the solution quality by ruling out incorrect hypotheses and, as a result, increase the frequency in which the correct solution is found by the minimal model approach. To measure the quality of a solution, we use a simple error measure: We count how often either two junction observations that correspond to different junctions have been mapped to the same node or two observations that correspond to the same junction have not been unified. To assure that even searching without pruning is possible in reasonable time, we used rather small problem instances varying the size of the environment between 4 and 16 junctions.

Table 1 shows the results of 15600 trials for each of the following settings and both the CompEnv and VisOnly variants of the minimal model algorithm: (1) only structural constraint, (2) structural constraint and planarity constraint, (3) structural constraint and cardinal direction constraints, (4) structural constraint, planarity constraint, and cardinal directions. In addition, Fig. 7 shows how the average error distances increase with the size of the correct model throughout the experiment.

As the average error distances show, the planarity constraint and in particular the direction constraints significantly improve the solution quality. For the

Table 1. Experimental results regarding the solution quality

Setting		Correct model found	Average error distance
CompEnv	structural only	4.77%	9.86
	structural, planarity	5.97%	7.27
	structural, card. dir.	50.62%	1.41
	structural, planarity, card. dir.	50.92%	1.18
VisOnly	structural only	59.00%	5.63
	structural, planarity	64.77%	4.07
	structural, card. dir.	97.92%	0.20
	structural, planarity, card. dir.	98.15%	0.17

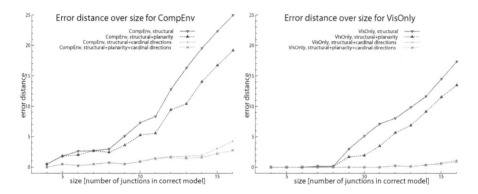

Fig. 7. Average error distance depending on the size of the correct model for the CompEnv and VisOnly variants

CompEnv variant, the planarity constraint achieves a 26.27% reduction of error distance, while direction information decreases the error distance by 85.70%. Combining both planarity and direction constraints only gives slightly better results than without applying the planarity constraint. The application of the constraints is highly beneficial but in most cases is not sufficient to resolve all ambiguities.

For the VisOnly case in which unvisited junctions are not included in the model, the improvements are even more drastic. The application of cardinal direction information leads to the correct model being found in 98.15% of all trials and has an extremely low average error distance of 0.20, or 0.17 when combined with the planarity constraint.

The experiment shows that the planarity constraint and in particular the cardinal direction constraints are able to resolve most of the model ambiguities remaining on the structural level leading to a largely increased solution quality.

Pruning efficiency. To investigate the effects of the individual settings on the size of the hypothesis space that has to be searched, we performed random experiments running the search until all solutions up to the size of the correct solution had been determined to mask out the effects of varying solution quality. We recorded (1) the number of expansion steps in which successors of a hypothesis are generated, (2) the average branching factor in the search tree, and (3) the maximal queue size occurring during the search. The number of expansion steps and the average branching factor give a good indication of the computational costs involved, while the maximal queue size tells us how many hypotheses were tracked simultaneously during the search and, hence, reflects the space-consumption.

The result of this experiment are summarized in Table 2. Fig. 8 shows how the number of expansions grows with increasing size of the environment (logarithmic scale is used for the y-axes).

Table 2. Results regarding the pruning efficiency

Setting		Expansions	Branch. factor	Max. queue size
CompEnv	structural only	2407.09	4.49	833.96
	structural, planarity	284.97	2.38	86.17
	structural, card. dir.	39.84	2.48	13.58
	structural, planarity, card. dir.	21.85	1.64	6.10
VisOnly	structural only	790.61	3.19	160.88
	structural, planarity	254.25	2.00	47.87
	structural, card. dir.	20.72	1.18	2.95
	structural, planarity, card. dir.	20.11	1.15	2.76

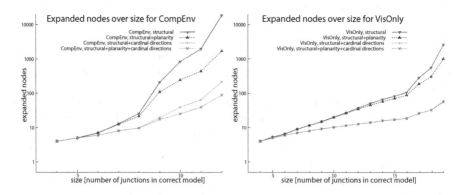

Fig. 8. Comparison of expansion steps depending on the size of the correct model for CompEnv and VisOnly

We clearly see that the CompEnv variant of the minimal model finding problem is much more complex than the VisOnly variant. The planarity constraint leads to an 88.16% decrease in expansion steps for CompEnv and 67.84% for VisOnly. The average branching factor has been decreased by 46.99% to 2.38 (CompEnv) and by 37.30% to 2.00 (VisOnly). For the cardinal direction constraints, we see a very high reduction of expansion steps of 98.35% for CompEnv and 97.38% for VisOnly. By combining both, an extreme reduction in expansion steps of 99.99% was achieved for CompEnv which corresponds to an average branching factor of 1.64. For VisOnly the cardinal direction constraints yields a 99.97% reduction (branching factor 1.18).

We conclude that the planarity assumption and the coarse direction information given by the qualitative cardinal relations lead to a much increased efficiency of the minimal model finding approach which would otherwise only be feasible for very small problem instances.

Absolute vs. relative direction information. One of the goals of our analysis was to compare the effects of employing absolute direction information (e.g., relations from the cardinal direction calculus) and relative direction information (e.g., $\mathcal{OPRA}_2$ relations). Therefore, we repeated the experiments for determining solution quality and pruning efficiency for both calculi using all constraints. Besides the already previously considered parameters, we distinguished the exact reasons for rejecting an hypothesis. The reasons are (1) direction ordering violation, (2) junction matching violation, and (3) global consistency violation as distinguished in Section 3.2. As also discussed there, direction ordering only plays a role for absolute direction information. In addition, it can only occur at unvisited junctions and, hence, when using the CompEnv variant.

Fig. 9 shows the diagrams for error distance and expansions steps for VisOnly. With regard to solution quality, the change from absolute to relative direction information increased the average error distance from 1.64 to 1.82 for CompEnv and from 0.44 to 0.68 for VisOnly. The average number of expansion steps increased from 49.12 to 82.60 (branching factor from 1.33 to 1.42) for CompEnv and from 12.79 to 13.93 (branching factor from 1.21 to 1.24) for VisOnly.

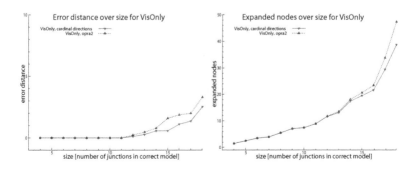

Fig. 9. Error distance and expansions for cardinal directions and $\mathcal{OPRA}_2$ (VisOnly)

Table 3. Reasons for rejection involving the direction information (CompEnv variant)

Setting	Violation	% of rejected hypotheses
cardinal direction calculus	direction ordering	0.59%
	junction matching	23.84%
	global consistency	75.57%
$\mathcal{OPRA}_2$ relations	direction ordering	—
	junction matching	23.09%
	global consistency	76.91%

The observed decrease in performance is not surprising as relative direction information in general allows for more perceptual aliasing. Taking this into account, the decrease in performance seems to be rather mild, especially for the VisOnly variant, and still much lower than for the less constrained settings investigated in the previous experiments.

When we look at the reasons for rejection that are summarized in Table 3 and Table 4, we see that direction ordering violation does not play an important role at all. Junction matching violations and global consistency violations show about the same rejection ratios for absolute and relative direction information in the case of the CompEnv variant. Global consistency violation occurs much more often than rejection caused by junction matching violations.

For VisOnly, the picture changes significantly. While for relative direction information global consistency violations still are the reason for about 44.41% of all rejections, only 21.27% are caused by global inconsistencies in the case of absolute direction information. The general increase of junction matching violations clearly results from the fact that for VisOnly complete information about all junctions is available. This allows rejection of many hypotheses early before global consistency is even tested. The difference between absolute and relative direction information in the case of VisOnly shows that absolute direction information reduces perceptual aliasing to a much higher degree and, hence, increases the predominance of rejections based on junction matching violations.

Table 4. Reasons for rejection involving the direction information (VisOnly variant)

Setting	Violation	% of rejected hypotheses
cardinal direction calculus	direction ordering	—
	junction matching	78.73%
	global consistency	21.27%
$\mathcal{OPRA}_2$ relations	direction ordering	—
	junction matching	55.59%
	global consistency	44.41%

Overall, as expected relative direction information is inferior to absolute direction information in terms of solution quality and pruning efficiency. The main advantage of relative information is that it often can be obtained more easily.

Overall computational costs. When investigating the pruning efficiency, we restricted ourselves to small problem instances which allowed to apply the model finding approach even without planarity and direction constraints. In addition, we focused on the effects of the different settings on the search space. For the complete minimal model finding approaches featuring all kinds of constraints, we further investigated how the approaches perform for larger problem instances. This investigation yielded two main results: First, even applying all constraints is not sufficient to conquer the combinatorial explosion for the CompEnv variant. Second, the computational costs of global consistency checking when employing the relative $\mathcal{OPRA}_2$ calculus quickly becomes excessive, making this approach infeasible for large environments for both variants.

As a result of the first observation, the CompEnv variant seems limited to scenarios with a rather small number of junctions in which the ability to predict the structure of unvisited parts is worth the increased computational costs. Taking a closer look at the second issue revealed that the computation times spent on global consistency checking for $\mathcal{OPRA}_2$ rise sharply, in some cases making up 90% of the overall computation time. We believe that there are two issues that contribute to this explosion in computational costs: the large number of base relations in the $\mathcal{OPRA}_2$ calculus which makes it impossible to store the complete composition table of general relations, and the size of constraint networks contains 2× the number of edges as variables. As a result, it seems that currently the VisOnly variant in combination with absolute direction information is the only one that scales sufficiently well to large environments.

5.2   Real-World Experiment

In a last experiment, we tested the combination of the minimal model finding approach with the Voronoi graph representation described in Sect. 4. The environment and the trajectory of the robot during the experiment is shown in Fig. 10(a)[1]. Due to the results from the previous experiments we only used the VisOnly variant. When employing the cardinal direction information, the local maps were manually aligned as the data set does not contain compass readings.

Fig. 10(b) shows the minimal model computed using the cardinal direction calculus which is indeed the correct graph model for this exploration run. For $\mathcal{OPRA}_2$, the resulting model was correct as well except for two wrongly merged nodes in a room that was entered via two different doors. However, while the computation took 16 seconds using cardinal directions, it took over 10 hours for $\mathcal{OPRA}_2$ because of the issues described in the previous section.

---

[1] The data set has been recorded at the Intel Research Lab, Seattle, and is available at (http://radish.sourceforge.net/), courtesy of D. Hähnel.

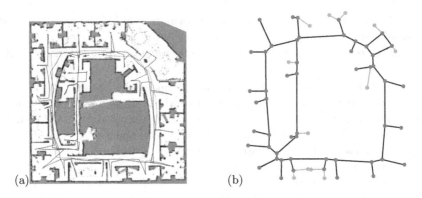

Fig. 10. (a) Environment of the real-world experiment, (b) computed route graph

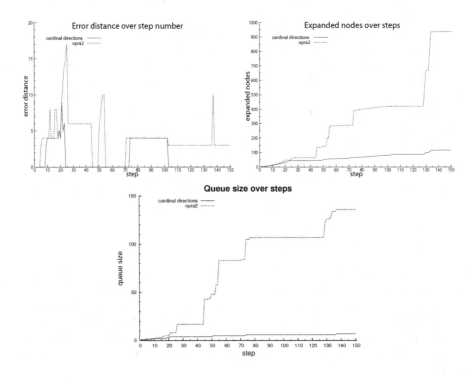

Fig. 11. Error distance and expansion steps for the real-world exploration experiment

Fig. 11 shows how error distance of the current minimal model and number of expansion steps develop over the 150 exploration steps for both spatial calculi. The diagram for the error distance shows that the variant using cardinal directions immediately settles for the correct hypothesis when the first loop traversal is completed in step 23, while this takes almost the entire second loop for $\mathcal{OPRA}_2$. We later see another increase in error distance caused by entering the new rooms. We also see that the number of expansion steps required and the number of tracked alternative hypotheses is significantly higher for $\mathcal{OPRA}_2$. However, the main reason for the hugely increased computation time again is the time spent on the global consistency checking.

6 Conclusions

We formulated topological mapping as the problem of finding a minimal route graph model that explains a sequence of observations and actions. Our solution consists of a search through the tree of possible graph hypotheses exploiting qualitative direction information (absolute or relative) and the planarity assumption. The experimental evaluation showed that this approach leads to a significantly reduced search space and improved solution quality. The approach has also been incorporated into a Voronoi-based mapping system and been applied to real exploration data. In contrast to typical topological mapping approaches only tracking a single hypothesis, our multi-hypothesis approach achieves a much higher degree of robustness as demonstrated by the successful mapping experiment for a large and complex indoor environment.

From the perspective of exploiting spatial information and reasoning, there is still a lot of room for improvements. The results with regard to spatial consistency checking based on qualitative direction information can be seen as a challenge for future research on qualitative spatial reasoning as none of the currently existing directional calculi ideally fits the demands arising from the problem. The cardinal direction calculus, for instance, while having good computational properties cannot express cyclic ordering information. For relative directional calculi like the $\mathcal{OPRA}_2$, the standard constraint reasoning techniques like algebraic closure are typically incomplete. In addition, even the application of the standard algebraic closure algorithm quickly became infeasible for $\mathcal{OPRA}_2$. This demonstrates the need for improved reasoning techniques and spatial calculi.

Finally, promising directions for future research also include the incorporation of other kinds of spatial constraints into the framework and investigating other search approaches, e.g., by limiting the number of simultaneously tracked hypotheses.

Acknowledgements. The author would like to thank Christian Freksa, Diedrich Wolter, and the reviewers for valuable comments. Funding by the Deutsche Forschungsgemeinschaft (DFG) under grant SFB/TR 8 Spatial Cognition and grant IRTG Semantic Integration of Geospatial Information is gratefully acknowledged.

References

1. Thrun, S.: Robotic mapping: A survey. In: Lakemeyer, G., Nebel, B. (eds.) Exploring Artificial Intelligence in the New Millenium. Morgan Kaufmann, San Francisco (2002)
2. Kuipers, B.: The Spatial Semantic Hierarchy. Artificial Intelligence (119), 191–233 (2000)

3. Remolina, E., Kuipers, B.: Towards a general theory of topological maps. Artificial Intelligence 152(1), 47–104 (2004)
4. Werner, S., Krieg-Brückner, B., Herrmann, T.: Modelling navigational knowledge by route graphs. In: Habel, C., Brauer, W., Freksa, C., Wender, K.F. (eds.) Spatial Cognition 2000. LNCS, vol. 1849, pp. 295–316. Springer, Heidelberg (2000)
5. Dudek, G., Jenkin, M., Milios, E., Wilkes, D.: Robotic exploration as graph construction. IEEE Transactions on Robotics and Automation 7(6), 859–865 (1991)
6. Dudek, G., Freedman, P., Hadjres, S.: Using multiple models for environmental mapping. Journal of Robotic Systems 13(8), 539–559 (1996)
7. Kuipers, B., Modayil, J., Beeson, P., MacMahon, M., Savelli, F.: Local metrical and global topological maps in the hybrid Spatial Semantic Hierarchy. In: Proceedings IEEE International Conference on Robotics and Automation 2004 (ICRA 2004), pp. 4845–4851 (2004)
8. Moratz, R., Nebel, B., Freksa, C.: Qualitative spatial reasoning about relative position: The tradeoff between strong formal properties and successful reasoning about route graphs. In: Freksa, C., Brauer, W., Habel, C., Wender, K.F. (eds.) Spatial Cognition III. LNCS, vol. 2685, pp. 385–400. Springer, Heidelberg (2003)
9. Moratz, R., Wallgrün, J.O.: Spatial reasoning about relative orientation and distance for robot exploration. In: Kuhn, W., Worboys, M.F., Timpf, S. (eds.) COSIT 2003. LNCS, vol. 2825, pp. 61–74. Springer, Heidelberg (2003)
10. Savelli, F., Kuipers, B.: Loop-closing and planarity in topological map-building. In: IEEE/RSJ International Conference on Intelligent Robots and Systems (IROS 2004), pp. 1511–1517 (2004)
11. Ligozat, G.: Reasoning about cardinal directions. Journal of Visual Languages and Computing 9, 23–44 (1998)
12. Moratz, R.: Representing relative direction as binary relation of oriented points. In: Proceedings of the 17th European Conference on Artificial Intelligence (ECAI 2006), August 2006, pp. 407–411 (2006)
13. Wallgrün, J.O.: Autonomous construction of hierarchical voronoi-based route graph representations. In: Freksa, C., Knauff, M., Krieg-Brückner, B., Nebel, B., Barkowsky, T. (eds.) Spatial Cognition IV. LNCS, vol. 3343, pp. 413–433. Springer, Heidelberg (2005)
14. Cohn, A.G., Hazarika, S.M.: Qualitative spatial representation and reasoning: An overview. Fundamenta Informaticae 46(1-2), 1–29 (2001)
15. Hopcroft, J., Tarjan, R.: Efficient planarity testing. Journal of the ACM 21(4), 549–568 (1974)
16. Wallgrün, J.O., Frommberger, L., Wolter, D., Dylla, F., Freksa, C.: Qualitative spatial representation and reasoning in the sparQ-toolbox. In: Barkowsky, T., Knauff, M., Ligozat, G., Montello, D.R. (eds.) Spatial Cognition 2007. LNCS, vol. 4387, pp. 39–58. Springer, Heidelberg (2007)
17. Mackworth, A.: Consistency in networks of relations. Artificial Intelligence 8(1), 99–118 (1977)
18. Montanari, U.: Networks of constraints: Fundamental properties and applications to picture processing. Information Science 7(2), 95–132 (1974)
19. Choset, H., Walker, S., Eiamsa-Ard, K., Burdick, J.: Sensor-based exploration: Incremental construction of the Hierarchical Generalized Voronoi Graph. The International Journal of Robotics Research 19(2), 126–148 (2000)

Comparing Relations with a Multi-holed Region

Maria Vasardani and Max J. Egenhofer

National Center for Geographic Information Analysis and
Department of Spatial Information Science and Engineering
University of Maine
Boardman Hall, Orono, ME 04469-5711, USA
{maria.vasardani,max}@spatial.maine.edu

Abstract. Relation models have treated multi-holed regions relations either the same as hole-free regions relations, loosing this way the peculiarities of the holed topology, or with methods dependent on the number of holes. This paper discusses a model of relations between a hole-free and a multi-holed region that departs from past approaches by using the frequencies of the relations in which the holes participate to summarize the relation. The model is independent of the number of holes and builds on the 23 topological relations between a hole-free and a single-holed region. With the help of a balanced algorithm the relation model is used in a method that compares relations for their topological similarity, by computing the cost of transforming one relation into the other. The placement of the holes in relation to the hole-free region is found to be of same importance as the placement of the host of the holes, for similarity comparisons.

1 Introduction

A plethora of geographic phenomena have discontinuities in the form of internal cavities. Such holes may occur as natural phenomena as in the case of kipukas—areas of land completely surrounded by one or more, younger lava flows, therefore appearing as holes within the younger lava area (Fig. 1a). Holes may also appear as artifacts in geosensor networks data streams (Stefanidis and Nittel 2004) in the form of *coverage* holes, that is, regions with a low density of working sensors (Ahmed *et al.* 2005) (Fig. 1b). The geometric representations of such phenomena shift the focus from the simple, homogeneously 2-dimensional regions that have been the prevailing objects of reference for most models of spatial relations in the plane (Egenhofer and Franzosa 1991; Randell *et al.* 1992), to regions with holes. This paper develops an analytical model of topological relations between a hole-free and a multi-holed region—a region with an arbitrary number of holes—that also enables the comparison of such relations for their topology.

The most fine-grained model of topological relations with holed regions explicitly enumerates the 9-intersection relations for all possible pairs comprising the host regions, a host region and a hole, or two holes (Egenhofer *et al.* 1994). The resulting relations are dependent on the number of holes, making this approach inefficient and painstaking as the number of holes increases. This dependence also prevents from

K. Stewart Hornsby et al. (Eds.): COSIT 2009, LNCS 5756, pp. 159–176, 2009.
© Springer-Verlag Berlin Heidelberg 2009

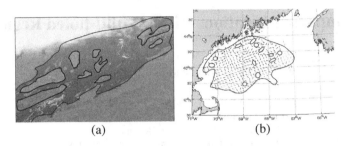

(a)   (b)

Fig. 1. Examples of holed spatial phenomena: (a) kipukas in younger lava and (b) coverage holes in the Gulf of Maine geosensors network

making comparisons between relations over multi-holed regions. Figure 2 shows a sketched topological scenario that requires such a comparison. Regions C and D represent underground sedimentary rock formations with concentrations of oil residing in the holes H1 through H8, respectively. Regions A and B are ground, hole-free regions that have been deemed appropriate for excavation to reach the oil deposits. Given that excavation from region A has already proved profitable, and that all holes in both underground formations are of equal profit importance, how different is the topological relation between regions A and C from that between B and D? If the two relations are topologically similar, excavation from region B may be considered comparably beneficial, but if they are very dissimilar, further oil-drilling profit estimations are needed, before it is decided whether excavations should proceed from region B.

To answer such questions, a model for reasoning with relations between a hole-free and a multi-holed region is needed. The relation model developed in this paper addresses such relations and is applicable to regions with varying numbers of holes. The approach is to summarize for each relation, which of the 23 topological relations hold between the hole-free and a single-holed region (Egenhofer and Vasardani 2007), selecting one hole at a time. This model also allows for relations between a hole-free and a multi-holed region to be compared for their topological similarity by applying a balanced algorithm that uses their summaries to transform one relation into the other. The cost associated with the transformation expresses a measure of the dissimilarity between the two relations, which is subsequently converted into their similarity (Nedas 2006).

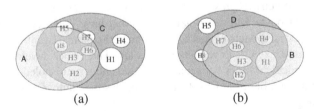

(a)   (b)

Fig. 2. Example sketched geological application with multi-holed oil-baring regions C and D and hole-free excavating regions A and B

The remainder of this paper is organized as follows: Section 2 discusses existing models of multi-holed regions. The formal definitions of a multi-holed region and of the relation between a hole-free and a multi-holed region, as used in this work, are given in Section 3. Section 4 describes the relation comparison method and demonstrates its use with an example, while Section 5 demonstrates the method's properties by evaluating the similarity among specifically chosen cases. In Section 6 the results from the similarity ranking of randomly generated topological relations between a hole-free and a multi-holed region are analyzed. The paper finishes with the concluding discussion on the relation model and the relation comparison method and with future work (Section 7).

2 Relation Models for Holed Regions

Holes in 2- and 3-dimensional objects have been extensively treated for their nature and variety from a philosophical viewpoint (Casati and Varzi 1994; Varzi 1996). The formalization of relations of 2-dimensional holed regions to meet the needs of spatial reasoning has been primarily approached with two methods: (1) the definition of topological relations based on the nine intersections of interiors, boundaries and exteriors (Egenhofer and Herring 1990) of the holed regions and (2) the use of the region-connection-calculus (RCC) (Randell *et al.* 1992).

2.1 Different Approaches to Modeling Relations

The vanilla 9-intersection has been applied for explicitly enumerating the binary relations for all pairs of homogenous regions (Egenhofer *et al.* 1994) and for describing relations of complex spatial objects, including holed regions (Schneider and Behr 2000). This approach may yield relations that correspond to different topological arrangements, but are, nevertheless, grouped together because they share the same 9-intersection. Such cases are distinct in the set of 23 relations between a single-holed and a hole-free region (Egenhofer and Vasardani 2007) and the set of 152 relations between two single-holed regions (Vasardani and Egenhofer 2008). Based on the 9-intersection as well, these relation models take into account the region's separation into different and mutually exclusive topological parts caused by the holes. While it addresses single-holed regions, the set of 23 relations between a single-holed and a hole-free region and its neighborhood graph (Section 2.2) acts as a building block for the relation comparison method described in this paper.

The region-connection-calculus (Randell *et al.* 1992) produces eight jointly exhaustive and pairwise disjoint (JEPD) topological relations (RCC-8) between two regions, based on a region definition that implicitly encompasses holed-regions as well. RCC-8 also groups under the same relation some topological scenarios that are clearly distinct, an issue that is partly addressed with an extended RCC relation set (Cohn *et al.* 1997).

2.2 Set of 23 Relations and Their Conceptual Neighborhood Graph

The 23 distinct relations between a hole-free and a single-holed region, the 23-t_{RRh}, are JEPD (Egenhofer and Vasardani 2007). Each relation has two constituent relations that

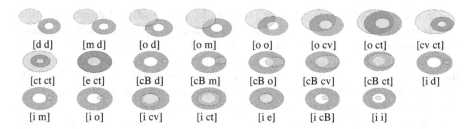

[d d]   [m d]   [o d]   [o m]   [o o]   [o cv]   [o ct]   [cv ct]

[ct ct]   [e ct]   [cB d]   [cB m]   [cB o]   [cB cv]   [cB ct]   [i d]

[i m]   [i o]   [i cv]   [i ct]   [i e]   [i cB]   [i i]

Fig. 3. Specifications of the 23-t_{RRh} and graphical examples with d=*disjoint*, m=*meet*, o=*overlap*, e=*equal*, cv=*covers*, cB=*coveredBy*, ct=*contains*, i=*inside*

are defined in a tuple [r_1 r_2]. Relation r_1 is called *principal* and represents the relation between the hole-free and the *generalized* region, while r_2 is the *refining* relation between the hole-free region and the hole. The generalized region is obtained by the union of the host region and the region that fills the hole. The 23-t_{RRh} may be viewed as refinements of the eight relations defined in the 9-intersection, i.e., the 8-t_{RR} (Egenhofer and Herring 1990), because a t_{RRh} offers more details about the location with respect to the hole than a t_{RR}. Graphical examples of the 23-t_{RRh} are depicted in Figure 3.

For a set of JEPD relations, the gradual topological transformation of each relation to a different one is followed on a conceptual neighborhood graph (CNG) (Freksa 1992). On such graphs, the number of edges along the shortest path connecting any two relations represents their conceptual distance—the steps of gradual change between them. The CNG for the 23-t_{RRh} is constructed by nesting together relations that share either one of their two constituent relations, each of which is one of the 8-t_{RR} binary relations between two hole-free regions. The nesting is achieved by linking

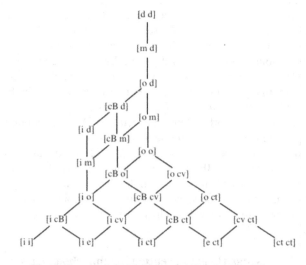

Fig. 4. Conceptual Neighborhood Graph of the 23-t_{RRh}

with an edge relations that either share the same principal relation and their refining relations are immediate neighbors on the 8-t_{RR} CNG, or their principal relations are neighbors on the 8-t_{RR} graph and their refining relation is the same (Fig. 4).

Two identical relations have a zero distance between them. A path of length one between two immediate neighbors carries the shortest non-zero distance, which reflects the minimum change that occurs in order to have one relation transform into another. The maximum distance d_{max23} is the shortest path of length eight between four pairs of relations: [d d]-[ct ct], [d d]-[i i], [i d]-[ct ct] and [i i]-[ct ct], capturing the biggest change that can happen to one t_{RRh} to transform it into another. The distances on the 23-t_{RRh} CNG offer a rationale for associating a cost of transforming a relation between a hole-free and a multi-holed region into another using the relations' summaries (Section 4), assessing this way the degree of topological similarity between them.

3   A Multi-holed Region and the Hole-Frequency Relation Model

A multi-holed region with n holes, denoted by n-B′, is what remains after subtracting from a hole-free region B, the union of the interior closures of n simple regions Hi, i=1…n. Region n-B′ comprises the generalized region B′* and n holes such that B′* *contains* each of the holes and all holes are pairwise *disjoint*. The generalized region B′* is defined as the union of the multi-holed region n-B′ and the regions filling its n holes Hi, i=1…n (Fig. 5).

Multi-holed region n-B′ with holes Hi, i=1…n:

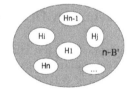

- n-B′ = B-$\bigcup\limits_{i=1}^{n}\overline{Hi}°$
- B′* = n-B′∪H1…∪Hn-1∪Hn
- ∀i:i=1…n : B′* *contains* Hi
- ∀i,j:i=1…n, j=1…n, i≠j : Hi *disjoint* Hj

Fig. 5. A multi-holed region n-B′

A topological relation between a hole-free region A and a multi-holed region n-B′, denoted by r(A, n-B′), is modeled as the union of the n relations between hole-free region A and each of the single-holed regions ℬi, i=1…n, as if each hole Hi was unique (Eqn. 1a). Because of this trait, the relation is named *multi-element* and each of the n *elements*, denoted by r(A, ℬi) is one of the 23-t_{RRh} between a hole-free and a single-holed region (Egenhofer and Vasardani 2007). Each single-holed region ℬi is equal to what remains after extracting Hi's interior closure from the generalized region B′* (Eqn. 1b). A depiction of a four-element topological relation is given in Figure 6.

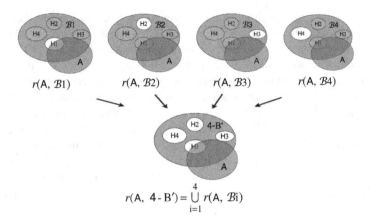

$$r(A, 4\text{-}B') = \bigcup_{i=1}^{4} r(A, \mathcal{B}i)$$

Fig. 6. Depiction of the different $r(A, \mathcal{B}i)$, i=1...4 that together make up the multi-element $r(A, 4\text{-}B')$ between a hole-free region A and multi-holed region n-B′

$$r(A, n\text{-}B') = \bigcup_{i=1}^{n} r(A, \mathcal{B}i), \text{ where } i=1...n \text{ and } \mathcal{B}i \in 23\text{-}t_{RRh} \tag{1a}$$

$$\forall i: i=1...n, \ \mathcal{B}i= B'^{*} - \overline{Hi}^{\circ} \tag{1b}$$

This model provides a summary of each multi-element relation. The summary consists of the occurrence frequencies for those of the 23-t_{RRh} that the holes participate in. It is, therefore, called the *Hole-Frequency Model* (HFM).

4   Frequency Distribution Method

It follows from the HFM that topologically different multi-element relations have different occurrence frequencies for their participating t_{RRh}s. The *Frequency Distribution Method* (FDM) achieves the transformation of one multi-element relation into another, using the concept of redistributing their summary frequencies. By assigning a cost to this transformation and translating the cost into a similarity measure, FDM enables the comparison of such relations for their topological similarity.

Two multi-element relations are used as reference for calculating the cost of relation transformation: Relation R_1 between hole-free region A and multi-holed region n-B′ and relation R_2 between hole-free region C and multi-holed region n-D′.

4.1   Conceptual Distance between Two Multi-element Relations

Relations R_1 and R_2 may be consisting of different t_{RRh}s as their elements. The normalized frequencies of R_1 and R_2's elements are summarized in vectors with m values each, m being the collective number of different t_{RRh}s present in R_1 and R_2 (Eqn. 2). Vector V_1 records the normalized frequencies $Fi_{i=1}^{m}$ of R_1, where Fi is the count of R_1's holes (fi) that participate in each of the m t_{RRh}s divided by the total number of holes n for normalization. Accordingly, vector V_2 records the normalized frequencies of R_2. Zero frequency values in a vector correspond to t_{RRh}s not present in the respective multi-element relation.

$$V = \begin{bmatrix} F1 \\ \vdots \\ Fm \end{bmatrix} \text{ where } \forall i = 1...m : \quad Fi = \frac{fi}{n} \tag{2}$$

The sum of a frequency vector V is defined as the sum of its m frequency values F (Eqn. 3). Since the frequency values in vectors V_1 and V_2 are normalized by the total number of holes, they are counted in frequency units and lie in the interval [0, 1]. A *frequency unit* is equal to the division of the unit by n (i.e., 1/n). Each frequency vector records all of the n elements of R_1 or R_2; therefore, the sums of both the vectors are equal to the unit (Eqn. 4).

$$\text{sum}(V) = \sum_{i=1}^{m} F_i \tag{3}$$

$$\text{sum}(V_1) = \text{sum}(V_2) = 1 \tag{4}$$

If the vectors differ in frequency values for corresponding elements, the multi-element relations R_1 and R_2 are different and have a *conceptual distance* between them that determines their dissimilarity. In the vectors have matching frequencies for corresponding elements, R_1 and R_2 do not differ conceptually and their topology is very similar. Nevertheless, R_1 and R_2 should not necessarily be regarded exactly the same, unless their graphic depictions verify so.

Definition 1: The *conceptual distance between two multi-element relations*, R_1 and R_2, is equal to the minimum cost of transforming frequency vector V_1 into vector V_2 by redistributing the normalized frequencies of V_1 so that they are identical to those of V_2.

The total cost assigned to this transformation is the weighted sum of the distances along the 23-t_{RRh} CNG between the elements, among which the redistribution of frequency units occurs. The number of frequency units moved between two relations represents the weight assigned to their distance on the 23-t_{RRh} CNG. The maximum cost is incurred when the maximum possible amount of frequency units is redistributed over the longest distance. The maximum amount of frequency units of a frequency vector V is sum(V), which is equal to the unit (Eqn. 4), and the longest distance on the 23-t_{RRh} CNG, d_{max23}, is eight (Section 2.1). Subsequently, the *maximum conceptual distance* $d_{max}^{R_1-R_2}$ between relations R_1 and R_2, or the maximum cost of transformation, is eight (Eqn. 5).

$$d_{max}^{R_1-R_2} = \text{sum}(V)*d_{max23} = 8 \tag{5}$$

In order to distinguish the amount of frequency units and the elements between which the units are redistributed, the *relation difference* is calculated.

Definition 2: The *relation difference* ($\Delta_{R_1 R_2}$) between two relations, R_1 and R_2, is a frequency unit vector, defined as the difference of the relations' normalized frequency vectors (Eqn. 6).

$$\Delta_{R_1 R_2} = V_1 - V_2 \qquad (6)$$

The values of the elements in $\Delta_{R_1 R_2}$ are the frequency units that need to be redistributed. There are both positive and negative values, with the positive corresponding to excess frequency units that will be transferred to balance the negative ones. Consequently, $\Delta_{R_1 R_2}$ carries the weights information necessary for computing the minimum cost of transforming V_1 into V_2, in the form of the exchanged frequency units. The values in $\Delta_{R_1 R_2}$ represent the difference units between the two relations' frequency vectors, which are normalized by the same number of holes n. Therefore, the elements in the relation difference cancel themselves out (i.e., the sum of the positive element values in $\Delta_{R_1 R_2}$ is equal to the sum of the negative element values) (Eqn. 7).

$$\mathrm{sum}(\Delta_{R_1 R_2}) = 0 \qquad (7)$$

4.2 Computing the Minimum Cost of the Relation Transformation

The determination of the minimum cost for transforming frequency vector V_1 into V_2 can be translated into a balanced transportation problem (Murty 1976; Strayer 1989), which is a case of the linear programming problem (Dantzig 1963). This section discusses how this translation is achieved, and the use of the transportation algorithm for acquiring a solution.

4.2.1 The Balanced Transportation Problem

An analogy of *warehouses* and *markets* is used for discussing the transportation problem. This problem, which considers the *supplies* of all the warehouses and the *demands* of all the markets, along with the *unit costs* for transportation between all pairs of warehouses and markets, is about finding the minimum cost for transporting all supplies from the warehouses to the markets so that all demands can be met. This analogy's components have their counterparts in the case of transforming one frequency vector into another. In particular, the p different elements with positive frequency differences in $\Delta_{R_1 R_2}$ correspond to warehouses, while the n different elements with negative frequency differences correspond to markets. The ith warehouse's (W_i) supply, denoted by s_i, is equal to the value of its corresponding positive frequency difference in $\Delta_{R_1 R_2}$. Similarly, the jth market's (M_j) demand, denoted by d_j, equals the value of the analogous negative frequency difference in $\Delta_{R_1 R_2}$.

The unit cost c_{ij} for moving a supply unit (i.e., a frequency unit) from W_i to M_j is the distance $d(W_i, M_j)$ on the 23-t_{RRh} CNG (Fig. 4) between the two t_{RRh} that correspond to W_i and M_j. The sum of the relation difference is zero (Eqn. 7); therefore, the sum of all supplies equals the sum of all demands (Table 2) and, due to this equality, this transportation problem is a *balanced* transportation problem.

Table 2. The balanced transportation problem for $\Delta_{R_1 R_2}$, with p warehouses ($W_{i=1}^p$) and p supplies ($s_{i=1}^p$), n markets ($M_{j=1}^n$) and n demands ($d_{j=1}^n$), and $p \times n$ unit costs ($c_{i,j=}^{p,n}$)

$\Delta_{R_1 R_2}$	M_1	M_2	$\ldots$	M_n	
W_1	c_{11}	c_{12}	$\ldots$	c_{1n}	s_1
W_2	c_{21}	c_{22}	$\ldots$	c_{2n}	s_2
$\vdots$	$\vdots$	$\vdots$	$\vdots$	$\vdots$	$\vdots$
W_p	c_{p1}	c_{p2}	$\ldots$	c_{pn}	s_p
	d_1	d_2	$\ldots$	d_n	

$$\text{sum}(\Delta_{R_1 R_2}) = 0: \quad \sum_{i=1}^{p} s_i = \sum_{j=1}^{n} d_j$$

If x_{ij} frequency units are to be transferred from warehouse (element) W_i to market (element) M_j, then the transportation problem is to determining those values of x for which the total cost z is a minimum (Eqn. 8) and the *supply* and *demand constraints* are satisfied. The supply constraint states that the sum of all frequency units distributed out from a certain warehouse must equal the total supply capacity of that warehouse (Eqn. 9a) and, similarly, the demand constraint states that the sum of all frequency units received by a market must equal the total demand of that market (Eqn. 9b). Cost z is then the minimum cost for transforming frequency vector V_1 into vector V_2.

$$z = \min\left(\sum_{i=1}^{p}\sum_{j=1}^{n} c_{ij} x_{ij}\right) \tag{8}$$

$$\forall i, \, i = 1 \ldots p: \quad \sum_{j=1}^{n} x_{ij} = s_i \tag{9a}$$

$$\forall j, \, j = 1 \ldots n: \quad \sum_{i=1}^{p} x_{ij} = d_j \tag{9b}$$

4.2.2 Relation Similarity Values Obtained from the Transportation Algorithm

Solutions to the balanced transportation problem of identifying the conceptual distance between two multi-element relations or the minimum cost for transforming one relation to the other can be obtained by employing the *transportation algorithm* (Murty 1976; Strayer 1989). This algorithm operates in two steps: First a basic feasible solution is found by randomly redistributing the frequency units in $\Delta_{R_1 R_2}$ so that the supply and demand constraints are met (Eqn. 9a-b). Since the basic solution might not be optimal as far as the minimum cost z of the transportation is concerned, the algorithm, in a second step, iteratively improves the redistribution, until an optimal solution is found (Strayer 1989). There may, in fact, be more than one ways to distribute the frequency units, so that the minimum value of z is obtained. The higher the final solution, the more costly is the redistribution of frequency units.

The computed value of z actually stands for the *dissimilarity* $\delta(R_1, R_2)$ between two multi-element relations—the conceptual distance between them. The similarity

$sim(R_1,R_2)$ and dissimilarity $\delta(R_1,R_2)$ of two relations are complement values; therefore, they add up to a constant (i.e., g) (Eqn. 10). Dissimilarity $\delta(R_1,R_2)$ lies in the interval $[0, d_{max}^{R_1-R_2}]$ and can be normalized in the closed interval $[0, 1]$ if divided by $d_{max}^{R_1-R_2}$ (Eqn. 11a). In this case, similarity is the dissimilarity's complement from the unit, and it also lies in the closed interval $[0, 1]$ (Eqn. 11b).

$$sim(R_1,R_2)+ \delta(R_1,R_2)=g \tag{10}$$

$$\delta(R_1,R_2) = \frac{z(R_1,R_2)}{d_{max}^{R_1-R_2}} \tag{11a}$$

$$sim(R_1,R_2) =1-\delta(R_1,R_2) \tag{11b}$$

4.2.3 Example

The similarity of the two multi-element relations depicted in the example sketched geological application of Section 1 (Fig. 2) is evaluated, as an example of applying FDM. The relations are: (1) $r(A, 8\text{-}C')$, between hole-free region A and multi-holed region 8-C', and (2) $r(B, 8\text{-}D')$, between hole-free region B and multi-holed region 8-D', or R_1 and R_2 for short (Fig. 7). Since *overlap* is the principal relation for both cases, five different 23-t_{RRh} elements are possible (Fig. 3). Table 3 displays these five t_{RRh} along with the single-holed regions that participate in each t_{RRh} with A or B ($C_{i=1}^8$ and $D_{i=1}^8$ respectively), according to the placements of the holes.

Table 3 reveals also the elemetns' frequencies for R_1 and R_2. It is, thus, possible to construct the normalized frequency vectors V_1 and V_2 (Eqn. 12). Since there are eight holes, normalization is achieved by dividing each frequency by eight (i.e., f/8, where f is the frequency of an element) and the frequency unit is 1/8.

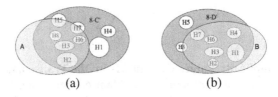

(a)   (b)

Fig. 7. Multi-element relations of example sketched geological application: (a) $r(A, 8\text{-}C')$ or R_1 and (b) $r(B, 8\text{-}D')$ or R_2

Table 3. The different t_{RRh} elements of the multi-element relations $r(A, 8\text{-}C')$ and $r(B, 8\text{-}D')$

	[o d]	[o m]	[o o]	[o cv]	[o ct]
A	C4	C1	C7 + C5	-	C2 + C3 + C6 + C8
B	D5	-	D8	D7	D1 + D2 + D3 + D4 + D6

$$
V_1 = \begin{bmatrix} [o\ d] & 0.125 \\ [o\ m] & 0.125 \\ [o\ o] & 0.25 \\ [o\ cv] & 0 \\ [o\ ct] & 0.5 \end{bmatrix} \quad \text{and} \quad V_2 = \begin{bmatrix} [o\ d] & 0.125 \\ [o\ m] & 0 \\ [o\ o] & 0.125 \\ [o\ cv] & 0.125 \\ [o\ ct] & 0.625 \end{bmatrix} \tag{12}
$$

Vectors V_1 and V_2 do not share the same frequencies for corresponding elements; therefore, the transportation algorithm is employed in order to produce a dissimilarity value between them. Equation 13 gives their relation-difference $\Delta_{R_1 R_2}$.

$$
\Delta_{R_1 R_2} = V_1 - V_2 = \begin{bmatrix} [o\ d] & 0 \\ [o\ m] & 0.125 \\ [o\ o] & 0.125 \\ [o\ cv] & -0.125 \\ [o\ ct] & -0.125 \end{bmatrix} \tag{13}
$$

By enforcing the sums constraint (Table 2), frequency units need to be redistributed to acquire a balance. $\Delta_{R_1 R_2}$ shows which elements with excess (positive) frequency differences offer units—[o m] and [o o] in this case—to the receiving elements with the negative frequency differences—[o cv] and [o ct]. The distances between the elements on the 23-t_{RRh} CNG (Fig. 4) provide the costs of the transportations. A basic feasible solution—which is also optimal for this simple example—is attained if 0.125 units are transferred from [o m] to [o cv] and 0.125 units are transferred from [o o] to [o ct]. The costs for these transfers are 2 and 2, respectively. Therefore, the value z for the cost of redistribution in $\Delta_{R_1 R_2}$ is 0.5 (Eqn. 14).

$$
z = 0.125 * d([o\ m] \rightarrow [o\ cv]) + 0.125 * d([o\ o] \rightarrow [o\ ct]) = 0.5 \tag{14}
$$

This value z is the cost of transforming V_1 into V_2. Using Equation 11a the dissimilarity $\delta(R_1, R_2)$ between the two multi-element relations is evaluated (Eqn. 15); subsequently using Equation 11b, the similarity $\text{sim}(R_1, R_2)$ between the two relations is estimated (Eqn. 16).

$$
\delta(R_1, R_2) = \frac{z}{d_{max}^{R_1 - R_2}} = \frac{0.5}{8} = 0.0625 \tag{15}
$$

$$
\text{sim}(R_1, R_2) = 1 - \delta(R_1, R_2) = 1 - 0.0625 = 0.9375 \tag{16}
$$

Converted to a percentage, this similarity value is 93.75%. However, FDM is for comparing relations for their similarity to another relation. The similarity between two relations is meaningless, unless one of them is the reference relation and similarity is evaluated between the reference and at least one more relation, to enable similarity comparisons (Janowicz et al. 2008). The similarity computed with FDM does not imply that two relations are as similar, percentage wise, as the calculated number indicates. Occasionally, multiple different frequency distributions yield the same minimal result. For more complex distributions, the second phase of the transportation algorithm is employed.

5   Properties of the Frequency Distribution Method

The FDM presents certain characteristics:

1.   The *number of holes* n is the same between query and random relations.
2.   There are three possible ways that two multi-element relations—between a hole-free and a multi-holed region—can differ: (1) the principal relation is the same, while the refining relations—those associated with the placement of the holes—are different, (2) the principal relation differs but the placement of the holes remains invariable, and (3) the principal and some, or all, of the refining relations differ.
3.   All edges in the 23-t_{RRh} CNG have the *same weight* (i.e., the unit), so that the distance between any two relations on the graph solely depends on the length of the shortest path between them.

This section examines in detail the tree ways that multi-element relations can differ, as well as how the weight constant affects the similarity evaluation procedure.

5.1   The Effect of the Principal Relation on the Similarity Evaluation

There are five principal relations—*disjoint, meet, equal, covers,* and *contains*—whose strong influence on the topology forces the hole-free region to be in the same relation with all the holes—*disjoint* for principal relations *disjoint* and *meet*, while for the remaining three principal relations the hole-free region *contains* the holes. Other multi-element relations may differ only in their principal relation, or both their principal and refining relations. They cannot, nonetheless, differ only in their refining relations—there are no neighbors on the 23-t_{RRh} CNG that share any one of these five principal relations (Fig. 8). As this information is imprinted in the 23-t_{RRh} CNG, in case two multi-element relations share any of these five principal relations, their topological equivalence is recognized in the FDM and no further calculations are performed.

For the remaining three principal relations—*overlap, coveredBy,* and *inside*—there are no restrictions; therefore, multi-element relations may differ in any of their constituent relations, or in both simultaneously. There are, however, a few exceptions for *inside*. While having eight possibilities for different hole placements, principal relation *inside* shares only five common refining relations with principal relations *coveredBy* and *overlap*—namely the refining relations *disjoint, meet, overlap, covers* and

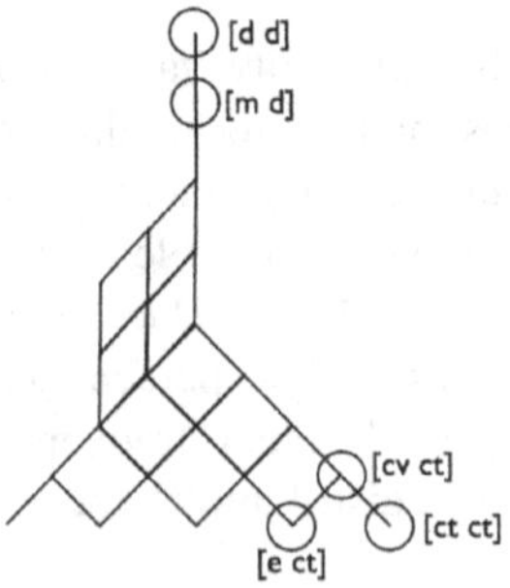

Fig. 8. The five t_{RRh}s with principal relations that when shared, force topological equivalence

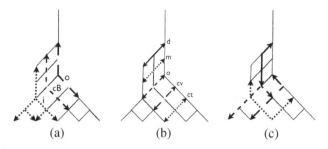

Fig. 9. Paths connecting multi-element relations with: (a) the same principal, but different refining relations, (b) different principal, but the same refining relations, and (c) different principal and some, or all, refining relations different

contains. When the principal relation is *inside* and the refining relation is one of *equal*, *coveredBy*, or *inside*, multi-element relations can vary only in their principal or both their constituent relations, but not solely in their refining relations. Figure 9 depicts various paths on the 23-t_{RRh} CNG between two different multi-element relations with *overlap*, *coveredBy*, or *inside* for their principal relation. These paths connect relations that have the same principal relation and different hole placements (Fig. 9a), or the same hole placements and different principal relations (Fig. 9b), or both principal and refining relations different (Fig. 9c).

5.2 Ramifications of the Weight Constant

The weight constant is related to the dependency of the method on the 23-t_{RRh} topological relations. In order to evaluate the implications of this constant on the FDM, the ranking of a few specifically chosen example relations against a query is examined. The choice of the unit as the weight of all the edges of the 23-t_{RRh} CNG equalizes the amount of change from one principal relation to a neighboring one while keeping the refining relation the same, with the amount of change of keeping the same principal relation and moving to a neighboring refining relation (Fig. 10). It implies that a change in the topological relation between the hole-free and the generalized region has the same topological importance as a change in the topological relation between the hole-free region and any one of the holes.

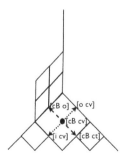

Fig. 10. Moving from [cB cv] to any of its four immediate neighboring relations causes the same amount of topological change

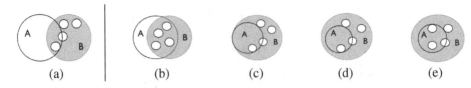

Fig. 11. Multi-element relations—(b) to (e)—to be assessed according to their similarity with the query relation (a)

The importance of the invariant distance unit weight is highlighted in the following examples. Figure 11a shows the query multi-element relation against which four other relations are evaluated for their degree of similarity. They are chosen so that the first one has the same principal but different refining relations (Fig. 11b), the second and third have different principal but the same refining relations (Fig. 11c-d), and the fourth one has different principal and some, but not all, different refining relations (Fig. 11e).

For the first relation (Fig. 11b), using the FDM to calculate the cost associated with transforming it to the query, four frequency units need to be transferred from the [o ct] element to each of the query's elements. The frequency unit is ¼—there are four holes in total—and the distances over which the units need to be transferred are $d([o\ ct] \rightarrow [o\ d]) = 4$, $d([o\ ct] \rightarrow [o\ m]) = 3$, $d([o\ ct] \rightarrow [o\ o]) = 2$ and $d([o\ ct] \rightarrow [o\ cv]) = 1$, as indicated from the 23-t_{RRh} CNG. The similarity sim(query, b) is calculated to be 68.75%. For the second relation (Fig. 11c), four frequency units need to be transferred again, but this time the principal relation changes, while the holes' placement is the same. Applying FDM gives sim(query, c) = 87.5%.

The third relation (Fig. 11d) has different principal and the same refining relations again, only this time the elements are farther apart from the query's respective relations on the CNG. The analogous calculations give sim(query, d) = 75%. For the fourth relation (Fig. 11e), the principal and some, but not all, refining relations are different. The redistribution of frequency units gives sim(query, e) = 62.5%.

While relation 11b exposes no change from the query with respect to the principal relation, it does not rank the highest in the similarity assessment. Instead relations 11c and 11d with different principal relations but with the same topological placement of their holes, rank higher. Last in ranking comes 11e, with both principal and refining relations different from the query's corresponding relations. Figure 12 presents the relations in sequence, according to their ranking.

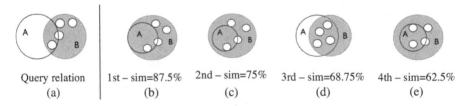

| Query relation | 1st – sim=87.5% | 2nd – sim=75% | 3rd – sim=68.75% | 4th – sim=62.5% |
| (a) | (b) | (c) | (d) | (e) |

Fig. 12. The ranking of the multi-element relations according to their similarity with the query relation: (b)=Fig. 11c, (c)=Fig. 11d, (d)=Fig. 11b and (e)=Fig. 11e

The weight constant is responsible for the specific ranking of the results. The 23-t_{RRh} CNG captures all topological changes among the 23-t_{RRh} possible elements of a multi-element relation. The path between two relations on the graph represents a path of least topological difference, implying that the closer the relations are on the graph, the smaller is the amount of change required in order to gradually transform one relation to the other (Bruns and Egenhofer 1996). Topological changes of the 23-t_{RRh}, however, have the characteristic of comprising two, not totally independent, variables: (1) change in the principal and (2) change in the refining relation. Neighboring relations on the graph may differ by a step of change in either of their constituent relations. Therefore, the placement of the elements on the 23-t_{RRh} CNG dictates the similarity ranking. Closeness between t_{RRh}s on the graph has a strong bearing on a multi-element relation's similarity to the query relation, even if their principal relations are different. In a different similarity model, where the two variables are treated separately by using a different principle than least topological change, the lengths between neighbors on the CNG would vary. In such a case, differences in their principal relation and differences in their refining relations would play distinct roles in the similarity assessment between relations over holed regions. In the model presented here, however, topological changes in both constituent relations are of the same importance.

6 Analysis of Random Query Experiments

With the help of a software prototype for randomly generating multi-element topological relations, experiments were conducted for ranking a set of ten such random relations (Fig. 13b-k) according to their similarity with a query multi-element relation (Fig. 13a), using FDM. The results of the similarity evaluation are given in Table 4.

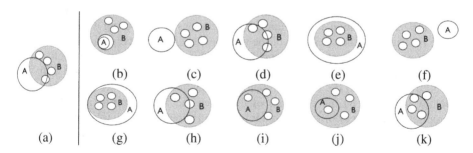

Fig. 13. Ten randomly generated archived topological scenarios (b-k), which are ranked against the query (a)

Table 4. Similarity evaluation for the ten randomly generated relations (Fig. 13b-k) with respect to the query relation (Fig. 13a)

Relation	13b	13c	13d	13e	13f	13g	13h	13i	13j	13k
Sim%	53.1	71.9	**90.6**	40.6	59.4	53.1	87.5	78.1	59.4	78.1

The similarity assessment results indicate that the relation topologically closest to the query is relation 13d with sim(13d) = 90.6%, followed by relation 13h with sim(13h) = 87.5%. In this case, the two most highly ranked scenarios not only share the same principal relation, but also display hole placements that resemble the topology of the query. Therefore, the elements of each multi-element relation appear very close to the query's corresponding elements on the 23-t_{RRh} CNG, and are responsible for the high similarity ranking. The scenarios that follow in the ranking present a combination of different principal and also quite different refining relations from the analogous relations of the query. The last rank is shared by two cases, 13b and 13g which have both elements located the farthest from the corresponding relations of the query, on the 23-t_{RRh} CNG. Specifically, [i ct] is the farthest from elements [o m] and [o o] of the query, while [cv ct] has distance four from element [o o] with the highest frequency, and three from element [o m]. Since these elements are located on central positions on the graph (Fig. 4), three and four respectively, are the second biggest distances that both elements can have from any other relation on the graph. Therefore, their placement on the 23-t_{RRh} CNG is responsible for placing 13g to the lowest ranking.

7   Conclusions and Future Work

This paper introduces a new method for modeling and comparing relations with multi-holed regions that is independent of the number of holes and recognizes the differences caused by the holes' topology. The method enables the assessment of the similarity of relations between a hole-free and a multi-holed region, providing a quantitative rationale to questions of finding the conceptually closer topological scenarios for either replacing or comparing with the query relation.

We found that it is both the placement of the holes in relation to the hole-free region and the interaction of the hole-free region with the host of the holes that affect the similarity between relations. The closer the elements of one multi-element relation are with the respective elements of another multi-element relation on the 23-t_{RRh} CNG, the more topologically similar are the relations. Neighboring elements have a strong influence on the similarity of the relations, even when their principal relations differ. Responsible for this influence is the FDM's weight constant, which places equal importance to a topological change caused by a change in the principal with that caused by a change in the refining relation. The method is, therefore, sensitive to the topology generated by the existence of the holes and identifies relations that could appropriately replace or offer valuable information about the query relation.

This approach to similarity evaluation creates new questions and opens up new research avenues. What happens, for example, when the participating multi-holed regions have different numbers of holes? If the same method is applied for these cases then what are the effects of dropping (or adding) holes to a region, in case of map generalization, for example? Furthermore, so far all holes have been considered equally, but in many real life scenarios, some holes or their contents have different importance. How does the similarity method handle holes differently and what happens in the case that holes of different importance are dropped? Broadening the similarity method's domain will certainly provide answers to a wider range of questions.

Finally, the case where both regions are multi-holed needs to be examined. Two different approaches are to either adapt the same method of using the 23-t_{RRh} and consider one region as hole-free and the other as singe-holed and then switch roles, or incorporate the 152-t_{RhRh} relations between two single-holed regions (Vasardani and Egenhofer 2008). In this case, both regions would be considered as single-holed until all the possible combinations with the different holes are exhausted. The target is to compare these two approaches.

Acknowledgements. This work was partially supported by the National Geospatial-Intelligence Agency under grant number NMA201-01-1-2003, a University of Maine Provost Graduate Fellowship and a University of Maine Graduate Research (CHASE) Award.

References

1. Ahmed, N., Kanhere, S.S., Jha, S.: The Holes Problem in Wireless Sensor Networks: A Survey. ACM SIGMOBILE Mobile Computing and Communications Review 9(2), 4–18 (2005)
2. Bruns, T., Egenhofer, M.: Similarity of Spatial Scenes. In: Kraak, M., Molenaar, M. (eds.) Seventh International Symposium on Spatial Data Handling (SDH 1996), pp. 4A. 31–42 (1996)
3. Casati, R., Varzi, A.: Holes and Other Superficialities. The MIT Press, Cambridge (1994)
4. Cohn, A., Bennett, B., Gooday, J., Gotts, N., Stock, O.: Representing and Reasoning with Qualitative Spatial Relations. In: Spatial and Temporal Reasoning, pp. 97–134. Kluwer Academic Publishers, Dordrecht (1997)
5. Dantzig, G.B.: Linear Programming and Extensions. Princeton University Press, Princeton (1963)
6. Egenhofer, M., Al-Taha, K.: Reasoning About Gradual Changes of Topological Relationships. In: Frank, A.U., Formentini, U., Campari, I. (eds.) GIS 1992. LNCS, vol. 639, pp. 196–219. Springer, Heidelberg (1992)
7. Egenhofer, M., Clementini, E., Felice, P.: Topological Relations between Regions with Holes. International Journal of Geographical Information Systems 8(2), 129–142 (1994)
8. Egenhofer, M., Herring, J.: A Mathematical Framework for the Definition of Topological Relationships. In: Brassel, K., Kishimoto, H. (eds.) Fourth International Symposium on Spatial Data Handling, pp. 803–813 (1990)
9. Egenhofer, M., Franzosa, R.: Point-set Topological Spatial Relations. International Journal of Geographical Information Systems 5(2), 161–174 (1991)
10. Egenhofer, M.J., Vasardani, M.: Spatial Reasoning with a Hole. In: Winter, S., Duckham, M., Kulik, L., Kuipers, B. (eds.) COSIT 2007. LNCS, vol. 4736, pp. 303–320. Springer, Heidelberg (2007)
11. Freksa, C.: Temporal Reasoning Based on Semi-Intervals. Artificial Intelligence 54, 199–227 (1992)
12. Janowicz, K., Raubal, M., Schwering, A., Kuhn, W.: Semantic Similarity Measurement and Geospatial Applications. Transactions in GIS 12(6), 651–659 (2008)
13. Murty, K.: Linear and Combinatorial Programming. John Wiley & Sons, Inc., Chichester (1976)
14. Nedas, K.: Semantic Similarity of Spatial Scenes. PhD Dissertation, Department of Spatial Information Science and Engineering, University of Maine (2006)

15. Randell, D., Cui, Z., Cohn, A.: A Spatial Logic Based on Regions and Connection. In: Nebel, B., Rich, C., Swartout, W. (eds.) Proceedings of the Third International Conference, Principles of Knowledge Representation and Reasoning KR 1992, pp. 165–176. Morgan Kaufmann, San Francisco (1992)
16. Schneider, M., Behr, T.: Topological Relationships between Complex Spatial Objects. ACM Transactions on Database Systems (TODS) 31(1), 39–81 (2006)
17. Stefanidis, A., Nittel, S.: Geosensor Networks. CRC Press, Boca Raton (2004)
18. Strayer, J.: Linear Programming and Its Applications. Springer, New York (1989)
19. Varzi, A.: Reasoning About Space: The Hole Story. Logic and Logical Philosophy 4, 3–39 (1996)
20. Vasardani, M., Egenhofer, M.J.: Single-Holed Regions: Their Relations and Inferences. In: Cova, T.J., Miller, H.J., Beard, K., Frank, A.U., Goodchild, M.F. (eds.) GIScience 2008. LNCS, vol. 5266, pp. 337–353. Springer, Heidelberg (2008)

The Endpoint Hypothesis: A Topological-Cognitive Assessment of Geographic Scale Movement Patterns

Alexander Klippel and Rui Li

Department of Geography, GeoVISTA Center
The Pennsylvania State University, University Park, PA, USA
{klippel,rui.li}@psu.edu

Abstract. Movement patterns of individual entities at the geographic scale are becoming a prominent research focus in spatial sciences. One pertinent question is how cognitive and formal characterizations of movement patterns relate. In other words, are (mostly qualitative) formal characterizations cognitively adequate? This article experimentally evaluates movement patterns that can be characterized as paths through a conceptual neighborhood graph, that is, two extended spatial entities changing their topological relationship gradually. The central questions addressed are: (a) Do humans naturally use topology to create cognitive equivalent classes, that is, is topology the basis for categorizing movement patterns spatially? (b) Are 'all' topological relations equally salient, and (c) does language influence categorization. The first two questions are addressed using a modification of the endpoint hypothesis stating that: movement patterns are distinguished by the topological relation they end in. The third question addresses whether language has an influence on the classification of movement patterns, that is, whether there is a difference between linguistic and non-linguistic category construction. In contrast to our previous findings we were able to document the importance of topology for conceptualizing movement patterns but also reveal differences in the cognitive saliency of topological relations. The latter aspect calls for a weighted conceptual neighborhood graph to cognitively adequately model human conceptualization processes.

1 Introduction

Humans think about space topologically. The question is: what does this statement actually mean? Is topology indeed the way to bridge the gap between formal characterizations of spatial information and spatial cognition? This topic has been addressed since Piaget (1955) published his theory of developmental stages. Piaget proposes that infants, as a first cognitive mechanism to make sense of their spatial environments, apply topological concepts. More recently, topology has been identified as the basic characteristic of image schemata (e.g., Lakoff, 1987) and Kuhn writes "Image schemas are often spatial, typically topological [...]" (2007, p. 155). These brief examples demonstrate the close relation of topology and (spatial) cognition. We have indeed witnessed tremendous research efforts that build on the assumption that humans reason and think about their spatial environments qualitatively (not only topologically though) and not quantitatively.

K. Stewart Hornsby et al. (Eds.): COSIT 2009, LNCS 5756, pp. 177–194, 2009.

While the majority of research (both cognitive and formal) has focused on static spatial relations, more recently (within approximately the last 20 years[1]) spatio-temporal research topics have gained prominence. For example, the formal characterization of movement patterns has developed into a central topic in geographic and related research (for a summary see Dodge, Weibel, & Lautenschütz, 2008). Paralleling the developments in spatial sciences, research in psychology is advancing our understanding of how the human cognitive system understands dynamic phenomena (termed events) (for an overview see, Zacks & Tversky, 2001; Shipley & Zacks, 2008). The need to relate these two research communities, in general, has long been acknowledged and addressing that need has lead to the development of qualitative spatial reasoning (Cohn, 1997; Freksa, Habel, & Wender, 1998) as a way to formally characterize human spatial cognition.

This paper is organized as follows. First, we review pertinent research on topology and similarity. Then we detail the setup of our behavioral experiments, analyze and discuss the results. Finally, we conclude with outlooks on how the results obtained can benefit research in the cognitive and geographic information sciences.

2 Topology

Since qualitative representation and reasoning strategies were first formally studied in the scientific community, fostered by the shift from classical categorization theories to prototype and resemblance theories (Rosch, 1975), work on fuzzy theories (Zadeh, 1965) and the Naïve Physics Manifesto (Hayes, 1978), topology has been amongst the top candidates able to bridge the gap between formal and cognitive characterizations of spatial relations. However, the number of papers evaluating the cognitive adequacy[2] of topological calculi for both representation and reasoning, as compared to the number of papers proposing topology based formalism, is rather small. Notable exceptions can be found in the work by Mark and Egenhofer (Mark & Egenhofer, 1994; 1995), Knauff and collaborators (Knauff, Rauh, & Renz, 1997; Renz, 2002), Riedemann (2005) and others (Zhan, 2002; Xu, 2007; Ragni, Tseden, & Knauff, 2007).

Our own behavioral research (Klippel, Worboys, & Duckham, 2008; Klippel, to appear) has addressed the role of topology in the spatio-temporal domain from a behavioral perspective[3]. We have addressed two questions so far. First, are paths through the conceptual neighborhood graph (see Figure 1) with the same start relation (DC) and three different ending relations (NTPP, NTPPi, EQ) basic to human category construction? Second, do ending relations at different positions in the conceptual neighborhood graph (e.g., DC, EC, PO, TPP, NTPP) influence categorization (as

---

[1] See, for example, the workshop on *Temporal GIS: the past 20 years and the next 20 years* that was held in conjunction with the GIScience 2008 conference.

[2] We use *cognitive adequacy* in the sense of modeling cognitive processes Strube, 1991.

[3] Please note that many approaches in spatial information science address this question from a cognitively inspired but formal perspective (e.g., Hornsby & Egenhofer, 1997; Worboys & Duckham, 2006; Galton, 2004; Mennis, Peuquet, & Qian, 2000; Peuquet, 2001).

generally proposed by Regier and Zhang 2007; formally by, for example, Camara & Jungert, 2007 and for the temporal domain by Lu & Harter, 2006). Surprisingly, to date we were not able to assert that topology has the same cognitive adequacy in the dynamic domain as previous research has shown it does for static spatial relations (Mark & Egenhofer, 1994; Knauff et al., 1997).

The results of previous experiments, as they relate to the findings reported here, will be detailed further in the discussion section, as they shed light on contextual factors on the conceptualization (category construction) of spatial relations[4].

Three concerns relate to this work: (a) what role does topology play in real world scenarios (not just geometric figures), (b) how similar are topological relations to each other and whether, for example, different weights should be assigned to edges in a conceptual neighborhood graph to improve the cognitive adequacy of topology based similarity measures, and (c) what are the linguistic influences on the categorization of movement patterns.

The first question is addressed by using a hurricane-peninsula scenario (as opposed to geometric figures, as done frequently in past research, including our own). The second question has been addressed previously in work by Knauff et al. (1997). Their experiments center both on the region connection calculus (RCC) and its two levels of granularity, RCC-8 and RCC-5 (Randell, Cui, & Cohn, 1992) and on the two levels of granularity that can be derived from Egenhofer's intersection models (Egenhofer & Franzosa, 1991). Knauff and collaborators summarize the outcome of their research in

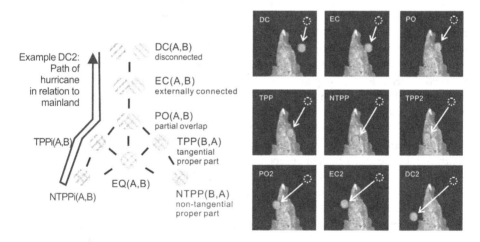

Fig. 1. Left: Conceptual neighborhood graph (Freksa, 1992; Egenhofer & Al-Taha, 1992). Right: Different paths of hurricanes distinguished by ending relations. Depicted is ONLY the ending relation of the hurricane movement. All hurricanes start in the upper right corner of each icon, disconnected from the peninsula (see also Figure 2).

---

[4] Please note: we use the term conceptualization and category construction in the sense that categories are not learned as part of the experiments Pothos & Chater, 2002; Medin, Watten-maker, & Hampson, 1987. Participants do not receive feedback whether their group-ing/categorization adheres to certain criteria.

the following way: First, that topological relations are relevant and that they are the dominating factor for categorizing spatial configurations; second, that RCC-8 captures the right level of granularity, while, third, RCC-5 and the medium granularity of Egenhofer's relations can be rejected as cognitively inadequate (on empirical grounds). We will come back to these results later.

The third question (the linguistic influence on categorization) has been controversially discussed in the literature (e.g., Boroditsky, 2001; January & Kako, 2007). Spatial relations such as direction concepts have been shown, however, to be susceptible to the influence of language (Crawford, Regier, & Huttenlocher, 2000; Klippel & Montello, 2007). Limited details will be provided, as our results show that in the case of topologically distinguished ending relations of movement patterns, there were no major differences between linguistic and non-linguistic conceptualization (see Results Section).

3   A Word on Similarity

Similarity assessment of geographic information has become a major research field, owing to a strong importance in human cognition (Ahlqvist, 2004; Schwering, 2008; Schwering & Kuhn, to appear; McIntosh & Yuan, 2005). The ultimate goal of similarity assessment is to assign meaning to data through similarity measures.

Yet, from a cognitive perspective similarity is controversial (Goldstone, 1994; Rips, 1989). Murphy and Medin (1985) make the argument that potentially everything could be as similar or dissimilar to everything else given a certain perspective. An extreme example is the similarity of plums and lawnmowers which are similar given that they both weigh less than 10,000 kg. The criticism is that similarity is regarded as being too flexible and that it cannot explain categorization (Bryant, 2000). Ultimately, we need a deeper level of explanation, often referred to as the *theory theory* (Laurence & Margolis, 1999) that allows for explaining why certain attributes are chosen over others on which a similarity rating can be based.

Contrasting these arguments, similarity has been given favor, especially from a perceptual perspective. For example, Gibson (1979) addressed the topic of *structural invariants* in the environment; properties of the environment that remain constant over time. Other researchers (Biederman, 1987; Yuille & Ullman, 1990) seek to identify perceptual aspects that can be, and are likely to be, used for both recognition and categorization. While it is acknowledged that similarity depends on cultural aspects, goals, and contexts, it is also believed that perceptual similarity is constrained enough to provide the groundwork for categorization (Biederman, 1987; Ahn, Goldstone, Love, Markman, & Wolff, 2005; Goldstone & Barsalou, 1998). Even in science and mature human beings, where potentially complex theoretical constructs could be applied to perform categorization tasks, a core of similarities that are perceptually invariant are used (Goldstone, 1994). Relevant to this work is the notion that perceptual properties, for example, those defined by topological relations, can provide a rich source of information for categorization (see, for example, Knauff et al., 1997). This way, through perceptual invariants characterized by topological relations, the gap between conception and perception might be closed.

4 Experimental Evaluation

Participants. 26 undergraduate geography students (six female, average age: 21.34) participated in the study.

Materials. Animated icons of different hurricane paths were constructed using Adobe Flash. The icons were 120 by 120 pixels in size (see Figure 2). The northernmost tip of Queensland (Australia) was used as inspiration but the color and shape were simplified so that it could not be easily recognized as a part of Australia[5].

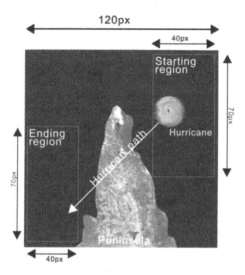

Fig. 2. Shown is the construction of animated icons for the case DC2, where the hurricane fully crosses the peninsula. Start and ending coordinates were randomly chosen from within the boxes labeled 'Starting region' and 'Ending region' satisfying additionally the constraints of paths through the conceptual neighborhood graph (see Figure 1).

The start and ending coordinates were randomized within the limits of topologically established equivalent classes. A region disconnected from the main land was chosen from which start coordinates were selected (see Figure 2). Likewise, a region for each ending relation was set up, from which randomly chosen coordinates, the ending position of the hurricane movement, were selected. In the cases of EC, PO, TPP relations only the vertical coordinates were randomized. Partial overlap (PO) was realized as a 50% overlap between hurricane and peninsula. Particular care was placed on two aspects in creating the animations: (a) that the ending relation was perceptually clear (see below); (b) that the velocity of the hurricane movements was identical. Each hurricane movement pattern animation lasted between two and five seconds depending on the length of the path. The ending relation was made salient by

---

[5] No participant made a comment that could be taken as an indication of recognizing Queensland. Technically, we should have referred to the moving low pressure system as a *cyclone* but kept the name *hurricane* , as US students are more familiar with that name.

a pause in the movement, before the movement pattern was repeated. The velocity of the hurricane movement was normalized such that all movement patterns had the same speed. This was achieved by normalizing the path length / frame ratio, that is, the longer the path the more frames were used to keep the velocity constant across all animated icons. Synchronous movement patterns were therefore automatically avoided due to different path lengths (the differences were the result of randomizing start and ending coordinates; the standardized speed led to a slight variation in the time the animation lasted).

The start relation for all movement patterns was DC (the hurricanes were disconnected/disjoint from and on the right side of the peninsula). This left the following nine possibilities for ending relations assuming a continuous movement of the hurricane (translation in terms of topology): DC (the hurricane does not make landfall), EC, PO, TPP, NTPP (see Figure 1). As the movement of the hurricane continued across the peninsula, the following topological relations were repeated and therefore indicated with a '-2': TPP2, PO2, EC2, DC2 (see Figure 1). Hence, we distinguished nine possible ending relations constituting different paths of hurricanes. For each ending relation (topologically equivalent class) we chose 6 randomizations (i.e., randomly chosen start and ending coordinates within the corresponding equivalent class regions) which led to a total of 54 animated icons. We chose a relatively small number of icons, as participants performed the grouping task twice (see below). We refrained from adding other factors to this experiment, such as different sizes of the hurricanes, so that the results focused on the topologically distinguished ending relations. All hurricanes also moved along a straight path. Given the scale of the overall region, this was seen as a reasonable decision[6].

Procedure. The experiments were organized as group experiments in a GIS lab at the Geography Department of Penn State. The lab could seat up to 16 participants at the same time. The lab is equipped with Dell Computers (Optiplex 755, Duo CPU E8200, 2.66GHz) that have 24'' Dell wide screen displays. All animated icons appeared on the left side of the screen, locations were randomly assigned (see Figure 3).

Participants performed the experiment in one session (with two grouping tasks). After arriving, the participants provided consent and entered personal information into the experiment software (see below). The instructions explicitly mentioned that participants were to group paths of hurricanes, but that how to create groups was completely up to the participants and that there were no right or wrong groupings. We used custom-made software (Klippel et al., 2008), an advanced version of the tool employed by Knauff et al. (1997), that allowed for: (a) collecting all relevant data from participants and the experiment; and (b) the presentation of animated icons (i.e., animated gifs). After collecting the participant information and providing participants with general instructions, participants were shown a short demonstration and performed a warm-up task. We used animal icons to introduce the simple handling interface of the grouping software. Icons could be placed into groups by simply using drag and drop. New groups could be created ('New Group' button) or deleted ('Delete Group' button). Only after all icons were placed into groups, a third button ('Finish') became active allowing participants to end the first part of the experiment and enter the second part.

---

[6] We did not find any comments indicating that this was perceived as an unnatural scenario.

In the second part, participants were presented with the groups that they created again and were asked to label the groups. This was an important aspect in the design of this experiment: labeling the groups required linguistic awareness on behalf of the participants. Afterwards, participants performed the same grouping task a second time, the second grouping was performed with this linguistic awareness. These two conditions will be referred to as NKL (no knowledge about the linguistic labeling task) and WKL (with knowledge about the linguistic labeling task). This linguistic awareness has been shown to influence the categorization of spatial relations (Crawford et al., 2000; Klippel & Montello, 2007).

It is important to note that participants were free to create as many groups as they deemed appropriate for the given stimulus. This kind of research setup is in line with research that addresses questions of unsupervised learning and category construction (Pothos & Chater, 2002; Medin, Wattenmaker, & Hampson, 1987). Participants initially do not receive any number of groups on the right side of the screen but have to explicitly create each group (see Figure 3).

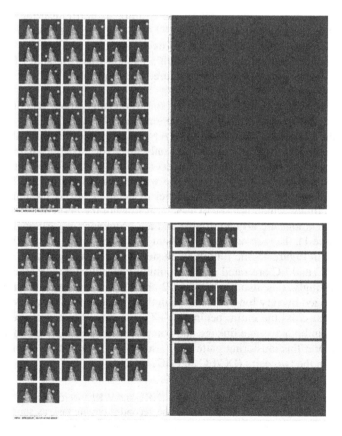

Fig. 3. Two screenshots: The top one is shows the initial screen that participants saw. The bottom one mimics an ongoing experiment.

5 Results

The experiment software collects the following data: the number of groups participants create[7], grouping time[8], and the linguistic labels for each group a participant
created. Most importantly, the software creates a similarity matrix that encodes the
results of the category construction for each participant (which animated icons are
placed together into the same group). This information is binary encoded, '0' for a
pair of icons that are not placed together into the same group and '1' for a pair of
icons that are placed together into the same group. The similarity matrix has 2916
cells. As similarity is symmetrically encoded—icon A is as similar to icon B as icon
B is to icon A—1431 cells are required for the assessment of the grouping behavior.
For each participant two matrices are created (as they perform the grouping twice,
NKL and WKL). To assess the overall similarity of animated icons across all participants (but within each condition), the matrices are added together. The highest similarity obtained for a pair of icons is therefore identical with the number of participants
in the experiment (N=26).

Once the overall similarity matrix is obtained, several analyses can be performed.
Cluster analysis can be used to reveal the overall, natural clustering structure. We
performed different cluster analyses using CLUSTAN (Wishart, 2004), and compared
the results to cross-validate our findings (Kos & Psenicka, 2000). Here, we used average and complete linkage, as well as Ward's method (increase sum of squares). All
methods yielded a similar clustering structure. The results of Ward's method for NKL
are shown in Figure 4.

A first observation is that in the NKL (no knowledge about labeling) condition,
most ending relations are present when we enforce a 9 cluster solution (which arguably is not the best cut of the dendrogram). Nonetheless, in this case all nine ending
relations can be clearly distinguished and only three (or 5.6%) animated icons are in
groups different from their topological peers (icon PP_1 ended up in a group together
with TPP2 icons, icon TPP2_3 in a group with NTPP icons, and EC2_4 with DC2
icons). The clustering structure at this level of detail is not as pronounced in other
clustering methods, which is to be expected (as it is not the best cut). Hence, while the
nine ending relations are all identifiable and are prominent in participants' groupings
(see also Table 1), they are not equally salient from a cognitive perspective.

Also seen in Figure 4, the following clusters in the dendrogram are clearly distinguishable: DC and EC are rated as more similar conceptually than other ending relations. Interestingly, this also holds for DC2 and EC2, that is, movement patterns that
are characterized by very long paths through the conceptual neighborhood graph, (i.e.,
hurricanes that cross the entire peninsula)[9]. This is indeed also a clustering structure
that is found in both average linkage and complete linkage clustering analysis. Hence,
we can interpret this clustering pattern as a strong one, indicating a conceptual similarity between the two pairs (DC/EC and DC2/EC2).

---

[7] No significant differences between conditions (NKL and WKL) were found.

[8] Participants took significantly less time for the second grouping task as shown by a paired-
sample t-test ($t(25) = 5.43$, $p<.001$). Participants were more familiar with the stimulus in the
second grouping task.

[9] Please note that this is validation that participants were able to properly assess and understand
the animations and that animations were watched through to the finish!

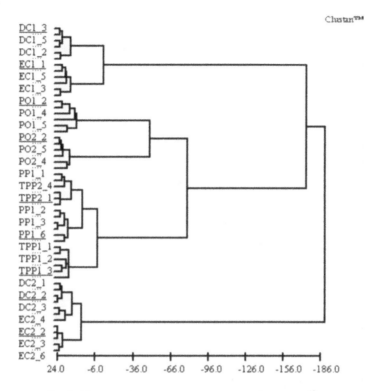

Fig. 4. Wards method, no knowledge on labeling (NKL)[10]

Likewise, across all clustering methods we find that the ending relations NTPP, TPP, and TPP2 are conceptually similar. In a 4 or 5 clustering solution, they even form a single cluster (see Figure 4).

Comparing these results for the NKL condition to the results for the WKL condition, the same clustering structure was present. For the current experimental setting, we found no differences in the clustering structure between condition 1 (participants were unaware of the labeling task, NKL) and condition 2 (participants performed the grouping task again after they provided linguistic labels for their first grouping, WKL)[11].

To perform further analysis that addressed the clustering structure and individual differences in greater detail, we used another custom software program, called KlipArt (Klippel, Hardisty, & Weaver, 2009), which is built in the Improvise visualization environment (Weaver, 2004). Figure 5 provides details on the interface and an example of all six animated icons of the category DC (hurricanes that do not make landfall).

---

[10] NTPP is abbreviated as PP.

[11] There are tiny differences in the DC2/EC2 group. While these groups are conceptually as close as in the NKL condition, the boundaries between them are less pronounced (two instead of one DC2 icon is grouped together with the EC2 icons).

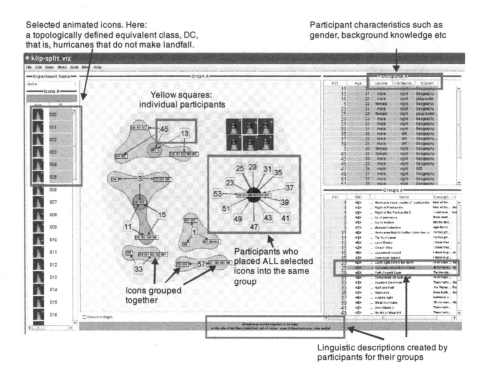

Selected animated icons. Here:
a topologically defined equivalent class, DC,
that is, hurricanes that do not make landfall.

Participant characteristics such as
gender, background knowledge etc

Yellow squares:
individual participants

Participants who
placed ALL selected
icons into the same
group

Icons grouped
together

Linguistic descriptions created by
participants for their groups

Fig. 5. A screenshot of KlipArt v4. Selected are all animated icons in which the hurricane does not make landfall (DC). The KlipArt tool allows for a very detailed analysis of the grouping behavior of participants (squares with numbers indicate individual participants), to select sub-groups of participants (e.g., male or female) and to analyze the linguistic descriptions participants provide for the groups they create.

KlipArt allows for the analysis of grouping behavior for each participant and groups of participants. It also allows the analysis of a subset of icons across all participants and the exploration of the linguistic descriptions provided by participants. One aspect we analyzed is whether there are differences in the number of participants who placed all animated icons of a topological equivalent class into the same group (a) across different equivalent classes, and (b) across the two conditions (NKL and WKL). Table 1 shows the results. Using chi-squared analysis, we did not find significant differences between NKL and WKL conditions except for PO. Additionally, both PO ending relations (PO and PO2) seem to be responsible for significant differences within each condition. These results, like the previously reported cluster analysis, indicate the special situation of the PO relations and support the notion that topological relations are conceptually salient to varying degrees. To better understand the grouping behavior of the participants we further explored these results with the KlipArt tool. We found that six participants grouped all but one animated icon (but not the same icon each time) from the PO group together (in the NKL condition). The spare icon was either placed into the TPP or the NTPP category.

Table 1. Raw counts of participants that placed all icons belonging to a topological equivalent class into the same group. χ^2 results (p values) are shown for (a) comparing all 9 ending relations at the same time within each condition (NKL and WKL), and (b) comparing each ending relation between NKL and WKL individually.

	DC	EC	PO	TPP	PP	TPP2	PO2	EC2	DC2	p value
NKL	16	12	6	12	16	14	11	17	17	0.042
WKL	19	15	14	8	18	16	9	14	16	0.029
p value	0.375	0.405	0.023	0.254	0.560	0.575	0.569	0.397	0.773	

Table 2. Comparing linguistic labels for two topological equivalent classes, DC and TPP

DC / NKL	TPP / NKL
Right side stopping circles	circles stopping in figure
Calm right before the storm	Hurricane _____ makes landfall !
hurricane stops short of land	hurricane lands on land
Path Doesn't Cross	Stick on Right Side
Completely off east coast	Completely over land
Right side	inside right
outside right	Land Hurricane
Weak Hurricane	Goes Over Land and Stops
Don't Make It	Within land
No Hit or Weak Hit	Continental
Before land	Hurricanes that partially crossed the landmass
Weak Hurricanes	Hurricane Landfalls
No landfall	
No landing	
hurricanes that never made it to shore	
Pre-landfall Hurricanes	

We also analyzed the linguistic descriptions participants created. While it would be possible to look into all linguistic descriptions for all icons (in total 5616), we decided to preselect the descriptions and only analyze those that were created by participants who placed all animated icons of a specific topological equivalent class into the same group (see Table 1). The descriptions for DC/NKL and TPP/NKL can be seen in Table 2. This analysis reveals several important aspects: while the perceptual characteristics of the stimulus material led participants to group icons in each topological equivalent class together, the linguistic descriptions are highly varied. This phenomenon is very well known and has been noted in research on object naming for HCI (Furnas, Landauer, Gomez, & Dumais, 1987). Compared to objects, the diversity of linguistic descriptions in dynamic situations is greater. We identified the following general principles in the DC/NKL case: the description of the endpoint of a movement pattern (e.g., *right side*, which is of course contextualized in this experiment), an inferred characteristic of the domain (e.g., *weak hurricane*), a description of the path characteristics (e.g, *path doesn't cross*), and an association with a domain characteristic (e.g., *no landfall*). Further distinctions could be made. The important point is that all descriptions we obtained for DC (except two) are directly describing the topological equivalent class, that is, that the hurricane does not reach the coast. If we compare these descriptions to the descriptions offered for the ending relation TPP/NKL, we find that the latter are much less tailored toward the topologic information that the participants used to create this group in the first place. This can be easily imagined if we asked another group of participants to draw sketch maps on the basis of the linguistic labels (see Table 2). It would be very unlikely that any participant would draw the hurricane/peninsula relation as TPP.

6 Discussion and Conclusions

One of the most important findings of this study is that the conceptual/perceptual salience of topological relations is not equal for all topologically defined ending relations. Already Mark and Egenhofer (1994) found that line-region relations characterized by the 9-intersection model may need aggregation to reflect cognitive conceptualizations of spatial relations. Knauff et al. (1997), although rejecting RCC-5, mentioned that certain topological relations (DC and EC) might be conceptually closer together than others in the case of static spatial relations. If the basic formal characterization is built on eight topological relations as identified in the RCC-8 formalism or by Egenhofer's intersection models, we need to add an additional description that assigns different weights to the edges in the conceptual neighborhood graph to reflect different cognitive-conceptual (and perceptual) saliencies.

We propose to use weights, instead of switching to a different topological calculus, as it is obvious that different granularity levels exist in the conceptualization of topological relations. These levels are visually present in the clustering structure shown in the dendrogram (see Figure 4). It is important to note that the clustering structure does not support RCC-8 as the cognitively most adequate solutions (Knauff et al., 1997). Different levels of granularity exist, as we would assume for a good model of a cognitive system (Hobbs, 1985).

Interestingly, though, we find (as in the static case) that DC and EC (and DC2 and EC2, respectively), are conceptually very close, which indicates that RCC-5 may be capturing the conceptualization of spatial relations cognitively more adequately then Egenhofer's intersections models. Yet, as we discuss below, this very well may be an effect of the domain (and/or the scale) we used in these experiments.

Weighting the edges in a conceptual neighborhood graph would be important for approaches that: use similarity measures to automatically infer the semantics of a data set, model human categorization through similarity measures, or work on design aspects of human-machine interfaces (Ahlqvist, 2004; Schwering, 2007; Nedas & Egenhofer, 2008; Furnas et al., 1987). One way to assign weights to the edges of the conceptual neighborhood graph would be to use *weighted fusion coefficients*. A fusion coefficient is a measure that indicates the distance at which two clusters merge. In Figure 4 the fusion coefficient is indicated as a measure of the horizontal axis of the dendrogram. For example, the fusion coefficient (normalized to 0) for the clusters DC and EC in Figure 4 is 36.8 and for EC2 and DC2 it is 18.61. In contrast, the fusion coefficient for DC/EC and PO/PO2 is 198.1. The fusion coefficients for TPP, NTPP, and TPP2 indicate the conceptual closeness between these relations (11 and 31.0, respectively).

More experiments and analyses (and potentially simulation Montello & Frank, 1996) would be required to derive appropriate weights (and, see below, these weights might not be identical for different domains/contexts). Most importantly, these results—additionally enforced by the more qualitative analysis of the linguistic labels that participants provided—show that different topologically distinguished ending relations are cognitively salient to different degrees and that using a conceptual neighborhood graph for assessing semantic similarity requires attaching weights to the edges to achieve cognitive adequacy.

An interesting question might be whether the weighting of the individual relations could be influenced by the domain. For example, for the stimulus material we used (paths of hurricanes), the underlying basis for placing DC and EC semantically closer together could be that hurricane boundaries are usually vaguely defined. Perceptually, the distinction between DC and EC was made clear as in case of the DC ending relation, where the hurricane stopped a radius away from the peninsula. That these relations were indeed perceptually distinguishable is expressed in the prominent distinction between the clusters containing DC icons and clusters containing EC icons. The semantic similarity between these two clusters compared to all other clusters might, however, best be explained by assuming that there was an influence based on knowledge about hurricanes (which is most likely not specific to geography undergraduate students in Pennsylvania). These assumptions of super-classes of topological relations in the characterization of movement patterns are also made in other approaches. A recent proposal in the spatio-temporal domain by Camara and Jungert (2007) groups the eight topological relations between two spatially extended entities (see Figure 1) into two categories: *distance* (DC) and *proximity* (all other relations). Camara and Jungert are primarily concerned with moving artificial entities on land (not movement patterns of natural entities such as hurricanes). However, it would be interesting to experimentally validate whether their assumption corresponds to human category construction.

Behavioral results analyzing Allen's (1983) intervals, in contrast, reflect the conceptualizations found in our experiments. Lu and Harter (2006) addressed the question of whether all of Allen's intervals are equally salient in the cognitive conceptualization of movement patterns (in their experiments, fish swimming in a tank). Their results indicate that relations that describe some kind of overlap (START, DURING, FINISHES, EQUAL) are distinguished from those relations that do not (BEFORE, MEET).

Topological relations are not only cognitively important but are also associated with distinguishable perceptual characteristics. This is a very important aspect that has been stressed in recent research on categorization. Work, especially by Barsalou and collaborators (1999; 2003; Goldstone & Barsalou, 1998) has shown how important it is to take into account the perceptual aspects of our environments to frame theories of categorization. In Gibson's (1979) terminology, topology could contribute to this framework by offering perceptual (and conceptual) invariants. Hence, topology might allow for perceptually grounding categorization, not only on the very abstract level of image schemata (Kuhn, 2007) but also on the level of direct perception. The critical question to ask is: why did topology (mostly) fail to guide the category construction in previous experiments (Klippel et al., 2008; Klippel, to appear)? One explanation can be found in research on category construction by Pothos and collaborators (Pothos & Chater, 2002; Pothos & Close, 2008). In cases where 'only' geometric figures are used, perceptual context effects dominate and, as Pothos states, the simplest one is singled out to guide category constructions. For example, while we distinguished the same nine different ending relations in a preceding experiment (Klippel, to appear), we also introduced greater variation as the stimulus consisted of geometric figures that are not constrained by background knowledge (although participants were encourage to think of geographic examples and did provide geographic scenarios as descriptions). Additional factors were different movement patterns (either one or both spatial entities moved) and size differences. Interestingly, the differences in size were singled out as the most prominent dimension for constructing categories. Size (or from a geographic perspective, scale) is an important factor. In the context of these experiments, size is the better solution from a cognitive-ergonomic perspective.

Overall, our research contributes to the rich body of knowledge that asserts topology in having a central role in human spatial cognition. Our research specifically contributes to endeavors that address the cognitive adequacy of topological calculi in the dynamic domain. The results offer a springboard for tailoring calculi toward cognitive adequacy, by providing insights into methods for weighting edges in a conceptual neighboring graph to better model the similarity of spatial relations. Additionally, we analyzed linguistic descriptions of conceptually similar movement patterns showing a variety of possibilities to characterize movement patterns linguistically. Systematically analyzing geographic scale event language will allow for informing the design of query languages and will improve our ability to automatically analyze large text corpora, interpreting the spatial information that is linguistically encoded.

Acknowledgements

We would like to thank all reviewers for valuable comments that helped to improve this paper. We also would like to thank David Mark for sharing with us ideas that developed out of Mark and Egenhofer's famous "The road across the park" example. We would like to acknowledge Thilo Weigel who implemented the grouping tool that Markus Knauff and collaborators used, which inspired our grouping tool. We sincerely thank Stefan Hansen for implementing our grouping tool and Frank Hardisty for his continuous support. We would also like to thank Chris Weaver for creating the KlipArt toolkit. This research has been supported by NSF and the National Geospatial-Intelligence Agency/NGA through the NGA University Research Initiative Program/NURI program. The views, opinions, and conclusions contained in this document are those of the authors and should not be interpreted as necessarily representing the official policies or endorsements, either expressed or implied, of the National Geospatial-Intelligence Agency or the U.S. Government.

References

Ahlqvist, O.: A parametrized representation of uncertain conceptual spaces. Transactions in GIS 8(4), 493–514 (2004)

Ahn, W.-K., Goldstone, R.L., Love, B.C., Markman, A.B., Wolff, P. (eds.): Decade of behavior 2000-2010. Categorization inside and outside the laboratory: Essays in honor of Douglas L. Medin. American Psychological Assoc., Washington (2005)

Allen, J.F.: Maintaining knowledge about temporal intervals. Communications of the ACM 26, 832–843 (1983)

Barsalou, L.W.: Perceptual symbol systems. Behavioral and Brain Sciences 22(4), 577–609 (1999)

Barsalou, L.W.: Abstraction in perceptual symbol systems. Philosophical Transactions of the Royal Society of London: Biological Sciences 358, 1177–1187 (2003)

Biederman, I.: Recognition-by-components: A theory of human image understanding. Psychological Review 94, 115–145 (1987)

Boroditsky, L.: Does language shape thought?: Mandarin and English speakers' conceptions of time. Cognitive Psychology 43, 1–22 (2001)

Bryant, R.: Discovery and decision: Exploring the metaphysics and epistemology of scientific classification. Fairleigh Dickinson University Press Associated Univ. Presses, Madison N.J (2000)

Camara, K., Jungert, E.: A visual query language for dynamic processes applied to a scenario driven environment. Journal of Visual Languages and Computing 18, 315–338 (2007)

Cohn, A.G.: Qualitative Spatial Representation and Reasoning Techniques. In: Brewka, G., Habel, C., Nebel, B. (eds.) Advances in Articial Intelligence KI 1997, pp. 1–30. Springer, Berlin (1997)

Crawford, L.E., Regier, T., Huttenlocher, J.: Linguistic and non- linguistic spatial categorization. Cognition 75(3), 209–235 (2000)

Dodge, S., Weibel, R., Lautenschütz, A.K.: Towards a taxonomy of movement patterns. Information Visualization 7, 240–252 (2008)

Egenhofer, M.J., Al-Taha, K.K.: Reasoning about gradual changes of topological relationships. In: Frank, A.U., Formentini, U., Campari, I. (eds.) GIS 1992. LNCS, vol. 639, pp. 196–219. Springer, Heidelberg (1992)

Egenhofer, M.J., Franzosa, R.D.: Point-set topological spatial relations. International Journal of Geographical Information Systems 5(2), 161–174 (1991)

Freksa, C.: Temporal reasoning based on semi-intervals. Artificial Intelligence 54(1), 199–227 (1992)

Freksa, C., Habel, C., Wender, K.F. (eds.): Spatial Cognition 1998. LNCS, vol. 1404. Springer, Heidelberg (1998)

Furnas, G.W., Landauer, T.K., Gomez, L.M., Dumais, S.T.: The vocabulary problem in human-system communication. Commun. ACM 30(11), 964–971 (1987)

Galton, A.: Fields and objects in space, time, and space-time. Spatial Cognition and Computation 4(1), 39–68 (2004)

Gibson, J.: The ecological approach to visual perception. Houghton Mifflin, Boston (1979)

Goldstone, R.: The role of similarity in categorization: Providing a groundwork. Cognition 52(2), 125–157 (1994)

Goldstone, R.L., Barsalou, L.W.: Reuniting perception and conception. Cognition 65, 231–262 (1998)

Hayes, P.: The naive physics manifesto. In: Michie, D. (ed.) Expert Systems in the Microelectronic Age, pp. 242–270. Edinburgh University Press, Edinburgh (1978)

Hobbs, J.R.: Granularity. In: Joshi, A.K. (ed.) Proceedings of the 9th International Joint Conference on Artificial Intelligence, Los Angeles, CA, pp. 432–435. Morgan Kaufmann, San Francisco (1985)

Hornsby, K., Egenhofer, M.J.: Qualitative representation of change. In: Frank, A.U. (ed.) COSIT 1997. LNCS, vol. 1329, pp. 15–33. Springer, Heidelberg (1997)

January, D., Kako, E.: Re-evaluating evidence for linguistic relativity: Reply to Boroditsky (2001). Cognition 104, 417–426 (2007)

Klippel, A., Hardisty, F., Weaver, C.: Star plots: How shape characteristics influence classification tasks. Cartography and Geographic Information Science 36(2), 149–163 (2009)

Klippel, A., Montello, D.R.: Linguistic and nonlinguistic turn direction concepts. In: Winter, S., Duckham, M., Kulik, L., Kuipers, B. (eds.) COSIT 2007. LNCS, vol. 4736, pp. 354–372. Springer, Heidelberg (2007)

Klippel, A., Worboys, M., Duckham, M.: Identifying factors of geographic event conceptualisation. International Journal of Geographical Information Science 22(2), 183–204 (2008)

Klippel, A.: Topologically characterized movement patterns – A cognitive assessment. In: Spatial Cognition and Computation (to appear)

Knauff, M., Rauh, R., Renz, J.: A cognitive assessment of topological spatial relations: Results from an empirical investigation. In: Frank, A.U. (ed.) COSIT 1997. LNCS, vol. 1329, pp. 193–206. Springer, Heidelberg (1997)

Kos, A.J., Psenicka, C.: Measuring cluster similarity across methods. Psychological Reports 86, 858–862 (2000)

Kuhn, W.: An image-schematic account of spatial categories. In: Winter, S., Duckham, M., Kulik, L., Kuipers, B. (eds.) COSIT 2007. LNCS, vol. 4736, pp. 152–168. Springer, Heidelberg (2007)

Lakoff, G.: Women, fire and dangerous things. Chicago University Press, Chicago (1987)

Laurence, S., Margolis, E.: Concepts and cognitive science. In: Margolis, E., Laurence, S. (eds.) Concepts. Core readings, pp. 3–81. MIT Press, Cambridge (1999)

Lu, S., Harter, D.: The role of overlap and end state in perceiving and remembering events. In: Sun, R. (ed.) The 28th Annual Conference of the Cognitive Science Society, Vancouver, British Columbia, Canada, pp. 1729–1734. Lawrence Erlbaum, Mahwah (2006)

Mark, D.M., Comas, D., Egenhofer, M.J., Freundschuh, S.M., Gould, M.D., Nunes, J.: Evaluation and refining computational models of spatial relations through cross-linguistic human-subjects testing. In: Kuhn, W., Frank, A.U. (eds.) COSIT 1995. LNCS, vol. 988, pp. 553–568. Springer, Heidelberg (1995)

Mark, D.M., Egenhofer, M.J.: Modeling spatial relations between lines and regions: Combining formal mathematical models and human subject testing. Cartography and Geographic Information Systems 21(3), 195–212 (1994)

McIntosh, J., Yuan, M.: Assessing similarity of geographic processes and events. Transactions in GIS 9(2), 223–245 (2005)

Medin, D.L., Wattenmaker, W.D., Hampson, S.E.: Family resemblance, conceptual cohesiveness, and category construction. Cognitive Psychology 19(2), 242–279 (1987)

Mennis, J., Peuquet, D.J., Qian, L.: A conceptual framework for incorporating cognitive principles into geographical database representation. International Journal of Geographical Information Science 14(6), 501–520 (2000)

Montello, D.R., Frank, A.U.: Modeling directional knowledge and reasoning in environmental space: Testing qualitative metrics. In: Portugali, J. (ed.) The construction of cognitive maps, pp. 321–344. Kluwer, Dodrecht (1996)

Murphy, G.L., Medin, D.L.: The role of theories in conceptual coherence. Psychological Review 92(3), 289–316 (1985)

Nedas, K.A., Egenhofer, M.J.: Integral vs. Separable attributes in spatial similarity assessments. In: Freksa, C., Newcombe, N.S., Gärdenfors, P., Wölfl, S. (eds.) Spatial Cognition VI. LNCS, vol. 5248, pp. 295–310. Springer, Heidelberg (2008)

Peuquet, D.J.: Making space for time: Issues in space-time data representation. GeoInformatica 5(1), 11–32 (2001)

Piaget, J.: The construction of reality in the child. Basic Books, New York (1955)

Pothos, E.M., Chater, N.: A simplicity principle in unsupervised human categorization. Cognitive Science 26(3), 303–343 (2002)

Pothos, E.M., Close, J.: One or two dimensions in spontaneous classification: A simplicity approach. Cognition (2), 581–602 (2008)

Ragni, M., Tseden, B., Knauff, M.: Cross-cultural similarities in topological reasoning. In: Winter, S., Duckham, M., Kulik, L., Kuipers, B. (eds.) COSIT 2007. LNCS, vol. 4736, pp. 32–46. Springer, Heidelberg (2007)

Randell, D.A., Cui, Z., Cohn, A.G.: A spatial logic based on regions and connections. In: Proceedings 3rd International Conference on Knowledge Representation and Reasoning, pp. 165–176. Morgan Kaufmann, San Francisco (1992)

Regier, T., Zheng, M.: Attention to endpoints: A cross-linguistic constraint on spatial meaning. Cognitive Science 31(4), 705–719 (2007)

Renz, J. (ed.): Qualitative Spatial Reasoning with Topological Information. LNCS (LNAI), vol. 2293. Springer, Heidelberg (2002)

Riedemann, C.: Matching names and definitions of topological operators. In: Cohn, A.G., Mark, D.M. (eds.) COSIT 2005. LNCS, vol. 3693, pp. 165–181. Springer, Heidelberg (2005)

Rips, L.J.: Similarity, typicality and categorisation. In: Vosniadou, S., Ortony, A. (eds.) Similarity and Analogical Reasoning, pp. 21–59. Cambridge University Press, Cambridge (1989)

Rosch, E.: Cognitive representations of semantic categories. Journal of Experimental Psychology: General 104(3), 192–233 (1975)

Schwering, A.: Evaluation of a semantic similarity measure for natural language spatial relations. In: Winter, S., Duckham, M., Kulik, L., Kuipers, B. (eds.) COSIT 2007. LNCS, vol. 4736, pp. 116–132. Springer, Heidelberg (2007)

Schwering, A.: Approaches to semantic similarity measurement for geo-spatial data: A survey. Transactions in GIS 12(1), 2–29 (2008)

Schwering, A., Kuhn, W.: A hybrid semantic similarity measure for spatial information retrieval. In: Spatial Cognition and Computation (to appear)

Shipley, T.F., Zacks, J.M. (eds.): Understanding events: How humans see, represent, and act on events. Oxford University Press, New York (2008)

Strube, G.: The Role of Cognitive Science in Knowledge Engineering. In: Schmalhofer, F., Strube, G., Wetter, T. (eds.) GI-Fachtagung 1991. LNCS, vol. 622, pp. 161–174. Springer, Heidelberg (1992)

Weaver, C.: Building highly-coordinated visualizations in improvise. In: Proceedings of the IEEE Symposium on Information Visualization 2004, Austin, TX (October 2004)

Wishart, D.: ClustanGraphics Primer: A guide to cluster analysis, 3rd edn. Clustan Limited, Edinburgh (2004)

Worboys, M., Duckham, M.: Monitoring qualitative spatiotemporal change for geosensor networks. International Journal of Geographical Information Science 20(10), 1087–1108 (2006)

Xu, J.: Formalizing natural-language spatial relations between linear objects with topological and metric properties. International Journal of Geographical Information Science 21(4), 377–395 (2007)

Yuille, A.L., Ullman, S.: Computational theories of low-level vision. In: Osherson, D.N., Kosslyn, S.M., Hollerback, J.M. (eds.) An invitation to cognitive science: Language, vol. 2, pp. 5–39. MIT Press, Cambridge (1990)

Zacks, J.M., Tversky, B.: Event structure in perception and conception. Psychological Bulletin 127(1), 3–21 (2001)

Zadeh, L.A.: Fuzzy sets. Information and Control 8, 338–353 (1965)

Zhan, F.B.: A fuzzy set model of approximate linguistic terms in descriptions of binary topological relations between simple regions. In: Matsakis, P., Sztandera, L.M. (eds.) Applying soft computing in defining spatial relations, pp. 179–202. Physica-Verlag, Heidelberg (2002)

Evaluating the Effectiveness and Efficiency of Visual Variables for Geographic Information Visualization

Simone Garlandini and Sara Irina Fabrikant*

Department of Geography, University of Zurich,
Wintherthurerstrasse 190,
CH-8057 Zurich, Switzerland
{simone.garlandini,sara.fabrikant}@geo.uzh.ch

Abstract. We propose an empirical, perception-based evaluation approach for assessing the effectiveness and efficiency of longstanding cartographic design principles applied to 2D map displays. The approach includes bottom-up visual saliency models that are compared with eye-movement data collected in human-subject experiments on map stimuli embedded in the so-called flicker paradigm. The proposed methods are applied to the assessment of four commonly used visual variables for designing 2D maps: size, color value, color hue, and orientation. The empirical results suggest that the visual variable size is the most efficient (fastest) and most effective (accurate) visual variable to detect change under flicker conditions. The visual variable orientation proved to be the least efficient and effective of the tested visual variables. These empirical results shed new light on the implied ranking of the visual variables that have been proposed over 40 years ago. With the presented approach we hope to provide cartographers, GIScientists and visualization designers a systematic assessment method to develop effective and efficient geovisualization displays.

Keywords: Geographic visualization, visual variables, eye movements, change blindness, empirical studies.

1 Introduction

The cartographic design process is about a systematic transformation of collected (typically multivariate) spatial data into a two-, three- or four-dimensional visuo-spatial display. This process is typically performed by applying scientific (i.e., systematic, transparent, and reproducible) cartographic design methods, as well as aesthetic expressivity. Principles and details of the map design process can be found in many of the well-established cartography textbooks (see for example Dent, 1999; Slocum et al., 2008). More recently, cartographers have not only been interested in "what looks good" or "what visually communicates well", but also increasingly how and why a particular design solution works well or not.

Although the seemingly intuitive design principles have been successfully used for hundreds of years, and some of them (e.g., "light is less–dark is more") have even

---

* Corresponding author.

K. Stewart Hornsby et al. (Eds.): COSIT 2009, LNCS 5756, pp. 195–211, 2009.

been internationally accepted as conventions, for example, in the statistics community (Palsky, 1999), very few of the proposed conventions have actually been tested systematically for their effectiveness and efficiency with human users. One such example is the well-known system of the seven visual variables proposed initially by the French cartographer Jacques Bertin (1967; and translated to English in 1983) and later extended by various cartographers, see for example, Morrison (1974) and MacEachren (1995). More recently, Bertin's work has also received attention in the information visualization literature (Mackinlay, 1989). The variables seem to work when employed logically, but designers are typically not certain why. Unfortunately, there is very little empirical evidence on the effectiveness and efficiency of these visual variables (MacEachren, 1995). How can GIScientists, geovisualizers, and cartographers be sure that their design decisions produce effective and efficient displays? Naïve users tend to extract information based on perceptual salience rather than on thematic relevance (Lowe, 2003; Fabrikant & Goldsberry, 2005). For this reason, an empirical evaluation of design principles, and a systematic look into the relationships between perceptual salience and thematic relevance in visualization design is needed (MacEachren & Kraak, 2001) to understand how and why certain displays are more successful for spatial inference and decision making than others.

2 Related Work

2.1 Visual Variables for Guiding Visual Attention

Bertin (1967/83) proposed a systematical approach to communicating information by visual means. He lists seven basic visual variables and presents effects of varying the perceptual properties of the visual variables in order to derive meaningful representations. There are two planar variables (the x and y position on the map plane), and five so-called "retinal" ones (size, color value, color hue, shape, and orientation), which we (and perhaps vision researchers) would probably translate as "pre-attentive" (Bertin, 1967/83). Although Bertin (1967/83) lists these variables individually, effective map representations can of course include a combination of various visual variables (MacEachren, 1995).

Bertin distinguishes *selective, associative, ordered* and *quantitative* visual variables. A visual variable is *selective* (e.g., color hue) and therefore fundamental for symbolization of data, if all symbols can be easily isolated (perceptually selected) to form a group of *similar* symbols based on this variable (e.g., *where* are the red signs compared to the green signs). Bertin contends that shape (for points, lines and areas) and orientation (only when applied to areas) are not selective. Conversely, a visual variable is called *associative* (e.g., shape) if it allows to perceptually group all categories or instances of symbols based on that particular visual characteristic (signs of the same shape with different sizes vs. signs of different sizes with the same shape). Only the visual variables *size* and *color value* are said to have perceptual *dissociative* characteristics (Bertin, 1967/83). With dissociative visual variables (e.g., size) it is easier to detect visual variations among the signs themselves, than to visually form groups of similar symbols across other visual variables. Dissociative variables can be *ordered* or *quantitative*. A visual variable is defined *ordered* if it is possible to perceptually rank symbols based on one particular visually varying characteristic (e.g., lighter vs.

darker shading). If it is possible to perceptually quantify the degree of variation of a visual symbol, the visual variable property is defined as *quantitative* (e.g., size). Bertin furthermore ranks visual variables in an explicit sequence: higher order variables (e.g., size) which possess a greater number of perceptual characteristics (i.e., quantitative, ordered, and dissociative), compared to lower order variables (e.g., orientation), that may only have associative characteristics (only for areas).

Ironically, Bertin does not cite any perceptual or psychophysical work that would provide empirical evidence to his design guidelines. In fact, his seminal volume on the *Semiology of Graphics* (1967/83) does not include any reference to any previous or related work. Bertin's contributions can be understood within the context of the work by Gestalt psychologists such as, Wertheimer and Koffka in the 1920s (reviewed by Gregory, 1987 and Goldstein, 1989) who posited that the arrangement of features in an image plane will influence the perceived thematic or group membership relations of elements (i.e., figure/ground separation). Bertin's proposals have been somewhat supported by later experimental evidence for classic visual search tasks (e.g., pop-out vs. conjunctive search) proposed by Treisman and colleagues (e.g., Treisman & Gelade, 1980). In a meta study summarizing several decades of visual search and attention work in psychology and neuroscience, Wolfe & Horowitz (2004) list color (hue), size and orientation as undoubted variables to guide visual attention (for static displays), and color value (luminance) and shape as probable cases. Interestingly, these variables are not congruent with the ordering that Bertin suggests. Most if not all of this empirical work, however, has been performed on highly controlled, and therefore simple graphic displays, typically containing only simple and isolated geometrical signs, thus not complex graphics such as commonly used maps, or other kinds of visualizations.

Visual search strategies in a geographic context have been studied on realistic looking scenes such as maps (Lloyd, 1997), aerial photographs (Lloyd et al., 2002), and on remotely sensed images (Swienty et al., 2007). Additional empirical evidence for the validity of the visual variable system in more complex cartographic displays have been provided in the context of weather maps (Fabrikant et al., in press), thematic map animations (Fabrikant & Goldsberry, 2005), or for depicting the distance-similarity metaphor in information spatializations (Fabrikant et al., 2004; Fabrikant et al., 2006). Visual attention guiding variables have also been employed for the construction of computational vision models, as will be discussed in the next section.

2.2 Computational (Bottom Up) Models of Visual Attention

Itti & Koch (2001) present a computational framework to model visual saliency, based on based on neurobiological concepts of visual attention (Itti et al., 1998). The aim of the various computational models of visual attention is to model and predict visual attention based on psychophysical and neurophysiological empirical findings with human subjects (Koch, 2004). Visual saliency models also allow investigating complex and dynamic situations like animations, and changing natural scenes (Rosenholtz et al., 2007). Hence, they seem to be promising candidates for evaluating map displays as well.

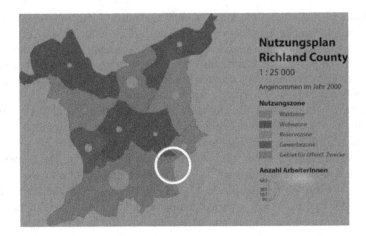

Fig. 1. Stimulus with predicted first eye fixation based on its saliency map

The Itti model is a neural-net based, neurobiologically plausible vision model. The goal of the model is to identify the focus of attention of a visual system (mammal or robot) based on the 'where' (e.g., perceptually salient characteristics), but not the 'what' (e.g., semantic characteristics, requiring cognition). In this model, three filters are applied to extract *color hue, color value* and *orientation* contrasts at several levels of image resolutions in a visual scene. Interestingly, these are three of Bertin's proposed visual variables. Three feature maps (one for each filter) are computed based on center-surround comparisons. Feature maps are additionally computed at several image resolutions and integrated to form a single conspicuity map for each feature type. A non-linear normalization is applied to each conspicuity map to amplify peaks of contrasts relative to noise in the background. In the final stage feature maps are combined to produce a single saliency map (SM). The saliency model also predicts a sequence of locations (ranked saliency peaks in the SM) that will attract a viewer's gaze in a scene (Parkhurst et al., 2002). The predicted initial eye fixation (white circle) is shown Figure 1. Lighter areas in Figure 1 identify image locations with higher saliency.

It is important to emphasize that the Itti saliency map does not reveal top-down components of visual attention. However, because we specifically employ a bottom-up approach within the flicker paradigm (see next section), and we are interested in evaluating the "retinal" (e.g., "pre-attentive") characteristics of map symbols, we contend this not to be a limitation for our study. Moreover, despite these limits, saliency map models appear to have already proven to be useful for cartographic purposes (Fabrikant & Goldsberry, 2005; Fabrikant et al., in press). We employ the visual attention model developed by Itti and colleagues (Itti et al., 1998) as a baseline to later compare human subject viewing behaviors collected with eye movement data. While visual variables are said to guide visual attention based on visual saliency, it is important to be aware of limitations or failures of the visual system, which we discuss in the next sections.

2.3 Failures in Visual Attention

Change blindness refers to a failure in the visual system in that observers often fail to detect even very salient and large changes in a scene when a blank field separates two alternating images. Change blindness is defined as *"the inability to notice changes that occur in clear view of the observer, even when these changes are large and the observer knows they will occur"* (Rensink, 2005: 76). According to Rensink (2005) change blindness occurs in different situations and under various conditions, thus it is a well-established phenomenon of human visual perception. Changes involving perceptually salient features are easier to detect than changes involving perceptually less salient features (Simons, 2000). As mentioned earlier, previous work has already demonstrated that visual attention and visual perception are tightly related (see review by Wolfe & Horowitz, 2004).

Rensink et al. (1997) introduced the flicker paradigm in order to investigate the phenomenon of change blindness. In the flicker paradigm *"an original image A repeatedly alternates with a modified image A', with brief blank fields placed between successive images"* (Rensink, 1997: 368).

Attention is characterized by bottom-up (stimulus-driven) and top-down (goal-driven) attentional control (Wright & Ward, 2008). The bottom-up component of attention is modeled in the flicker paradigm asking observers to detect the change as quickly as possible (Rensink, 2005). As a consequence, the memory impact on the experiment is reduced, but not completely inhibited (Rensink, 2005).

The dependent variable that can be measured under flicker conditions is the response time (Rensink, 2005). An observer is asked to solve three kinds of tasks: 1) change detection (what?), 2) change localization (where?), and 3) change identification (how?) (Rensink, 2002). Experimental results report that the identification task is typically the most complex task to handle (Rensink, 2002).

3 Experiment

In a controlled experiment we empirically investigated the relationships between the perceptual salience and thematic relevance in static 2D map displays. We employed a systematic bottom-up evaluation approach using the flicker paradigm (Rensink et al., 1997), in combination with the eye movement data collection method. In our experiment we focus specifically on those visual variables (i.e., size, color value, color hue and orientation) that according to Wolfe & Horowitz (2004) have been proven in psychophysical studies not only to guide visual attention, but are also used in a state-of-the-art visual saliency models (Itti et al., 1998).

In order to test the efficiency and effectiveness of these visual variables with users we prepared thirty-two thematic 2D map stimuli varying the visual variables size, color value, color hue and orientation (within-subject independent variables), embedded in a flicker display. The experiment consisted in solving three kinds of tasks: change detection, change localization, and change description. We hypothesize that the most efficient visual variable is detected *faster* in a flicker display than less efficient ones. Moreover, the more *effective* a visual variable, the more accurate participants' responses will be in a flicker display, compared to a less effective visual variable. To investigate these two hypotheses, the dependent variables *time of response* and *accuracy of response* are measured. In addition to the traditional success measures we additionally collect procedural data in the form of participants' eye movements when solving the experiment tasks. In this way, we hope to not only identify which visual variable works best, but also *how*. Finally, we derived saliency maps of the stimuli using a bottom-up computational model of visual attention (Itti et al., 1998). These saliency maps provide additional information about the saliency effects of the employed visual variables, and permit validation with the collected eye movement data.

Participants: Twenty participants (9 females and 11 males), recruited from the University of Zurich (UZH) and from the Swiss Federal Institute of Technology (ETH) Zurich, took voluntarily part in this study. They were not given any recompensation for participation. Participants were on average 29 years old, and no one indicated to be color-blind. Participants were selected to represent a range of professional backgrounds, without any experience regarding the flicker paradigm and its implications. On average the participant pool has a low to average training in geographic information science, such as cartography, geographical information systems, including the general familiarity with and usage of spatial data. Participants had a low or average level of training in computer science and related fields.

Materials: Sixty-four 2D map stimuli were designed in AdobeIllustrator and embedded in thirty-two flicker animations using AdobeFlash, according to the guidelines proposed by Rensink et al. (1997). The animations were embedded in a web page that could be automatically loaded by the eye tracker management software during the experiment. The flicker animations include four types of maps (i.e., eight flicker animations for each type) systematically varying the visual variables color *hue*, color *value*, *size*, and *orientation* (within-subject independent variables). To keep the map design consistent across trials, the stimuli included graduate circles and choropleths, as depicted in Figure 2 below. For the size stimuli, circle sizes changed, while the uniform area fills in the choropleth map was held constant. For the other three variables the area fills were affected by change, while the cirlces sizes were held constant. Fgure 2 shows a map stimulus used in the experiment.

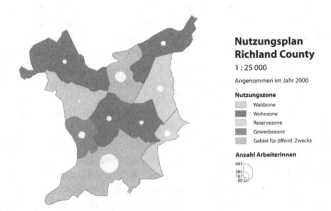

Fig. 2. Sample map stimulus evaluated in the study (color hue)

The maps in the flicker animation depict a set of randomly selected Swiss municipalities at a scale of 1:100,000. The geometry of the maps was systematically rotated in steps of forty-five degrees to assure that participants would not recognize the location, and therefore are able to focus their attention entirely on the change detection tasks. Based on the data characteristics (shown in the legend), we selected the appropriate visual variable for each thematic map stimulus, applying Bertin's (1967/83) design guidelines. Only the map portion of the graphic stimulus exhibits change between two consecutive displays. The change locations were also systematically varied so that areas in the center and various periphery locations in the map would change. Map title and legend never changed. An arbitrary map title was chosen by randomly selecting a county name in the U.S.A. (unknown to Swiss participants). The chosen name does not match the shown geometry. The legend includes a map scale (i.e., randomly selected representative fraction), a map symbol key, and respective attribute information. The maps do not contain any other map elements, such as author information, data source, or copyright sources. We reduced the design to a necessary (ecologically valid) minimum, in order to minimize cognitive load, and thus not further distract participants from the change detection tasks.[1]

Setup: The experiment took place in a windowless office, specifically designed and used to run eye movement experiments. It was administered on a Dell Precision 390 Windows workstation. The Tobii Studio software was employed to display the map stimuli and test questions on a 20-inch flat panel display, at 1024 by 768 pixels screen resolution. A standard mouse and keyboard were used for input. Participants' eye movements were recorded using a Tobii X120 eye tracker, at 60 Hz resolution. We employed a fixation filter with radius of 50 pixels, and minimal fixation duration at 100ms to collect participants' eye movements. Response time was measured as the elapsed time in milliseconds between the trial display appearing on the screen and the participant hitting a designated key on the keyboard to proceed to the next screen containing test questions.

Procedure: At the beginning of the test session participants were welcomed to the eye-tracking lab, signed a consent form, and filled out a background questionnaire. Participants were then asked to sit comfortably in front of the experiment computer connected to the eye tracker. Information on the testing procedure was displayed on the screen. Participants first performed two change detection trials to get comfortable with the test instrument, without having their eyes tracked. Following the practice trials participants' eye movements were calibrated with the eye tracker. Participants were again informed to sit comfortably, but as still as possible during the experiment, to improve calibration accuracy and consequently the eye tracking accuracy for the experiment.

For each flicker animation, participants were asked to hit the F10 key as soon as they saw a change. After the animation stopped and the stimulus disappeared, an answer screen appeared displaying a black and white reference map including area labels. Participants were asked to answer three questions. Firstly, if they had seen a

---

[1] Stimuli and experimental questions are available at: http://www.geo.uzh.ch/~sgarland/master/.

visual change (detection task); secondly, where they had seen the change (localization task); and finally, to describe the change (identification task). Participants responded to the test questions orally by refering to area labels on the reference map and the experiment leader recorded their answers using a digital microphone, and by typing responses into a digital file. After answering the three questions, participants launched the next flicker animation by hitting the F10 key. If participants did not see any change, the animation stopped automatically after 60 seconds. Participants were then asked to continue to the next trial by hitting the F10 key. The display order of the stimuli was randomized to avoid any potential learning bias. After completing the on-screen experiment participants were debriefed, and thanked for participation.

4 Results

Figure 3 shows participants' response times (efficiency) for the change detection task on the four tested visual variables. On average, participants took more time to detect a change in a map display varying the visual variable orientation (M=1.94s, SD=1.08s) compared to the other tested visual variables. The variable size yielded the shortest response time (M=0.65s, SD=0.21s), followed by color hue (M=0.92s, SD=0.73s) and color value (M=1.00s, SD=0.33s).

A repeated measures ANOVA (including a Bonferroni correction) reveals a significant overall effect for the (within-subject) "visual variables" factor, $F(25.805) = .000$, $p < .05$, indicating that there is a significant efficiency difference between the visual variables under study. Pairwise comparisons reveal that the variable orientation is indeed the least efficient visual variable for detecting a change. For maps containing this visual variable people take significantly longer to detect a change than for all other maps. Furthermore, while the variable size is the fastest of all tested visual variables, it is only significantly faster than orientation and color value. The speed advantage to color hue is not significant. There are no significant speed differences between color hue and color value.

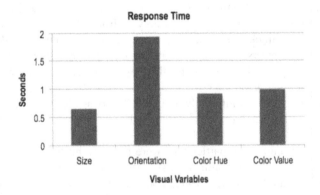

Fig. 3. Response time values in seconds

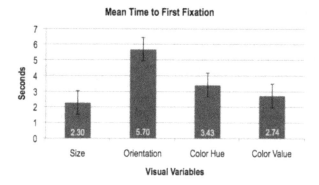

Fig. 4. Mean *"time to first fixation"*

We additionally investigated the efficiency (detection speed) of the visual variables by examining participants' eye movement behavior. For each stimulus, we delineated an area of interest (AOI) where a change occurs in the map. The efficiency metric *time to first fixation* (Goldberg & Kotval, 1999) can be employed to identify how long participants take to first fixate that particular AOI. This metric is negatively correlated with the potential degree of saliency of a region. High values of time to first fixation denote low degrees of saliency (Jacob & Karn, 2003). Figure 4 depicts the average length (in seconds), until participants fixated the relevant AOI for the first time during a trial. Again, people are slowest to first fixate on orientation changes, compared to color hue, color value, or size changes (fastest).

A repeated measures ANOVA reports a significant main effect for the four tested visual variables, $F(6.623) = .004$, $p < .05$. Size is significantly faster compared to orientation, but there are no significant differences between size and color hue or color value. Orientation is significantly slower than all the other tested variables, except compared to color hue. Size (fastest) and orientation (slowest) are at the extreme ends of the efficiency spectrum. There are no clear winners between color hue and color value.

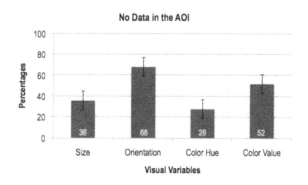

Fig. 5. Percentages of changes detected without looking explicitly at the change AOI

As Irwin (2004) notes, it is likely that the area of visual attention is larger than the location to where the fovea is pointing during a fixation. Evidence for this can be found in Figure 5. This Figure shows percentages of change that participants were able to detect correctly, without even fixating in the respective AOI. It is notable, that in 68% of the orientation trials (thus more than just by guessing) participants detected change without even fixating the respective "change" AOI. The percentages for the other trials are: color value (52%), size (36%) and color hue (28%), respectively.

To further look into the attention guiding potential or saliency of a visual variable we computed a ratio between the fixation duration within an AOI placed in the visual center of the map and the fixation duration within a "change" AOI. If this ratio provides lower values, observers' eyes were less attracted to the target AOI compared to "staring" into the center of the map. Higher ratio values might suggest that people's gazes moved around the map more or were attracted more readily to other attention guiding regions of the display. Size and color value (both 1.59) have the highest ratio, compared with orientation (1.29), and color hue (1.20). This measure qualitatively confirms the results depicted in Figure 4. Size and color value seem to have attracted participants gazes more than color hue and orientation.

We now turn to change localization. Regardless of the visual variable, people generally performed very well on the change localization tasks. This might be due to the stimuli having relatively low complexity. The size changes were localized practically error free (99%), followed by color hue and color value (both M=.994 SD=.028), and finally orientation with the lowest score (M=.925, SD=.143).

A repeated measures ANOVA for the change localization task provides evidence that there are significant differences among the visual variables, F(7.589) = .002, p < .05. Analog to the efficiency outcome for the change detection task, the variable orientation (least accurate localization) differs significantly from size (most accurate localization). No significant effects seem to exist between the other visual variables.

Figure 6 above also shows the percentage of correctly described types of changes. There is little difference in people's accuracy describing the change for size (99%), color hue (92%) and color value (97%) displays. However, changes in orientation seemed to have been much harder for people to describe accurately (69%).

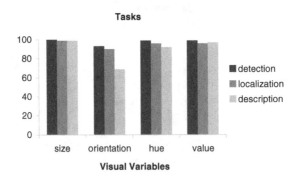

Fig. 6. Percentages of correct change detection, localization and description

According to a repeated measures ANOVA there seems to be a significant difference in the change description accuracy across the visual variables, $F(15.227) = .000$,
$p < .05$. The variable orientation differs significantly from to the other three visual
variables, yielding the least accurate results. Size scores are highest again, with 100%
description accuracy; significantly better than color hue and orientation. Color value
does not differ significantly from size and color hue.

4.1 Computational Saliency Evaluation

We additionally evaluated the animated flicker displays with previously mentioned
Itti saliency maps, using specifically the saliency model for dynamic visual scenes. In
addition to contrasts in color hue, color value and orientation (for static scenes), the
dynamic model also takes movement variables into consideration to compute the
resulting saliency map. The additional dynamic variables considered are: change in
location (motion up/down/right/left) as well as flicker (i.e., appearance and disappearance at a location).

We compared the location of highest saliency computed by the model and its respective predicted eye fixation pattern with the actual change locations and our own
collected eye movement data. The region of the change is indeed predicted by the
model to be the most salient region in the saliency map. The model seems to work
particularly well for the size displays. Comparing the predicted saliency maps of the
map stimuli across the four tested variables, it is notable that the model yields a few
highly concentrated areas of high saliency for the size stimuli, but less so for the other
variables, where salient areas are more spread out and less crisp. On average, color
hue has more salient locations in its saliency maps than the other tested variables.
Consequently, one would expect that observers would be attracted to a larger number
of locations competing for saliency (e.g., distractors), which might make the detection
("pop out") of a changing area more difficult. Based on this, one might further argue
that the variable color hue would yield the worst results in a change detection task.
However, our empirical results do not support this hypothesis. Participants had greater
difficulty and took significantly longer to detect a change in an orientation map than
for the other maps. Perhaps orientation maps do not provide enough visual contrast
between the enumeration areas. The linear pattern of the zone boundaries is harder to
isolate, due to the linear fill pattern within the zones. The individual enumeration
units seem to form larger homogeneous regions with little figure-ground contrast.
Henderson & Ferreira (2004) note that uniform regions are characterized by low fixation counts and consequently they do not draw visual attention. On average, orientation provided fewer fixation counts in the "change AOI" than the other three visual
variables.

As we used animated graphic stimuli for the assessment of the visual variables, we
need to also consider the effectiveness and efficiency of the visual variables for animated, or dynamic (e.g., interactive) visualizations. The overall advantage of size and
(to a lesser extent) color value in the change description task can perhaps be explained
by the additional influence of the dynamic variables (also computed for the saliency
map). Figures 7-8 show samples of overall saliency maps for the four tested visual
variables, overlaid on top of a map stimulus (upper left panel). The lighter the shade

("spot light") the higher the saliency. The white circle in the map stimulus is the predicted first gaze point (location of highest saliency). All the saliency attributes contributing to the overall saliency map are placed to the right and below of the map stimulus (panels with black background). Both size (Figure 7a) and color value (Figure 8b) yield areas of high saliency that are highly localized, compact, of small extent, and with crisp boundaries (especially for the size variable). This is perhaps due to optimal correlation of the visual variables (hue, value and orientation) with the dynamic ones such as, flicker (on/off) and motion (left, down, up, and right). The hue maps (Figure 8a) and orientation maps (Figure 7b), showing a much more dispersed pattern in their saliency maps, for both the static (visual) and dynamic variables, seem to be less effective at guiding people's attention to the relevant areas of change.

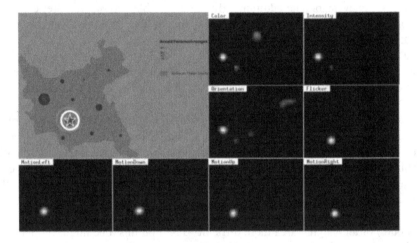

Fig. 7. Saliency maps for the visual variable size

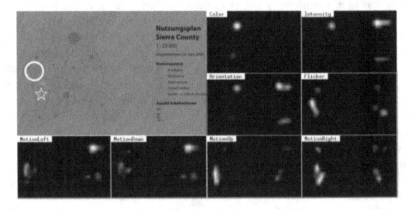

Fig. 8. Saliency map for the visual variable orientation

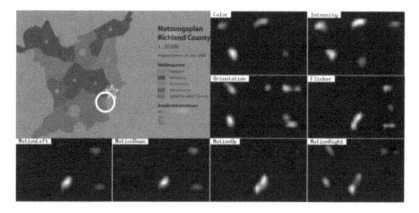

Fig. 9. Saliency map for the visual variable color hue

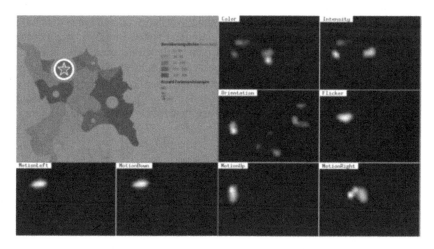

Fig. 10. Saliency map for the visual variable color value

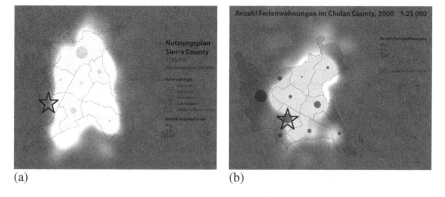

Fig. 11. Fixation concentrations across all participants for (a) orientation and (b) size

We now contrast (predicted) model results with our collected eye movement data. Figure 9 depicts two sample stimuli with aggregated eye fixations of all our participants. The lighter the display the higher the fixation concentration and magnitude.

It is striking (but somewhat counter intuitive) that the model correctly predicts (as shown in Figure 7b), and confirmed by our empirical data (Figure 9), that the largest of the graduated circles in the size display is attended the least. Both the center of the map and the smallest symbols receive most attention in the size display. It seems that (center-surround) contrast changes (modeled explicitly in the saliency maps) are indeed attention guiding. The smaller circles offer more "contrast-changes" against a homogeneous background than larger circles. Interestingly, only a small portion of the relevant change AOI (marked with a star symbol in Figure 9) was fixated in the orientation stimulus (compare with Figure 7b). This might be explained by the corner of the AOI being closest to the center of the map. Furthermore, if center-surround contrasts are relevant, then the concentration of stable boundary lines converging in a corner offer perhaps more contrast opportunities compared to directional changes of a linear fill pattern.

Overall, the model results and empirical result are very encouraging for cartographers, because they suggest that commonly employed visual variables, when correctly applied, are indeed able to effectively and efficiently guide observers' attention to relevant information. As Lowe (2003) suggests, congruently displaying thematically relevant information in a perceptually salient manner is one of the key challenges for designing effective and efficient map displays. However, empirical results presented in Figures 5 and 9, also provide some evidence that foveal attention and saliency are not always located in the same location.

5 Discussion

Summarizing our results we find that the selected four tested visual variables (Bertin, 1967/83) are indeed attention guiding, as people performed significantly above chance (e.g., 50%) in detecting, localizing and describing a change in the display. This is in accordance to the summary of results presented in Wolfe & Horowitz (2004)'s meta study on attention guiding attributes. These authors list color, motion, orientation, and size as "undoubted attributes" to guide visual attention. However, unlike Bertin (1967/83), Wolfe & Horowitz (2004) do not provide a ranking of the attributes. Our empirical results do provide some evidence for the implied ordering of Bertin's visual variables. We find the visual variable size to be the most efficient and effective variable to guide viewers' attention in thematic 2D maps, under flicker conditions. Perhaps this can be explained by the size displays being visually the least complex (e.g., having fewer visual distractors), according to the computational saliency model shown in Figure 7. According to Bertin, size is the only visual variable that has quantitative, ordered, selective (the signs perceived as different), and dissasociative characteristics (the signs are perceived as not similar). In fact, Bertin attributes size most "dissassociativeness". Since size emphasizes sign *difference* (e.g., change), one might argue from an information theoretic encoding perspective that difference or change could be an aspect of "interestingness", and thus, a very useful quality to guide attention. Since early eye movement studies on visual displays (Buswell, 1935;

Yarbus, 1967), it has been known that people concentrate their fixations on *interesting* and *informative* scene regions (Henderson & Ferreira, 2004).

The visual variable orientation appeared to be least effective and efficient of the four tested visual variables. As Bertin (1983: 93) writes: "in *area representation* variation in orientation is the easiest to construct, but it is at the same time the least selective" [of all seven visual variables]. Bertin assigns orientation only one attention guiding characteristic (i.e., associativity). He argues that with orientation (in areas) it is harder to isolate an area of change, as the variable emphasizes similarity, thus has a more uniform or homogeneous appearance. The computed saliency maps and our collected gaze data seem to support this idea.

For the color value and color hue variables the result pattern is not as clear. While color hue and color value yielded similar results, color value seems to have a slight (but non significant) advantage. In Bertin's system, color value differs from size only in the lack of a quantitative characteristic, thus one would have expected color value to perform better than hue for change detection. These results might support Wolfe & Horowitz (2004)'s questioning of luminance polarity (e.g., contrast in brightness or color value) as an attention-guiding attribute. They suggest it might be a subset of color, that is, the luminance axis of a three-dimensional color space.

6   Conclusion

This paper presents a systematic empirical evaluation approach to assess the effectiveness and efficiency of four commonly employed visual variables (size, color value, color hue and orientation) for the design of 2D map displays (Bertin, 1967/83). The proposed evaluation approach combines the application of visual saliency models developed in research on human vision with the assessment of change under flicker conditions by combining traditional performance measures (accuracy and speed) with eye movement recordings. We find that the visual variable size performs most effectively (accurately) and most efficiently (fastest) under flicker conditions. Conversely, the visual variable orientation seems to be least effective and efficient in our change detection experiment. For color hue and color value the results pattern are not as clear. Our results suggest validity to the implied ordering of the visual variables proposed by cartographer Jacques Bertin (1967/83) over 40 years ago. This study also shows that both the saliency map approach and the measurement of eye fixations under flicker conditions can be employed to systematically assess the utility of Bertin's (1967/83) system of seven visual variables widely used in cartography, and also discovered in information visualization (Mackinlay, 1989). The visual variable system was developed specifically to help cartographers better control the visual salience of symbols on maps. However, until today it lacked in systematical validation procedures, which we hope to have provided with this contribution.

Acknowledgments. We would like to thank our participants who were willing to participate in our research and are grateful for Mary Hegarty's continued insightful feedback on all things related to the eye movement data collection method. We also thank Alan MacEachren for his valuable feedback on an earlier draft of this manuscript.

References

Bertin, J.: Semiology of Graphics: Diagrams, Networks, Maps. University of Wisconsin Press, Madison (1983) (French edn., 1967)

Buswell, G.T.: How People Look at Pictures. University of Chicago Press, Chicago (1935)

Mackinlay, J.D.: Automating the Design of Graphical Presentations of Relational Information. ACM Transactions on Graphics 5(2), 110–141 (1986)

Dent, B.D.: Cartography. Thematic Map Design, Wm. C. Brown, Dubuque, IA (1999)

Fabrikant, S.I., Goldsberry, K.: Thematic Relevance and Perceptual Salience of Dynamic Geovisualization Displays. In: Proceedings, 22th ICA/ACI International Cartographic Conference, A Coruña, Spain, July 9-16 (2005)

Fabrikant, S.I., Montello, D.R., Ruocco, M., Middleton, R.S.: The Distance-Similarity Metaphor in Network-Display Spatializations. Cartography and Geographic Information Science 31(4), 237–252 (2004)

Fabrikant, S.I., Montello, D.R., Mark, D.M.: The Distance-Similarity Metaphor in Region-Display Spatializations. IEEE Computer Graphics & Application, 34–44 (2006)

Fabrikant, S.I., Rebich-Hespanha, S., Hegarty, M.: Cognitively Adequate and Perceptually Salient Graphic Displays for Efficient Spatial Inference Making. Annals of the Association of American Geographers (in press)

Goldberg, J.H., Kotval, X.P.: Computer Interface Evaluation using Eye Movements: Methods and Constructs. International Journal of Industrial Ergonomics 24, 631–645 (1999)

Goldstein, E.B.: Sensation & Perception, 2nd edn., Wadsworth, Belmont, CA (1989)

Gregory, R.L. (ed.): The Oxford Companion to the Mind, pp. 491–493. Oxford University Press, Oxford

Griffin, A.L.: Visual Variables. In: Kemp, K. (ed.) Encyclopedia of Geographic Information Science, pp. 506–509. SAGE Publication, Thousand Oaks (2008)

Henderson, J.M., Ferreira, F.: Scene Perception for Psycholinguists. In: Henderson, J.M., Ferreira, F. (eds.) The Integration of Language, Vision, and Action: Eye Movements and the Visual World, pp. 1–58. Psychology Press, New York (2004)

Irwin, E.: Fixation Location and Fixation Duration as Indices of Cognitive Processing. In: Henderson, J.M., Ferreira, F. (eds.) The Integration of Language, Vision, and Action: Eye Movements and the Visual World, pp. 105–134. Psychology Press, New York (2004)

Itti, L., Koch, C.: Computational Modeling of Visual Attention. Nature Reviews Neuroscience 2(3), 194–203 (2001)

Itti, L., Koch, C., Niebur, E.: A Model of Saliency-Based Visual Attention for Rapid Scene Analysis. IEEE Transactions on Pattern Analysis and Machine Intelligence 20(11), 1254–1259 (1998)

Jacob, R.J.K., Karn, K.S.: Eye Tracking in Human-computer Interaction and Usability Research: Ready to Deliver the Promises. In: Hyönä, J., Radach, R., Deubel, H. (eds.) The Mind's Eye: Cognitive and Applied Aspects of Eye Movement Research, pp. 573–605. Elsevier, Amsterdam (2003)

Koch, C.: Selective Visual Attention and Computational Models (2004), http://www.klab.caltech.edu/cns186/PS/attention-koch.pdf

Lloyd, R.: Visual Search Processes Used in Map Reading. Cartographica 34(1), 11–12 (1997)

Lloyd, R., Hodgson, M.E.: Visual Search for Land Use Objects in Aerial Photographs. Cartography and Geographic Information Science 29(1), 3–15 (2002)

Lowe, R.K.: Animation and Learning: Selective Processing of Information in Dynamic Graphics. Learning and Instruction (13), 157–176 (2003)

MacEachren, A.M.: How Maps Work. Representation, Visualization, and Design. Guilford Press, New York (1995)

MacEachren, A.M., Kraak, M.-J.: Research Challenges in Geovisualization. Cartography and Geographic Information Science 28, 3–12 (2001)

Morrison, J.L.: A Theoretical Framework For Cartographic Generalization with the Emphasis on the Process of Symbolization. International Yearbook of Cartography 14, 115–127 (1974)

Muller, J.C.: Bertin's Theory of Graphics/A Challenge to North American Thematic Cartography. Cartographica 18(3), 1–8 (1981)

Palsky, G.: The Debate on the Standardization of Statistical Maps and Diagrams (1857-1901). Elements for the history of graphical language. Cybergeo: European Journal of Geography (85) (1999) (March 16, 1999),
`http://www.cybergeo.eu/index148.html`

Parkhurst, D., Law, K., Niebur, E.: Modeling the Role of Salience in the Allocation of Overt Visual Attention. Vision Research 42, 107–123 (2002)

Rensink, R.A., O'Regan, J.K., Clark, J.J.: To See or Not to See: The Need for Attention to Perceive Changes in Scenes. Psychological Science 8, 368–373 (1997)

Rensink, R.A.: Change Detection. Annual Review of Psychology 53, 245–277 (2002)

Rensink, R.A.: Change Blindness. In: Itti, L., Rees, G., Tsotsos, J.K. (eds.) Neurobiology of Attention, pp. 76–81. Elsevier, San Diego (2005)

Rosenholtz, R., Li, Y., Nakano, L.: Measuring Visual Clutter. Journal of Vision 7(2), 1–22 (2007)

Simons, D.J.: Current Approach to Change Blindness. Visual Cognition 7(1/2/3), 1–15 (2000)

Slocum, T.A., McMaster, R.B., Kessler, F.C., Howard, H.H.: Thematic Cartography & Geographic Visualization, 3rd edn. Prentice Hall, Upper Saddle River (2008)

Swienty, O., Kurz, F., Reichenbacher, T.: Attention Guiding Visualization in Remote Sensing IIM Systems. Photogrammetrie, Fernerkundung, Geoinformation 4, 239–251 (2007)

Treisman, A.M., Gelade, G.: Feature-Integration Theory of Attention. Cognitive Psychology 12, 97–136 (1980)

Wolfe, J.M., Horowitz, T.S.: What Attributes Guide the Deployment of Visual Attention and how do they do it? Nature Reviews Neuroscience 5(6), 1–7 (2004)

Wright, R.D., Ward, L.M.: Orienting of Attention. Oxford University Press, New York (2008)

Yarbus, A.L.: Eye Movements and Vision. Plenum, New York (1967)

SeaTouch:
A Haptic and Auditory Maritime Environment for Non Visual Cognitive Mapping of Blind Sailors

Mathieu Simonnet[1,*], Dan Jacobson[2], Stephane Vieilledent[1],
and Jacques Tisseau[1]

[1] European Center for Virtual Reality (CERV), Britain European University (UEB)
25 rue Claude Chappe, 29280 Plouzane, France
{mathieu.simonnet,stephane.vieilledent}@univ-brest.fr,
jacques.tisseau@enib.fr
http://www.cerv.fr
[2] Investigating Multi Modal Representations of Spatial Environments (IMMERSE),
Department of Geography, University of Calgary
2500 University Drive NW, Calgary, AB, T2N 1N4, Canada
dan.jacobson@ucalgary.ca
http://www.immerse.ucalgary.ca

Abstract. Navigating consists of coordinating egocentric and allocentric spatial frames of reference. Virtual environments have afforded researchers in the spatial community with tools to investigate the learning of space. The issue of the transfer between virtual and real situations is not trivial. A central question is the role of frames of reference in mediating spatial knowledge transfer to external surroundings, as is the effect of different sensory modalities accessed in simulated and real worlds. This challenges the capacity of blind people to use virtual reality to explore a scene without graphics. The present experiment involves a haptic and auditory maritime virtual environment. In triangulation tasks, we measure systematic errors and preliminary results show an ability to learn configurational knowledge and to navigate through it without vision. Subjects appeared to take advantage of getting lost in an egocentric "haptic" view in the virtual environment to improve performances in the real environment.

Keywords: Navigation, spatial frames of reference, virtual reality, haptic, blind, sailing.

1 Introduction

Since Vygotsky's "theory of mediated activity"[1], the psychological community has paid particular attention to psychological tools. For example, in the spatial domain maps and compasses provide support for humans to think about

---

* Currently a postdoctoral fellow at the Naval Academy Research Institute in Lanvoc-Poulmic, France.

K. Stewart Hornsby et al. (Eds.): COSIT 2009, LNCS 5756, pp. 212–226, 2009.
© Springer-Verlag Berlin Heidelberg 2009

space. Thorndyke and Hayes-Roth [2] revealed that map learning is superior to environmental navigaton for judgements of relative locations and straight line distances among objects. Thus, learning a configurational layout does not necessarily require entire body displacement. Clearly, there are many differences between consulting a map and walking in an environment. While the map consultation unfolds in an allocentric spatial reference frame, independent of the perceiver position and orientation, the walking sequence takes place in an egocentric frame of reference which is directly relative to the current state of the body [3]. Thinus-Blanc and Gaunet [4] showed the necessity to coordinate these two views using invariants as a form of mediated common frame of reference.

The development of virtual reality techniques have provided researchers with promising tools to investigate how individuals learn space in controlled environments. Tlauka and Wilson [5] compared the spatial knowledge of subjects after a computer simulated navigation in an egocentric reference and the consultation of a map in an allocentric reference. Referring to the results found by Thorndyke and Hayes-Roth [2], the authors concluded that virtual and real navigation lead to the potential for building equivalent spatial knowledge. Similar outcomes were found by Richardson, Montello and Hegarty [6]. However, most of the time, during navigation in virtual environments, subjects predominantly accessed only visual information. This is in contrast to environmental navigation where increased acces to visual cues and displacements of the entire body provide people with inertial, kinesthetic and proprioceptive information. However, numerous studies from spatial virtual environments found that learning large-scale environments could be done effectively from purely visual sources and did not require body based information at all [7,8].

As vision is the spatial sense par excellence and virtual environments are both predominantly visual and predicated by the use of vision, the utility and the ability of blind people to generate meaningful spatial information *via* exposure to virtual environments has not been widely explored (See Lahav and Mioduser [9] for an exception). Consequently the question of what the role of virtual environments are able to play as a tool to aid the spatial knowledge learning of blind people remains largely unknown. For centuries there have been philosophical and experimental debates as to the capacity of people without sight to acquire a functional and holistic view of geographic space (See Ungar [10] for a review). The "difference theory" contends that blind people have no deficit in spatial processing but need more time to develop it [11]. The difficulties experienced by blind individuals in traversing an environment derive in part from their perceptual inability to gather distal (out of touch) spatial information [12]. While micro navigation, that which is proximal to the body, is taught as obstacle avoidance in traditional orientation and mobility training, and the environmental learning of an area is most commonly acquired through sequential and procedural exploration. The generation of configurational (macro level) knowledge of an environment remains problematic and is only achieved through intensive exploration of an environment [13,14]. However, spatial learning can be facilitated by the use of other spatial representations as tactile maps and models [15], digital touch and auditory interfaces to information [13,16], haptic and auditory displays [17] or personal guidance systems [18,19]. In any case, the ability to integrate, translate and move between experiences and representations relative to the egocentric and allocentric spatial frames of reference critical.

Blind people are able to successfully and independently navigate urban environments, where necessary using long canes and guide dogs as obstacle avoidance tools at the micro level. Access to the tools listed above is rare at best [20,21], however the acquistion of route, procedural and landmark knowledge, remains possible, providing enough information for blind people to locate and orientate themselves relative to the surroundings. In a built environment, there is a wide array of environmental cues, for example, sidewalks, road edges and junctions, auditory and olfactory landmarks, that facilitate the generation of spatial knowledge of an area and the ability to recreate routes. The guideing question behind the research is the complex issue of how do blind individuals locate themselves in natural environment that does not provide conventional urban cues for navigation, or the ability to spatially update through vestibular, kinaesthic, proprioceptive or inertial processes [22]. With an environmental tabula rasa, unable to access the cues above how would blind individuals be able to locate, orient and comprehend their spatial environmnet. Such an environment is provide by the ocean. Here, are the use of tactile maps and vocal compasses sufficient to connect egocentric and allocentric frames of reference?

In Brest (France), blindsailors were able to helm the sailboat in a straight line due to wind sensations [23]. In this case, the wind direction appeared to be a key feature to hold one's course and became the main directional reference. However, avoiding distant obstacles like rocks and locating themselves on the map remains a complex spatial task for blind sailors. In a maritime environment the type and variability of potential environmental information available is very different from a land based situation. These include, wind (speed and direction), boat displacement (speed, heading, pitch, roll, yaw), and potentially spatial updating via path integration based upon monitoring tacking, from a combination of the above information. All of this information is spatially and temporally highly variable. In order to assess these questions and in an applied manner to provide a navigational solution, we developed "SeaTouch". This application allows blind people to explore a representation of a maritime environment by means of a haptic and auditory interface. This enables virtual navigation with egocentric and allocentric "haptic views". Practically, this system aims to help blind people to master sailing navigation during real voyages, and in this manner is analogous to a land based personal guidance system [19].

After the description of the SeaTouch functionalities, we present an experiment to compare the performances obtained by blind sailors after virtual training sessions perfomed in egocentric or allocentric conditions.

2   SeaTouch

2.1   General Description

SeaTouch software and hardware aim to provide for blind people's cartographic needs, in a maritime environment using haptic sensations, vocal announcements and realistic sounds (Fig. 1). SeaTouch allows blind sailors to prepare their maritime itineraries. The digital maritime charts used in the development of SeaTouch conform to the S-57 International Hydrographic Office (IHO) exchange format, ensuring opportunities for interoperability. The digital charts contain many relevant geographic objects; "Handinav" software was developed to transform the S-57 data into XML structured files. Thus, objects of particular salience can be chosen to be displayed or not: sea areas, coastlines, land areas, beacons, buoys, landmarks were used in our research. Additional data contained in this maritime XML format is retained for potential future use. The position of the boat can be selected by entering coordinates in the simulator when it is started. The simulated weather conditions, such as the direction and the speed of the wind are modifiable. When simulation is on, the speed of the boat results from the interaction of the direction and speed of the wind with the orientation of the boat, generating a new heading speed. These calculations are based upon ocean

Fig. 1. A visualization of the SeaTouch virtual maritime environment. A participant's hand is interacting with the stylus of the Phantom haptic mouse. The land area, coastline, and maritime features are displayed.

based data collection, from an 8 meter sailboat "Sirius". Blind sailors choose the boat's heading during the entire simulation by using the right and left arrows of the computer keyboard. When the boat hits the coast, the simulation stops, this is indicated to the users *via* an auditory "crash".

2.2   Haptic Contacts and Constraints

The representational workspace is in the vertical plane, 40 centimetres wide, 30 centimetres high and 12 centimetres deep. Using a *Phantom Omni* force-feedback device, *via* a haptic cursor, calibrated to the representational workspace, blind participants explore the scene (Fig. 1). They touch different objects on the maritime maps as 2D-extruded haptic features. The salient features are sea surface, coastline, land area, navigational beacons, buoys and landmarks. The sea surface and land area are formed by two flat surfaces separated by two centimetres. Between the land and sea, the coastlines form a perpendicular wall, analogous to a cliff face, that allows users to follow it with the Phantom. The display of coastlines uses the "contact haptic force feedback", a virtual wall. By contrast, for beacons, buoys and landmarks, we apply a "constraint haptic force feedback" to a spring of one centimeter diameter. This spring is an active force feedback field that maintains the cursor inside of the object with a 0.88 Newton force, analogous to a "gravity well". In order to move outside of the spring, participants have to apply a stronger force. The position of the boat is displayed by the same haptic spring effect. It can be located from anywhere in the workspace by pressing the first button of the Phantom stylus, then the haptic cursor is relocated to the current position of the boat.

2.3   Sonification

In the sonification module, as soon as the stylus is in contact with virtual geographic objects audible naturalistic sounds are played. When in contact with the sea, a looping water sound is played. When the haptic stylus is touching the coastline, a virtual cliff face, the sounds of seabirds are played, and when land areas are in contact with the stylus the sounds of land birds are played. Prior testing confirmed clear discriminability of the sea and land birds sonification signals. If participants push through the sea surface, they hear the sound that a diver would make. If the cursor is wandering in the air, a wind sound is played. It is possible to touch the wake of the boat by hearing wash sounds. Our intention is that the redundancy and overlap between haptic and auditory information make this virtual environment as intuitive as possible.

2.4   Vocalization

Using "Acapela" vocal synthesis, a text to speech software, auditory information can be automatically spoken by SeaTouch. When the Phantom cursor enters in a beacon, buoy or landmark field, the nature and the name of these are announced.

Alternatively, blind participants can ask for information about distances and directions between the boat and the beacons or between two beacons. The format of this information can be preselected by the user, distance in nautical miles, or in kilometers, or in time relative to the current speed of the boat. In the allocentric representation mode, directions can be vocalized in the cardinal system (North, South, etc.) or in numeric cardinal degrees (0 - 360). In the egocentric representation mode directions can be obtained in hours relative to the boat orientation (noon is in front of the boat and 6 o'clock is behind), or in port and starboard numeric degrees relative to the boat orientation. For instance 90 degrees to starboard would be equivalent to 3 o'clock.

2.5   Virtual Navigation and Spatial Frames of Reference

SeaTouch software allows virtual interaction with two modes of haptic perspective : the allocentric ("bird's eye view") or the egocentric ("on board view") perspectives.

Northing / allocentric mode. The northing view provides a conventional presentation format of the scene, where the map remains invariant in a fixed frame of reference, aligned north up. The subject faces the north and the boat moves over the map. Thus exploration unfolds in an allocentric frame of reference (Fig. 2).

Heading / egocentric mode. By contrast, the heading view takes place in an egocentric frame of reference. Although the participant remains in an orthographic map view the participants' view of the map is continually re-orientated to always face the heading of the ship aligned to the top of the scene. This means that the ship does not rotate in the workspace, but the map rotates to maintain the heading of the ship to the top of the scene. Thus, the scene is dynamic and shifts as the blind sailor explores the auditory haptic representation (Fig. 3). In this condition the Phantom is comparable to a sort of "long maritime white cane" that the subject can use either to touch the boat itself or to localize the obstacles in the scene. The scale of the scene clearly facilitates exploration beyond the proximal in the micro environment, affording a "map-like" overview of the area.

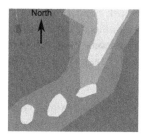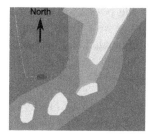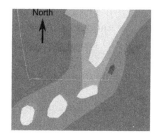

Fig. 2. The northing mode (allocentric) of SeaTouch: while changing boat directions, the boat moves on the map but the orientation of the map stays stable

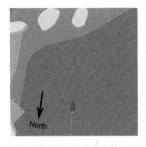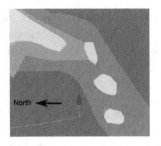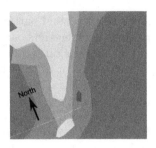

Fig. 3. The heading mode (egocentric) of SeaTouch: while changing boat directions, its position and orientation in the workspace stay stable but the map orientation moves to ensure a "boat up" view

Aim of the study. The heading and northing possibilities offered by SeaTouch to navigate in a virtual environment raise the question of their respective impact to the construction of an efficient non visual spatial representation at sea. The research focuses on investigating the difference between information gained in either of these frames of reference and the participants ability to transfer this information to a sailing activity in the ocean environment. Ultimately assessing whether virtual navigation in a heading condition (egocentric) is more efficient that in a northing condition (allocentric) to help facilitate blind sailors to train, locating the landmarks, or beacons, in the environment during a real voyage.

3   Method

3.1   Subjects, Cartographic Material and Experimental Organization

Subjects. As a preliminary study two blind sailors performed our experiment. They both navigate at sea regularly, are familiar with maps and computers.

Cartographic Material. During the learning tasks of each condition, the subjects were asked to explore SeaTouch maps including a configuration of six named beacons to set up their itinerary. This was composed of five ordered directions between these named anchor points (e.g. "rock" "raft", "spot", "pole", "net", "buoy"). The configuration of the points are the same in both conditions (egocentric and allocentric) except that we applied a mirroring symmetry translation along the north-south axis and changed the names of the beacons to avoid any learning effects (Fig. 4). In addition subjects performed the tasks of the two conditions in opposite orders, in a cross over design.

Experimental organization (Fig. 5). In order clarify each condition (heading versus northing), the virtual learning task can be divided in two sub-learning phases. The first phase consisted in exploring the map in a static setting; the participants familiarized themselves with the equipment, the interface and its operation until they were satisfied, and then they acquired their itinerary. The second learning

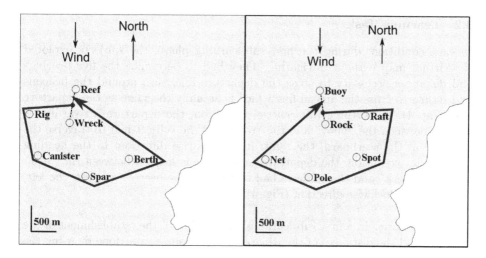

Fig. 4. Northing and heading maps. On the left the map we used in the northing condition, on the right the one we used in the heading condition. These two maps are similar except that, for illustrative purposes a symmetric transformation was applied to the right map relative to the vertical axis crossing the central point. The dark blue lines represent the trajectories of each course.

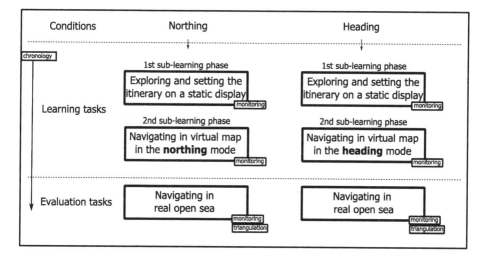

Fig. 5. The experimental organization

phase was made up of navigating through the virtual environment attempting to follow their itinerary between the named anchor points. Then, after completing these two sub-learning tasks that we describe more precisely below, subjects had to actualize their learning by navigating in the open sea evaluation task depicted in further below (See subsection 3.3).

3.2   Learning Task

In each condition, during the first sub-learning phase, the subjects explored the virtual map without navigating. They had to determine the five bearings and distances necessary to cross the departure line, turn around the beacons and return to cross the arrival line which is actually the same as the departure line (Fig. 4). In northing (allocentric) condition, the departure/arrival line is located between the "buoy" and the "raft" and the subject had to traverse the beacons on the starboard, the right, in a clockwise direction. In the heading (egocentric) condition, the departure/arrival line is located between the "reef" and the "wreck" and the subjects had to traverse the beacons to port, the left, in a counter-clockwise direction (Fig. 4).

Itinerary Setting. In a first sub-learning phase (Fig. 5), the establishment of the bearings and the distances of the itinerary of the course were done by using the "measure command" of SeaTouch. Here, subjects placed the haptic cursor to an initial point and said "origin" in the microphone, a voice answered "origin" Then subjects moved the cursor to a destination point and said "measure". The voice answered them the bearing (in cardinal numeric) and distance (in kilometers) between these two points. They set the all itineraries in this manner. This learning sub-task is exactly the same in the both conditions except for the positions of the landmarks.

Virtual Navigation. By contrast, in a second sub-learning phase (Fig. 5), subjects performed a virtual navigation in a heading or northing mode respectively corresponding to the egocentric and allocentric conditions. Here they could actively direct the orientation of the ship by speaking in increments of 1 or 10 up to 90 degrees to starboard or to port in the microphone. The speed of the boat only depended of its interaction with the wind. This blew from the north. If the angle between the ship direction and the wind orientation was less than 45 degrees, it could not proceed, replicating conditions in the open water. During this virtual navigation, blind sailors could ask for the bearings (in cardinal numeric degrees) and the distance (in kilometers) of the nearest beacon by saying "beacon". They also could ask for the current speed and heading. During the virtual navigation, we recorded the exploration displacements of the haptic cursor and of the trajectory of the boat. We called this record the monitoring (Fig. 5).

This procedure was replicated with the participants repeating the experiment in the alternate condition, that is participant who explored in the heading (egocentric) mode now explored the virtual environment in the northing (allocentric) mode and *vice versa*.

3.3   Evaluation Task

Here, we assessed how precisely subjects were able to locate themselves on the sea when navigating aboard a real sailboat encountering all of the usual constraints of sailing, such as tacking. After the virtual navigation, the blind sailors navigated in an open ocean environment aboard Sirius, an 8 meters long sailboat of the Orion association. The wind also blew from the north (±15 degrees). Subjects were asked to follow their set of named beacons, gathered from the virtual explorations in SeaTouch. In the real navigation, they managed Sirius by the same commands (1, 10 or 90 degrees) which were applied to an "automatic pilot". The automatic pilot is an electrical system, including an electronic compass, that movers the tiller to maintain a magnetic heading of the boat without assistance. In essence the participants were sailing "hands free", but encountering the complete array of other ocean based cues, such as, wind, swell and the roll, pitch and yaw of the boat. As in virtual navigation, subjects could ask information about the nearest beacon, the speed and heading of the sail boat. However, they did not have the haptic interface at their disposal so had no access to other distal cues.

To assess how precisely blind sailors could locate themselves, we asked them to point out the directions of three beacons in the middle of each of their five segments of the itinerary and we picked up these directions with a bearing's compass. So, we obtained fifteen angular errors about the directions of the landmarks. Directional pointing data was used to apply the projective convergence technique [24]. This triangulation provided us with five error triangles. Their areas informed us about the consistency of the responses (Fig. 6).

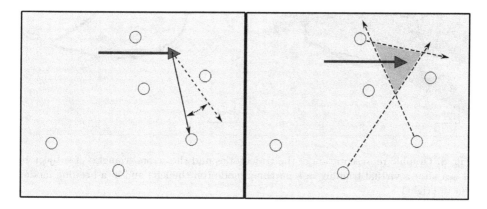

Fig. 6. Examples of angular error (on the left) and error triangle (on the right). The yellow circles are the beacons. The small red triangle is the sailboat. The large blue line is the track of the ship. The dotted lines are the directions estimated. On the left figure, the angle between the plain and dotted lines constitutes the angular error. On the right figure, we report estimations from the beacons to the ship. Where the dotted lines cross, the large grey triangle drawn is the error triangle.

4   Results

Due to the non normal distribution and measurement level of the data we used the non parametric Wilcoxon paired test to evaluate differences in the egocentric and allocentric frames of reference.

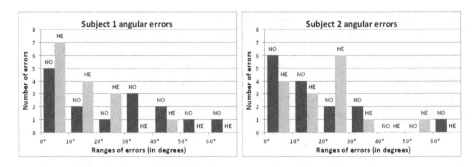

Fig. 7. Range of angular errors of the subjects 1 and 2 at sea after training in northing (NO) and heading (HE) conditions

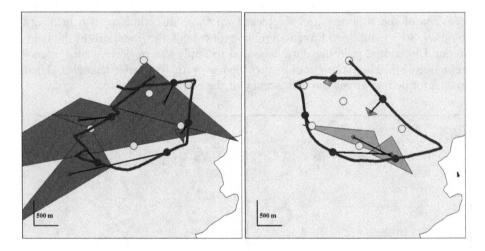

Fig. 8. Graphic representation of the trajectories and the errors triangles of subject 1 at sea after a virtual training in a northing mode (on the left) and in a heading mode (on the right)

The comparison of the angular errors of subject 1 after performing virtual training in heading (egocentric) and northing (allocentric) conditions revealed a significant difference (p=0.02). After heading virtual training, 14 responses among 15 were under 30 degrees of error (Fig. 7). After northing virtual training, only 8 estimations were under 30 degrees. So, the angular errors of subject 1 were significantly less important in the heading condition. By contrast the comparison of the angular errors of the subject 2 in heading and northing conditions did not reveal any significant difference (p=0.82) (Fig. 7).

These results were confirmed when we compared the areas of error triangles. For subject 1, all areas of error triangles were under 0.4 km^2 after virtual training in heading condition. By contrast, after virtual training in northing condition, 3 responses among 5 were over 0.5 km^2. So, relative to the areas of the error

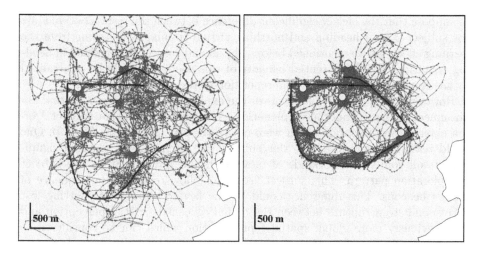

Fig. 9. The heading haptic patterns of exploration. On the left is the recorded pattern of exploration of the subject 1. On the right is the recorded pattern of exploration of the subject 2.

triangles, subject 1 performed significantly (p=0.04) better at sea after training in heading (egocentric) condition (Fig. 8). Subject 2, replicating the situation for angular errors, no significant difference was found between the error triangles in northing and heading conditions.

The better performances of subject 1 could not come from the learning effect because he firstly performed the heading condition. So these results are reinforced by the order of the experiment. The qualitative results provided with the monitoring allow us to see that the exploration movements of subjects 1 and 2 were different, especially in heading condition (Fig. 9). This report suggests an interesting discussion point, as to what is the role of "getting lost"? In other words what are the effects and differences in acquiring levels of spatial knowledge, does exploring an environment searching for information necessarily lead to higher more configural spatial knowledge rather than following a pre-learned prescribed route.

5   Discussion

The present study aimed at assessing the influence of egocentric and allocentric representations in virtual training, and their effect on participants abilities to locate and navigate in a true maritime environment without vision. The evaluation task was performed during true navigation that obviously took place in an egocentric spatial frame of reference. An explanation of the better results of subject 1 in the heading (egocentric) condition could be that the transfer of spatial knowledge is favored by the similarity of the spatial frames of reference involved in the learning and evaluation tasks. In this case, it would be reasonable to suppose that the lack of significant difference between the results obtained by the subject 2 after heading and northing virtual training could come from the learning effect. As we mentioned before, the second subject first trained in heading condition. So, the potential benefits of this condition could be neutralized by the learning undertaken by the repetition of the experiment.

However, the qualitative analysis of the exploratory strategies in virtual environment suggested another explanation. Actually, it seems that subject 1 often explored in directions which were out of the set up itinerary (Fig. 9). One could reasonably propose that this subject got often lost, or traveled significantly "off route". These kinds of movements did not appear on the subject 2 exploration pattern. This subject seems to follow regularly the sequence of named beacons. This difference could be the key to our results. Getting lost corresponds to a rupture between the actual egocentric spatial perception and the previously more global spatial representation which can be considered as allocentric [25]. So, when subject 1 got lost, he had to mentally try and connect the actual nearest beacon direction with the configuration of the anterior memorized map. This cognitive process could be an efficient training mechanism if we refer to Thinus-Blanc [4] who showed that mastering space consists in coordinate egocentric and allocentric spatial frames of reference using the concept of invariant. For these authors, an invariant would be a common entity of the egocentric field of perceptions and the allocentric spatial representation. In this case, subject 1 had to mentally identify an invariant when getting lost in the virtual world whereas subject 2 did not need to process this spatial reasoning because he did not encounter any difficulties in following the itinerary previously learned. This could explain why subject 2 did not benefit from the heading navigation.

This small prelimary case study shows that it could be beneficial to get lost in a virtual environment in an egocentric view in order to practice connecting egocentric and allocentric frames of reference. Our results should be interpreted with caution due to the small sample size (n=2), and the inherent possibility for a large role to be played by individual differences. Inspite of this small sample size the data collected are able to demonstrate that clearly different strategies may be utilised by people using the SeaTouch. However the experimental structure and methodology used is able to provide useful, insightful information. Our results highlight the need for further investigations with more subjects. In more general terms we have an environment and methodolody for exploring the role of egocentric and allocentric, haptic and auditory learning in virtual environments for people without vision and the transfer of this knowledegde to maritime surroundings. This affords us the opportunity to investigate further the role of frames of reference, map alignment, and a user's perspective on spatial information, in combination with issues of multimodal spatial data representation. The role of each of these factors transcends beyond the maritime or vision impaired community and are central to understanding how we learn from virtual environments and then utilise this information in the real world.

References

1. Vygotsky, L.S.: Mind of society. Harvard University Press, Cambridge (1930)
2. Thorndyke, P., Hayes-Roth, B.: Differences in spatial knowledge acquired from maps and navigation. Cognitive Psychology 14, 560–589 (1982)
3. Klatzky, R.L.: Allocentric and egocentric spatial representations: Definitions, distinctions, and interconnections. In: Freksa, C., Habel, C., Wender, K.F. (eds.) Spatial Cognition 1998. LNCS, vol. 1404, pp. 1–17. Springer, Heidelberg (1998)
4. Thinus-Blanc, C., Gaunet, F.: Representation of space in blind persons: vision as a spatial sense? Psychological Bulletin 121(1), 20–42 (1997)
5. Tlauka, M., Wilson, P.: Orientation-Free Representations from Navigation through a Computer-Simulated Environment. Environment and Behavior 28(5), 647–664 (1996)
6. Richardson, A., Montello, D., Hegarty, M.: Spatial knowledge acquisition from maps and from navigation in real and virtual environments. Memory and Cognition 27(4), 741–750 (1999)
7. Rossano, M., West, S., Robertson, T., Wayne, M., Chase, R.: The acquisition of route and survey knowledge from computer models. Journal of Environmental Psychology 19(2), 101–115 (1999)
8. Waller, D., Greenauer, N.: The role of body-based sensory information in the acquisition of enduring spatial representations. Psychological Research 71(3), 322–332 (2007)
9. Lahav, O., Mioduser, D.: Haptic-feedback support for cognitive mapping of unknow spaces by people who are blind. International Journal of Human-Computer Studies 66, 23–35 (2008)
10. Ungar, S.: Cognitive mapping without visual experience. In: Kitchin, R., Freundschuh, S. (eds.) Cognitive Mapping: Past, Present and Future, pp. 221–248. Routledge, London (2000)
11. Kitchin, R., Blades, M., Golledge, R.: Understanding spatial concepts at the geographic scale without the use of vision. Progress in Human Geography 21(2), 225–242 (1997)
12. Millar, S.: Models of Sensory Deprivation: The Nature/Nurture Dichotomy and Spatial Representation in the Blind. International Journal of Behavioral Development 11(1), 69–87 (1988)
13. Jacobson, D.: Cognitive mapping without sight: Four preliminary studies of spatial learning. Journal of Environmental Psychology 18, 189–305 (1998)
14. Jacobson, R., Lippa, Y., Golledge, R., Kitchin, R., Blades, M.: Rapid development of cognitive maps in people with visual impairments when exploring novel geographic spaces. Bulletin of People-Environment Studies 18, 3–6 (2001)
15. Casey, S.: Cognitive mapping by the blind. Journal of Visual Impairment and Blindness 72, 297–301 (1978)
16. Jacobson, R.D.: Representing Spatial Information Through Multimodal Interfaces: Overview and preliminary results in non-visual interfaces. In: 6th International Conference on Information Visualization: Symposium on Spatial/Geographic Data Visualization, pp. 730–734. IEEE Proceedings, London (2002)
17. Rice, M., Jacobson, R.D., Golledge, R.G., Jones, D.: Design Considerations for Haptic and Auditory Map Interfaces. Cartography and Geographic Information Science 32(4), 381–391 (2005)

18. Golledge, R.G., Loomis, J.M., Klatzky, R.L., Flury, A., Yang, X.L.: Designing a personal guidance system to aid navigation without sight: Progress on the GIS component. International Journal of Geographical Information Systems 5, 373–396 (1991)

19. Golledge, R.G., Klatzky, R.L., Loomis, J.M., Speigle, J., Tietz, J.: A geographical information system for a GPS based personal guidance system. International Journal of Geographical Information Systems 12, 727–749 (1998)

20. Rowell, J., Ungar, S.: The world of touch: an international survey of tactile maps. Part 1: production. British Journal of Visual Impairment 21(3), 98–104 (2003)

21. Rowell, J., Ungar, S.: The world of touch: an international survey of tactile maps. Part 2: design. British Journal of Visual Impairment 21(3), 105–110 (2003)

22. Simonnet, M.: Virtual reality contributions for the coordination of spatial frames of reference without vision. SeaTouch, a haptic and auditory application to set up the maritime itineraries of blind sailors. PhD Thesis. University of Brest (France), Department of Sports (2008)

23. Simonnet, M.: Sailing and blindness: The use of a sensorial perceptions system of a blind helmsman. Graduate Report, University of Brest (France), Department of Sports (2002)

24. Hardwick, D.A., McIntyre, C.W., Pick Jr, H.L.: The Content and Manipulation of Cognitive Maps in Children and Adults. Monographs of the Society for Research in Child Development 41, 1–55 (1976)

25. Wang, R.F., Spelke, E.S.: Human spatial representation: insights from animals. Trends in Cognitive Sciences 6, 376–382 (2002)

Assigning Footprints to Dot Sets: An Analytical Survey

Maximillian Dupenois and Antony Galton

School of Engineering, Computing and Mathematics, University of Exeter, UK

Abstract. While the generation of a shape, or *footprint*, from a set of points has been widely investigated, there has been no systematic overview of the field, with the result that there is no principled basis for comparing the methods used or selecting the best method for a particular application. In this paper we present a systematic classification of footprints, algorithms used for their generation, and the types of applications they can be used for. These classifications can be used to evaluate the suitability of different algorithms for different applications. With each algorithm is associated a vector of nine values classifying the footprints it can produce against a standard list of criteria, and a similar vector is associated with each application type to classify the footprints it requires. A discussion of, and a method for, the assessment of the suitability of an algorithm for an application is presented.

1 Introduction

While the generation of a shape, or *footprint*, from a set of points has been widely investigated, there has been no systematic overview of the field, with the result that there is no principled basis for comparing the methods used or selecting the best method for a particular application. In this paper we present a systematic classification of the footprints, the algorithms used, and the types of applications they can be used for. Our classification of footprints bears some similarity to the set of criteria proposed by Galton and Duckham [9] for evaluating the footprints produced by different algorithms. However, here we provide a more detailed analysis and propose a method for choosing an algorithm appropriate for a given context. It should be noted that this paper makes no attempt to evaluate these footprints, only to classify them. Using this classification we evaluate the suitability of algorithms for general applications, but this has no bearing on how 'good' a footprint is. For a discussion on the quality of a footprint with regard to perceived shape or any other cognitive criteria see [8]. The classifications of the footprints, algorithms and applications are all linked but for the sake of clarity are declared separately, with appropriate relations discussed later. While we only cover two-dimensional footprints, much of the analysis should carry over to the three-dimensional case; but it is likely that three-dimensional footprints have additional properties not covered by the present classification.

K. Stewart Hornsby et al. (Eds.): COSIT 2009, LNCS 5756, pp. 227–244, 2009.

2   Definitions

The general problem under consideration is that of assigning a region-like entity to a collection of point-like entities in space. In this paper we call the former a 'footprint' and the latter 'dots'. Here we provide a brief justification for this choice of terminology.

- **Dots** We refer to dots rather than points because when considering the various algorithms it became apparent that, in addition to coordinates, the point-like entities may possess attributes such as shape, area, or velocity, any of which might be of relevance to an algorithm.
- **Footprint** What we are here calling footprints have variously been called outlines, shapes, hulls, and regions. We reject 'outline' and 'shape' as being too focussed on the boundary; while 'region' seems too general, with nothing to indicate any special relationship to the dots. 'Hull' has often been chosen to reflect the idea of a footprint as a generalisation of the convex hull, and indeed many of the algorithms are modified forms of convex-hull algorithms (e.g., Concave Hull [13]); however, the definition of a hull operator in computational geometry requires all the dots to lie within the hull, which must itself be connected [12], both of which conditions may be violated by footprint algorithms.[1] 'Footprint' is used in various fields to denote the impression of some entity,[2] and this connotation seems appropriate here.

3   Background

As mentioned in the introduction, much work has been done on the generation of footprints, but there has been surprisingly little by way of comparative analysis. Four types of analysis largely absent from the literature on footprints are:

- Analysis of the types of footprint.
- Analysis of the methods used in the algorithm, and how these methods limit the footprints produced.
- Analysis of the algorithm type, e.g., whether it requires some form of pre-processing on the data set.
- Analysis of the context that the algorithm was created for.

An early, and much-referenced, paper on the subject is by Edelsbrunner et al. [5], who present a method for creating footprints from a point set.[3] The method produces straight-line graphs called α-shapes, obtained from a generalisation of the convex hull. For a set S the convex hull can be considered to be the intersection of all closed half-planes that contain all the points of S. The α-*hull*

---

[1] A hull operator is also required to be idempotent; it is not clear what this could mean in the case of an operator which generates a region from a finite set of points.

[2] E.g. the memory space a piece of software uses, the actual footprint of an animal, the carbon footprint.

[3] Point set and not dot pattern as the method only uses the coordinates.

is obtained by using closed discs of radius $1/\alpha$ instead of half-planes; the α-shape is derived from this in a straightforward way. The authors do not discuss any principled way to choose the appropriate α for the type of shape required.

Chaudhuri et al. [3] present two methods for generating a footprint, called the *external shape*, from a dot pattern. Although they use the term 'dot pattern' they make no distinction between points and dots. For the first method, a grid of squares of side-length s is drawn on the plane, and the union of all grid-squares containing at least one of the dots is returned as the footprint, called the *s-shape*. For the r-shape they inscribe a disc of radius r round each dot, and draw an edge connecting any pair of dots whose discs intersect in a point not contained in any of the other discs. These edges provide an outline which, in our terms, may be regarded as the boundary of the footprint. As with the α-shape, no principles are given for selecting appropriate values of r or s.

Garai and Chaudhuri [10] propose a 'split and merge' method for generating footprints. This method starts from the convex hull and attempts to refine it to a shape more closely resembling what they refer to as the *underlying shape*. The method consists of three separate algorithms (four if the convex hull algorithm is included): *splitting*, *isolation*, and *merging*. This is one of the few algorithms that provides a way of *aiming* for a particular shape without having to re-run the algorithm with different parameters, so long as the user is able to identify a desired maximum area or number of sides just from a cursory examination of the dot pattern. Again the authors say little about the quality or type of footprint they generate.

Alani et al. [1] developed the *Dynamic Spatial Approximation Method* (DSAM). This system takes in both the dot pattern of the region to be found and the dot pattern of the area known to exist outside the region. It builds a Voronoi diagram based on these coordinates and takes the union of all the cells which contain an 'interior' point as its footprint. This work pays more attention than many in the area to the quality of footprint produced; this can be assessed in terms of how closely the region found fits the expected region. The existence of a contextually determined target shape differentiates this paper from others in the field.

Arampatzis et al. [2] follow on from Alani et al. [1]. However, they adapt DSAM to use Delaunay triangulations in conjunction with a system for finding point locations using web queries. They call this adaptation *the recolouring algorithm* and use it to generate boundaries for imprecise regions. Much like the DSAM this system has a target shape and, as such, this paper has more analysis of the footprint found than much of the field.

Galton and Duckham [9] propose two methods for finding footprints. The first method is a generalisation of the Jarvis March ('gift-wrapping') algorithm for convex hulls. The idea behind the Jarvis March is simple. From an origin point outside the dot set a radial half-line is swung in an arbitrary direction until it meets one of the dots. This dot is made the new origin point from which a radius is swung in the same direction as before until it meets another dot. This is repeated until the first dot is encountered again; the sequence of dots encountered in this way form the vertices of the convex hull. Dots are removed from consideration if they have already been marked as being on the convex hull or if they lie within the area enclosed by the dots encountered so far. The 'Swinging Arm' algorithm is similar except that it uses a line-segment of some predetermined length instead of a half-line. The second method starts with the Delaunay triangulation and successively removes the longest external edge, subject to constraints of maintaining connectedness and regularity, until either some predetermined minimum length is reached, or no more edges can be removed. The authors note that there can be no uniquely 'optimal' footprint when the application context is considered to be general. The paper proposes nine criteria which may be used for evaluating footprint algorithms with respect to different application contexts, although little is said about any actual applications. Some of these criteria are used by the classification developed in the present paper.

Moreira and Santos [13] present a 'Concave Hull' algorithm. Like the Swinging Arm, Concave Hull is also derived from the Jarvis March algorithm, its difference being that it always selects the next vertex from the k nearest neighbours of the current vertex. This is the crux of the algorithm's effectiveness: by having a non-contextual integer as the variable that restrains the hull algorithm, they have a default base value from which they can run the algorithm (i.e. $k = 3$); if this fails to produce a footprint that satisfies the criteria (having no intersecting lines and containing all the points) then the algorithm is run with increasing values of k till such a footprint is created. Like most of the other authors they pay little attention to the quality of the footprint in relation to any application type, though they do mention the criteria given in [9]. Like the split and merge method [10], the Concave Hull algorithm requires some pre-processing of dots, using the Shared Nearest Neighbour (SNN) algorithm to determine any separable groupings in the dot pattern prior to running the algorithm. Like Garai and Chaudhuri they do not take account of this pre-processing algorithm in determining the computational complexity of their own.

Duckham et al. [4] provide a fuller account of the Delaunay-based method introduced in [9], now called the χ-algorithm. This paper includes a discussion of the footprint's properties, and how these are directly tied to the method by which it is created. More attention is paid to the choice of the length parameter l. There are practical limits on l for any triangulation (if it is too large then no lines will be removed, if it is too small too many will be removed) and consequently l can be normalised. Duckham et al. propose using this normalised parameter (λp) to find a starting value which should achieve what they call a *characteristic shape* for many, if not all, dot patterns. While they conclude that there is no λp that always produces a "good" characterization, the fact that they spend time considering this is unusual within the field. Unlike Moreira and Santos [13] and Garai and Chaudhuri [10], Duckham et al. do not discount the pre-processing (in this case computing the Delaunay triangulation and sorting the edges) when determining the complexity of the algorithm.

Galton [8], instead of proposing an algorithm, searches for objective criteria for evaluating the acceptability of any proposed footprint in relation to the 'perceived' shape of a dot pattern. The paper notes that in most of the published work, "while lip-service is generally paid to the fact that there is no objective definition of such a 'perceived shape', little is said about how to verify this, or indeed, about exactly what it means". Restricting attention to footprints in the form of *polygonal hulls*, simple polygons having vertices selected from the dot pattern, all the other dots being within the interior, the paper presents evidence that while a dot pattern may have several equally acceptable perceived shapes, they all represent optimal or near-optimal compromises between the conflicting goals of simultaneously minimising both the area and the perimeter of the hull.

4   Classifications

4.1   Footprints

Before embarking on the classification, we make some preliminary observations of a general nature.

1. While a footprint is, considered in itself, a region with a shape, what makes it a footprint is the relationship it bears to the dot pattern from which it is derived. For this reason, it must be emphasised that our classification does not attempt to be a general classification of *shapes*; it only considers those aspects of shape which are relevant to the role of being a footprint.
2. Our classification criteria can be divided into *intrinsic* criteria, which concern properties of the footprint in itself, without reference to the dots, and *relational* criteria which concern the relationship between the footprint and the dots.

Intrinsic footprint criteria

[C] **Connected** *The footprint consists of a single connected component.*
Figure 1 shows examples of connected and disconnected footprints for the same dot pattern. Some algorithms will always generate a single connected component, implicitly assuming that any clustering has been done beforehand, with the algorithm being applied to individual clusters (e.g., Concave Hull [13], χ-shape [4]); others can yield footprints with multiple components (e.g., Swinging Arm [9]). The desirability or otherwise of multiple components is application-dependent, e.g., if only connected footprints are appropriate, use an algorithm guaranteed to produce such components.

[R] **Regular** *The footprint is topologically regular.*
Assuming the footprint is topologically closed, this criterion amounts to whether or not the footprint contains boundary elements that do not bound the footprint's interior, such as the linear 'spike' in Figure 2(b) or isolated linear component in Figure 2(c).

[P] **Polygonal** *The boundary of the footprint is made up of only straight lines.*
For a polygonal footprint the boundary is made up entirely of straight line-segments as opposed to curves. (Figure 3).

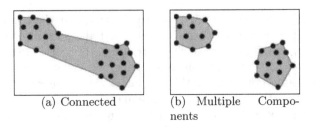

(a) Connected   (b) Multiple Components

Fig. 1. Connectedness

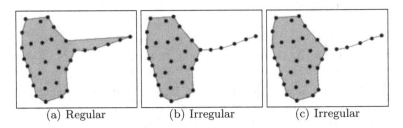

(a) Regular   (b) Irregular   (c) Irregular

Fig. 2. Regular

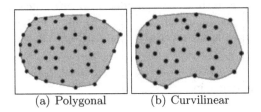

(a) Polygonal   (b) Curvilinear

Fig. 3. Polygonal

[JC] Jordan Components *Each component of the footprint has a Jordan boundary.*

A Jordan boundary is a boundary which is a Jordan curve, i.e., homeomorphic to a circle. Such a boundary does not meet itself, so it is possible to traverse the entire boundary passing through each of its points only once. (Figure 4(a)). In Figure 4(b) the component with a non-Jordan boundary is represented as a 'bow tie' shape; of course this is not the only way the Jordan property can fail.[4]

[SCC] Simply Connected Components *Each component of the footprint is simply connected.*

---

[4] In relation to the 'bow-tie' configuration, if the footprint is formed by tracing out its boundary, then the constriction point may be either a *self-intersection*, where the boundary actually crosses itself, or a *pinch point*, where the boundary touches itself without crossing. An intersection or pinch-point may or may not occur on one of the dots; examination of the algorithms suggests that a self-intersection is more likely to occur away from a dot, whereas the opposite is true for a pinch point.

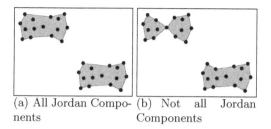

(a) All Jordan Compo- (b) Not all Jordan
nents   Components

Fig. 4. Jordan Boundary

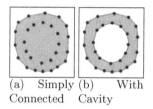

(a) Simply (b) With
Connected Cavity

Fig. 5. Simply Connected

A component that is not simply connected contains a 'hole' (Figure 5(b)). In two dimensions this means that the boundary is disconnected, with one of the boundary components facing the 'outside', and each other component bounding an internal cavity.[5]

Relational footprint criteria

[CED] Curvature Extrema At Dots *All curvature extrema of the footprint boundary coincide with dots.*
Very often a footprint is constructed by tracing its boundary through some or (more rarely) all the dots of the dot pattern. In such cases it is typical for the dots to mark curvature extrema of the outline; this is the normal situation when the outline is polygonal, with the dots at its vertices (Figure 6(a)), and is always found in the case of the convex hull.

Note that this criterion is independent of whether all, some, or none of the dots occur on the boundary (which is given by criteria [ADB] and [NDB] introduced next), as shown by Figure 6, where each value for one criterion can co-occur with each value of the other. However, [CED] ∧ [NDB] (all curvature extrema are dots and all dots are off the boundary) can only be true if the footprint is circular, in which case there are no curvature extrema, so [CED] is true by default.

---

[5] In three dimensions there are more varieties of connectivity to consider, e.g., the distinction between an internal cavity and a perforation. We shall not discuss these further here.

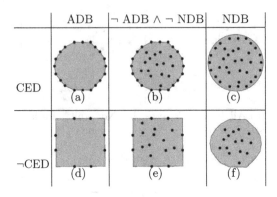

Fig. 6. Curvature Extrema and Dots On/Off Boundary

[ADB] All Dots on Boundary *All of the dots lie on the boundary of the footprint.*
In general we would not expect footprints to satisfy this criterion, but in some applications the dots are specifically intended to represent boundary points, and in such cases this criterion is appropriate. As mentioned above [ADB] is linked to, but distinct from, whether or not the curvature extrema coincide with dots (Figure 6).

[NDB] No Dots on Boundary *None of the dots lie on the boundary of the footprint.*
Criteria [ADB] and [NDB] cannot be simultaneously satisfied, and they are not independent. As with [ADB] it is linked to, but distinct from, whether or not the curvature extrema coincide with dots (Figure 6). Some algorithms (e.g., the Voronoi-based method of [1]) create footprints by amalgamating 'areas of influence' surrounding the dots. In such cases the dots typically all lie in the interior of the footprint, and hence off the boundary.

[FC] Full Coverage *All of the dots are included in the closure of the footprint.*
It is possible that a footprint algorithm may be able to distinguish certain dots from the pattern as 'noise', and as such it may wish to exclude them from the footprint. We call such dots *outliers* (Figure 7).

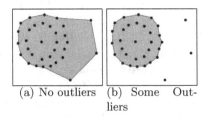

(a) No outliers   (b) Some   Outliers

Fig. 7. Full Coverage

Table 1. Classification of Footprint Examples

Footprint Examples	C	R	P	JC	SCC	CED	ADB	NDB	FC
Example 1 [Figure 8(a)]	+	+	+	+	+	+	−	−	+
Example 2 [Figure 8(b)]	+	−	+	−	−	+	−	−	−
Example 3 [Figure 8(c)]	−	+	−	+	+	+	−	+	+
Example 4 [Figure 8(d)]	−	+	+	+	+	−	−	−	+

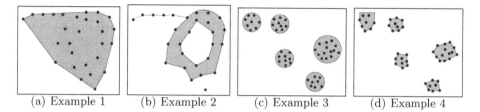

(a) Example 1   (b) Example 2   (c) Example 3   (d) Example 4

Fig. 8. Footprint Examples

The criteria listed here can be combined together to give an overall classification of any particular footprint. For compactness, we shall represent such a classification using the abbreviations introduced above. For a footprint F we might present this in the form of a list such as

$$F\{C, R, P, JC, SCC, CED, ADB, \neg NDB, FC\}$$

but when comparing footprints it is convenient to present the data in tabular form as in Table 1, which presents the classifications for the footprints illustrated in Figure 8. In Table 1 the + indicates a true value and − a false.

Note that Example 1 is the convex hull of its dot pattern, and the only properties for which it is not classified as + are [ADB] (all dots on the boundary) and [NDB] (no dots on the boundary). In general, the convex hull of any dot pattern will satisfy the criteria C, P, SCC, ¬NDB, and FC. Unless the dots are collinear, it will also satisfy R and JC. It may or may not satisfy ADB.

4.2   Algorithms

There are three types of algorithm classification criteria, relating to the nature of input expected, the process and the output. For the purposes of this paper we are most interested in the output produced, but we shall first briefly describe the other two types.

With regard to the input there are three key questions to ask: Does the algorithm require pre-processing? Does it require any data on the dots besides the coordinates (e.g. velocity or area)? Does it require an input parameter (e.g., line-length in Swinging Arm [9], α in α-shapes [5])?

The process has two criteria. First, does the algorithm build the footprint incrementally from the dot pattern or does it create the footprint by decrementally removing elements from some initial shape (usually the convex hull)? Second, does the algorithm optimise the footprint with respect to some pre-defined criteria designed to produce some form of best-fit shape (e.g., based on some general idea of what the final shape should look like). This optimisation can be performed in two ways. The first method is to produce a footprint, analyse it and then, if the footprint fails to match some pre-determined criteria, to iterate the algorithm with some internal value changed, as in Concave Hull [13]. The alternative is to examine the footprint to see if performing the next step of the algorithm invalidates some criteria, e.g., the χ-Hull algorithm [4] checks whether removing a line will make the footprint no longer a simple polygon. The two types may be called *remedial* and *preventive* optimisation respectively.

Output

This makes use of the footprint classification but with the added complication that the terms become modalised. There are many possible footprints for any dot set and infinitely many possible dot sets. We cannot predict all the footprints an algorithm can produce, but sometimes we can state whether a particular algorithm can ever produce a footprint with certain properties. Some properties are *necessary* (e.g., the output from Swinging Arm [9] is always polygonal), some are *unconstrained* (e.g., an α-shape [5] may have a cavity, but need not) and some are *impossible* (e.g., an s-shape [3] cannot have all its curvature extrema on dots). We can denote these cases using positive, null and negative respectively. Below, Table 2 shows the value system chosen to represent these.

Table 3 gives a few examples of algorithms as classified by the footprints they produce. As in the footprint examples we begin with the convex hull, using the Jarvis March algorithm. Any convex hull algorithm will be classified in the same way as they all produce the same footprint for a given dot pattern. Note that

Table 2. Possible Modal Values

Value	Description
1	All of the footprints produced by the algorithm satisfy the criterion
0	Some, but not all, of the footprints produced by the algorithm satisfy the criterion
−1	None of the footprints produced by the algorithm satisfy the criterion

Table 3. Algorithm Examples

Algorithm Examples	C	R	P	JC	SCC	CED	ADB	NDB	FC
Jarvis March [11]	1	1^-	1	1^-	1	1	0	−1	1
Swinging Arm Algorithm [9]	0	1^-	1	1^-	1	1	0	−1	1
α-shape [5]	0	1^-	1	1^-	0	1	0	−1	0
Concave Hull [13]	1	1^-	1	1^-	1	1	0	−1	1
DSAM [1]	1	1	1	0	0	−1	−1	1	1

when assessing criteria R and JC, we discount cases where the input dots are all collinear, for which most algorithms will produce a straight line segment as the footprint. If in all other cases the footprint satisfies R or JC then, we shall record this using 1^- here (to mean 'all footprints for non-collinear inputs') rather than 0. Later, when using these values, 1^- will be treated as 1.

In Table 3 it can be seen that the outputs from the Concave Hull algorithm and the Jarvis March algorithm are classified in the same way. However, while they produce the same footprint types by the classification, they may be very different from the point of view of the quality evaluation criteria, as mentioned in §1.

4.3 Application Types

Little work has been done on relating the various methods to application types. It is understandable that there is a desire to abstract the theory away from the application, so that a general method for finding a footprint can be created and used in any circumstance. However, existing methods can produce very different results with a variety of different computational complexities. As such it seems that linking a method to, at least, a general application type would help the understanding of the aims behind each method and when it is best used. But before this can be done the application types have to be classified.

This is meant to be a broad description of the types of applications that these algorithms can be used for, not an in-depth study of each actual application. The classification is laid out with the idea that there are certain types of application and within each type there are optional requirements for each instance.

Pattern Recognition. Finding a particular shape from an image is obviously linked to footprint generation, particularly in a digital field where the dots represent pixels. Within pattern recognition the relevance of associated data is apparent, so it is likely the algorithm will need data for the dots beyond their coordinates, e.g. colour values.

Classification. In this type of application, the dot patterns do not exist in physical space, but represent positions in some abstract quality space, representing the attributes of different entities, each of which is assigned to some known class. The problem is to determine the region of quality space occupied by entities of that class. An example is shown in Figure 9, where instances of two different classes of entity (represented by the red and blue[6] colours) are shown. It is clear that there is a distinct difference between the groupings and a footprint algorithm might be used to find these. There is a presumption that the dots representing entities in the same class will form distinct groupings in the quality space — if this is not the case then that quality space has been inappropriately selected for the classification task.

---

[6] Darker and lighter if viewed in black and white.

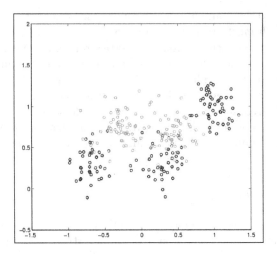

Fig. 9. Application Type: Classification (synthetic data from [6])

Region approximation. Applications in this category are required to generate a geographical region on the basis of sample points within it. We distinguish three subcategories: simplification, estimation, and precisification. This represents a distinction in aims: the methods used in each case could well be the same.

Simplification occurs when the region itself is exactly known, but, perhaps for reasons of storage economy, the generation of a simpler version of its outline from a set of sample points within it is desired. An example might be that of approximating a county boundary from a collection of points known to lie inside it, as exemplified by the methods described in [2].

Estimation occurs when the region has a determinate boundary but this is incompletely known, e.g., if we try to locate the border of a city from the buildings within it. There is a definite boundary to the city but by using the footprint given by dot set (in which the dots represent a large number of the buildings) we might be able to approximate it.

Precisification Here there is no definite region to start with, for example the border of a forest. We have certain tree locations and we can use these to provide a sharp boundary to represent what may in reality be an indeterminate transition zone. The term 'precisification' is used in Supervaluation Theory [7], where it denotes the replacement of a category with incompletely determinate membership by one with fully determinate membership.

Aggregation. This category is concerned with the creation of an aggregate entity from possibly scattered components, e.g., for the purpose of map generalisation, where what appears on a large scale map as a collection of individual buildings may need to be represented as a single connected built-up area when the scale is reduced. This is distinct from region approximation in that there need be no presumption that what is being represented is a region distinct from the buildings themselves; rather, what we want to represent *is* the buildings themselves, but at a coarse level of granularity where they cannot be distinguished as individuals.

With any of these categories one might wish to track changes over time as the entities represented by the dots appear, move, or disappear: in this case the application type becomes *dynamic* rather than *static*. This implicitly means we are treating our data as more than just coordinate locations as we have added a 'history' to each dot. The linkages between footprint types and application types may vary according as the latter are regarded as static or dynamic. However, for the scope of this paper we will restrict ourselves to the static form.

5   Classification Relations

As already mentioned, algorithm types are linked to the footprints they produce. Moreover, application types may be linked to the footprints types they require (§5.1).

5.1   Relating Application Types to Footprints

For a given application type, a particular footprint property may be required, permitted, or forbidden. For this reason we can use the same modal type classification as we used for describing the outputs from footprint algorithms (Table 2). This will also aid in choosing which algorithms can be used for a specific application type. The tables below give our initial thoughts concerning the footprint requirement of the various application types; more detailed analysis will be required for a more definitive statement of these, as described below in §6.

5.2   Application Types to Algorithms

Just as an application type can have requirements on the footprints, it may also be constrained as to what data it can provide to an algorithm. However, this kind of constraint may not be relevant at the level of generality with which we are treating application types here.

There is one algorithm criterion which is of particular interest. Whether or not the dot pattern contains only coordinate data is likely to be heavily affected by making the data dynamic. When running whichever loop is required to allow the algorithm to generate the footprint over the 'frames' that make up the data (live or pre-set) it would be naïve not to keep track of the 'past' associated with each point. Without this the main body of the algorithm needs to be re-run for each frame, instead of using the knowledge of the current footprint and likelihood of movement for each dot to cut down on processing time.

Table 4. Application Types Related to Footprints

Application Type	C	R	P	JC	SCC	CED	ADB	NDB	FC
Pattern Rec.	0	0	0	0	0	0	0	0	1
Classification	1	1	0	1	0	−1	−1	1	1
Simplification	1	1	0	1	0	0	0	−1	0
Estimation	0	1	0	0	0	0	0	0	1
Precisification	0	1	0	0	0	−1	−1	1	0
Aggregation	1	1	0	1	0	0	0	0	1

Algorithm Suitability. By comparing the classification of an algorithm with the application-type classification, we can endeavour to determine the suitability of different algorithms for various applications. Earlier we introduced a value system for our modalities. The footprint criteria were given a value of 1 if it was necessarily true that it holds for all elements, −1 if it was necessarily true that it does not hold for any elements and 0 if the criterion was unconstrained. Application contexts have a vector of these values for their required footprints and the algorithms have vectors for their possible output footprints.

With the goal of having a clear, systematic and quantitative approach we looked at several ways of treating the results. The first method that presented itself was to treat these vectors as points in nine dimensional space, allowing us to find the Euclidean distance between them and thereby assess, numerically, the suitability of the algorithm. As the maximum possible distance between any of the values is 2 we can have a maximum total distance of $\sqrt{9 \cdot 2^2} = 6$. If the two vectors are equal we have the minimum distance of 0. This allows us to present the 'quality' of the algorithm, with regards to the application, on a scale of 0–1, where 1 is best fit and 0 is worst, using the formula $1 - (x/6)$ where x is the distance between the two classification vectors.

However this assumes that all the footprint criteria should be treated as of equal weight. It may well be the case that an algorithm is shown to be close to a best fit but that the criterion it fails is the most important for a specific context. There is also the issue that vectors with many small single value differences are the same distance apart as vectors with few large differences (four 1 value differences to each 2 value). This goes some way to depicting the importance large value differences have over small ones and as a result may be of benefit, but the degree of increase in worth is simply a product of the way the distance is found and therefore may be inappropriate.

Despite these concerns the 'real' issue with the Euclidean distance is how to interpret the result. When the method was used on the previously mentioned algorithms against the application vectors the results were all in the range of 0.38 to 0.76, with most of the values being very close for each application. These close results made it difficult to determine which algorithms would perform better than others, particularly as assigning the qualitative value of 'good' to a strictly quantitative measure seemed presumptious, even pseudoscientific.

Further to the Euclidean distance we looked at using the scalar product and the angle between the vectors, but both these methods suffered from the same problems as the distance.

The facts that the value system indicates that all the criteria are equally weighted and that the resultant values are so similar are clear indicators that the three values may not be enough. One approach to correcting this would be to relate applications to footprints using a continuous scale from -1 to 1, allowing the values to be ranked by importance. This could also be applied to the algorithms: on the original discrete scale an algorithm which is connected in all but the rarest cases would still be marked as a 0, whereas it may be more appropriate to assign a value of, for example, 0.9, indicating that the algorithm mostly produces connected footprints. This continuous scale could also lead to greater differences between the distances so assessing them would be easier. However there is still the problem of judging the result in a systematic manner and the added problem of assigning the original values, for example how do you judge if an algorithm is '0.45' on a scale of producing a regular footprint? Even if you take it as a likelihood for how often an algorithm produces a type of footprint then you have an issue, if an algorithm only produces an irregular footprint when the dots are collinear there are still infinitely many dot patterns which will cause this.

The duality in meaning of 0 is also worth noting. The value 0 can mean two separate things for an application: (1) the algorithm is required to be able to produce both extremes; or (2) the algorithm does not care which value is produced. Thus if the application vector has 0 indicating 'uncaring' in a certain position, this should be allowed as a good match with any of 1, 0, and -1 in the algorithm vector, whereas a 0 indicating a requirement of both values should only match to a 0 on the algorithm.

Seeing that there were more than a few issues with the system it seemed sensible to go over some specific application fields and attempt to find their appropriate vectors. In doing this we hoped that the answers to the above issues would present themselves. However after just a few it became apparent that even an application as specific as 'removing outliers from a spatial distribution' gave a vector composed almost entirely of 0's, it often being the case that a dot pattern can be envisioned for each possible value for a criterion. These 'weak' results would indicate that the dot pattern heavily influences the requirements for the application.

Based on the influence the dots have it appears that an exact method for assessing the suitability of the algorithm requires a change in the classification approach. Instead of simply stating that the application always/sometimes/never requires footprints of a given type as output, we might now additionally try to characterise the inputs for which outputs of that type are to be produced, thereby moving towards providing a potential specification for an algorithm satisfying that application. A similar system could be applied to the algorithm, by specifying the types of output associated with different types of input. These two classifications could then be compared to check for suitability. Unfortunately the

Table 5. Assesing Swinging Arm [9] for suitability to the application of removing outliers from a spatial distribution. Special cases on the algorithm: R and JC are -1 when dots are collinear.

	C	R	P	JC	SCC	CED	ADB	NDB	FC
Algorithm:	0	1^-	1	1^-	1	1	0	-1	1
Application:	0	0	$\sim$	0	0	1	0	-1	-1

process for creating such a system is not known. It would require having a complete set of dot pattern classifiers and a way of assigning values to them. Then the system would have to be created by considering which of the dot pattern descriptors affect which footprint classifiers and in what manner. For example; if the minimum and maximum distances between any two points differ largely from the average distances does this affect the likelihood that the footprint will need to be connected? This would also need to be done for the algorithm in terms of how the dot pattern affects the footprint the algorithm produces. It is likely that such a process would be complicated and prone to assigning values as arbitrarily as in the suggestion for continuous values.

After much work it became apparent that we needed to 'tighten' our focus. Our goal could be achieved much more simply and straightforwardly using direct comparison. This still leaves us with the problem of the dual meaning of 0 for the applications, and the fact that an algorithm which will produce one type of footprint except in special circumstances (i.e. collinearity) will strictly be given a value of 0. As such we will re-introduce the 1^- and, if necessary, add -1^+ to both algorithms and applications; these indicate that except in special circumstances, which should be described, the algorithm produces, or the application requires, a value of 1 or -1 respectively. We will also strictly define the 0 on an application as requiring both, if the application has no preference for the result the value is a $\sim$ sign. These extended values allow for very easy comparison, users can concern themselves only with values about which they care. They can also clearly see which special circumstances can occur and decide if they are applicable to their application. An example of the way this could be set out is shown in Table 5.[7] Running through this we can see that the Swinging Arm algorithm [9] fails completely on SCC and FC and only satisfies R and JC if the special cases are likely to come up regularly in the application field. As such we can say that the Swinging Arm algorithm would be unsuitable for this application.

6   Conclusion

Before discussing the additions this work has made to the field the shortcomings must be explained. The values given on the application types are easily

---

[7] For the sake of the example we have made the assumption that, for this application, curvature extrema are required at dots even though this is not stated by the application title.

debatable, and perhaps some of the values could even be changed to their polar opposite while keeping in line with the application descriptions. However, we chose the values while attempting to keep in mind the aim of maintaining a very general view. Even should there be any significant disagreement with the values the goal of this paper is unaffected, our aim being to show the possibility of developing a systematic value system that can be used to rate algorithms against applications.

We noted in §3 that current literature has little to say about the types of footprint generated or required. We advocate the use of a classification system along the lines presented here, though we expect that further work will refine the details considerably. There is also a noticeable deficiency with regard to applying the algorithms to any applications. Even from the very general definitions given in this paper it should be apparent that the contexts can differ hugely and may have conflicting requirements. Given this diversity of applications, it seems strange to present a footprint algorithm without linking it to an application context, without which little can be said about the suitability of the footprints generated. The disparity also means a truly general algorithm is unlikely. The algorithms considered here were largely produced without specific applications in mind; it is interesting to speculate how different they might have been had they been created with particular contexts in mind.

Aside from the above mentioned shortfalls the paper presents a systematic way of rating the appropriateness of algorithms for applications by classifying the footprints they create. This rating uses a clear nomenclature which is easily repeatable and therefore usable by others. In further work we plan to refine the list of criteria, to examine the dot pattern types more closely, and to consider how input and process criteria for the algorithms relate to application types.

References

[1] Alani, H., Jones, C.B., Tudhope, D.: Voronoi-based region approximation for geographical information retrieval with gazetteers. International Journal of Geographical Information Science 15(4), 287–306 (2001)

[2] Arampatzis, A., van Kreveld, M., Reinacher, I., Jones, C.B., Vaid, S., Clough, P., Joho, H., Sanderson, M.: Web-based delineation of imprecise regions. In: Computers, Environment and Urban Systems, vol. 30, pp. 436–459. Elsevier, Amsterdam (2006)

[3] Ray Chaudhuri, A., Chaudhuri, B.B., Parui, S.K.: A novel approach to computation of the shape of a dot pattern and extraction of its perceptual border. In: Computer Vision and Image Understanding, vol. 68, pp. 257–275. Academic Press, London (1997)

[4] Duckham, M., Kulik, L., Worboys, M., Galton, A.: Efficient generation of simple polygons for characterizing the shape of a set of points in the plane. In: Pattern Recognition, vol. 41, pp. 3224–3236. Elsevier, Amsterdam (2008)

[5] Edelsbrunner, H., Kirkpatrick, D.G., Seidel, R.: On the shape of a set of points in the plane. In: Computer Vision and Image Understanding, vol. IT-29, pp. 551–559. IEEE, Los Alamitos (1983)

[6] Fieldsend, J.E., Bailey, T.C., Everson, R.M., Krzanowski, W.J., Partridge, D., Schetinin, V.: Bayesian inductively learned modules for safety critical systems. In: Computing Science and Statistics, vol. 35, pp. 110–125 (2003)

[7] Fine, K.: Vagueness, truth and logic. Synthese 30(3-4), 265–300 (1975)

[8] Galton, A.: Pareto-optimality of cognitively preferred polygonal hulls for dot patterns. In: Freksa, C., Newcombe, N.S., Gärdenfors, P., Wölfl, S. (eds.) Spatial Cognition VI. LNCS, vol. 5248, pp. 409–425. Springer, Heidelberg (2008)

[9] Galton, A., Duckham, M.: What is the region occupied by a set of points? In: Raubal, M., Miller, H.J., Frank, A.U., Goodchild, M.F. (eds.) GIScience 2006. LNCS, vol. 4197, pp. 81–98. Springer, Heidelberg (2006)

[10] Garai, G., Chaudhuri, B.B.: A split and merge procedure for polygonal border detection of dot pattern. In: Image and Vision Computing, vol. 17, pp. 75–82. Elsevier, Amsterdam (1999)

[11] Jarvis, R.A.: On the identification of the convex hull of a finite set of points in the plane. In: Information Processing Letters, vol. 2, pp. 18–21. North-Holland Publishing Company, Amsterdam (1973)

[12] Klette, R., Rosenfeld, A.: Digital Geometry: Geometric Methods for Digital Picture Analysis. Morgan Kaufmann, San Francisco (2004)

[13] Moreira, A., Santos, M.Y.: Concave hull: A k-nearest neighbours approach for the computation of the region occupied by a set of points. In: International Conference on Computer Graphics Theory and Applications GRAPP (2007)

Mental Tectonics - Rendering Consistent μMaps

Falko Schmid

Transregional Collaborative Research Center SFB/TR 8 Spatial Cognition,
University of Bremen, P.O. Box 330 440, 28334 Bremen, Germany
`schmid@sfbtr8.uni-bremen.de`

Abstract. The visualization of spatial information for wayfinding assistance requires a substantial amount of display area. Depending on the particular route, even large screens can be insufficient to visualize all information at once and in a scale such that users can understand the specific course of the route and its spatial context. Personalized wayfinding maps, such as μMaps are a possible solution for small displays: they explicitly consider the prior knowledge of a user with the environment and tailor maps toward it. The resulting schematic maps require substantially less space due to the knowledge based visual information reduction. In this paper we extend and improve the underlying algorithms of μMaps to enable efficient handling of fragmented user profiles as well as the mapping of fragmented maps. Furthermore we introduce the concept of mental tectonics, a process that harmonizes mental conceptual spatial representations with entities of a geographic frame of reference.

1 Introduction

The visualization of geographic data for wayfinding assistance on limited display resources is a demanding task: ideally we have to show the complete route on a level of detail such that all decisive elements (e.g. involved streets, turns, start and destination, landmarks, etc.) are clearly recognizable and easy to recall. Additionally, we have to prevent visual clutter, thus unnecessary visual elements which are known to affect the cognitive processing of visual information [1]. Depending on the area actually covered by a route (and of course depending on the available screen size) it can be hard to display the general course of a route and the details on street level at once. If we have the possibility to display all information at once in a suitable scale, we still have to face the problem of supporting the the cognitive processing of the information, e.g. focusing on the crucial elements of a route, which are typically decision points.

Current mapping services usually choose an output scale which ensures the complete coverage of the queried route on the available target display. If the route is not comparable short and/or the available display exceptionally large, this will typically result in afterward interaction with the generated map: the user will have to zoom-in and zoom-out either to understand the crucial details of the route or to understand the larger spatial context in which it is embedded. It has been shown that this interaction is problematic since both, map size

K. Stewart Hornsby et al. (Eds.): COSIT 2009, LNCS 5756, pp. 245–262, 2009.

and completeness of geographic information is crucial for successful knowledge acquisition and problem solving [2, 3].

A solution for these problems are schematic wayfinding maps; they visualize geographic data explicitly for the wayfinding task by considering cognitive spatial representations [4, 5] and interaction principles. Two examples for schematic maps are LineDrive Maps and Focus Maps. LineDrive Maps [6] reflect an activity based schematization for routes, based on the observation that routes often incorporate long parts where no decision activity (like turning or changing a road) is required. An example is driving for a long period on highways. When we visualize the geographic region in accordingly scale on a map, these parts can require a significant amount of the available limited interface space. Agrawala and Stolte propose to adapt the scale of the particular route elements to the corresponding wayfinding activity: a high degree of required activity will lead to a more detailed view of the involved entities; a low degree of required activity will lead to a highly schematized view. The result is a strip map which needs significantly less display area if the route incorporates large parts with no decisive wayfinding activity.

Focus maps, developed by Zipf and Richter introduce a different form of schematization [7, 8]. The primary aim is not the compression of the visual representation of a route, but to facilitate the extraction and processing of a route and its context from a rich map. FocusMaps highlight the route by schematizing and fading out map features depending on their proximity to the route. I.e., the closer a feature is to the actual route, the higher is its level of detail and the intensity of its color. This concept reflects the observation that a larger spatial context is helpful during wayfinding, but not all spatial regions are of equal interest for the given task.

1.1   Why Maps at All?

In times of GPS navigation, why should we use maps at all? GPS based step-by-step instructions are known to be superior in performance [9]. However, there are increasing indications that step-by-step assistance prevents people from learning the environment; further they advance and amplify an individual feeling of insecurity during navigation. Studies showed that users of turn-by-turn instructions made more stops than map users and direct-experience participants, made larger direction estimation errors, and drew sketch maps with poorer topological accuracy ([9, 10, e.g.]). These are strong indicators that people do not learn the environment properly and seem not to trust the assistance. We are currently at the edge of a technological evolution and can observe a significant change in how people access geographic information: cars are delivered with build in navigation devices, geographic information is accessed via Internet services. So far it is unclear how a possible life-long learning of the environment with rather context-free representations will affect the formation of a mental map.

In contrast to the negative side-effects of turn-by-turn assistance, maps enable people to learn complex configurations of the environment and allow them to navigate without assistance once they learned it [11, e.g.]. To improve future navigation assistance, we will have to reconsider the communication of route information. We will have to create a sense of place, an awareness for spatial context beyond the route (similarly to maps) within a reduced representation. A promising approach is the combination of the effectiveness of GPS-based assistance and the individual enabling of map based representations. However, displaying a dot on map is not enough, as we still have to consider the visualization problem for geographic data on small displays.

1.2   The Difficulty of Transforming Geographic Information

A fundamental set of human activity depends on a visualization of our environment. Tasks like wayfinding, spatial planning or thematic information visualization require conceptually veridical visualization of spatial information, which can require large display areas (just think of a detailed map of a city, a country, or even the world). In the context of this work the term *veridical* has to be understood as *geographically truthful*, the correct correspondence between represented entities and the entities within a representation. This demand is increasingly problematic as the the access to spatial data is currently migrating to mobile devices. I.e., we will have to adapt the visualization of geographic information for small screens by focusing on the specific task to be solved [12]. However, the algorithmic transformation of geographic data is computationally a hard task. Geographic data is in a fragile equilibrium: there are many implicit and explicit constraints which have to be considered in order to keep the results consistent with the real world. Straightening a curvy road might disturb topological relations of other entities (e.g. a building can be placed at the wrong side of the road afterward). Altering the size of selected elements can have similar effects (a region can suddenly contain more or less elements as in the real world). The aggregation of features or to omit features from the real world can cause semantic conflicts. This means, we cannot just demagnify some elements and magnify others – we will always have to check the consistency of all visible elements.

Mental Conceptual Consistency: At this point we have to distinguish between spatial/geographic consistency and mental conceptual consistency. *Spatial* consistency describes mutual configurational correctness between represented and real entities: all constraints, e.g. topology and size have to be satisfied relatively to each other. A survey map like a general city map is spatially consistent as it depicts all elements in their relative correct dimensions and orientations. In contrast to that, conceptual consistency describes the mutual correspondency between *mental* concepts of constraints amongst features and the elements of real world. I.e., conceptual consistency allows for explicit distortion of spatial constraints if they reflect human conceptual primitives and/or systematic distortions in the mental map (for an overview see [13]): the visual representation can be still understood as the distortions meet the conceptualization of and the expectations of the user in the real environment. Examples for maps of these kinds are the schematic maps discussed in Section 1.

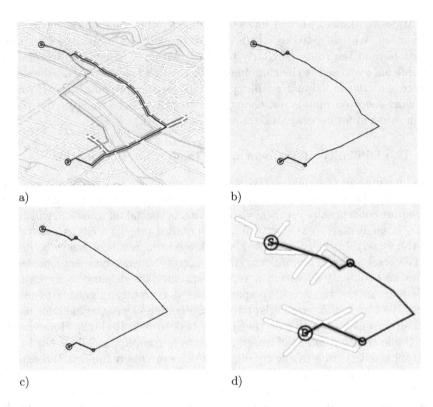

a)   b)

c)   d)

Fig. 1. Generating a μMap: The map in a) shows the map annotated with previous knowledge (bold magenta/dark gray lines), the shortest path from S to D (red/light gray), and the path across prior knowledge (black). In b) we see the unfamiliar segments of the route connected by the path across previous knowledge. In c) the prior knowledge path is schematized. In d) the familiar, schematized path is minimized by means of the convex hull distance, and the unfamiliar segments of the route are moved towards each other. The final μMap in d) is significantly smaller than the original map in a).

2   Personalized Mapping with μMaps

One promising approach to effectively compress the visualization of geographic data is to tailor maps to the individual prior spatial knowledge of an user. The idea is following: with available technology (like GPS enabled mobile phones) it is possible to record and analyze the trajectories of an user and to constantly build-up a spatial user model [14, 15] consisting of the set of historically visited personally meaningful places and frequently traveled paths (the *prior knowledge* as referred to in the following). This user model serves as an input for the generation of e.g. μMaps [16]: if a user queries a route to an unknown destination (or from an unknown origin), the route planning incorporates the prior knowledge and tries to compute the course of the route along known elements. μMaps only display the unfamiliar segments of the route in full detail, the familiar segments are highly schematized and minimized. If a significant part of the actual route can be directed across familiar parts of the environment, the map can be compressed to only a fraction of the size required by traditional maps. Another benefit of μMaps is the abnegation of assistance where it is not required: the user is not cluttered with unnecessary information, and new knowledge is always related to existing knowledge (which facilitates spatial learning). The identified routes are cognitively "lightweight": as the user knows the familiar segment of the route, these parts of the route do not introduce additional decision points. In the following we will use the term *familiar segments* when we refer to the familiar parts of the environment incorporated by a route; likewise we will refer to the unfamiliar parts as *unfamiliar segments*. Unfamiliar segments have to be understood as the unfamiliar part of the route plus additional contextual information (e.g. parts of the street network, see Figures 1, 2, 7).

2.1 Routing Across Knowledge Fragments

Routing Across Coherent Knowledge: The current algorithmic framework for generating μMaps considers coherent user profiles generated from idealistic trajectories. I.e., it assumes complete and error-free sensory information. However, especially GPS enabled mobile devices are known for noisy and fragmented data acquisition. Their handling is contrary to the requirements of a GPS device: they are usually carried at places with weak signal reception, e.g. in jackets, trouser pockets, or bags. This massively reduces the quality of the received signals and causes signal loss and therewith data loss. If the user moves while the device has no reception, it will result in data-gaps in the trajectory and finally leads to the fragmentation of the captured prior knowledge. The previously proposed route

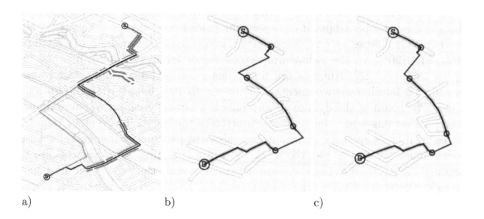

a)   b)   c)

Fig. 2. Fragmented μMap. The map in a) shows the street network annotated with fragmented prior knowledge (magenta/dark gray), the shortest path between S and D (red/light gray), and the path across fragmented prior knowledge (black). b) shows the corresponding μMap, c) the chorematized μMap (see Section 3.2).

search algorithm implemented an explicit planning strategy by trying to minimize the unfamiliar segments of an route and at the same time maximizing the familiar segments (see [16]). This implies the identification of plausible shortest paths from the starting point toward, across, and from the prior knowledge to the destination, by at the same time optimizing the cumulative length of the route. This procedure ensures the overall shortest route for the given policy: maximizing the familiar parts and minimizing the visual output, ensuring the smallest possible maps. However this procedure is costly as the familiar segment of the route can be accessed at n points (depending on the number of topological street network vertices), which affects the length across the prior knowledge. Furthermore it does not integrate into existing routing frameworks easily. Similarly to the route search algorithm, the basic rendering algorithm was limited to basic configurations of familiar and unfamiliar segments.

Routing Across Fragmented Knowledge: To improve the mentioned points, we developed a Dijkstra [17] based algorithm, which is illustrated in the following. In the geo-data corresponding with the search for an optimal route, we require the edges to be annotated with a familiarity measure. The annotation itself takes place during map-matching, i.e. when the positioning information is matched with the geo-data. When the currently traveled street is identified, the corresponding topological edges of the underlying street network data are annotated as *known* to a certain degree. When the user now queries for a route between any places P_1, P_2, each familiar edge E_i is weighted differently to unfamiliar edges. Instead of using the geographic distance as the crucial weight, we implement a dynamic reduction factor d: to enforce the algorithm to enter prior known edges, we increase d temporally up to 50% of the geographic distance of the incorporated edges, i.e. we will virtually shorten them by the value of d. Edges which have not been priorly visited are not altered, thus their actual geographic lenght is used for the computation. As a result the algorithm prefers the assumed shorter edges, if incorporating prior knowledge is an option at all. As soon as it enteres an annotated vertex, we decrease d stepwise by 10% (thus, 40%, 30%, 20%) to the behavioral detour factor for in-situ route planning of up to 10%, see [18]. Humans are no perfect route planners and select routes in complex familiar environments that are up to 10% longer than the optimal route. The result of the dynamic shortening is a virtually deformed environment (the familiar edges are shorter), which attracts the algorithm to enter familiar edges by at the same time guaranteeing a shortest path under human behavioral route choice heuristics. I.e., the identified route is assured in the worst case to be only slightly longer than a route a human would select (only the first four steps of the shortening produce longer paths). The runtime complexity of the underlying Dijkstra is not affected as the distance modifications can be processed in linear time. An additional benefit is the implicit routing across fragmented knowledge: if the incorporation of multiple fragments is geographically plausible, the algorithm will prefer the selection of them, otherwise they will be not or only partially integrated (see e.g. Figure 7a).

Algorithm 1
COMP-FRAG-μMAP$(G, R, dist)$

| **Input** | : A graph G consisting of vertices and edges of the street network, a route R consisting of vertices in G, and the distance threshold $dist$ ensuring that functional components will be visually separated. |
| **Output** | : Will return the μMap for R |

1 $vec \leftarrow \emptyset$
2 $C \leftarrow \emptyset$
3 $S \leftarrow$ segment route R into familiar/unfamiliar segments s_i
4 **forall** $s_i \in S$ **do**
5   **if** $s_i \equiv unfamiliar$ **then**
6        $s_i \leftarrow$ extend street network around vertices of $s_i \in G$
7   **else**
8        $s_i \leftarrow$ schematize(s_i)
9     $c_i \leftarrow$ getConvexHull(s_i)
10    $C \leftarrow$ add$(C,\ c_i)$
11 **forall** $s_{i>1} \in S$ **do**
12   **if** $s_i \equiv familiar$ **then**
13        $c_{pre} \leftarrow$ getConvexHull$(\{c_1 \cup, ..., \cup c_{i-1}\} \in C)$
14        $c_{suc} \leftarrow$ getConvexHull$(\{c_{i+1} \cup c_{i+2} \cup, ..., \cup c_n\} \in C)$
15        $s_i \leftarrow$ scale s_i with the maximal possible minimization factor according to $dist$ between c_{pre} and c_{suc}
16        $vec \leftarrow$ get displacement vector for the minimization factor
17   **forall** $(s_j \equiv c_j \in S) \vee c_j \in c_{suc}$ **do**
18           $s_j, c_j \leftarrow$ translate elements with $vec \times s_j, vec \times c_j$
19 **return** $\{s_1 \cup ... \cup s_n\} \in S$

2.2 Rendering Fragmented Routes

As μMaps are visual representations of the environment, we have to consider principle rendering issues. The rendering of μMaps across fragmented knowledge works in its core as follows. The familiar segments between unfamiliar segments of the environment are minimized by a convex hull based distance optimization: by computing a convex hull around each unfamiliar segment and minimizes the distance between their two closest points (the unfamiliar segments are moved towards each other, see Figure 1d). This method ensures geographic veridicality, as it preserves the spatial relationships amongst the unfamiliar segments. We now extend the basic algorithm, in order to treat fragmented routes, as discussed above. Figure 3 illustrates the functional core of Algorithm 1, Figure 2 shows a generated example. In the following we give a detailed description of COMP-FRAG-ROUTE, the algorithm to compute the schematization of μMaps for fragmented prior knowledge (see Algorithm 1).

    The algorithm requires the street network graph G, the route R, and the distance threshold $dist$, the minimal distance to be kept between all visual

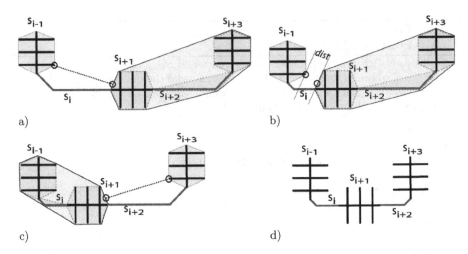

Fig. 3. Fragmented μMizing: a) illustrates the three *unfamiliar* segments $s_{i-1}, s_{i+1}, s_{i+3}$ and the *familiar* segments s_i, s_{i+2}. s_i will be minimized, all other segments are surrounded with their convex hulls. The green/light gray region is c_{suc}. The dashed line between the two circles is the shortest distance between c_{pre} the convex hull of s_{i-1} and c_{suc}. In b) s_i is minimized to the distance threshold *dist* between c_{pre}, c_{suc}. In c) s_{i+2} is minimized, the green region is c_{pre}, c_{suc} is the convex hull around s_{i+3}. d) illustrates the result of the fragmented minimization.

elements. We initialize the algorithm by creating a variable for the displacement vector *vec*, C the list of all convex hulls for the list of segments S of R.

We segment R into familiar and unfamiliar segments $s_0, ..., s_i$ and store them in S (steps 1-3). A segment consists of a list of vertices of R. For each s_i, independent if it is familiar or unfamiliar we compute the convex hull c_1 and store it in C (steps 4-10). We require the convex hull to check the consistency of topological constraints between the segments. Figure 3a illustrates the convex hulls as gray regions around the familiar and unfamiliar segments of the route (the convex hull of s_i is not shown as it is not required in this step). For each familiar s_i in S we compute two complex convex hulls (steps 11-14) $c_{pre} = \{c_0 \cup, ... \cup c_{i-1}\}$ and $c_{suc} = \{c_{i+1} \cup, ... \cup c_n\}$ (s_i and c_i are to be interpreted as corresponding objects). In Figure 3a c_{pre} is identical with the convex hull of the unfamiliar segment s_{i-1}, c_{suc} is the illustrated yellow region. Only the corresponding convex hull c_1 of the familiar segment s_i is not a member of either c_{pre} or c_{suc}. c_{pre} contains all convex hulls of the segments of R before the current segment s_i; c_{pre} the convex hulls of the remaining segments after s_i. When we minimize a s_i, we have to ensure that no other functional element of the map is interfered. I.e., we have to avoid touching or intersections of elements as an unwanted side-effect of the schematization. The convex hulls serve as a approximation of the shape of the segments of the map, and *dist* serves as the distance-to-keep between the convex hulls. In order to minimize s_i we compute the minimal possible distance according to *dist* between the closest points between c_{pre}, c_{suc}. We apply the minimization factor to all elements in s_i and apply the corresponding displacement vector vec to all elements in c_{suc} (steps 15-18) (see also Figure 1). Figure 3a shows the two closest points between c_{pre} and c_{suc} (dashed line). Figure 3b illustrates the result of the operation.

3   Mental Tectonics - Supporting Mental Prototypical Configurations

As denoted in 1.2, human spatial memory is not a veridical representation of the real world (see e.g., [13]). There is a number of systematic distortions introduced by the mental conceptual processing and encoding of spatial aspects, like the representation of direction concepts. When people are asked to draw sketch maps of routes or to verbalize them, they discretize the angular information to a high degree. Instead of drawing or verbalizing precise angles, they make use of prototypical patterns in both language and drawing. They say "turn left" instead of "turn for 281 °" or draw a 270 ° angle instead of the 281 ° angle. It is assumed that the turn-based encoding of the environment is responsible for this effect: when we learn an environment from navigating through it, we have to take a series of turning at decision points (e.g. we turn "left" or "right" at an intersection), but at the same time we have a very limited vocabulary for describing these actions.

Based on these observations, Tversky an Lee proposed verbal and pictorial toolkits for generating route directions and maps [20]. The idea behind these toolkits is to support the mental encoding of turns appropriately with matching

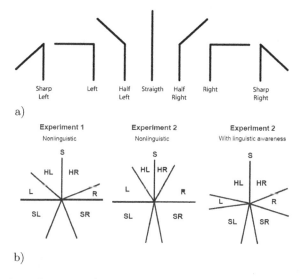

a)

b)

Fig. 4. Wayfinding choremes: a) shows the seven wayfinding choremes. b) The discretization sectors as described in [19], (Figure taken from [19]).

external representations. Klippel further formalized the mental conceptualization of turning actions [21] with wayfinding choremes as prototypical representations as depicted in figure 4a. In [19] the authors identified different sector models for the mental discretization of angles at intersections into choremes. Depending if the situation implied communication (linguistic awareness) or not, the sectors to organize a set of turns significantly differed (see Figure 4b).

The conceptualization of turns has direct implications on how relations between entities of the environment are stored in the mental map and later recalled. When we travel along a route, we encode and store the incorporated turns by the corresponding wayfinding choreme. When we later recall a route, we recall the chorematized route instead of the route with real angular information. As a consequence, all involved elements (e.g. places, streets, landmarks) will be rearranged to be consistent with the currently processed route; the route and the mental locations of its elements are then prototypically arranged.

3.1   Mental Tectonics

Our goal is to visualize the familiar segments of the route as they are mentally represented and to arrange the unfamiliar segments according to them. In [22] the authors describe how to chorematize a strip map by means of using prototypical angles at intersections instead of the real angles, the so-called *chorematization*. Chorematization can easily result in maps violating the real spatial relations between origin and the destination; depending on the specific sequence of turns, the chorematized course of the route can be far from the real situation. However, as we relate to mental concepts and communicate new knowledge (which will be stored in the mental map), we have to provide geographic veridicality, i.e. a truthful allocentric configuration which is consistent with the real spatial situation. The placement of the new elements has to be carefully balanced between conceptual mental arrangements and real spatial constraints – a process we term *Mental Tectonics*. Otherwise the map will not meet the expectations and will introduce substantial distortions in the mental map.

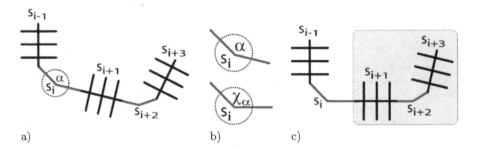

Fig. 5. Chorematizing a μMap: illustration a) depicts the original situation. s_i, s_{i+2} are *familiar* segments, all others are *unfamiliar*. b): α is replaced by the corresponding choreme χ_α. c) depicts the result of the chorematization step. Note the rearrangement of all entities in the gray region.

Mental Tectonics is a method to generate personally meaningful µMaps with the aim to communicate survey knowledge. In order to facilitate the correspondency between the mental representation of the familiar segments, we represent the decision points along the route by means of wayfinding choremes (see Figure 5). However, in contrast to strip maps, µMaps visualize multiple complex parts of an environment; those parts need to be anchored correctly within the environment to preserve the correct mental embedding. Otherwise, due to the specific sequence of turns, different routes would relate the same parts of the environment mutually differently to each other within the same allocentric frame of reference (e.g. one route relates the segment A as "North" of B, a different route across the same segments A, B could relate A "South" of B).

3.2   Computing Mental Tectonics

The computation of cognitively and at the same time geographically veridical µMaps works as described in the following. First of all we require a sound discretization function for angles at intersection. As the discretization of angles is obviously context dependent (see [19]), and as it is unclear how the available empirical results scale to the mental encoding of familiar paths, in our implementation we used a pragmatic definition of the sector model. Especially the sector for "Straight" is in the empirical results under all conditions organized very strict. As this is contrary to our everyday experience where "Straight" is a more flexible relation, we decided to extend "Straight" to a sector of $30\,°$ ($\pm 15\,°$ around $0\,°$ (in an egocentric frame of reference), see Figure 6a for details.

The algorithm requires a route R, consisting of the familiar and unfamiliar segments and the contextual extensions of the unfamiliar segments (e.g. parts of the adjacent street network). Furthermore we require the set of choreme mappings X which allow to map a given angle to a respective choreme, as well as the error correction parameters e expressed as sectors around the borders of the choreme sectors (see Figure 6a+b). This parameter enforces geographic compensation at an early stage: whenever the global error plus the error of the current replacement results are above or below e, we select the neighbored and also plausible choreme to minimize the error. This method prevents the change of the configurational concept between the unfamiliar segments (within the egocentric reference frame of the wayfinding choremes). In our implementation we use a sector of $25\,°$ relative to the border as , thus $\pm 12.5\,°$ around the border. I.e., the global error is reduced to $\|\chi_{\sphericalangle max}\| - \frac{e}{2}$ ($\|\chi_{\sphericalangle max}\|$ denotes the maximal angular sector of X, in our implementation the sectors of "Half-Left" and "Half-Right").

We initialize the accumulative global error variable $globalErr$ with 0 (we do not have a displacement error so far), we further need to keep track of the local error, this will be represented by $localErr$ later in the algorithm. We further split R into its familiar and unfamiliar segments and store them in the list S (steps 1-2). When we replace each angle along a route by a choreme, the accumulated displacement can be arbitrarily high (depending on the layout of the route): each replacement introduces a specific rotation and transition to the remaining parts.

Algorithm 2
COMP-MENTAL-TECTONICS(R, X, e)

Input	: A route R consisting of vertices v, and the set of wayfinding
	choremes mappings X, and the global error reduction parameter e
Output	: The chorematized and geographically veridical R

1 $globalErr \leftarrow 0$
2 $S \leftarrow$ segment route R into familiar/unfamiliar segments s_i
3 **forall** $s_i \in S$ **do**
4   **if** $s_i \equiv familiar$ **then**
5          $s_i \leftarrow$ schematize(s_i)
6   **forall** $v_j \in s_i, v > 1$ **do**
7              $\alpha \leftarrow$ getAngle(v_{j-1}, v_j, v_{j+1})
8              $\chi_\alpha \leftarrow$ get corresponding choreme $\chi_k \in X$
9   **if** $\chi_\alpha < 0$ **then**
10                 $\chi_\alpha \leftarrow \|\chi_\alpha\|$
11             $localErr \leftarrow (\chi_\alpha - \|\alpha\|)$
12   **if** $localErr > 0 \wedge globalErr + localErr \geq e$ **then**
13                 $\chi_\alpha \leftarrow$ get χ_{k-1}
14                 $localErr \leftarrow (\chi_\alpha - \|\alpha\|)$
15   **else if** $localErr < 0 \wedge globalErr + localErr \leq e$ **then**
16                 $\chi_\alpha \leftarrow$ get χ_{k+1}
17                 $localErr \leftarrow (\chi_\alpha - \|\alpha\|)$
18             $globalErr \leftarrow globalErr + localError$
19             $vec \leftarrow$ compute rotation and displacement vector for χ_α
20   **forall** $v_{j>1} \in s_i \wedge s_{k>i} \in S$ **do**
21                 $v_{j>1}, s_k \leftarrow vec \times v_{j+2}, vec \times s_k;$

22 **return** $\{s_1 \cup ... \cup s_n\} \in S$

To tackle this problem, first of all we limit the numbers of chorematized turns to the most significant ones in the familiar segments by schematizing the path before chorematization (steps 3-5).

We now iterate through the remaining vertices $v_j \in s_i$. We compute the egocentric angle α from the vertices v_{j-1}, v_j, v_{j+1}. We select the corresponding choreme χ_k for α and compute the local error by subtracting α from χ_k (steps 6-9). Note that we express the error in deviation from the actual choreme by using absolute values for α and the choreme χ_α if it is smaller than 0. If the angle is exactly the angle of the choreme, the introduced error is 0. Otherwise it is the positive or negative difference (see Figure 6a+b). If the local error is 0, we apply the choreme without further checking, as the current global error is not affected. Otherwise we select the choreme $\chi_{k\pm1}$ to minimize the overall error (local + global) effectively. + refers to the choreme in positive direction, - in negative direction (see Figure 6a+b). $globalErr$ is updated accordingly (steps 10-14). $globalErr$ expresses the error as the deviation from the original geographic configuration and uses the reference frame of the egocentric choremes to optimize

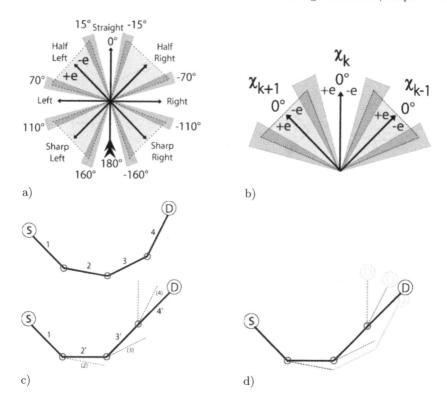

a)

b)

c)

d)

Fig. 6. Error minimization: a) the sector model with discretization borders in degrees relative to the egocentric frame of reference; the dark gray sectors illustrate the error correction parameter e around the choreme sector borders. b) if an angle is exactly the angle of the choreme χ_k, the error is 0. If it is larger than $+e$ or smaller than $-e$ the neighbored choreme $\chi_{k\pm1}$ is selected to minimize the global error. c) and d) see text in section 3.2 for details.

it. After replacing the choreme, we apply the actual displacement caused by replacement of for α with χ_α to all concerned elements (15-16). Finally, if all segments are treated accordingly, we return the route.

The route, or rather the map, still needs to be minimized to be a μMap. Obviously, we can nest Algorithm 2 within Algorithm 1. However the order of the execution is important: the chorematization should be computed prior to the minimization as the minimization requires the final layout as an input; Mental Tectonics modifies the layout of the route and implicitly changes the spatial constraints between the segments of the route.

Figure 6c and d) illustrates the compensation: in c, the top image shows the original path from S to D. In the bottom illustration we see the corresponding chorematized path (bold lines) and the original course (dashed lines). The accumulated error is the difference between the original edges and the chorematized edges. The error introduced with edge 4 is larger than e, thus we select

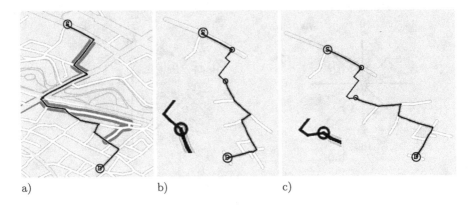

a)   b)   c)

Fig. 7. Corrected μMap. The map in a) shows the street network annotated with prior knowledge (magenta lines/dark gray in b/w), the shortest path between S and D (red/light gray in b/w), and the path across fragmented prior knowledge (black). Note that the computation of the route did not enforce a path via prior knowledge (see further Section 2.1). b) shows the corresponding chorematized μMap. The resulting map violates geographic veridicality: the destination environment is located "South" of the start environment (cardinal directions). c) shows the corrected but still chorematized μMap. The destination environment is now veridically arranged "South-East" of the start environment (cardinal directions). Note the embedded enlarged parts of the crucial decision points.

the choreme that minimizes the error most effectively; in this case we select "Straight" instead of "Half-Left" (dashed bold gray line). in d) the overlay of the routes illustrates the effectiveness: although all edges are replaced by choremes, the geographic veridicality is preserved. D (dark circle) still meets the geographic constraints as no change of concept of D within the cardinal and the egocentric reference frame was enforced; compare with the alternative locations (light gray circles). Figure 7a-c shows a μMap generated with and without using Mental Tectonics. Compare the original situation in a), the chorematization without the correction in b), and c) the geographic veridical chorematization.

4   Discussion

When μMaps route across highly fragmented prior knowledge they can (depending on the route) generate a high number of familiar and unfamiliar segments. I.e., we have to identify a plausible number of fragments, and a minimum size for a familiar fragment. If a fragment is too small (e.g. only one street segment between two intersections), visualizing it will possibly cause more cognitive load than concealing its existence. Each transition from an familiar segment to an unfamiliar segment of a route (and vice versa) implicitly means the switch of the frame of reference: from the geographic frame of reference (unfamiliar segments) to the personal frame of reference (familiar segments). Each switch implies a certain cognitive load and will influence the effectiveness of the maps. In his classic publication Miller [23] proposes "7±2" information chunks as a general guideline for information representation, as this number demarcates a processing limit (which is of course not applicable to all domains similarly). If we follow this proposal as a rule of thumb and interpret the familiar and unfamiliar segments of the route each as chunks, we can integrate 2-3 familiar segments within the route. This depends on the actual configuration of familiar and unfamiliar segments; however, as between each familiar part must be one unfamiliar part, we have either 3-5 overall fragments when we allow 2 familiar segments, or 5-7 segments if we allow 3 familiar segments. The communication and processing of the change of the frame of reference can be enhanced by integrating personally meaningful anchor points of the personal frame of reference within the familiar segments of μMaps. In [24] we demonstrate the enhanced accessibility of personalized maps by integrating a selection of personally meaningful places according to their relevance for the course of the route (places are selected when they are highly familiar and/or located at significant locations along the route). They serve as a key to the compressed representation of μMaps and help to understand the scale and the actual course of the route.

The rendering of the familiar segments of the route benefits from chorematization as it supports the mental processing. However, this practice can disturb geographic relations up to a high degree. We showed that it is possible to use choremes to describe the route by at the same time preserving geographic consistency. Our proposed approach is only a local optimization and does not consider the global composition of angles along the route. A global approach could select specific decision points to control the effect of the correction.

The degree of schematization has great influence on the remaining geometry of a segment and is largely responsible for the introduced error after chorematization. A very strict schematization will result in a straight line, a moderate one will keep eventually too much details. The automatic identification of a suitable parameter is a hard task, as we usually have no operationizable parameters at hand. The preservation of the geographic veridicality after chorematization could be such a parameter: we can select the degree of schematization in dependency of the global error after the chorematization of the schematized segments. We found a suitable granularity if the global chorematization error is minimized.

Further, we implemented one fixed error correction for all choremes. This parameter could be more adaptive to the relationships among choremes: so far we implied that users will accept selection of a neighbored concept uniformly. However, the studies of [19] clearly show that the borders between choremes are varying, which also indicates context dependent flexibility in accepting changes of concepts (as different conditions resulted in different discretization sectors).

5   Conclusions and Outlook

The visualization of geographic data for wayfinding assistance on small displays is a critical task. Either the map has a very small scale and is hard to read, or the information is only partially visible. This leads to a substantially increased cognitive effort to process and to successfully understand the presented information. A possible solution are µMaps. µMaps transform the requested geographic space according to the familiarity of an individual user. The individual knowledge can be present in inconsistent or fragmented user profiles. To cope with this problem, we need to extend the existing algorithmic framework of µMaps. We present an Dijkstra based routing algorithm with dynamic weights considering human behavioral route choice heuristics. This results in improved routing and connects fragmented knowledge, if it is geographically and behaviorally plausible.

Based on this algorithm and potentially fragmented routes across several familiar segments, we introduced a rendering algorithm to optimize the representation of a µMaps accordingly. The minimization of familiar links between unfamiliar parts of the environment optimizes the distance locally for every pair of configuration of familiar and unfamiliar segments. Although µMaps explicitly address mental spatial concepts, they do not only have to support the mental representation and processing of familiar routes, but they have to support the mutual geographic veridicality of all spatial elements as well. I.e., they have to support mental conceptual consistency and spatial consistency at the same time. In this paper we introduce a method we call *Mental Tectonics* to balance mental conceptual configurations with geographic veridicality. This allows for relating new spatial knowledge to prior knowledge by integrating it in existing mental layouts. Mental Tectonics qualifies as a general method to integrate chorematized representations into a geographic frame of reference.

The integration of multiple prior knowledge fragments raises the question for the maximal and optimal amount of familiar and unfamiliar segments in a route. Although we can limit the number of allowed fragments pragmatically, only empirical studies will shed light on the cognitive demands and limits of understanding µMaps. A question which definitely requires an answer.

The interplay of route choice, schematization and chorematization has great influence on the layout of the resulting map. When the route incorporates large familiar segments, the applied human route choice heuristic usually allows for the selection of different paths (as it accepts up to 10% detour, which can result in multiple candidates). Since each route has a different layout, the selection of the actual path might not only dependent on a familiarity measure. Especially among equal choices, we could select the path with the best layout properties. An automated adjustment of the crucial parameters of the schematization algorithms under the consideration of geographic veridicality and map compactness would additionally foster the generation of compact and at the same time accessible maps, which is the ultimate goal of µMaps.

Acknowledgments

This research is carried out as part of the Transregional Collaborative Research Center SFB/TR 8 Spatial Cognition. Funding by the Deutsche Forschungsgemeinschaft (DFG) is gratefully acknowledged.

References

[1] Rosenholtz, R., Li, Y., Nakano, L.: Measuring visual clutter. Journal of Vision 7(2), 1–22 (2007)

[2] Dillemuth, J.: Map size matters: Difficulties of small-display map use. In: International Symposium on LBS and TeleCartography, Hong Kong (November 2007)

[3] Tan, D.S., Gergle, D., Scupelli, P., Pausch, R.: Physically large displays improve performance on spatial tasks. ACM Trans. Comput.-Hum. Interact. 13(1), 71–99 (2006)

[4] Freksa, C.: Spatial aspects of task-specific wayfinding maps - a representation-theoretic perspective. In: Gero, J.S., Tversky, B. (eds.) Visual and Spatial Reasoning in Design, University of Sidney, Key Centre of Design Computing and Cognition, pp. 15–32 (1999)

[5] Klippel, A., Richter, K.F., Barkowsky, T., Freksa, C.: The cognitive reality of schematic maps. In: Meng, L., Zipf, A., Reichenbacher, T. (eds.) Map-based Mobile Services - Theories, Methods and Implementations, pp. 57–74. Springer, Berlin (2005)

[6] Agrawala, M., Stolte, C.: Rendering effective route maps: Improving usability through generalization. In: SIGGRAPH 2001, Los Angeles, California, USA, pp. 241–249. ACM Press, New York (2001)

[7] Zipf, A., Richter, K.F.: Using focus maps to ease map reading — developing smart applications for mobile devices. KI Special Issue Spatial Cognition 02(4), 35–37 (2002)

[8] Richter, K.F., Peters, D., Kuhnmünch, G., Schmid, F.: What do focus maps focus on? In: Freksa, C., Newcombe, N.S., Gärdenfors, P., Wölfl, S. (eds.) Spatial Cognition VI. LNCS (LNAI), vol. 5248, pp. 154–170. Springer, Heidelberg (2008)

[9] Parush, A., Ahuvia, S., Erev, I.: Degradation in spatial knowledge acquisition when using automatic navigation systems. In: Winter, S., Duckham, M., Kulik, L., Kuipers, B. (eds.) COSIT 2007. LNCS, vol. 4736, pp. 238–254. Springer, Heidelberg (2007)

[10] Ishikawa, T., Fujiwara, H., Imai, O., Okabe, A.: Wayfinding with a GPS-based mobile navigation system: A comparison with maps and direct experience. Journal of Environmental Psychology 28, 74–82 (2008)

[11] Ishikawa, T., Montello, D.R.: Spatial knowledge acquisition from direct experience in the environment: Individual differences in the development of metric knowledge and the integration of separately learned places. Cognitive Psychology 52, 93–129 (2006)

[12] Reichenbacher, T.: Mobile Cartography: Adaptive Visualization of Geographic Information on Mobile Devices. PhD thesis, University of Munich, Munich, Germany (2004)

[13] Tversky, B.: Distortions in cognitive maps. Geoforum 2(23), 131–138 (1992)

[14] Schmid, F., Richter, K.F.: Extracting places from location data streams. In: UbiGIS 2006 - Second International Workshop on UbiGIS (2006)

[15] Schmid, F.: Formulating, identifying, and analyzing individual spatial knowledge. In: Seventh IEEE International Conference on Data Mining Workshops (ICDMW 2007), vol. 0, pp. 655–660. IEEE Computer Society Press, Los Alamitos (2007)

[16] Schmid, F.: Knowledge based wayfinding maps for small display cartography. Journal of Location Based Services 2(1), 57–83 (2008)

[17] Dijkstra, E.W.: A note on two problems in connexion with graphs. Numerische Mathematik 1, 269–271 (1959)

[18] Wiener, J., Tenbrink, T., Henschel, J., Hoelscher, C.: Situated and prospective path planning: Route choice in an urban environment. In: Proceedings of the CogSci 2008, Washington (2008) (to appear)

[19] Klippel, A., Montello, D.R.: Linguistic and nonlinguistic turn direction concepts. In: Winter, S., Duckham, M., Kulik, L., Kuipers, B. (eds.) COSIT 2007. LNCS, vol. 4736, pp. 354–372. Springer, Heidelberg (2007)

[20] Tversky, B., Lee, P.U.: Pictorial and verbal tools for conveying routes. In: Freksa, C., Mark, D.M. (eds.) COSIT 1999. LNCS, vol. 1661, pp. 51–64. Springer, Heidelberg (1999)

[21] Klippel, A.: Wayfinding choremes. In: Kuhn, W., Worboys, M.F., Timpf, S. (eds.) COSIT 2003. LNCS, vol. 2825, pp. 320–334. Springer, Heidelberg (2003)

[22] Klippel, A., Richter, K.F., Hansen, S.: Wayfinding choreme maps. In: Bres, S., Laurini, R. (eds.) VISUAL 2005. LNCS, vol. 3736, pp. 94–108. Springer, Heidelberg (2006)

[23] Miller, G.A.: The magical number seven, plus or minus two: some limits on our capacity for processing information. Psychological review 101(2), 343–352 (1994)

[24] Schmid, F.: Enhancing the accessibility of maps with personal frames of reference. In: Jacko, J.A. (ed.) INTERACT 2009. LNCS, vol. 5612. Springer, Heidelberg (2009)

To Be and Not To Be:
3-Valued Relations on Graphs

John G. Stell*

School of Computing, University of Leeds
Leeds, LS2 9JT, U.K.
j.g.stell@leeds.ac.uk

Abstract. Spatial information requires models which allow us to answer 'maybe' to questions asking whether a location lies within a region. At the same time, models must account for data at varying levels of detail. Existing theories of fuzzy relations and fuzzy graphs do not support notions of granularity that generalize the successes of rough set theory to rough and fuzzy graphs. This paper presents a new notion of three-valued relation on graphs based on a generalization of the usual concept of three-valued relation on sets. This type of relation is needed to understand granularity for graphs.

1 Introduction

The stark choice – to be or not to be – that faced the Prince of Denmark is not forced upon us in the world of spatial information. Whether a location is among places having some attribute or is not among such places is a choice we have learned to sidestep. Once we admit vagueness, uncertainty and related factors into our spatial models we allow locations which are in a sense both inside and outside regions. The question of being vs not being is no longer a categorial distinction, but has engendered a new possibility: both being and not being.

As Lawvere observes [Law86], this apparent contradiction reveals the presence of a boundary. The algebraic formulation of this boundary was noted as relevant to spatial information in an earlier COSIT by Stell and Worboys [SW97], but for graphs this account was restricted to the crisp case. In the present paper this restriction is removed, and the algebraic framework is generalized to three-valued subgraphs. Building on this, a novel theory of three-valued relations on graphs is developed as the next stage in an ongoing programme of granularity for both crisp and fuzzy graphs.

The structure of the paper is as follows. In Section 2 the basics of three-valued subsets are reviewed. This account is extended to three-valued relations in Section 3, in which we find that the usual definition is inadequate and a more general (fuzzier) notion is required. In Section 4 we consider subgraphs, initially reviewing the algebra of crisp subgraphs and then showing how this can

---

* Supported by the EPSRC (EP/F036019/1) and Ordnance Survey project *Ontological Granularity in Dynamic Geo-Networks*.

K. Stewart Hornsby et al. (Eds.): COSIT 2009, LNCS 5756, pp. 263–279, 2009.

be generalized to three-valued subgraphs. This generalization introduces two apparently novel operations in the lattice of three-valued subgraphs and their expressiveness is demonstrated by examples in this section. Section 5 shows how to define relations on three-valued subgraphs.

2   Three-Valued Sets

Spatial information has as one of its core activities the assignment of locations to propositions and of attributes to locations. While locations may be idealized points at some level of detail, they are more generally regions of space. Such regions divide space into region and non-region making a precise distinction. It has often been observed that such models fail to cope with the richness of human conceptualization of the world in which besides 'in' and 'out' we also have the judgement 'maybe'. This leads to models where we classify entities by three values instead of two. Specific examples include *'flou sets'* [Wyg96], *'egg-yolk regions'* [CG96], and regions with *'broad boundaries'* [CD96], the latter two sources being both in the book [BF96].

First it must be noted that there is no such thing as a three-valued set; there are only three valued subsets. That is, we must have some universe of entities which we value or classify with three values. The three values will be denoted here by the set $\mathbf{3} = \{\top, 0, \bot\}$, where $\top$ should be though of as true, $\bot$ should be thought of a false, and 0 as the middle value 'maybe' or 'possibly'. The values are ordered $\bot < 0 < \top$. There are many other ways to denote the three values in the literature, and Łukasiewicz [Łuk70, p87] originally used $\{0, 2, 1\}$ before settling on $\{0, \frac{1}{2}, 1\}$.

When a set X is given, a $\mathbf{3}$-valued subset of X, or just a $\mathbf{3}$-subset, is a function $A : X \to \mathbf{3}$. The set of all $\mathbf{3}$-subsets of X is denoted $\mathcal{P}_3 X$ and the algebra in this structure gives us a model in which we can talk about parthood ($A \leqslant B$ iff $\forall x \in X \cdot Ax \leqslant Bx$) as well as unions and intersections of $\mathbf{3}$-subsets. The $\mathbf{3}$-subset $x \mapsto \bot$ will be denoted just by $\bot$. Similarly the $\mathbf{3}$-subset which takes the value 0 everywhere is just 0 and likewise $\top$. The most interesting aspect of the structure of $\mathcal{P}_3 X$ is that it supports three distinct notions of outside as shown in Figure 1.

For $A \in \mathcal{P}_3 X$, the $\mathbf{3}$-subset $-A$ is the mapping $x \mapsto -(Ax)$ where $-\top = \bot$, $-0 = 0$, and $-\bot = \top$. The operators $\neg$ and $\dashv$ are the pseudocomplement and its dual, which following Lawvere [Law86], will be called the supplement. These are described in [SW97] in the equivalent context of two-stage sets. The value for theories of spatial information of the operations $-$, $\neg$, and $\dashv$, lies in the way that when combined with each other and with $\vee$ and $\wedge$ they allow the formation of expressions which describe practically useful concepts. Three of these are illustrated in Figure 2. The boundary consists of those points that are both possibly inside and possibly outside the set. The core consists of those places for which it is possible that they are outside the places possibly outside, that is they must be definitely inside the set. The co-core is the double pseudocomplement, and consists of elements possibly not outside the region.

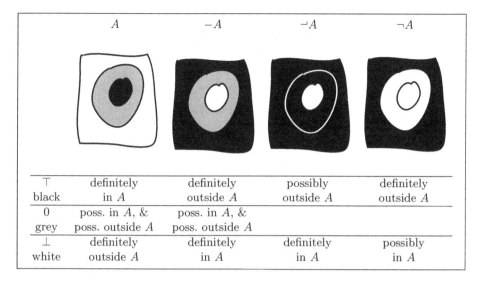

	A	$-A$	$\neg A$	$\neg A$
$\top$ black	definitely in A	definitely outside A	possibly outside A	definitely outside A
0 grey	poss. in A, & poss. outside A	poss. in A, & poss. outside A		
$\bot$ white	definitely outside A	definitely in A	definitely in A	possibly in A

Fig. 1. Three kinds of outside and their interpretation for a **3**-subset

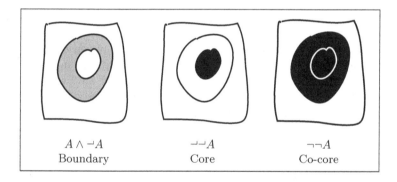

$A \wedge \neg A$	$\neg\neg A$	$\neg\neg A$
Boundary	Core	Co-core

Fig. 2. Boundary, Core and Co-core for a **3**-subset

While the algebraic operations are valuable for these **3**-subsets, the expressiveness becomes more apparent when we apply them to graphs in Section 4. Before doing so, we examine what should be meant be a three-valued relation on a set.

3 Fuzzy Relations: The 3-Valued Case

On a set X, a relation R can be drawn as a set of arrows as in Figure 3. Each arrow in the diagram goes between two elements of X (possibly the same element). The fact that for elements $x, y \in X$ we have $x \, R \, y$ (that is visually that we have an arrow from x to y) can model various situations such as x is similar to y or that x is near to y or that y is accessible from x etc.

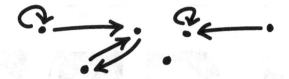

Fig. 3. A crisp relation on a crisp set

Having a relation R on X can be thought of as a viewpoint from which X is seen. The idea here is that instead of the elements $x \in X$ we can only see granules of the form

$$\{x\} \oplus R = \{y \in X \mid x \, R \, y\}$$

where $\oplus$ denotes the morphological dilation operation. This operation is discussed, along with its counterpart erosion, in the context of relations in [Ste07]. We can describe a subset $A \subseteq X$ in terms of these granules in various ways. The most obvious two possibilities being the set of granules wholly within A and the set of granules which intersect A. In the case of R being an equivalence relation, these would be respectively the lower and upper approximations in rough set theory.

It is natural to qualify the extent of the relatedness. For example, we may say that x is very similar to y, but that y is only slightly similar to z. In another example, perhaps a is very near to b but c is not that close to d although we wouldn't want to go so far as to say that c is not in any sense near d. These natural examples suggest we need some kind of fuzzy relations, in which instead of two items either being related or not, we admit some kind of intermediate relatedness. There is a well-established notion of fuzzy relation [Zad65] taking values in the interval $[0,1]$ or more generally in some lattice of truth values Ω. This notion takes an Ω-valued relation on a set X to be simply a function $R : X \times X \to \Omega$.

In the current paper I am only considering the case of three truth values where $\Omega = \{\bot, 0, \top\}$. This means that an Ω valued relation on X can be presented visually by drawing black arrows between elements of X which are related to extent $\top$, and by drawing grey arrows between elements which are only related to the medium extent 0. This is illustrated in Figure 4. When there is no relatedness (or rather the relatedness is given the value $\bot$) we draw no arrows, but we can think of drawing white arrows instead.

Fig. 4. Classical concept of a 3-valued on a crisp set

We shall see shortly that although the idea of a function $X \times X \to \mathbf{3}$ has been very widely accepted as the 'right' notion of a $\mathbf{3}$-valued relation, it is inadequate to model the requirements of a theory of granularity. What a function $R : X \times X \to \mathbf{3}$ can actually do is to take a crisp subset $A \subseteq X$ and produce a $\mathbf{3}$-subset of X. That is, $A \oplus R$ is the function $X \to \mathbf{3}$ where $(A \oplus R)(y)$ is the maximum value in $\{R(a, y) \in \mathbf{3} \mid a \in A\}$.

To give a concrete example, suppose that X is a set of places, and that for each ordered pair of places we have one of three values for how near the first place is to the second. We can think of these as being, *very near*, *somewhat near* and *not near*. Such $\mathbf{3}$-valued nearness relations have been discussed in the literature [DW01] and might arise from the opinion of a single human subject, from data collected from many individuals, or by some form of objective measurement. Given a single place $x \in X$ we can consider the set of places to which x is near. This provides not (in general) a crisp subset of X but a $\mathbf{3}$-valued subset, $\{x\} \oplus R$, in which x is *very near* some places, *somewhat near* others, and *not near* others. That is, for each $y \in X$ the value of $(\{x\} \oplus R)(y)$ is the extent to which x is near y.

This scenario extends readily to any crisp subset $A \subseteq X$, with $(A \oplus R)(y)$ being the greatest extent to which anything in A is near y. However, suppose we start not with a crisp subset of X but with a $\mathbf{3}$-valued subset. For example, we might want to model the nearness of places to a region that was only vaguely defined. To be specific, take a question such as "where is the forest near to?". If we ask this of a variety of subjects we can expect that there will be places x and y where some people will say that x is definitely near y but that it is not clear that x is in the forest. In such a situation we may not be able to say that somewhere in the forest is definitely near to y, because the fuzzy nature of nearness is compounded by the fuzzy nature of the forest itself. This means that we need our fuzzy relations to be able to operate on a fuzzy subset (e.g. the forest) and produce a fuzzy subset as a result (e.g. where the forest is near to). The classical notion of a $\mathbf{3}$-valued relation on X as a function from $X \times X$ to $\mathbf{3}$ is unable to do this, and a fuzzier kind of relation is required.

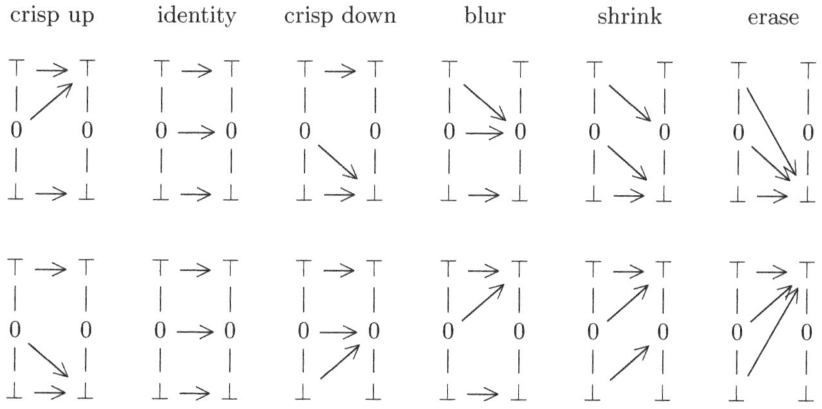

Fig. 5. The six join-preserving mappings $\mathbf{3} \to \mathbf{3}$ and their right adjoints

From an algebraic perspective, the root of the problem is that functions $X \times X \to \mathbf{3}$ are not equivalent to join-preserving functions on $\mathcal{P}_3 X$. This can be seen by taking $X = \{x\}$, in which case there are just three functions $X \times X \to \mathbf{3}$, but there are six elements of $\mathcal{P}_3 X$. These six are obtained from the upper row of Figure 5 by interpreting each of $\top$, 0, and $\bot$ as the value assigned to x.

The names given to the six functions in Figure 5 reflect the operations they induce on $\mathbf{3}$-subsets by composition. For example, if $\alpha : \mathbf{3} \to \mathbf{3}$ is the crisp up function, then for any $A : X \to \mathbf{3}$, the composite $\alpha A : X \to \mathbf{3}$ is the crisp subset of X obtained by marking as definite all the merely possible elements of A. The lower part of Figure 5 shows the right adjoint of each function. This can be thought of informally as a 'best approximation' to an inverse for the function. A brief account of adjunctions between posets (also called Galois connections) can be found in [Ste07] and more details can be found in [Tay99]. In the present context the significance of the adjoints is that they are needed to understand converses of relations.

Having rejected the usual notion of three-valued relation, the fuzzier kind of relation we require is as follows.

Definition 1. *A* $\mathbf{3}$*-relation on a set X is a mapping $R : X \times X \to \mathbf{3}^3$, where* $\mathbf{3}^3$ *is the lattice of join-preserving mappings on* $\mathbf{3}$.

Given $\mathbf{3}$-relations R and S on X their composite is $R \,;\, S$ where

$$x \, R \,;\, S \, z = \bigvee \{(x \, R \, y)(y \, S \, z) \in \mathbf{3}^3 : y \in X\},$$

in which $(x \, R \, y)(y \, S \, z)$ is the composition of join-preserving functions (in the order where $x \, R \, y$ is applied first) and the join is that in the lattice shown in Figure 6.

There is a bijection between $\mathbf{3}$-relations on X and join-preserving functions on $\mathcal{P}_3 X$. To the relation R is associated to the function $\_ \oplus R : \mathcal{P}_3 X \to \mathcal{P}_3 X$ where

$$(A \oplus R)x = \bigvee_{y \in X} (yRx)(Ay).$$

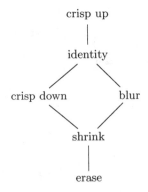

Fig. 6. The lattice of six functions from Figure 5

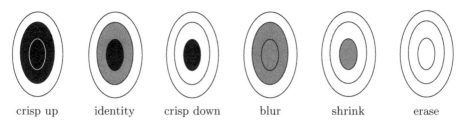

| crisp up | identity | crisp down | blur | shrink | erase |

Fig. 7. The six functions from Figure 5 as egg-yolks

This provides the dilation operation, and erosion is obtained as its right adjoint. To the function $\alpha : \mathcal{P}_3 X \rightarrow \mathcal{P}_3 X$ is associated the relation R where

$$(x\ R\ y)\omega = (\alpha(\{x\}_\omega))y,$$

and where for $\omega \in \mathbf{3}$, the ω-singleton set $\{x\}_\omega$ is the **3**-subset of X where

$$\{x\}_\omega y = \begin{cases} \omega & \text{if } y = x, \\ \bot & \text{otherwise.} \end{cases}$$

With **3**-relations defined as in Definition 1, it appears difficult at first to visualize specific examples. Instead of colouring arrows black or gray as in Figure 4 we have to label the arrows by the six functions found in Figure 5. However, we can think of the arrows we require as being "egg-yolk arrows", where the three parts (the yolk, the white, and outside the egg) are not spatial regions but the three truth values ($\top$, 0, and $\bot$), and each of these "regions" is occupied to an extent which is again one of these three truth values. This view is illustrated in Figure 7.

The diagrams of Figure 7 lead to one way of visualizing the six types of egg-yolk arrow as shown on the left of Figure 8 for the identity case. This pencil-like view is unnecessarily elaborate, because the region outside the egg (the outermost layer of the pencil) is always white. By dropping this outermost layer, we are led to the simpler visualization of the identity type of arrow shown on the right of Figure 8. The remaining types of arrow may be drawn similarly.

For ordinary relations, the idea of taking the converse is both easy to understand and of practical importance. The converse is visualized simply by reversing the direction of all the arrows in any diagram for the relation. In the classical conception of a **3**-valued relation, where things are related to one of just three degrees, the converse is again a matter of reversing the coloured arrows. For **3**-relations where we have six kinds of arrow instead of just three, the converse is more subtle. Surprisingly, taking the converse is not simply a matter of reversing the arrows; in two cases the reversed arrow changes its colouring.

In detail, if (y, x) is labelled by λ then in the converse (x, y) is labelled by the mapping $\omega \mapsto -(\lambda_* - \omega)$ where λ_* is the right adjoint of λ. The right adjoints of each possible λ are shown in Figure 5, and calculating $-(\lambda_* - \omega)$ in each case shows that to find the converse of a **3**-relation we need to label (x, y) by λ' where

Fig. 8. Possible visualizations of egg-yolk arrows

λ labels (y, x), and $\lambda \mapsto \lambda'$ is the involution of the lattice in Figure 6 obtained by reflection in the vertical axis.

4   The Algebra of Subgraphs

4.1   Graphs and Subgraphs

We recall the definition of a graph used in [Ste07] which followed [BMSW06].

Definition 2. *A **graph** consists of a set G with source and target functions $s, t : G \rightarrow G$ such that for all $g \in G$ we have $s(t(g)) = t(g)$ and $t(s(g)) = s(g)$.*

These graphs allow loops and multiple edges. The underlying set G makes no distinction between nodes and edges, but using the functions s and t the nodes can be described as those elements g for which $sg = g = tg$. The subgraphs of G in this setting are those subsets of G which are closed under the source and target functions.

For a graph G we can consider both subsets and subgraphs. The notation $\mathcal{P}G$ will be used to denote the lattice of subsets of G, and $\mathcal{G}G$ will be used to denote the lattice of subgraphs of G. By forgetting the source and target data, we can produce a set $|G|$ from the graph G. In the opposite direction, there are two ways that a subset $A \subseteq |G|$ can be made into a subgraph of G. To describe the properties of these constructions we again need to make use of adjunctions between posets.

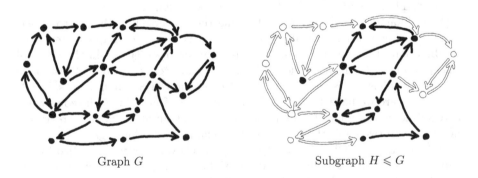

Graph G Subgraph $H \leqslant G$

Fig. 9. A Graph and a Crisp Subgraph

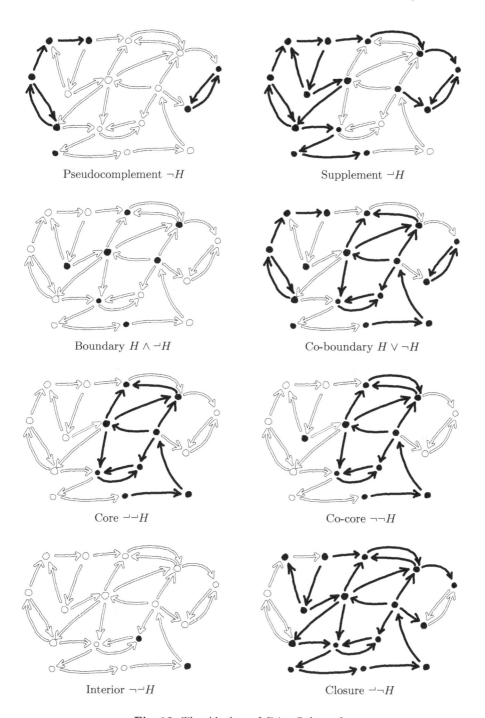

Pseudocomplement $\neg H$

Supplement $\dashv H$

Boundary $H \wedge \dashv H$

Co-boundary $H \vee \neg H$

Core $\dashv\dashv H$

Co-core $\neg\neg H$

Interior $\neg\dashv H$

Closure $\dashv\neg H$

Fig. 10. The Algebra of Crisp Subgraphs

Lemma 1. *The mapping* $|\_| : \mathcal{G}G \to \mathcal{P}G$ *has both a left and a right adjoint.*

$$[\_]^\wedge \dashv |\_| \dashv [\_]_\vee$$

For any $A \subseteq |G|$, *the graph* $[A]^\wedge$ *is the smallest subgraph containing the set* A, *and* $[A]_\vee$ *is the largest subgraph contained within* A. *Explicit constructions are given by*

$$[A]^\wedge = A \cup \{sa \in A : a \in A\} \cup \{ta \in A : a \in A\} = \bigwedge\{H : |A| \leqslant |H|\}$$

$$[A]_\vee = \{a \in A : sa \in A\} \cap \{a \in A : ta \in A\} \qquad = \bigvee\{H : |H| \leqslant |A|\}.$$

The boolean complement of subsets gives an isomorphism $- : \mathcal{P}X \to (\mathcal{P}X)^{\mathrm{op}}$ between the powerset and its opposite (that is the poset with the same elements, but with the converse partial order). For a subgraph $H \leqslant G$ simply taking the complement of the set of elements in H may not produce a subgraph, but we can ask for the greatest subgraph within this set or for the least subgraph containing it. This leads to two operations playing the same algebraic role in $\mathcal{G}G$ as $\dashv$ and $\neg$ do in the lattice of **3**-subsets of a set.

Lemma 2. *The operations* $\dashv$ *and* $\neg$ *on subgraphs defined by* $\dashv H = [-|H|]^\wedge$, *and* $\neg H = [-|H|]_\vee$ *are respectively the supplement and pseudocomplement in* $\mathcal{G}G$.

These operations and some constructions using them appear in Figure 10.

4.2   3-Valued Subgraphs

The value of the operations on subgraphs was already noted in [SW97], but the extension of this to subgraphs where some nodes and arrows may not be unequivocally inside or outside the subgraph does not seem to have been explored before. First we need to define these fuzzy subgraphs.

Definition 3. *A* **3**-*subgraph of a graph* (G, s, t) *is a* **3**-*subset,* $H : G \to \mathbf{3}$ *such that* $H(g) \leqslant H(sg) \wedge H(tg)$ *for all* $g \in G$.

We can draw examples of **3**-subgraphs as in Figure 11, with elements shown by outlines when they have value $\bot$, in grey when they have value 0, and in solid black when they have value $\top$. The definition requires that arrows must be present at least as strongly as their sources and targets.

The set of all **3**-subgraphs of G will be denoted $\mathcal{G}_3 G$. It is straightforward to check that this is a lattice, with the meets and joins given by the intersections and unions of the **3**-subsets underlying the graphs. In the crisp case, we used the pseudocomplement and the supplement in the lattice $\mathcal{G}G$ to construct the boundary, the core, and other important notions.

Before introducing the corresponding operations in $\mathcal{G}_3 G$, we state the following result, which is needed to justify their properties. In this the operation $\uparrow : \mathcal{G}_3 G \to \mathcal{G}_3 G$ makes a crisp version of a **3**-subgraph by colouring any grey nodes or edges black. Similarly, $\downarrow$ replaces grey nodes or edges by white ones.

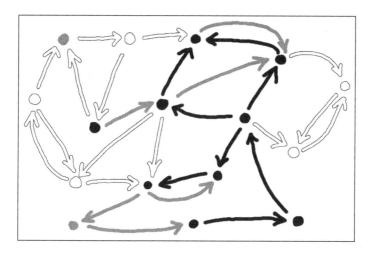

Fig. 11. Example of a **3**-Subgraph

Lemma 3. *For all* $H, K \in \mathcal{G}_3 G$ *the conditions* $H \wedge K = \bot$ *and* $\uparrow H \wedge K = \bot$ *are equivalent. Also, the conditions* $H \vee K = \top$ *and* $\downarrow H \vee K = \top$ *are equivalent.*

The lattice $\mathcal{G}_3 G$ has a pseudocomplement and a supplement, defined using $\neg$ and $\neg$ for crisp subgraphs as follows.

Lemma 4. *If* H *is a* **3**-*subgraph of* G *then the operations* $\neg$ *and* $\neg$ *defined by* $\neg H = \neg(\uparrow H)$ *and* $\neg H = \neg(\downarrow H)$, *provide the pseudocomplement and supplement in the lattice* $\mathcal{G}_3 G$.

Proof. To justify the pseudocomplement we have to show that $H \wedge \neg(\uparrow H) = \bot$, and that if $H \wedge K = \bot$ then $K \leqslant \neg(\uparrow H)$. As $\neg \uparrow H \leqslant \neg |H|$ the first of these is immediate from properties of $\neg$ in $\mathcal{G}G$. By Lemma 3 $H \wedge K = \bot$ iff $\uparrow H \wedge \uparrow K = \bot$, so working in $\mathcal{G}G$ we get $\uparrow K \leqslant \neg \uparrow H$ and the result follows as $K \leqslant \uparrow K$. A dual argument establishes the supplement property.

The supplement and pseudocomplement are illustrated in Figure 12.

Although $\mathcal{G}_3 G$ has a pseudocomplement and a supplement, these are both crisp subgraphs which conceptually does not seem the most appropriate way to

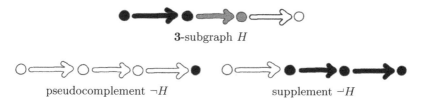

3-subgraph H

pseudocomplement $\neg H$ supplement $\neg H$

Fig. 12. A **3**-subgraph and crisp models of its outside

Fig. 13. Fuzzy interior and boundary concepts

model operations on **3**-fuzzy graphs. This can be seen by considering Figure 13 which shows that the notions of boundary, interior and exterior naturally classify their nodes into three classes: the definitely in, the possibly in, and the definitely not in. That is, we should be able to model boundaries, interiors and other derived subgraphs as **3**-subgraphs which will not be crisp in general, but using the pseudocomplement and the supplement in $\mathcal{G}_3 G$ cannot do this.

It would be possible to use pairs of crisp subgraphs, for example obtaining a fuzzy version of $\neg$ as a pair $(\neg(\uparrow H), \neg(\downarrow H))$. This would be similar to the use of upper and lower approximations in rough set theory, where an essentially fuzzy entity is modelled as a pair of crisp ones. However, there there are advantages in being able to think of a region as being just that: a single region of a new kind rather than a pair of ordinary regions.

To achieve this we need to generalize the operations $[\_]^\wedge$ and $[\_]_\vee$ from crisp subsets of a graph to **3**-valued ones. A full account of the technical details is beyond the scope of the present paper, but it can be shown that for any **3**-subset A of the elements of a graph G there are well-defined **3**-subgraphs $[A]^\wedge$ and $[A]_\vee$. These are respectively the smallest **3**-subgraph of G containing A, and the largest **3**-subgraph of G contained within A. This allows us to make the following definition.

Definition 4. *For $H \in \mathcal{G}_3 G$ the **3**-supplement, ∇H, and **3**-complement, $\triangle H$ are the **3**-subgraphs defined by $\triangle H = [-|H|]_\vee$ and $\nabla H = [-|H|]^\wedge$.*

It follows by standard techniques that these operations are adjoint to their opposites,

$$(\mathcal{G}_3 G)^{\mathrm{op}} \underset{\nabla}{\overset{\nabla^{\mathrm{op}}}{\underset{\longleftarrow}{\overset{\longrightarrow}{\perp}}}} \mathcal{G}_3 G \underset{\triangle^{\mathrm{op}}}{\overset{\triangle}{\underset{\longleftarrow}{\overset{\longrightarrow}{\perp}}}} (\mathcal{G}_3 G)^{\mathrm{op}}$$

and from these adjunctions several basic properties of these operations follow. The main properties are summarized in the following lemma.

Lemma 5. *The following hold for all $H, K \in \mathcal{G}_3 G$.*

1. *$H \leqslant K$ implies $\triangle K \leqslant \triangle H$ and $\nabla K \leqslant \nabla H$.*
2. *In $\mathcal{P}_3 G$ we have $|\triangle H| \leqslant -|H| \leqslant |\nabla H|$.*
3. *If H is crisp then $\neg H = \triangle H$ and $\neg H = \nabla H$.*

4. $\nabla\nabla H \leqslant H \leqslant \triangle\triangle H$.
5. $\nabla\nabla\nabla H = \nabla H$ and $\triangle\triangle\triangle H = \triangle H$.

While ∇H is defined as the smallest **3**-subgraph of G which contains the subset $-|H|$, and $\triangle H$ is the largest **3**-subgraph contained in $-|H|$, more explicit descriptions of these **3**-subgraphs are useful in practice.

When k is an edge, then $[k]^\wedge$ will be this edge together with its end-points. The next lemma essentially characterizes ∇H as obtained by replacing the value of each element g by $-Hg$ and then choosing higher values for any nodes which are needed to support the edges. Dually, $\triangle H$ is obtained by leaving values of all nodes as $-Hg$ and choosing the highest possible value for each edge which can be supported by these nodes.

Lemma 6. *For $H \in \mathcal{G}_3 G$ and for any $g \in |G|$,*

1. the subgraphs ∇H, and $\triangle H$ satisfy

$$\nabla Hg = \bigvee\{-Hk : g \in [k]^\wedge \text{ and } k \in |G|\}, \text{ and}$$
$$\triangle Hg = \bigwedge\{-Hk : k \in [g]^\wedge \text{ and } k \in |G|\}.$$

2. If g is an edge then $\nabla Hg = -Hg$, and if g is a node then $\triangle Hg = -Hg$.

In Figure 14 several examples are shown of subgraphs that can be constructed using the operations $\triangle$ and ∇ when applied to the **3**-subgraph H of Figure 11. These demonstrate the expressiveness of the language for describing fuzzy subgraphs that is available with these new operations.

5   Relations on 3-Subgraphs

A relation on a graph is just an ordinary relation on the underlying set which interacts in the right way with the source and target. The details of this treatment for crisp relations are given in [Ste07]. The generalization to **3**-relations is as follows.

Definition 5. *A **3**-relation, R, on the set of elements of a graph G is **graphical** if for all $g, h \in G$, it satisfies*

1. $g \, R \, h \leqslant (g \, R \, sh) \wedge (g \, R \, th)$, and
2. $(sg \, R \, h) \vee (tg \, R \, h) \leqslant g \, R \, h$.

Lemma 7. *Let R, S be graphical **3**-relations on G. Then the composite $R \, ; S$ is graphical.*

Proof. For any k we have $k \, S \, h \leqslant k \, S \, sh$ so $(g \, R \, k)(k \, S \, h) \leqslant (g \, R \, k)(k \, S \, sh)$. This gives $g \, R \, ; S \, h \leqslant g \, R \, ; S \, sh$ by taking the join over all $k \in G$. Similarly, $g \, R \, ; S \, h \leqslant g \, R \, ; S \, th$.

To show that $(sg \, R \, ; S \, h) \vee (tg \, R \, ; S \, h) \leqslant g \, R \, ; S \, h$, use the definition of compostion of relations and the fact that for any $k \in G$ we have $(sg \, R \, k) \vee (tg \, R \, k) \leqslant g \, R \, k$.

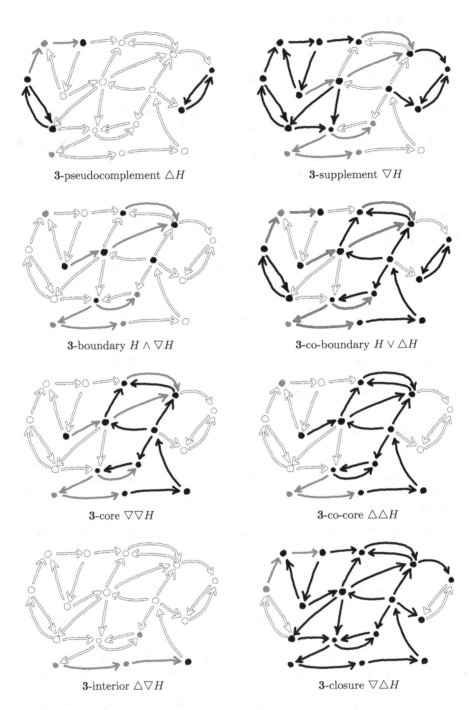

3-pseudocomplement $\triangle H$

3-supplement ∇H

3-boundary $H \wedge \nabla H$

3-co-boundary $H \vee \triangle H$

3-core $\nabla\nabla H$

3-co-core $\triangle\triangle H$

3-interior $\triangle\nabla H$

3-closure $\nabla\triangle H$

Fig. 14. Various **3**-Subgraphs constructed from H of Figure 11

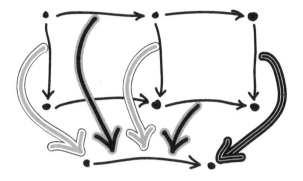

Fig. 15. Presentation for a graphical **3**-relation

So far, in order too keep matters as simple as possible, we have dealt with relations on a single set or a single graph. It is routine to extend this to the more general case where we have a relation going from one structure to another. We assume this more general setting in the following simple example shown in Figure 15. In this example, we have two graphs, the upper one with seven edges and six vertices, and the lower one with two vertices and a single edge. The relation between these graphs is shown using the egg-yolk technique on the right hand side of figure 8. To avoid too many arrows it is assumed that arrows required to make this relation graphical are implicitly included (strictly speaking the figure provides a presentation for a relation, rather than the relation itself).

By providing a **3**-relation from the one graph to a second graph, we specify how **3**-subgraphs of the first graph are to be viewed as **3**-subgraphs of the second graph. This is expressed formally in terms of dilation. Given a graphical relation R on G we can dilate any $H : G \to \mathbf{3}$ to obtain a **3**-subset $H \oplus R : G \to \mathbf{3}$. It is straightforward to check that when H is a subgraph that $H \oplus R$ is a subgraph too, and similarly for the erosion $R \ominus H$. Examples of dilation are provided in Figure 16. The upper part of each of the diagrams shows a subgraph of the graph which is the domain of the relation presented in Figure 15. The lower part shows the result of dilating the subgraph by the relation.

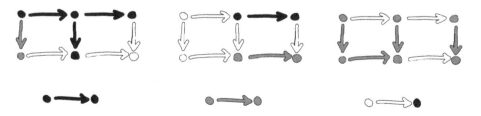

Fig. 16. Three examples of dilation using the relation from Figure 15

We have already seen that the converse of a **3**-relation on a set is not as simple as just reversing the direction of all the arrows, since two of the types of arrow also change their labelling. In the case of graphs the converse becomes more complex still, as already noted in the case of crisp graphical relations in [Ste07]. While there is insufficient space for anything beyond a brief discussion here, it is worth noting that the operations ∇ and $\triangle$ developed above do play an important role in understanding the converse of graphical **3**-relations. As graphical **3**-relations are determined by their dilation operations we are able to define the converse of a relation R as the relation where the effect of dilating a **3**-subgraph H is given by $\nabla(R \ominus \triangle^{\mathrm{op}} H)$. The following diagram summarizes the construction, and the symbols $\perp$, $\dashv$ and $\vdash$ in it denote adjunctions following the practice in [Tay99].

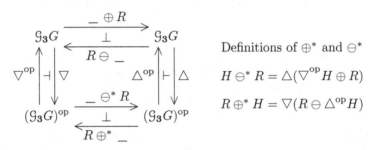

Definitions of $\oplus^*$ and $\ominus^*$

$$H \ominus^* R = \triangle(\nabla^{\mathrm{op}} H \oplus R)$$

$$R \oplus^* H = \nabla(R \ominus \triangle^{\mathrm{op}} H)$$

6   Conclusions and Further Work

This paper has introduced two important new features as part of an ongoing programme to understand the granularity aspects of graphs. These two features are the pair of operations ∇ and $\triangle$, and the definition of a **3**-relation on a graph. The value of both of these has been demonstrated by additional constructions built on them, such as the fuzzy notions of boundary and interior illustrated in Figure 14 and the means to obtain the converse of a graphical **3**-relation.

The main direction in further work is to apply the theoretical constructions presented here in various practical examples. It is significant that the relations described here have arisen from considerations of mathematical morphology. This connection means there is good evidence, in view of the relationship to rough set theory, to support the view that continuing the programme of work is likely to produce an account of granularity for both fuzzy and crisp graphs which will have analogous practical applications to those enjoyed by rough sets.

References

[BF96]   Burrough, P.A., Frank, A.U. (eds.): Geographic Objects with Indeterminate Boundaries. GISDATA Series, vol. 2. Taylor and Francis, Abingdon (1996)

[BMSW06]   Brown, R., Morris, I., Shrimpton, J., Wensley, C.D.: Graphs of Morphisms of Graphs. Bangor Mathematics Preprint 06.04, Mathematics Department, University of Wales, Bangor (2006)

[CD96]   Clementini, E., Di Felice, P.: An algebraic model for spatial objects with indeterminate boundaries. In: Burrough and Frank [BF96], pp. 155–169

[CG96]   Cohn, A.G., Gotts, N.M.: The 'egg-yolk' representation of regions with indeterminate boundaries. In: Burrough and Frank [BF96], pp. 171–187

[DW01]   Duckham, M., Worboys, M.F.: Computational structures in three-valued nearness relations. In: Montello, D.R. (ed.) COSIT 2001. LNCS, vol. 2205, pp. 76–91. Springer, Heidelberg (2001)

[Law86] Lawvere, F.W.: Introduction. In: Lawvere, F.W., Schanuel, S.H. (eds.) Categories in Continuum Physics. Lecture Notes in Mathematics, vol. 1174, pp. 1–16. Springer, Heidelberg (1986)

[Łuk70] Łukasiewicz, J.: Selected Works. North-Holland, Amsterdam (1970)

[Ste07] Stell, J.G.: Relations in Mathematical Morphology with Applications to Graphs and Rough Sets. In: Winter, S., Duckham, M., Kulik, L., Kuipers, B. (eds.) COSIT 2007. LNCS, vol. 4736, pp. 438–454. Springer, Heidelberg (2007)

[SW97]   Stell, J.G., Worboys, M.F.: The algebraic structure of sets of regions. In: Frank, A.U. (ed.) COSIT 1997. LNCS, vol. 1329, pp. 163–174. Springer, Heidelberg (1997)

[Tay99] Taylor, P.: Practical Foundations of Mathematics. Cambridge University Press, Cambridge (1999)

[Wyg96] Wygralak, M.: Vaguely Defined Objects. Kluwer, Dordrecht (1996)

[Zad65] Zadeh, L.A.: Fuzzy sets. Information and Control 8, 338–353 (1965)

Map Algebraic Characterization of Self-adapting Neighborhoods

Takeshi Shirabe

Department of Geoinformation and Cartography
Vienna University of Technology
1040 Vienna, Austria
shirabe@geoinfo.tuwien.ac.at

Abstract. A class of map algebraic operations referred to as "focal" characterizes every location as a function of the geometry and/or attribute of all locations
that belong to the "neighborhood" of that location. This paper introduces a new
type of map algebraic neighborhood whose shape is unspecified but required to
have a specified size. This paper explores how the use of such neighborhoods
affects the design of focal operations, as well as their implementation and application. It is suggested that the proposed operations can contribute to site selection analyses, which are often subject to size restriction (e.g. due to limited
budgets or environmental concerns), but needs more theoretical investigation
for them to be fully operational in practice.

1 Introduction

Map algebra is a high-level computational language that has been implemented in a
number of geographic information systems (GIS). More generally, it is a set of conventions for the organization of cartographic data and the processing of those data in a
manner that attempts to facilitate their interpretation with both clarity and flexibility
[32]. To do so, it presents a set of data-processing operations on single-attribute maps
referred to as "layers." Since all of these operations accept layers as input and generate layers as output, they are easily combined by simply using the output of one as input to another. Furthermore, each operation can be defined in terms of its effect on a
single, typical location with an understanding that all locations will in fact be subject
to the identical effect.

Over the last decades map algebra both as spatial information theory and as spatial
information tool has been repeatedly explored in the literature. This reflects the
importance of the philosophy underlying this methodology as well as its applicabil-
ity/adaptability to a wide range of subjects. Related works include those that formalize the structure of map algebra [12, 14, 30], design its visual interfaces [5, 13,], and
adapt its principle to other data types [20, 22]. The present paper, too, is one such attempt—more precisely, a possible extension of some map algebraic operations that
explicitly take into account geometric relations between locations.

K. Stewart Hornsby et al. (Eds.): COSIT 2009, LNCS 5756, pp. 280–294, 2009.
© Springer-Verlag Berlin Heidelberg 2009

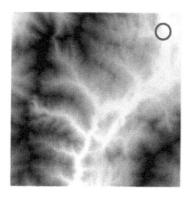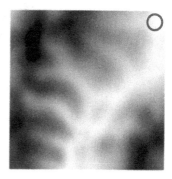

Fig. 1. A convolution operation and its input layer (left), window (delineated by a bold line), and output layer (right). Note that the darker a location is shaded, the higher its value is.

To begin, imagine that we seek a site for building a large-scale facility, say, a cricket field in a given study area, which is characterized by a layer representing the construction cost associated with each location. If the site is supposed to take the form of a circle of a certain diameter, we can use a single "convolution" [6] operation to create a layer that represents the cost of each potential site. The operation places a "window" [6] of the specified dimension on every location (except where too close to edges), adds the values of all locations that fall in that window, and records the result on a new layer. Figure 1 illustrates an example, which uses a circular window of diameter 160m and transforms the layer on the left-hand side (encompassing a 1800m-by-1800m rectangular area) to the one on the right-hand side. The output layer tells us that darker locations are more costly to place the center of the new cricket field.

A circular window was chosen in this particular example, but it could be any other shape such as oval, square, wedge, or annulus depending on the context. In any case, once some window is placed at every location, the convolution operation is easy to carry out.

Now what if a different context required the window used above to have a fixed size but not a fixed shape? At first sight, this assumption would make the effect of the convolution operation ambiguous (if not meaningless), as the window can take on an arbitrary shape. To avoid this ambiguity, one condition is imposed: at each location, the window is to be shaped in a way that minimizes the sum of those values that fall in that window. Then, the computing procedure can be intuitively understood such that during its execution each location lets its window evolve for an optimal (i.e. minimizing) shape. The convolution operation with this condition was applied to the same input layer used above, and its output layer is presented in Figure 2.

In the vocabulary of map algebra, the convolution operation is classified as "focal," and the window placed on each location is referred to as the "neighborhood" of that location. It is easy to conceptualize and implement focal operations if the shapes and sizes of all neighborhoods are completely specified. As seen above, however, it is possible to design more flexible focal operations—flexible, i.e. in the sense that each neighborhood can take on any shape subject to some size restriction. Neighborhoods of this kind are referred to here as "self-adapting," as they are to adapt their shapes to

Fig. 2. Layer generated by a convolution operation with "yet-to-be-shaped" windows of 20100 (approximately $80 \cdot 80 \cdot \pi$) m^2 and a window shaped optimal for a certain location (delineated by a bold line)

achieve some goals. The major objective of this paper is then to explore how the introduction of self-adapting neighborhoods affects the concept, implementation, and application of focal operations.

The rest of the paper is organized as follows. Section 2 reviews the basic concept of map algebra. Section 3 introduces a new design of focal operations. Section 4 illustrates an example of their use. Section 5 concludes the paper.

2 Map Algebra

This section summarizes some ideas of map algebra relevant to the present context. While the theory of map algebra is not necessarily raster oriented, its data-processing capabilities can be relatively easily implemented in raster form [33]. Hence, in this paper, it is assumed that all layers are encoded in a (common) raster format and all operations process them as such, and a location and a cell are interchangeably used.

As mentioned earlier, a map algebraic operation takes one or more existing layers as input and returns a new layer as output. It does so by applying the same computing rule to each individual location. This "worm's eye" perspective on map transformation gives rise to three major types of map algebraic operation. First are "local" operations that calculate an output value for each location as a specified function of one or more input values associated with that location. Second are "zonal" operations. These calculate an output value for each location as a specified function of all values from one input layer that occur within the same "zone" as that location on a second input layer, where a zone is a set of like-valued locations. The third group of operations includes those that calculate an output value for each location as a specified function of input values that are associated with neighboring locations. These are called "focal" operations in reference to the fact that each location is regarded as the center or "focus" of a "neighborhood" from which its output value is to be computed. Each location within such a neighborhood is defined as such by virtue of its distance and/or direction to/from the neighborhood focus. A neighborhood may include,

Fig. 3. Neighborhoods (lightly shaded) that are bounded by various distance and directional ranges from their focus (darkly shaded)

for example, all locations with 100 meters or all locations within 100 meters due North. The distance and/or directional relationship between a neighborhood location and the neighborhood focus, however, can also become more sophisticated. A neighborhood may be restricted, for example, to only those locations lying "within sight" or "upstream" of the neighborhood focus. It is important to note that neighborhoods may vary in shape from one focus to another (see Figure 3). Such variation can be controlled, for example, through one or more layers that give each location a unique distance and/or direction value for defining its neighborhood [32, p. 231-232].

Earlier implementations of map algebra assume that one layer characterizes each location with a single numerical value, but this restriction, in theory, can be relaxed. For example, Li and Hodgson [20] proposed a map algebra for vector (as opposed to scalar) layers. A vector layer assigns each location two values and effectively models a spatial phenomenon involving magnitude and direction (e.g. wind). A three-dimensional extension of map algebra designed by Mennis, Viger, and Tomlin [22] is another example that employs multi-valued layers. Their "cubic map algebra" can be used to analyze time-varying spatial phenomena (e.g. temperature) by spending two dimensions for representing locations and the third dimension for recording a time series at each location. Also, a series of work done by Heuvelink (e.g. [17, 18]) suggest that a random variable rather than a constant value should be associated with each location, when the differences between reality and data are not negligible but are only known in probabilistic terms.

As far as the present paper is concerned, a more notable extension of map algebra is one proposed by Takeyama [30, 31]. Takeyama's "Geo-Algebra" employs a new data type called "relational map." A relational map takes the form of a conventional layer, and expresses any arbitrary neighborhood for each location in zero-one terms (i.e. locations of value 1 are included in the neighborhood and locations of value 0 are not). The collection of all locations' relational maps amounts to a "meta-relational map" and may be regarded as yet another type of map algebraic layer.

3   Focal Operations with Self-adapting Neighborhoods

This section presents an idea of how conventional focal operations can be modified by self-adapting neighborhoods, and discusses a rudimentary approach to their implementation.

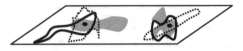

Fig. 4. Potential forms (lightly shaded, enclosed by a bold line, enclosed by a dotted line) of self-adapting neighborhoods and their focuses (darkly shaded)

3.1 Conceptual Design

The configuration of a self-adapting neighborhood relies on two topological properties: adjacency and contiguity (or connectedness). Since topological properties in raster space are often ambiguous [19, 35, 36, 27, 28], we clarify these concepts here. In general, two cells are said to be adjacent to each other if they share a "boundary." A boundary may be a cell corner or a cell side. This makes a further distinction between two types of adjacency: 4-adjacency and 8-adjacency. A 4-adjacency relation is attributed to a pair of cells sharing a cell side, and an 8-adjacency relation to a pair of cells sharing either a cell corner or a cell side. Once adjacency is defined in either way, contiguity is defined such that a set of cells is contiguous if any two cells in that set can be connected by at least one sequence of adjacent cells. For ease of discussion, we will employ the 4-adjacency in the remainder of the paper. But this choice is arbitrary.

A self-adapting neighborhood is an unspecified, contiguous set of locations associated with (and including) a specific location designated as its focus. It is additionally required to contain a specified number of locations. Thus, a self-adapting neighborhood can potentially take any shape as long as it is contiguous and correctly sized (Figure 4).

Since self-adapting neighborhoods are more like unknown variables than given parameters, we propose a new set of focal operations to optimally "solve" (i.e. assigns locations to) all such neighborhoods *and* compute new values for their focuses. Optimally—that is, in the sense that all the resulting focal values are either **maximum** or **minimum** possible depending on the user's explicit intent.

The idea of letting neighborhoods unknown was already implied in Tomlin's early foundational work [32]. To see this, consider a generic focal operation with conventional neighborhoods specified by a distance range in terms either of Euclidean distances or of non-Euclidean (e.g. cost-weighted) distances. In the former case, all neighborhoods are immediately found equally circular. In the latter case, however, their ultimate shapes (which are likely to differ from one another) are not known at the outset, but need to be revealed (possibly by computing the shortest cost-weighted distances from each focus to all other locations) before their focal values are computed.

The idea can be extended to deal with other cases in which the search of neighborhoods cannot be done separately from (or prior to) the computation of focal values because the two processes are dependent on each other. As seen below, such cases are found when focal operations involve descriptive statistical functions including: sum, mean, maximum, and minimum.

The conventional version of a focal operation involving any one of the aforementioned functions generates a new layer in which each location is assigned a value computed as the chosen function of all values found in that location's neighborhood (see Figure 5).

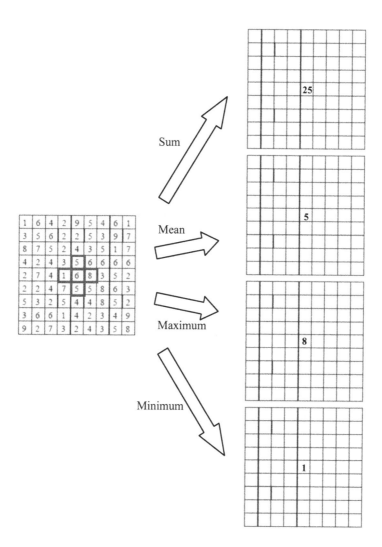

Fig. 5. *FocalSum/Mean/Maximum/Minimum* operations with fixed "+"-shaped neighborhoods (encircled by bold lines). These take the layer on the left as input and generate one of the layers on the right depending on the specified function (i.e. sum, mean, maximum, or minimum). The figure shows how a new value is computed for a typical cell (located in the center), and suppresses all other information

A focal operation that adopts self-adapting neighborhoods finds for each location a (not necessarily uniquely determined) neighborhood such that the value computed as the chosen function of all values found in that neighborhood is no smaller (or greater) than that for any other possible neighborhood, and generates a new layer in which each location is assigned the (uniquely determined) optimal value (see Figure 6). To distinguish the proposed focal operations from the existing ones, the term "maxi" or "mini," which indicates the type of optimization, has been added to their names.

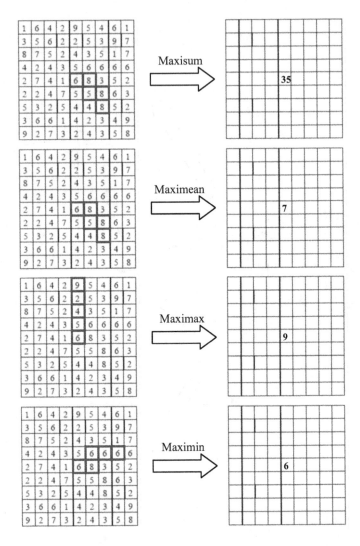

Fig. 6. *FocalMaxisum/Maximean/Maximax/Maximin* operations with five-cell self-adapting neighborhoods. These take the layer on the left as input and generate one of the layers on the right depending on the specified function. The figure shows how an optimal neighborhood (encircled by a bold line) and a new value are computed for a typical cell (located in the center).

3.2 Implementation

Here we sketch a strategy for implementing each of those focal operations described above. For ease of presentation, we employ the following notation. Let I denote the set of cells in the given layer, r each focus cell, $A(i)$ the set of cells adjacent to cell i, c_i the attribute value of cell i, k the number of cells to be included in cell r's neighborhood, and x_i a (unknown) variable that equals 1 if cell i is included in cell

r's neighborhood and 0 otherwise. y_{ij} is an auxiliary variable associated with a pair of cells i and j that takes on a nonnegative continuous value.

FocalMaxi(Mini)sum

These operations can be seen as equivalent to solving, for each cell r, a mathematical optimization problem formulated as follows.

$$\text{Maximize/Minimize} \quad \sum_{i \in I} c_i x_i \tag{1}$$

$$\text{Subject to} \quad x_r = 1 \tag{2}$$

$$\sum_{j \in A(i)} y_{ij} - \sum_{j \in A(i)} y_{ji} = x_i \qquad \forall i \in I : i \neq r \tag{3}$$

$$\sum_{j \in A(i)} y_{ji} \leq (k-1)x_i \qquad \forall i \in I \tag{4}$$

$\sum_{i \in I} c_i x_i$ indicates the sum of all values found in cell r's neighborhood. When it is maximized (or minimized), it becomes the focal value to be assigned to cell r. Equation (2) states that cell r is included in its neighborhood. Equations (3) and (4) are taken from [29] to ensure that the neighborhood is contiguous.

The problem formulated above is generally referred to as mixed integer programs (MIPs). Although there are general-purpose algorithms for MIPs (e.g. the branch-and-bound algorithm), none of them are efficient (or non-polynomial time). For example, Shirabe [29] reported that MIP (1)-(4) involving more than 200 cells could not be solved in reasonable time. Williams [34] gave the problem a different MIP formulation but experienced similar computational difficulty. There are other MIPs that might be utilized (or modified) for implementing the present operation (see, e.g. [37, 16, 11, 2, 3, 9]). In any case, however, considering that a (presumably hard) MIP needs to be solved *for every single cell*, the MIP-based implementation does not seem to meet practical standards.

A heuristic is currently under development. It is based on an observation that any connected set of cells can be numbered in a way that the removal of each cell in that order will never make the remaining set disconnected (Figure 7). This implies that any connected set of k cells can be created by adding one cell to a connected set of $(k-1)$ cells. A dynamic-programming procedure has been experimentally designed which successively derives sub-optimal (but expected-to-be-good) neighborhood from smaller sub-optimal neighborhoods. Informal preliminary experiments suggest that when the heuristic is applied to small grids (10-by-10), it constantly finds a good sub-optimal solution whose deviation from the optimal solution is within several percents. However, any statement should be taken merely as an optimistic hypothesis until a formal study (in)validates it.

Fig. 7. A numbering of a connected set of cells (shaded)

FocalMaxi(Mini)mean

These operations can make use of the same approach discussed above, because by definition each cell's output value of the *FocalMaxi(Mini)mean* operation is that of the *FocalMaxi(Mini)sum* operation divided by k (i.e. neighborhood size).

FocalMaxi(Mini)max

The *FocalMaximax* operation is relatively easy to implement. A simple procedure scans all cells that can be reached from each cell r through a connected sequence of k cells, finds the maximum value, and assigns it to cell r.

The *FocalMinimax*, on the other hand, seems more difficult to implement efficiently. Still, as with the case of the *FocalMinisum* operation, it is amenable to a MIP formulation as follows:

Minimize z_r (5)

Subject to

$c_i x_i \leq z_r$ $\forall i \in I$ (6)

Equations (2)-(4)

z_r indicates the greatest value found in cell r's neighborhood. When it is minimized, it will be assigned to cell r. Equation (6) allows no cell in cell r's neighborhood to have a value greater than z_r.

Like the MIP for the *FocalMinisum* operation, the present MIP can be solved by any general-purpose MIP algorithm in theory, but will suffer from computational difficulty in practice. Thus, further research is expected to explore efficient heuristic methods (such as region growing [4], simulated annealing [1], evolutionary algorithm [38], or dynamic programming)[1].

FocalMaxi(Mini)min

These operations can take the approach discussed above, since the output of the *FocalMaximin* and *FocalMinimin* operations are obtained by applying the *FocalMinimax* and *FocalMaximax* operations, respectively, to the inverted (i.e., multiplied by −1) input layer, and inverting the resulting layer.

---

[1] A reviewer of the paper has suggested a simple procedure for this problem. It iteratively revises a temporary region, R, which contains only a focus cell at the outset. At each iteration, add to R (a necessary number of) those cells that are connected to R and whose values are not larger than the largest value found in R. If R is still short of k cells, add to R the smallest-valued cell among those cells that are connected to R. Continue this until R has k cells.

4 Example

One important application area of the proposed focal operations is conservation of wildlife habitats [24, 21, 8, 15, 23, 7, 25, 26]. A typical problem is concerned with identifying a specific size of contiguous region suitable for conservation. If a wildlife manager is interested not just in a single presumably optimal region but in many good alternatives, (s)he would benefit from a tool that relates one alternative region to each location with the understanding that that location must be contained by that region. Linking a location and a region (or an object) this way is exactly the same idea that Cova and Goodchild [10] originally proposed. And, as illustrated below, this is the kind of task that focal operations with self-adapting neighborhoods can perform.

A highly simplified case is considered here using data pertaining to a small locality (encompassing 3600m-by-3600m rectangular area) called "Browns Pond" located in Massachusetts, USA. The study area is characterized by four layers representing attitudes of elevation, vegetation, hydrology, and development (Figure 8).

To evaluate the suitability for wildlife habitat conservation at each location of the study area, a simple "cartographic modeling" [32] approach is taken. It first gives each location a score indicating its suitability for conservation with respect to one attribute (which may be derived from existing and/or other derived layers) at a time, then weights these individual scores according to the relative importance of their corresponding attribute, and finally, aggregates them into a composite score. It is assumed that a location closer to water body, farther from buildings and roads, less steep, and covered with more woods is more suitable for conservation. Then some sequence of conventional map algebraic operations is applied to one layer after another until a layer representing conservation suitability over the study area is obtained as illustrated in Figure 9. I acknowledge that this is a highly contrived scenario, and avoid discussing its details and validity in this paper.

Based solely on the information obtained through the above procedure, one may attempt to seek regions of 40000 m^2 suitable for conservation. To do so, the *FocalMaxisum* operation is applied to the Suitability layer using self-adapting neighborhoods of 40000 m^2. The resulting layer (shown on the left of Figure 10) depicts the total suitability score of an optimal neighborhood found for each location. Such neighborhoods (a sample of which is shown on the right of Figure 10) are an important by-product of the proposed focal operation. The wildlife manager can take them as candidate regions for conservation, and compare them in terms of any factors that were not incurporated in the original suitability analysis. As discussed by Cova and Goodchild [10], the integration of the location-to-alternative linking with geographic information systems facilitates the user to explore a decision space.

From a map algebraic perspective, a challenging issue is how to deal with materialized self-adapting neighborhoods. They take on irregular shapes and thus cannot be specified by geometric relations (such as distance and direction) only. Like Takeyama's meta-relational map, it would be an idea to introduce a new type of "layer" in

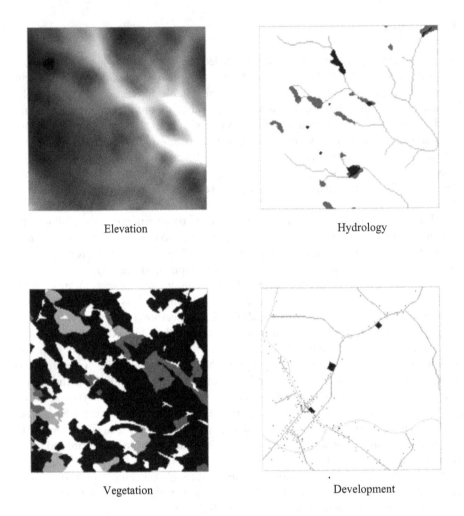

Elevation   Hydrology

Vegetation   Development

Fig. 8. Data on Browns Pond Study Area. The top-left layer indicates height above sea level in meters at each location. The top-right layer indicates the type of surface water at each location. The bottom-left layer indicates the type of vegetation at each location. The bottom-right indicates the type of human artifact at each location. Note that each shade of gray corresponds to some (qualitative or quantitative) value, although it is not revealed here.

which each location is assigned a neighborhood possibly in the form of a set of pointers pointing to its member locations. If this approach is adopted, at least two questions arise: how different types of layers are perceived and manipulated by geographic information systems and their users, and whether they can (or should) transform from one type to the other in a manner consistent with the convention of map algebra.

Fig. 9. Suitability for wildlife habitat conservation. Darker locations are more suitable for conservation.

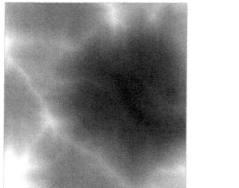 

Fig. 10. Output of *FocalMaxisum* of Suitability layer with neighborhoods of 40000 m² (left), and optimal neighborhoods associated with selected cells (those at the top-left corner, at the top-right corner, at the bottom-left corner, at the bottom-right corner, and at the center (right)

5   Conclusion

This paper presents a recent attempt to extend map algebra by proposing a new concept of neighborhood (called self-adapting neighborhood) that has no fixed shape but has a fixed size. It is theoretically interesting as three basic cartographic properties, namely, distance, direction, and area (size) are now included in the configuration of neighborhoods. Also it has practical applications including region selection in the context of wildlife conservation. It has been found, however, that some operations in this extension scheme are difficult to implement efficiently. The difficulty relates to combinatorial complexity caused primarily by the contiguity requirement imposed on every self-adapting neighborhood. Thus it is well anticipated that other focal operations that are not considered in this paper (e.g. those concerning product, range, diversity, frequency of neighbors' values) would encounter the same issue. Therefore, while the present paper focuses on a theoretical account of how focal operations behave with self-adapting neighborhoods, it would be practical to use heuristic methods for their implementation although they may not produce exact output.

Acknowledgments

I would like to thank C. Dana Tomlin for his valuable comments on an earlier draft of this paper, as well as for helping me understand his work. I also thank the anonymous reviewers for their detailed comments for improving the paper. Still all misunderstandings or incorrect statements found in this paper are my own responsibility.

References

1. Aerts, J.C.J.H., Heuvelink, G.B.M.: Using simulated annealing for resource allocation. International Journal of Geographical Information Science 16, 571–587 (2002)
2. Benabdallah, S., Wright, J.R.: Shape considerations in spatial optimization. Civil Engineering Systems 8, 145–152 (1991)
3. Benabdallah, S., Wright, J.R.: Multiple subregion allocation models. ASCE Journal of Urban Planning and Development 118, 24–40 (1992)
4. Brookes, C.J.: A parameterized region-growing programme for site allocation on raster suitability maps. International Journal of Geographical Information Science 11, 375–396 (1997)
5. Bruns, T., Egenhofer, M.: User Interfaces for Map Algebra. Journal of the Urban and Regional Information Systems Association 9(1), 44–54 (1997)
6. Burrough, P.A., McDonnell, R.A.: Principles of Geographical Information Systems, 2nd edn. Oxford Press, Oxford (1998)
7. Cerdeira, J.O., Gaston, K.J., Pinto, L.S.: Connectivity in priority area selection for conservation. Environmental Modeling and Assessment 10(3), 183–192 (2005)
8. Church, R.L., Gerrard, R.A., Gilpin, M., Sine, P.: Constructing Cell-Based Habitat Patches Useful in Conservation Planning. Annals of the Association of American Geographers 93, 814–827 (2003)
9. Cova, T.J., Church, R.L.: Contiguity Constraints for Single-Region Site Search Problems. Geographical Analysis 32, 306–329 (2000)
10. Cova, T.J., Goodchild, M.F.: Extending geographical representation to include fields of spatial objects. International Journal of Geographical Information Science 16, 509–532 (2002)
11. Diamond, J.T., Wright, J.R.: Design of an integrated spatial information system for multiobjective land-use planning. Environment and Planning B 15, 205–214 (1988)
12. Dorenbeck, C., Egenhofer, M.: Algebraic Optimization of Combined Overlay Operations. In: Mark, D., White, D. (eds.) AutoCarto 10: Technical papers of the 1991 ACSM-ASPRS annual convention, Baltimore, Maryland, USA, pp. 296–312 (1991)
13. Egenhofer, M.J., Bruns, H.T.: Visual map algebra: a direct-manipulation user interface for GIS. In: Proceedings of the third IFIP 2.6 Working Conference on Visual Database Systems, Lausanne, Switzerland, pp. 235–253 (1995)

14. Frank, A.U.: Map Algebra Extended with Functors for Temporal Data. In: Akoka, J., Liddle, S.W., Song, I.-Y., Bertolotto, M., Comyn-Wattiau, I., van den Heuvel, W.-J., Kolp, M., Trujillo, J., Kop, C., Mayr, H.C. (eds.) ER Workshops 2005. LNCS, vol. 3770, pp. 194–207. Springer, Heidelberg (2005)
15. Fischer, D., Church, R.L.: Clustering and Compactness in Reserve Site Selection: an Extension of the Biodiversity Management Area Selection Model. Forest Science 49, 555–565 (2003)
16. Gilbert, K.C., Holmes, D.D., Rosenthal, R.E.: A multiobjective discrete optimization model for land allocation. Management Science 31, 1509–1522 (1985)
17. Heuvelink, G.B.M.: Error Propagation in Environmental Modelling with GIS. Taylor & Francis, London (1998)
18. Heuvelink, G.B.M., Burrough, P.A., Stein, A.: Propagation of errors in spatial modelling with GIS. International Journal of Geographical Information Systems 3, 303–322 (1989)
19. Kovalevsky, V.A.: Finite topology as applied to image analysis. Computer Vision, Graphics, and Image Processing 46, 141–161 (1989)
20. Li, X., Hodgson, M.E.: Data model and operations for vector fields. Geographic Information Sciences and Remote Sensing 41(1), 1–24 (2004)
21. Mcdonnell, M.D., Possingham, H.P., Ball, I.R., Cousins, E.A.: Mathematical Methods for Spatially Cohesive Reserve Design. Environmental Modeling and Assessment 7, 107–114 (2002)
22. Mennis, J., Viger, R., Tomlin, C.D.: Cubic Map Algebra Functions or Spatio-Temporal Analysis. Cartography and Geographic Information Science 32(1), 17–32 (2005)
23. Nalle, D.J., Arthur, J.L., Sessions, J.: Designing Compact and Contiguous Reserve Networks with a Hybrid Heuristic Algorithm. Forest Science 48, 59–68 (2003)
24. Önal, H., Briers, R.A.: Incorporating Spatial Criteria in Optimum Reserve Network Selection. In: Proceedings of the Royal Society of London: Biological Sciences, vol. 269, pp. 2437–2441 (2002)
25. Önal, H., Briers, R.A.: Designing a conservation reserve network with minimal fragmentation: A linear integer programming approach. Environmental Modeling and Assessment 10(3), 193–202 (2005)
26. Önal, H., Wang, Y.: A Graph Theory Approach for Designing Conservation Reserve Networks with Minimum Fragmentation. Networks 51(2), 142–152 (2008)
27. Roy, A.J., Stell, A.J.: A qualitative account of discrete space. In: Egenhofer, M.J., Mark, D.M. (eds.) GIScience 2002. LNCS, vol. 2478, pp. 276–290. Springer, Heidelberg (2002)
28. Shirabe, T.: Modeling Topological Properties of a Raster Region for Spatial Optimization. In: Fisher, P. (ed.) Developments in Spatial Data Handling: Proceedings of the 11th International Symposium on Spatial Data Handling, pp. 407–420. Springer, Heidelberg (2004)
29. Shirabe, T.: A Model of Contiguity for Spatial Unit Allocation. Geographical Analysis 37, 2–16 (2005)
30. Takeyama, M.: Geo-Algebra: A mathematical approach to integrating spatial modeling and GIS. Ph.D. dissertation, Department of Geography, University of California at Santa Barbara (1996)
31. Takeyama, M., Couclelis, H.: Map dynamics: integrating cellular automata and GIS through Geo-Algebra. International Journal of Geographical Information Science 11, 73–91 (1997)
32. Tomlin, C.D.: Geographic information systems and cartographic modeling. Prentice-Hall, Englewood Cliffs (1990)

33. Tomlin, C.D.: Map algebra: one perspective, Landscape and Urban Planning 30, 3–12 (1994)
34. Williams, J.C.: A Zero-One Programming Model for Contiguous Land Acquisition. Geographical Analysis 34, 330–349 (2002)
35. Winter, S.: Topological relations between discrete regions. In: Egenhofer, M.J., Herring, J.R. (eds.) SSD 1995. LNCS, vol. 951, pp. 310–327. Springer, Heidelberg (1995)
36. Winter, S., Frank, A.U.: Topology in raster and vector representation. Geoinformatica 4, 35–65 (2000)
37. Wright, J., Revelle, C., Cohon, J.: A multipleobjective integer programming model for the land acquisition problem. Regional Science and Urban Economics 13, 31–53 (1983)
38. Xiao, N.: An evolutionary algorithm for site search problems. Geographical Analysis 38, 227–247 (2006)

Scene Modelling and Classification Using Learned Spatial Relations

Hannah M. Dee, David C. Hogg, and Anthony G. Cohn

School of Computing,
University of Leeds,
Leeds LS2 9JT, United Kingdom
{hannah,dch,agc}@comp.leeds.ac.uk

Abstract. This paper describes a method for building visual scene models from video data using quantized descriptions of motion. This method enables us to make meaningful statements about video scenes as a whole (such as "this video is like that video") and about regions within these scenes (such as "this part of this scene is similar to this part of that scene"). We do this through unsupervised clustering of simple yet novel motion descriptors, which provide a quantized representation of gross motion within scene regions. Using these we can characterise the dominant patterns of motion, and then group spatial regions based upon both proximity and local motion similarity to define areas or regions with particular motion characteristics. We are able to process scenes in which objects are difficult to detect and track due to variable frame-rate, video quality or occlusion, and we are able to identify regions which differ by usage but which do not differ by appearance (such as frequently used paths across open space). We demonstrate our method on 50 videos making up very different scene types: indoor scenarios with unpredictable unconstrained motion, junction scenes, road and path scenes, and open squares or plazas. We show that these scenes can be clustered using our representation, and that the incorporation of learned spatial relations into the representation enables us to cluster more effectively.

1 Introduction and Motivation

Through observing the way in which people and vehicles move around, we can learn a lot about the structure of a scene. Some regions will have very constrained motion (a lane of a road will usually feature motion in just one direction) and other regions will have unconstrained motion (chairs and benches are associated with motion in all directions as people shift and gesticulate). We aim to model the variation in motion patterns across a range of very different videos, and show that even in scenes which are superficially unconstrained, such as plazas, we can detect patterns in the way in which people use the spaces.

1.1 Related Work

Whilst there is a large literature on modelling spatial regions using appearance e.g. [13] the current paper falls in the category of scene modelling *from automated*

K. Stewart Hornsby et al. (Eds.): COSIT 2009, LNCS 5756, pp. 295–311, 2009.

analysis of video. This form of scene modelling has thus far been dominated by systems based upon trajectory analysis, object detection or on optical flow patterns.

Trajectory analysis systems take as their input the position over time of individuals moving through the scene. By recording the trajectories over a period of time, clustering can then be used to determine entrance and exit points, popular routes, and other aspects of scene geography. Johnson and Hogg [11] use vector quantization to learn behaviour vectors describing typical motion. Stauffer and Grimson [23] take trajectories and perform a hierarchical clustering, which brings similar motion patterns together (such as activity near a loading bay, pedestrians on a lawn). Makris and Ellis [19] learn scene models including entrances and exits, and paths through the scene, and use these for tracker initialisation and anomaly detection; a similar approach is used in [20]. The rich scene models obtained from trajectory modelling have been used to inform either about the observed behaviour (e.g., typicality detection as in [11]) or about the scene (e.g., using models of trajectory to determine the layout of cameras in a multi camera system as in [12]), or to detect specific behaviour patterns of interest (e.g., elderly people falling down [20]). These systems often require significant quantities of training data. As they rely upon tracking systems, they are also susceptible to tracking error, which can be caused by occlusion or segmentation errors, particularly for busy scenes.

In scene modelling the aim is to describe the way in which people and vehicles tend to move around a scene and to generate a representation tied to geography, specifying which types of motion tend to happen in which parts of the scene. Activity recognition however, works on a smaller scale, determining whether a particular video sequence contains a specific activity (running, jumping, or more fine grained activities such as particular tennis shots). These systems typically use "moving pixels" (such as optical flow) as their inputs, rather than trajectories of entire moving objects. Efros et al., in [7], present early work on activity modelling using normalised optical flow to compare input video with a database of labelled sequences. Laptev and colleagues [15,17,16], use "Space-Time Interest Points", which are spatio-temporal features developed from the 2D spatial-only Harris interest point operators to learn various actions, from surveillance style static camera videos and more recently from cinema movies. Dalal et al. in [5] use similar techniques (based upon histograms of gradients and differential optical flow) for the detection of humans in video; their human detector combines appearance and motion cues so can be seen as using activity to aid detection. Wang et al [25] use a hierarchial Bayesian approach to detect activities occuring in specific places, and as such combine scene modelling with activity recognition.

Understanding the 3D structure of a scene from images or video is another related task; it involves estimating camera geometry and the surface structure of visible objects. Hoiem and colleagues [9,10] have used estimates of building geometry and detections of known objects (people and cars) to estimate the structure of a scene from a single image. Rother et al [21] use the tracks of people and height estimates to work out camera calibration details, but not scene structure. More recently, Breitenstein et al [3] use person detectors to work out what areas of a scene are "walkable", and go on to estimate the orientation of these regions.

We model scenes through observation of motion, and try to determine the way in which the space is used by the people and vehicles moving around within it; in this way our work is similar to the scene modelling work from trajectories. However our approach differs from previous scene modelling techniques in two principal ways: we use as our input features which are more akin to those used in the activity modelling field (which allows us to determine more fine-grained motion), and we use these features over a short time window. Because of the implications of these differences we are able to model a wider variety of scenes than those systems based upon object tracking or background subtraction, in terms of both the range of modelled activities and the quality of the input video. This paper represents a significant extension of our previous work[6], in which we introduced the idea of modelling scenes using directional histograms and presented indicative results on three video datasets.

2   Experimental Dataset

The 50 videos used in our experiments cover a range of different scene types, frame-rates and image qualities. The majority of these have been saved from Internet webcams, and as such have a frame-rate affected by external factors such as server and connection speed. For these videos the frame rate is impossible to determine accurately, but it appears to vary from 5 frames per second (FPS) to a frame per minute, sometimes within the same video sequence. In addition to these webcam videos we use a number of standard frame-rate videos (25 FPS) which are either publicly available or locally captured. Table 1 summarises the datasets.

Table 1. Summary of the 50 videos used in our experiments

Short descriptions	Width	Height	Framerate
4 full frame rate videos: Indoor coffee room; large plaza; underground station; roundabout	720	576	25 FPS
1 full frame rate video of a road	352	288	25 FPS
28 webcams including roads, junctions, foyer scenes, offices, computer clusters, and outdoor plaza scenes	640	480	variable
9 webcams including roads, plazas, junctions, office and computer cluster scenes	704	480	variable
4 webcams featuring roads and plazas	352	288	variable
Outdoor plaza	768	576	variable
Outdoor plaza	704	576	variable
Junction	800	600	variable
Road	320	240	variable

The aim of this work is to characterise these video datasets based upon the observed motion within them. The datasets have significant variation in both content and technical difficulty (from a computer vision perspective); this latter issue restricts the vision tools available to us for processing the visual data. Thus we tackle the interesting challenge of characterising diverse motion patterns whilst dealing with low-frame rate, variable frame rate and poor resolution input data. The webcam streams used in this experiment were chosen from an online site which indexes webcam data and from search engine results[1]. The webcams selected were chosen to provide a cross-section of types of scenes with approximately equal numbers in 4 classes: Roads (15), Plazas or open spaces (11), Indoor office or computer cluster scenes (13), and Junctions (11). These categories are somewhat arbitrary, and emerged through observation of the varieties of video stream people choose to put on the Internet.

3 Feature Detection and Tracking

In order to build models of such varied scenes, we need to use a very robust measure of motion. Different scales and frame rates mean that estimates of velocity are unreliable. Thus our scene model is built upon histograms of motion direction – the only inputs to our system are counts of motion in particular directions. We use the "KLT" tracker [22,18,24] to provide this. KLT is based upon the insight that feature selection for tracking should be guided by the

| 25 frame tracklets | 3 frame tracklets |
| Dataset 4 | Dataset 7 |

Fig. 1. Single frame of video showing tracklets: these give a robust indication of motion in the image plane without committing to any object segmentation or scale

[1] Webcam index site address is http://www.drakepeak.net/index.php?content=livecams; to find webcam data using a search engine, search for axis-cgi/jpg/image.cgi within a page's URL. To save webcam streams as a sequence of numbered image files we use a simple shell script.

tracker itself: the features we select for tracking should be precisely those features which are easy to track. It does not use any motion predictor and is therefore excellent at handling unconstrained scenes with unpredictable motion. To avoid problems of occlusion and frame jumps, we use *tracklets*: by reinitialising the tracker every M frames we get tracks which are reliable but long enough to provide a descriptive representation of feature motion. We divide our datasets into two classes based upon frame rate, and for those videos with full (25 FPS) frame rate set $M = 25$, and for the variable frame rate webcam data set $M = 3$. Thus the duration of our features is short – whilst it is impossible to determine the frame rate of some of our datasets, our intuitions estimate the average to be in the region of three frames per second. For tracklets where distance travelled between first and last points exceeds a threshold (2 pixels, in our experiments) the direction of travel is quantized into 8 directions: *up, up-right, right, down right,* .... Figure 1 shows sample tracklets for two of our scenes.

4   Representing Videos by Their Motion Patterns

Each input scene is broken down into non-overlapping blocks with $N \times N$ pixels. In the current implementation, N is 16, regardless of input video dimensions and regardless of perspective effects; we have carried out preliminary experiments varying N and find that the resultant segmentations are fairly robust to different values. The procedure we follow to create our representation is detailed in Sections 4.1 and 4.2, and summarised in Figure 2.

4.1   Spatial Quantization: Histograms of Features

For each $N \times N$ pixel block in the image, we create an 8 dimensional feature vector $\mathbf{f}$ by counting the number of tracklets that pass through that square in each of our 8 canonical directions. These counts are thresholded and bins with small counts (fewer than 10 instances per video) are set to zero in order to remove noise in regions of the scene which see little motion. The resulting counts are then normalised by the maximum count for that particular block. We can visualise these histograms as "star diagrams", which have arrows of length proportional to the count in the corresponding direction, and give an intuitive visualisation of the overall motion pattern found in each $N \times N$ grid square. A sample grid and some of the accumulated histograms can be seen in Figure 3.

These feature vectors are computed for each of the the blocks across all 50 of our input videos: the total across all scenes is 59,402. This forms our feature set $\{\mathbf{f_j}\}$. We then learn a set of K prototype vectors using K-means from a subset of 20 videos consisting of 5 scenes chosen randomly from each of our 4 rough scene classes (*Junction, Indoor, Plaza* and *Road*). The size of this training set varies depending on the choice of videos, but typically this procedure gives us around 22,000 training vectors which represent various types of observed motion, from which we learn a codebook $\{\mathbf{p_k}\}$ of K prototypes. Each input square is labelled with its nearest neighbour in the codebook which provides us with an

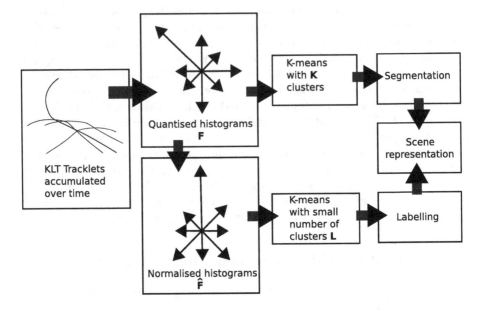

Fig. 2. Flow diagram summarising the scene representation

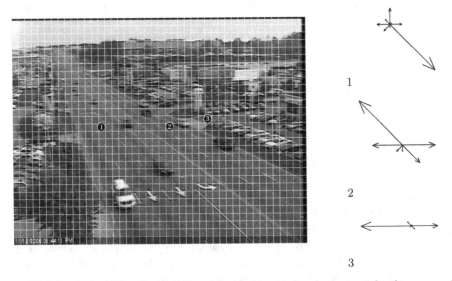

Fig. 3. Screenshot of video 31 with grid superimposed; column to right shows sample histograms represented as "star diagrams". This scene is a large road with carparks to either side and several junction regions in which cars enter and exit the road area. The star diagrams show a predominantly unidirectional road region (1) a region around a junction (2) and the road to a carpark where vehicles both enter and exit (3).

Image of scene   K=15   K=20   K=30

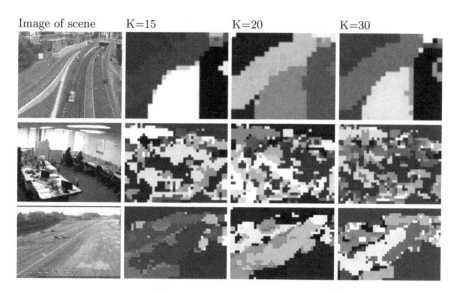

Fig. 4. Figures showing raw K-means segmentation for three of the input scenes (From top to bottom, videos 10, 19 and 39). Note that higher K (rightmost column) can split nearground and farground in road regions. The office scene appears very noisy, but it is possible to determine that table/desk regions tend to be coloured similarly, as do chair regions.

initial partitioning of each input scene. Some typical K-means segmentations for different values of K are shown in Figure 4. The choice of K has to be large enough to capture the common patterns of motion, but small enough to provide meaningful segmentations in which similar regions are classified in a similar way.

To create smoother segmentations we use a Conditional Random Field formulation [2,14,1] and find the MAP solution using graph cuts; we define energy as a function of both the input frequency histogram (the *data term*) and the cost of labelling adjacent squares differently (the *smoothness term*). We use a smoothness term which penalises the labelling of adjacent squares with different classes – i.e., that encourages large regions of the same class. The advantage of the CRF framework is that it creates smooth segmentations which preserve sharp boundaries where they exist. The effect of varying the relative influence of the smoothness term and the data term is shown in Figure 5; we choose a smoothness proportion of 0.025 as this removes small noise regions whilst preserving scene detail.

4.2   Direction Invariant Representation

The descriptor developed in the previous section captures the gross pattern of motion observed in a scene square, but does not capture the way in which we often find similar behaviours but in different directions. Consider a road with two carriageways, going in opposite directions: these will be classified as two

Input $C_s = 0$ $C_s = 0.025$ $C_s = 0.1$ $C_s = 0.5$

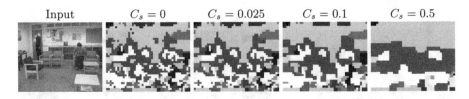

Fig. 5. The effect of varying the smoothness constant in the MRF formulation; input image at left, increasing smoothness constant going from left to right. C_s is the weight given to the smoothness term: 0.5 gives smoothness and data terms equal weight, 0 removes the smoothness term altogether.

separate types of motion using the previous descriptor. Our next step allows us to classify these as the same *type* of motion, just in different directions. We do this by transforming our input descriptors to a direction invariant motion descriptor, and then through a second application of K-means.

The direction invariant motion descriptors $\{\hat{\mathbf{F}}_j\}$ are created by "rotating" the input feature vectors $\{\mathbf{F}_j\}$ so that their largest entry is in the first position. These

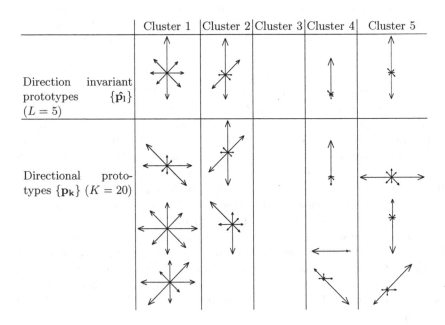

Fig. 6. Top row shows direction invariant histograms, and the lower rows show some of the corresponding directional histograms. (We do not show all 20 corresponding directional histograms due to space constraints.) Segmentation is performed using the directional histograms, and regions thus obtained are then labelled using the direction invariant histograms. Note that prototype three is not an empty column, but the prototype corresponding to very little motion.

Fig. 7. Sample multilevel segmentations. The first two scenes show roads. The second scene shows one of our more challenging videos with very low frame rate and large amounts of cloud motion; however the road in the foreground has been classified as having the same variety of motion as the road in the first scene. Note that the chair regions in the coffee room and the computer cluster scene are highlighted in the same class (but also that the challenging road scene has some spurious foreground motion classified similarly). These figures were generated with $K = 20$ and $L = 5$.

are then clustered using K-means into L clusters, which gives us a codebook $\{\hat{\mathbf{p}}_\mathbf{l}\}$ of direction-invariant motion prototypes. Obviously, the clusters obtained vary with L, but we find most values of L lead to a cluster which represents very little motion, clusters representing both uni- and bi-directional motion, a cluster representing motion in all directions, and in the case of higher L we often find clusters representing "crossroads" patterns representing motion on orthogonal axes.

To build our scene model we use directional prototypes $\{\mathbf{p_k}\}$ with $K = 20$ to segment the scene into regions, and then for each of these we rotate its prototype and find the nearest direction invariant prototype in the set $\{\hat{\mathbf{p}}_\mathbf{l}\}$, which we use as the label for that region. The representations obtained show clear similarities with the input video for many of our videos. These similarities are at their clearest in scenes with well-defined roads and with good video quality; however even in our most challenging datasets some similarities are apparent. The star diagrams in Figure 6 show a codebook of direction invariant histograms learned when $L = 5$, and the corresponding directional histograms which are used to segment the input image into regions. Figure 7 shows some example scene models.

5   Classifying and Summarising Motion

These segmentations give us a model of the scene in terms of the types of motion pattern we can expect to see in a particular scene region. We would like to be able to draw conclusions about the structure of the scene by considering the way in which regions of different class are found together – for example we would like to be able to say that regions with predominantly unidirectional motion are often found together (as we see with roads and roundabouts).

5.1   Learning Spatial Relationships

We adopt an approach similar to that described in [8] and learn pairwise spatial relationships from our set of learned regions. The relations we want to capture for spatial reasoning are things like *"above"*, *"below"* and *"overlapping"*. We capture metric information about the pairwise spatial relationships between learned

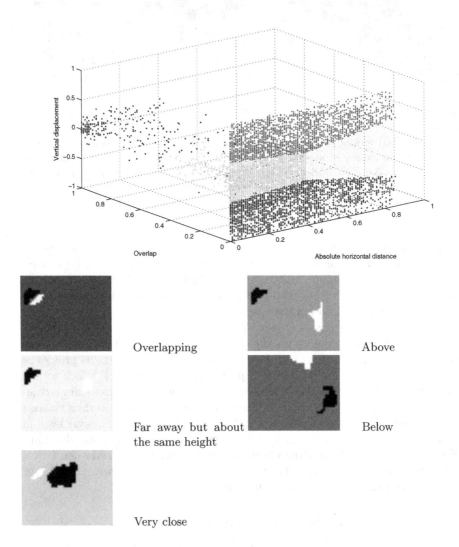

Fig. 8. Spatial relationships learned from image data. Scatterplot shows qualitative spatial relations obtained from the 3D relational descriptor as regions in feature space. The images below are each an example of a pair of objects identified as instantiating one of the 5 spatial relations indicated in the scatter plot, with the black region being in relation X to the white region.

image regions using three measures extracted from the pair of regions' bounding boxes (i.e. minimal enclosing rectangle): the vertical displacement, the absolute horizontal displacement, and the degree of overlap. These dimensions are summarised in Equation 1, in which Y and X represent the vertical and horizontal position of an object's bounding box centre respectively, and B represents the set of pixels inside the object's bounding box.

$$\left(Y_i - Y_j, |X_i - X_j|, \frac{|B_i \cap B_j|}{|B_i|} \right) \tag{1}$$

These measures are recorded for each pair of regions within each scene, then clustered using K-means into 5 clusters to learn spatial relations. The relations we learn correspond roughly to the concepts of "above", "below", "very close", "about the same height and far", and "overlapping". The classification of points in this three dimensional *shape space* and the classification of relationships between some example pairs of shapes are shown in Figure 8.

We also use a variant on these spatial relations only considering pairwise spatial relationships between objects which are *near* to each other. This is because we expect the relationships between those objects that are close to each other to be more significant than relationships between objects at opposite sides of the scene. Interestingly, the relations learned when only considering near objects are the same, just with fewer members in the "farther away but at the same height", "above" and "below" categories.

Combining these learned spatial relations with the segmentations learned from data we can generate a compact representation of each video by building a histogram of pairwise relations between region classes. For the $L = 5$ representation this results in a 125 element histogram (5 region classes x 5 region classes x 5 spatial relations). These histograms capture the fact that regions are in particular spatial relations to each other, but do not preserve information about region size or other metric information.

6   Evaluation

There are various ways in which we can evaluate this work. Some of it has to be informal, as there are things the scene modelling technique tells us that we didn't know before. This is particularly true of plaza scenes in which we find common paths across otherwise uniform open space; observation of people moving in this scenes would tell us that the paths exist but the details only emerge when seen through a system such as ours. We can evaluate against two forms of "ground truth": one is the simple four class (*Plaza, Junction, Road, Indoor*) scene categorisation, and the other is a more detailed classification in which each scene is hand-segmented into conceptual regions: road, junction, foliage, pedestrian crossing, plaza, chair, sky, and "null space", which we expect to be empty. This hand segmentation is then summarised as a normalised histogram with the proportion of video in each class. Images representing this ground truth are shown in Figure 9 alongside the $L = 5$ segmentations and a video screenshot for four sample scenes.

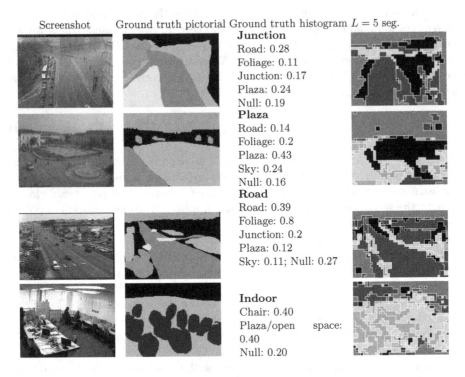

Screenshot   Ground truth pictorial Ground truth histogram $L = 5$ seg.

Junction
Road: 0.28
Foliage: 0.11
Junction: 0.17
Plaza: 0.24
Null: 0.19
Plaza
Road: 0.14
Foliage: 0.2
Plaza: 0.43
Sky: 0.24
Null: 0.16
Road
Road: 0.39
Foliage: 0.8
Junction: 0.2
Plaza: 0.12
Sky: 0.11; Null: 0.27

Indoor
Chair: 0.40
Plaza/open   space:
0.40
Null: 0.20

Fig. 9. Pictorial (column 2) and histogram (column 3) ground truth shown alongside segmentations and screen shots for sample scenes. The histogram ground truth is represented as text with empty categories omitted for space reasons. The top row shows a scene classed "junction", but the junction part of the scene comprises 17%, and the video also includes elements of road and plaza. This shows that single-class ground truth must be an approximation.

6.1   Informal Evaluation of Segmentation Quality

Our learned segmentations show a superficial similarity to the structure we see in the scene, particularly in the office and the road scenes where we can often find clear images of the roads and chairs in the learned segmentation (see, for example, the first column in Figure 7 in which the carriageways of the road are clearly segmented and identified as being the same type). Chair and bench regions are often identified as regions of increased motion in all directions.

If we consider plaza regions – areas of geographically undifferentiated open space – we find that our motion-based segmentations show us these are not actually undifferentiated at all. As any architect will confirm, people moving around open space do not use the whole space and informal paths often form (much to the annoyance of gardeners, in grassed areas). Figure 10 shows segmentations learned from open spaces: from these it is clear that pedestrians choose to follow different paths through open space and that usage is far from uniform.

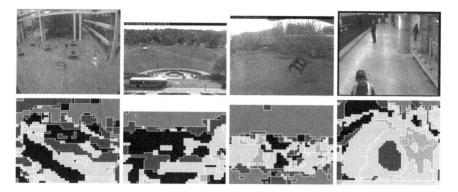

Fig. 10. Multilevel segmentations for some of the plaza scenes. The first two scenes show roads. Black regions are largely bidirectional; dark regions unidirectional. These images show that even in open space certain regions are effectively "paths". In particular, the scene depicted in the first column has unidirectional regions heading from right-to-left and from left-to-right. These figures were generated with $L = 5$

6.2 Clustering against a Single Class "Ground Truth"

Our input scenes fall into four broad classes, which provide us with a rough "ground truth" about the sort of things we can find in the scene. Clearly the difference between a *road* and a *junction* is imprecise, however, we can still use these categories as an approximate guide to evaluate our unsupervised clusterings. The way in which we do this is to compute a range of summary feature vectors for each scene representing the segmentation and then to cluster each of these vectors into four categories. If we have a perfect match, all of the roads will be in one class and all of the indoor scenes in another. As this is an unsupervised classification task we cluster the scenes and then find the optimal one-to-one assignment of each group to one of our four classes. We investigate the use of five different feature vectors, which are summarised in Table 2.

The misclassification rates for these clusterings are shown in Figure 11: note that chance performance here would be a misclassification rate of 74%[2] It is clear from this figure that the incorporation of spatial relations in the clustering lowers the misclassification rate significantly. To obtain these results we performed the experiment ten times with a different random selection of 20 videos for training each time: we present the mean of these ten runs.

An investigation of the detail behind these results indicates that we can deal with the "easy" scenes but have difficulty with the harder ones: many scenes fall into more than one of our rough ground truth classes (e.g. the plaza with a road across the back, or the office with a road clearly visible through the window). Perhaps most interestingly, these results show clearly that spatial information can assist in scene categorisation.

---

[2] This is due to the slight discrepancy between the size of each class in the ground truth – we have a few more roads and therefore random performance would be slightly less than 75%.

Table 2. Summary of feature vectors used for evaluation against ground truth. All of these feature vectors are normalised by dividing by the maximum per scene before clustering.

Label	Size	Summary
Block counts	L	Number of blocks of each region class in the scene.
Region counts	L	Number of regions of each class in the scene.
Nearness alone	$L \times L$	Region class of each pairwise combination, counting only those pairs where the regions are nearby.
Spatial relationships	$L \times L \times 5$	Spatial relationship and region class for each pairwise combination.
Spatial relationships near only	$L \times L \times 5$	Same as "spatial relationships", but only counting regions which are also near.

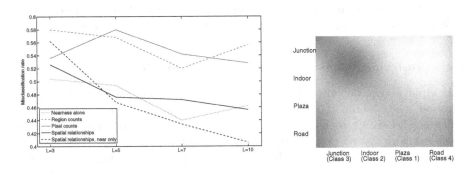

Fig. 11. Left: Overview of misclassification rates for clusterings with "pixel" counts, region counts, and spatial information. Right: Confusion matrix for the $L = 10$ spatial relationships with near (the best performing). Note that the highest values (darkest colours) are predominantly on the main diagonal, and that the majority of confusion is between the road, plaza and junction categories.

6.3   Information Retrieval

Another way to evaluate our representations is to treat each scene in turn as a "target" scene, and find the other scenes which are closest to that scene as in information retrieval. Using the same feature vectors as for clustering (see Table 2) we can find the nearest few scenes to any particular input scene. Using this technique and comparing it to the approximate single class ground truth (road, junction, plaza or indoor) we get the results shown in Table 3, which shows the percentage of scenes of the same class found when we retrieve 5 scenes from our dataset.

Using our second type of ground truth, we can obtain an ordering of all scenes in terms of similarity (so for scene 1, we can rank the rest of the videos in terms of similarity of content). We do this by treating the ground truth histogram as a feature vector and ordering the scenes according to Euclidean distance in feature

Table 3. Information retrieval results: count of % retrieved in same class as target using single class ground truth; 5 scenes retrieved

Type	$L = 3$	$L = 5$	$L = 7$	$L = 10$
Pixel counts	43	46	45	48
Region counts	35	29	23	27
Nearness	37	47	49	51
Spatial relations	32	38	37	47
Spatial relations and nearness	37	46	43	53

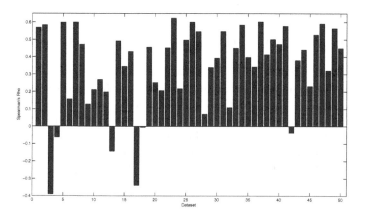

Fig. 12. Correlation results between rankings derived from histogram ground truth and rankings derived from spatial relationships using $L = 10$

space from the target scene. We can perform the same kind of ordering with our representations. Given two rank orderings these can be compared statistically using a rank order correlation, for example Spearman's Rho[4], and these correlations can then be tested for significance. Spearman's Rho is calculated as in Equation 2, in which n is the number of videos, and d is the difference between matched pairs of ranks.

$$r_s = 1 - \frac{6\sum_{i=1}^{n} d_i^2}{n(n^2 - 1)} \tag{2}$$

Figure 12 shows correlations between the rankings derived from ground truth and those derived from "Spatial relationships and nearness" with $L = 10$ (other correlations are omitted due to lack of space, but show a similar pattern). 35 out of 50 correlations have $p < 0.05$; this shows that our representation generally correlates strongly with the ground truth in ranking datasets. Those correlations which were particularly low (indeed, negative) indicate datasets upon which our method does not match the ground truth well; these are datasets 3, 4, 13, 17, 18 and 42. Dataset 17 is a road scene with large amounts of cloud motion, and considerable camera shake (shown in the second column of Figure 7). All of the other problematic datasets are plaza scenes, and we have already described the issues surrounding with evaluation in these cases.

7   Conclusions and Future Directions

This paper has described a novel way of learning a spatial representation directly from video, in an unsupervised fashion. We do this without needing to detect or track moving objects, and the accumulated short tracklets provide a robust representation that withstands the ambiguity of cluttered scenes and variable frame rates. These representations tell us how space is used by the people and vehicles which move around within it, and can be used for clustering videos and ordering video streams in terms of structural similarity between the depicted scenes. We have shown that incorporating learned spatial relationships into these representations provides us with a compact but powerful way of characterising video in terms of regions of motion and the relations between these regions.

A multi-scale approach, perhaps allowing for perspective effects, could be advantageous and this is certainly something to consider in future work. It would also be interesting to include models of temporal variation within the scenes; the work described here essentially integrates all motion observed within the time frame of the video and does not attempt to model any temporal variation.

Acknowledgements

This work was supported by EPSRC project LAVID, EP/D061334/1.

References

1. Boykov, Y., Kolmogorov, V.: An experimental comparison of min-cut/max-flow algorithms for energy minimization in vision. IEEE transactions on Pattern Analysis and Machine Intelligence (PAMI) 26(9), 1124–1137 (2004)
2. Boykov, Y., Veksler, O., Zabih, R.: Efficient approximate energy minimization via graph cuts. IEEE transactions on Pattern Analysis and Machine Intelligence (PAMI) 20(12), 1222–1239 (2001)
3. Breitenstein, M.D., Sommerlade, E., Leibe, B., Van Gool, L., Reid, I.: Probabilistic parameter selection for learning scene structure from video. In: Proc. British Machine Vision Conference, BMVC (2008)
4. Clarke, G.M., Cooke, D.: A basic course in statistics, 3rd edn. Edward Arnold, London (1992)
5. Dalal, N., Triggs, B., Schmid, C.: Human detection using oriented histograms of flow and appearance. In: Leonardis, A., Bischof, H., Pinz, A. (eds.) ECCV 2006. LNCS, vol. 3952, pp. 428–441. Springer, Heidelberg (2006)
6. Dee, H.M., Fraile, R., Hogg, D.C., Cohn, A.G.: Modelling scenes using the activity within them. In: Freksa, C., Newcombe, N.S., Gärdenfors, P., Wölfl, S. (eds.) Spatial Cognition VI. LNCS, vol. 5248, pp. 394–408. Springer, Heidelberg (2008)
7. Efros, A.A., Berg, A.C., Mori, G., Malik, J.: Recognizing action at a distance. In: Proc. International Conference on Computer Vision (ICCV), Nice, France (2003)

8. Ommer, B., Buhmann, J.M.: Object categorization by compositional graphical models. In: Rangarajan, A., Vemuri, B.C., Yuille, A.L. (eds.) EMMCVPR 2005. LNCS, vol. 3757, pp. 235–250. Springer, Heidelberg (2005)
9. Hoiem, D., Efros, A.A., Hebert, M.: Putting objects in perspective. In: Proc. Computer Vision and Pattern Recognition (CVPR), pp. 2137–2144 (2006)
10. Hoiem, D., Efros, A.A., Hebert, M.: Closing the loop in scene interpretation. In: CVPR (2008)
11. Johnson, N., Hogg, D.C.: Learning the distribution of object tractories for event recognition. Image and Vision Computing 14(8), 609–615 (1996)
12. KaewTraKulPong, P., Bowden, R.: Probabilistic learning of salient patterns across spatially separated, uncalibrated views. In: Intelligent Distributed Surveillance Systems, pp. 36–40 (2004)
13. Kaufhold, J., Colling, R., Hoogs, A., Rondot, P.: Recognition and segmentation of scene content using region-based classification. In: Proc. International Conference on Pattern Recognition, ICPR (2006)
14. Kolmogorov, V., Zabih, R.: What energy functions can be minimized via graph cuts? PAMI 26(2), 147–159 (2004)
15. Laptev, I.: On space-time interest points. International Journal of Computer Vision 64(2/3), 107–123 (2005)
16. Laptev, I., Marszalek, M., Schmid, C., Rozenfeld, B.: Learning realistic human actions from movies. In: Proc. Computer Vision and Pattern Recognition, CVPR (2008)
17. Laptev, I., Pérez, P.: Retrieving actions in movies. In: Proc. International Conference on Computer Vision (ICCV), Rio de Janeiro, Brazil (2007)
18. Lucas, B.D., Kanade, T.: An iterative image registration technique with an application to stereo vision. In: International Joint Conference on Artificial Intelligence, pp. 674–679 (1981)
19. Makris, D., Ellis, T.: Learning semantic scene models from observing activity in visual surveillance. IEEE Transactions on Systems, Man and Cybernetics 35(3), 397–408 (2005)
20. McKenna, S.J., Charif, H.N.: Summarising contextual activity and detecting unusual inactivity in a supportive home environment. Pattern Analysis and Applications 7(4), 386–401 (2004)
21. Rother, D., Patwardhan, K.A., Sapiro, G.: What can casual walkers tell us about a 3D scene?. In: Proc. International Conference on Computer Vision (ICCV), Rio de Janeiro, Brazil (2007)
22. Shi, J., Tomasi, C.: Good features to track. In: Proc. Computer Vision and Pattern Recognition (CVPR), pp. 593–600 (1994)
23. Stauffer, C., Grimson, E.: Learning patterns of activity using real-time tracking. IEEE transactions on Pattern Analysis and Machine Intelligence (PAMI) 22(8), 747–757 (2000)
24. Tomasi, C., Kanade, T.: Detection and tracking of point features. Technical Report CMU-CS-91-132, Carnegie Mellon (1991)
25. Wang, X., Ma, X., Grimson, W.E.L.: Unsupervised activity perception in crowded and complicated scenes using hierarchical Bayesian models. IEEE transactions on Pattern Analysis and Machine Intelligence (PAMI) 31(3), 539–555 (2009)

A Qualitative Approach to Localization and Navigation Based on Visibility Information

Paolo Fogliaroni[1], Jan Oliver Wallgrün[1], Eliseo Clementini[2],
Francesco Tarquini[2], and Diedrich Wolter[1]

[1] Universität Bremen, Department of Mathematics and Informatics
Enrique-Schmidt-Str. 5, 28359 Bremen, Germany
{paolo,wallgruen,dwolter}@informatik.uni-bremen.de
[2] Department of Electrical Engineering, University of L'Aquila
67040 Poggio di Roio (AQ), Italy
{eliseo.clementini,francesco.tarquini}@univaq.it

Abstract. In this paper we describe a model for navigation of an autonomous agent in which localization, path planning, and locomotion is performed in a qualitative manner instead of relying on exact coordinates. Our approach is grounded in a decomposition of navigable space based on a novel model of visibility and occlusion relations between extended objects for agents with very limited sensor abilities. A graph representation reflecting the adjacency between the regions of the decomposition is used as a topological map of the environment. The visibility-based representation can be constructed autonomously by the agent and navigation can be performed by simple reactive navigation behaviors. Moreover, the representation is well-qualified to be shared between multiple agents.

1 Introduction

Navigation has been defined as coordinated and goal-directed movement through the environment [1] and is deemed to be one of the fundamental abilities for an autonomous physical agent like a mobile robot. Navigation comprises several important subprocesses like *localization, path planning*, and *locomotion*. Navigation is also intrinsically linked to the agent's internal representation of the environment, often simply called the agent's *map*, which needs to be learned and maintained based on sensor information gathered over time (a process referred to as *mapping*).

The predominant approach to perform mapping and navigation on a mobile robot is to have the robot use a map in which spatial features of the environment are specified based on metric coordinates within a global frame of reference. Localization then comprises tracking the exact position of the robot within this coordinate system (see [2] for an overview on state-of-the-art techniques following this approach). However, there have also been a number of approaches in which localization is performed in a rather qualitative way based on distinct places derived from the environment which can be recognized and distinguished by their

K. Stewart Hornsby et al. (Eds.): COSIT 2009, LNCS 5756, pp. 312–329, 2009.

respective sensory input [3,4]. The resulting spatial representation is typically a graph representation called a *topological map* in which vertices stand for places or views and edges connect vertices that are connected or adjacent [5]. This second approach is often inspired and motivated by biological systems, for instance by results on human navigation and spatial representations which show that knowledge about the geometric layout of an environment is developed last and that the metric information is usually distorted [6,7].

One approach to deriving places from environmental information is based on the visibility of particular salient objects called *landmarks* and the spatial relations holding between them as it is known that landmarks play an important role in human navigation [8,9]. One early approach of a landmark-based navigation system for an autonomous agent is the work by Levitt and Lawton [10] which has later been adapted by Schlieder [11] and Wagner [12]. Levitt and Lawton use the lines connecting pairs of point landmarks to decompose the environment into regions which constitute the topological map. The regions differ in the cyclic order of landmarks which an agent will perceive from any location within the region. In contrast, to other topological mapping approaches like the view graph approach [4] in which the structure of the graph depends on the starting point and exploration path of the agent, the representation here is directly induced by the environment. As a result, the same map can be generated from the perceptions of a roaming agent as well as from an allocentric geometric representation of the environment.

In this work, we describe an approach that is similar in spirit to the landmark-based approach. However, in our case the decomposition of space is derived by combining a novel relational model of visibility and occlusion with cyclic order information. One main innovation is that our model deals with extended convex objects instead of being restricted to point landmarks. As (partial) occlusion of objects is the typical case rather than the exception in most environments, basing the spatial representation of an agent on these concepts and allowing for extended objects makes our approach very generally applicable and well-qualified for scenarios in which the representation should be shared between multiple agents. Since the regions in our decomposition approach can be distinguished by agents with very limited sensor abilities and due to the qualitative nature of our approach which does not rely on precise sensor measurements, localization, navigation, and map learning algorithms can also be implemented rather easily.

The remainder of this paper is structured as follows: we first develop a relational model for visibility and occlusion between extended objects in Sect. 2. Based on this model, we define a decomposition of navigable space and derive a topological map representation from it (Sect. 3). Identification of the regions from the egocentric perspective of an autonomous agent is described in Sect. 4. In Sect. 5 we explain how navigation and map learning are obtained in our approach. Finally, we describe an extension which leads to a finer grained decomposition of space in Sect. 6.

2   Visibility and Occlusion between Extended Objects

In the following we will first introduce a universal model for visibility and occlusion between objects (partially described in [13]) and then adapt the model to an agent-based approach to represent visibility of extended objects from the observation point of the agent. We model visibility and occlusion as ternary relations between convex objects. Concavities can still be treated by decomposing objects into convex parts or, simply considering the convex hull instead. As a result, we obtain a much more compact and elegant theory.

In contrast to other work on visibility and occlusion [14,15], our model only differentiates relations which can be distinguished by the observing agent (e.g., we do not consider different cases of complete occlusion like in [14]) where 14 relations are distinguished corresponding to the topological RCC-8 relations [16]. Another difference is that we use ternary relations which allows us to switch between an allocentric perspective and the egocentric perspective of an observing agent.

In the following we will use x, y, z, etc. to refer to points. Regions are simple bounded point-sets and will be indicated with capital letters X, Y, Z, etc. Lines are specified by two points lying on them, e.g., $\overline{xy}$. A closed interval on the line is written as $[\overline{xy}]$, where x and y are the extreme points of the interval. An open interval is indicated by $(\overline{xy})$.

2.1   Visibility among Points

In this section we define a ternary visibility relation among points based on collinearity which is a basic concept of projective geometry. Given a configuration of three points $x, y, z \in \mathbb{R}^2$ where x is the primary (or observed) object and y and z are the reference objects (or, in order, the obstacle and the observer), point x is visible from point z if and only if point y does not lie on segment $[\overline{xz}]$ as depicted in Fig. 1(a). Hence, we define the ternary projective relation

$$Visible(x, y, z) \iff y \notin [\overline{xz}] \tag{1}$$

On the other hand, when point y lies on segment $[\overline{xz}]$ the $Visible$ relation is not satisfied and we introduce another relation called $Occluded$ for this case (see Fig. 1(b)):

$$Occluded(x, y, z) \iff y \in [\overline{xz}] \iff \neg Visible(x, y, z) \tag{2}$$

The way Occluded is defined, also holds when the obstacle coincides with either the observer or the observed object. We note that the $Visible$ and $Occluded$ relations are invariant with regard to the permutation of first and third objects which will be relevant later for changing between an allocentric and egocentric perspective:

$$\begin{aligned} Visible(x, y, z) &\iff Visible(z, y, x) \\ Occluded(x, y, z) &\iff Occluded(z, y, x) \end{aligned} \tag{3}$$

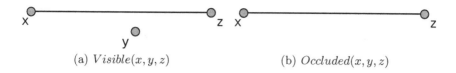

(a) $Visible(x, y, z)$ (b) $Occluded(x, y, z)$

Fig. 1. Visibility relations among points

2.2   Visibility among Regions

We will now define ternary relations $r(A, B, C)$ among regions where A is again the primary (or observed) object, and B and C are the reference objects (obstacle and observer). The extension of the visibility model from points to regions is basically an application of the visibility relations to the point sets belonging to the three regions. However, some new aspects arise. The *Visible* and *Occluded* relations remain and coincide with the instance where the whole observer can see the whole observed object and with the case where no part of the observed object can be seen by any part of the observer, respectively. However, when the *Visible* relation is true for some points of C but not for others, we have a case of partial visibility. We assume that the observer is able to determine which object is the occluding one (the one which is closer along the line of sight that passes through both objects), which means that several cases of partial occlusion can be distinguished. Therefore, we introduce new relations $PartiallyVisible^{Left}$, $PartiallyVisible^{Right}$, and $PartiallyVisible^{Joint}$. We define these by giving the respective acceptance areas for A given the reference frame formed by the observer C and the obstacle B (see Fig. 2).

The acceptance areas are constructed by tracing the internal and external tangents between B and C (an idea taken from [17,18]) which splits the plane into zones. As shown in Fig. 2(a) and 2(b), two cases can arise depending on whether the external tangents intersect behind B or not. We will in the following use the notation $\lambda_{C,B}^{ext,right}$ to refer to the external tangent passing C and B on the right as seen from C and similar notations for the other tangents (cf. Fig. 2). Each tangent has three parts: The middle part connecting B and C is a line segment and will be called $middle(\lambda)$ for a tangent λ while the other two parts are half-lines referred to as *half-tangents* and we will write $behind(\lambda, B)$ for the half-tangent of λ that lies behind B (from the perspective of C).

By tracing the tangents we obtain a *LightZone* (LZ) and a *ShadowZone* (SZ) and also, as said, acceptance areas for partial visibility: *LeftTwilightZone* ($TZ^{Left}(B, C)$) and a *RightTwilightZone* ($TZ^{Right}(B, C)$) can be distinguished based on whether the occluding object is to the left or the right of the occluded one. When the *ShadowZone* is limited (Fig. 2(a)) another zone named *JointTwilightZone* ($TZ^{Joint}(B, C)$) originates which means the occluded objects appears to both sides of the occluding one.

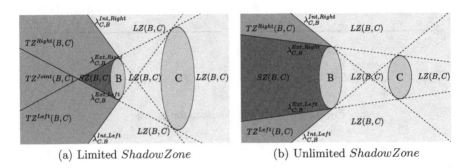

(a) Limited *ShadowZone*   (b) Unlimited *ShadowZone*

Fig. 2. Visibility acceptance areas for regions

Our set of elementary visibility relations among regions is defined based on these acceptance areas:

$$Visible(A, B, C) \iff A \subseteq LZ(B, C) \tag{4}$$

$$PartiallyVisible^{Left}(A, B, C) \iff A \subseteq TZ^{Left}(B, C) \tag{5}$$

$$PartiallyVisible^{Right}(A, B, C) \iff A \subseteq TZ^{Right}(B, C) \tag{6}$$

$$PartiallyVisible^{Joint}(A, B, C) \iff A \subseteq TZ^{Joint}(B, C) \tag{7}$$

$$Occluded(A, B, C) \iff A \subseteq SZ(B, C) \tag{8}$$

Lastly, we want to point out that because the observed object A is an extended region, it can overlap more than one acceptance zone simultaneously. We will refrain from introducing new relation for these cases and instead we will for instance write $A \cap LZ(B, C) \neq \emptyset \wedge A \cap TZ^{Right}(B, C) \neq \emptyset$ if we need to describe a configuration in which A overlaps both $LZ(B, C)$ and $TZ^{Right}(B, C)$.

2.3   Hybrid Visibility: Point-Region

In this section, we will briefly discuss the hybrid case where the observer object is a point (as is suitable for an agent with monoscopic perception) and the obstacle is a region. We can look at it as an extreme case of visibility between regions (Sec. 2.2) in which the observer object collapses into a point. As a result, the internal and external tangents will coincide so that the *TwilightZone* and the *PartiallyVisible* relations (Right, Left and Joint) will not exist. Consequently, the point case properties (Eq. 3) are applicable for this case as well:

$$\begin{aligned} Visible(A, B, c) &\iff Visible(c, B, A) \\ Occluded(A, B, c) &\iff Occluded(c, B, A) \end{aligned} \tag{9}$$

Furthermore, the following new property arises:

$$A \cap LZ(B, c) \neq \emptyset \wedge A \cap SZ(B, c) \neq \emptyset \iff PartiallyVisible^*(c, B, A) \tag{10}$$

where by $*$ we mean that one of the three partial visibility relations can be true (Right, Left and Joint).

3   Space Subdivision and Topological Representation

In a Cartesian 2-dimensional frame of reference, a point is identified by a pair
of real numbers (x,y). But looking at humans, it becomes clear that such a
precision is really not required in order to be able to navigate successfully. In
the following, we develop a low-precision qualitative coordinate system in which
the qualitative coordinates are expressed as "what we can see from a certain
position". The approach should be able to deal with extended and point-like
objects. We start by subdividing the plane into a finite set of regions based on
our visibility model for extended objects from the previous section. A method
for identifying the regions will be developed in Sec. 4. An example showing the
subdivision for a configuration of three extended objects is given in Fig. 3(a).
We will use this example throughout the remainder of this text.

To give a definition of the subdivision, we first introduce a function $cut(\mu)$
that for a half-tangent $\mu = behind(\lambda, X)$ yields the segment of μ that connects
X with the closest point on the boundary of another object which intersects
μ, or the entire half-tangent μ if no such object exists. We need this function
because if a half-tangent intersects another object, the two points defining the
tangent will not be visible simultaneously on the other side of this object and,
hence, cannot serve as a region boundary there.

Now given the set $\mathcal{O}$ of objects in the environment and the set $\mathcal{T}$ containing
all tangents $\lambda_{X,Y}^{ext,left}$, $\lambda_{X,Y}^{ext,right}$, $\lambda_{X,Y}^{int,left}$, and $\lambda_{X,Y}^{int,right}$ for each pair of objects
$X, Y \in \mathcal{O}$, the set of boundaries demarcating the regions in our subdivision
consists of:

1. the boundaries of each object $O \in \mathcal{O}$
2. all $cut(behind(\lambda_{X,Y}, X))$ and $cut(behind(\lambda_{X,Y}, Y))$ with $\lambda_{X,Y} \in \mathcal{T}$ for which
   there is no object which intersects $middle(\lambda_{X,Y})$

The regions of the subdivision are the holes in the mesh generated by these
boundaries except for those holes which correspond to objects from $\mathcal{O}$. In the
following $\mathcal{R}$ stands for the regions contained in the subdivision induced by the
objects in the environment as described above. The restriction in point 2 above is
needed because an object crossing the middle part of a tangent means that both
points defining the tangent will never be observable simultaneously from a point
behind X or Y. This case occurs in the example from Fig. 3(a) in which object
B intersects one of the external tangents between A and C. As a result, the
half-tangents $behind(\lambda_{C,A}^{ext,right}, A)$ and $behind(\lambda_{C,A}^{ext,right}, C)$ do not contribute to
the boundaries in the subdivision.

One consequence of the subdivision approach described here is that the actual
number of regions in the subdivision depends on the actual configuration of
objects. The strength of our subdivision approach is that the regions can be
identified from the egocentric perspective of an autonomous agent with very
limited sensor abilities as will be explained in Sect. 4.

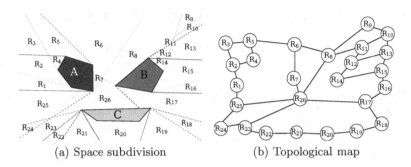

(a) Space subdivision   (b) Topological map

Fig. 3. The subdivision and topological map based on the visibility model

Given a subdivision $\mathcal{R} = \{R_1, R_2, ..., R_n\}$ it is possible to build a topological map in form of an undirected graph $\mathcal{G} = (\mathcal{V}, \mathcal{E})$:

- Every vertex $v_i \in \mathcal{V}$ represents one of the regions we obtained from the space subdivision ($\mathcal{V} \equiv \mathcal{R}$)
- An edge $e_{ij} = (v_i, v_j) \in \mathcal{E}$ exists between vertices v_i and v_j iff one can move between v_i and v_j by crossing a single boundary line.

The topological map corresponding to the example depicted in Fig. 3(a) is shown in Fig. 3(b).

4   Identification of the Regions

In this section we show how it is possible for an agent to distinguish the regions obtained by the space subdivision. To do that we developed a system to map from an allocentric to an egocentric representation of the space based on cyclic ordering information and visibility of objects.

By definition the regions we obtain from the space subdivision (Sec. 3) correspond to regions of the environment from which it is possible to either completely or partially see reference objects in the plane. Hence, one could expect that it is possible to detect the agent's position within the environment by considering the objects it can see around it. In the following, we introduce what we call Visual Cyclic Orderings (VCOs) and make use of the properties in Eq. 3 and Eq. 9, assuming that the observing agent can be considered a point.

4.1   Relating Visibility Zones and Viewed Objects

Since the space subdivision is obtained by merging the visibility reference frames built for every pair of objects, specific visibility relations have to hold between a generic point of a region and every object. According to the visibility relations defined in Sec. 2.2, our agent is either $Visible$, $PartiallyVisible^{Left}$, $PartiallyVisible^{Right}$, $PartiallyVisible^{Joint}$, or $Occluded$ from every object. Recurring to properties Eq. 3 and Eq. 9, we can deduce which relations hold between the agent and all objects. As an example, let us imagine an agent a standing in a region obtained by overlapping $LZ(A, B)$ and $TZ^{Right}(B, C)$. In this case, we can make the following inferences:

$$Visible(a, A, B) \Rightarrow Visible(B, A, a)$$
$$PartiallyVisible^{Right}(a, B, C) \Rightarrow C \cap LZ(B, a) \neq \emptyset \wedge C \cap SZ(B, a) \neq \emptyset$$

This way it is possible to constrict the agent's position within the environment by what it can observe (or not observe). Nevertheless, it is still possible that the same visibility relations hold for multiple disjoint regions in the subdivision meaning that considering only the visibility relations is not sufficient to determine the agent's location. To remedy this problem, we will in the following not only consider the subsisting visibility relations but also the cyclic order in which obstacles appear.

4.2   The Visual Cyclic Ordering

For every observation point $p_o \in \mathbb{R}^2$, it is possible to obtain a VCO label based on the projection of the visible objects on a circle centered in p_o (cmp. Fig. 4(a)). The cyclic order of objects in the projection is denoted using a sequence delimited by angular parenthesis in which objects may occur multiple times, e.g., $\langle A\,B\,A\,C \rangle$. It serves as a first step of deriving the VCO.

Let us now assume that agent a stands in region R_{11} of the environment depicted in Fig. 3(a), obtained by intersecting $LZ(B, A)$, $LZ(C, A)$, $LZ(A, B)$, $LZ(C, B)$, $LZ(A, C)$, and $SZ(B, C)$. Applying Eq. 9 we can then infer the following:

$$Visible(a, B, A) \Rightarrow Visible(A, B, a)$$
$$Visible(a, C, A) \Rightarrow Visible(A, C, a)$$
$$Visible(a, A, B) \Rightarrow Visible(B, A, a)$$
$$Visible(a, C, B) \Rightarrow Visible(B, C, a)$$
$$Visible(a, A, C) \Rightarrow Visible(C, A, a)$$
$$Occluded(a, B, C) \Rightarrow Occluded(C, B, a)$$

that is, the agent can see all objects except C which is occluded by B. This yields the VCO shown in Fig. 4(a). Let us now suppose the agent stands in region

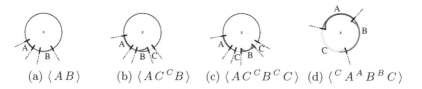

(a) $\langle A\,B \rangle$ (b) $\langle A\,C^C B \rangle$ (c) $\langle A\,C^C B^C C \rangle$ (d) $\langle {}^C A^A B^B C \rangle$

Fig. 4. Cyclic view of objects perceived by agent

$R_{10} = LZ(B,A) \cap LZ(C,A) \cap LZ(A,B) \cap LZ(C,B) \cap LZ(A,C) \cap TZ^{Right}(B,C)$
from where the first five relations still hold while the sixth by applying Eq. 10 becomes:

$$PartiallyVisible^{Right}(a,B,C) \Rightarrow C \cap LZ(B,a) \neq \emptyset \wedge C \cap SZ(B,a) \neq \emptyset$$

Focusing on C we can observe that a part of it is $Visible$ from a while another one is not because it is occluded by B. The agent will perceive no gap between C and B (see Fig. 4(b)) and the latter one is perceived as closer which means B is the occluding object (this is depicted in the iconic VCO representation by indenting thepart of B bordering C towards the center). Moreover, B is on the right side of C. Whenever such a case of occlusion exists, the VCO is modified by adding the occluded object as a superscript on the corresponding side of the occluding one. When the agent lies in a $JointTwilightZone$ the occluded object will appear on both left and right side of the occluding one which is indicated by two superscripts in the VCO. Thus, if we assume that the agent is within region $R_9 = LZ(B,A) \cap LZ(C,A) \cap LZ(A,B) \cap LZ(C,B) \cap LZ(A,C) \cap TZ^{Joint}(B,C)$ for which the sixth relation stays the same as for R_{10}, but the VCO changes as depicted in Fig. 4(c).

The VCO description as introduced above is also able to describe more complex configuration as depicted in Fig. 4(d) where the following main visibility relations and inferences subsist:

$$PartiallyVisible^{Right}(a,B,A) \Rightarrow A \cap LZ(B,a) \neq \emptyset \wedge A \cap SZ(B,a) \neq \emptyset$$
$$PartiallyVisible^{Right}(a,C,B) \Rightarrow B \cap LZ(C,a) \neq \emptyset \wedge B \cap SZ(C,a) \neq \emptyset$$
$$PartiallyVisible^{Right}(a,A,C) \Rightarrow C \cap LZ(A,a) \neq \emptyset \wedge C \cap SZ(A,a) \neq \emptyset$$

4.3   Identifying Regions

The VCO labels can be derived by the agent based on the cyclic view perceived from a position within a region as indicated above and it can use the labels to distinguish the regions and localize itself. The VCOs, however, are not completely unique as ambiguous cases may occur under very specific circumstances: Fig. 5(a) again shows the subdivision for the example configuration used throughout this text, this time with all regions labeled by their respective VCOs. There are two different zones labeled by the same VCO $\langle AC \rangle$ in the left part of the figure. This occurs when a region obtained by n objects is split into several subregions because of the presence of the $(n+1)^{th}$ landmark. However, due to the fact that the new regions induced by the $(n+1)^{th}$ object are labeled uniquely, it is sufficient to consider the VCOs of two neighboring regions to resolve the ambiguities.

If a geometric representation of the environment is available, it can be used to derive a topological map in which the regions are labeled with their respective VCOs by first using simple computational geometry to derive the cyclic view for an arbitrary point in each region and then construct the VCO in the same way as the agent would. The corresponding labeled topological map is shown in Fig. 5(b).

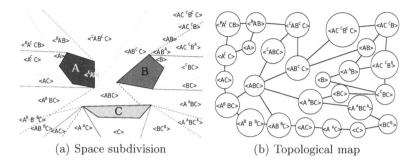

(a) Space subdivision   (b) Topological map

Fig. 5. Identifying regions based on their VCOs

5   Navigation and Exploration

As discussed in the previous section, regions can typically be uniquely identified by their VCOs. In some cases, the VCO of a neighboring region needs to be taken into account to tell apart regions which have the same VCO. This, together with the agent's ability to determine the VCO from its current position means that localization becomes trivial. In this section, we describe how navigation with an existing topological map of the environment can be realized simply by making use of the agent's ability to localize itself and by implementing elementary reactive navigation behaviors for crossing the boundaries in the subdivision. In addition, we explain how the agent can incrementally build up the topological map from its egocentric perspective while exploring the environment.

5.1   Path Planning and Navigation

The topological map offers a compact representation of the navigable space and explicitly describes the connectivity between different places. Hence, navigation from the agent's location to a goal region can be realized in two steps: First, standard graph search techniques like Dijkstra's shortest path algorithm [19] or A* [20] can be used, potentially exploiting rough distance estimates if those are available. The resulting path consists of a sequence of vertices standing for adjacent regions which need to be traversed in order. Second, the agent executes the plan by moving from region to region as specified.

The question then is how locomotion between two adjacent regions can be realized in order to be able to execute the plan. Our first observation is that for two regions which share a boundary—meaning they are connected by an edge in the map—their respective VCOs differ in the relation holding between objects X and Y only if the boundary is part of one of the half-tangents induced by X and Y. The exact difference depends on what kind of half-tangent the common boundary belongs to.

For instance, if we look at the regions with the VCOs $\langle B\,C\rangle$ and $\langle A^A\,B\,C\rangle$ on the right of Fig. 6(a), the fact that in the first region A is not visible (i.e., it is part of $SZ(B,A)$) while in the latter A is partially occluded from the right by

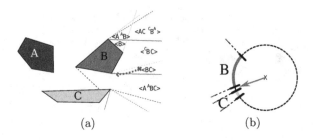

Fig. 6. Applying a navigation behavior to move from a *ShadowZone* to a *RightTwilightZone*: (a) the resulting trajectory, (b) the view of the agent at the starting position

B (i.e., it is part of $TZ^{Right}(B, A)$) means that the boundary separating these two regions has to be part of the half-tangent $behind(\lambda_{A,B}^{ext,right}, B)$.

As a consequence of this observation, it is possible to distinguish a finite set of boundary crossing cases, two for each pair of adjacent acceptance areas in the underlying visibility model (e.g., SZ to TZ^{Right} and TZ^{Right} to SZ), and implement reactive navigation behaviors for handling this particular cases.

We here restrict ourselves to describing the navigation behavior for the example above of moving from a *ShadowZone* to a *RightTwilightZone*. An important point hereby is that these navigation procedures should be easily realizable by our sensorically limited agent and should not require local geometric information about the objects surrounding the agent.

Fig. 6(b) shows the agent's view for the position marked by a cross in Fig. 6(a). Since moving from a *ShadowZone* to a *RightTwilightZone* means that A has to appear on the left of B, the navigation behavior for these particular cases should always move the robot towards a point directly left of B until A appears resulting in the trajectory shown in the figure.

For the reverse direction (*RightTwilightZone* to *ShadowZone*) A has to disappear behind B so that it is obviously a good strategy to move to a point directly to the right of B until A actually disappears.

Due to not being based on geometric information, the result of applying a navigation behavior is not deterministic in the sense that the agent may end up in a different region than the intended one. For instance, when moving from the region $\langle {}^C B C \rangle$ to $\langle A C^C B^A \rangle$ the agent may cross the half-tangent $behind(\lambda_{C,B}^{ext,right}, B)$ first and end up in $\langle B \rangle$. If this happens, the agent will notice it based on the unexpected VCO it perceives and can trigger the computation of a new plan. So far, our simulation experiments indicate that the agent never has to traverse more than one additional region for every pair of connected regions in the original plan. In the case that some kind of local geometric information is available to our agent, it should be possible to improve the navigation behaviors so that unintentionally moving into other regions is avoided and in a way that leads to shorter trajectories overall.

5.2   Exploration and Construction of the Topological Map

As a result of the fact that the regions making up the subdivision of space are distinguishable from the agent's egocentric perspective of space, building up the topological map autonomously while exploring the environment is straightforward.

A basic exploration algorithm simply does the following:

1. Generate the VCO l for the starting region from the egocentric view that is available from the start point; add a first vertex to the still empty map and label it with l.
2. While moving around constantly compute a temporary VCO l_{cur} from the current perception.
3. As soon as l_{cur} differs from the previously computed description, the agent knows that it has entered a new region:
   (a) If a region with the label l_{cur} is already contained in the map and if there is not already an edge connecting this region with the region the agent was in previously, add a new edge connecting these two.
   (b) If instead no region with label l_{cur} is contained in the map, add a new vertex with label l_{cur} and connect this one with the vertex of the previous region.
4. Repeat ad finitum.

If the VCOs for all regions are unique, it is guaranteed that this basic exploration algorithm will have constructed the correct and complete topological map of the environment, once the agent has passed between every pair of adjacent regions. The algorithm can easily be modified to deal with ambiguous VCOs by also storing the label of the previously visited region. However, we need the agent to at least be able to recognize when the map is complete and even better have a way to systematically explore the environment in a way that constructs the map efficiently. Furthermore, it seems likely that not all regions and crossings between regions have to be traversed but that parts of the map can be simply inferred. These issues still need to be investigated more closely in future research.

6   A Finer Grained Subdivision Based on the 5-Intersection Model

As we mentioned previously, our approach has a lot in common with the top-level representation in the outdoor navigation system described by Levitt and Lawton [10], even though their approach is restricted to point landmarks and uses the cyclic order of landmarks (called a panorama) to identify regions (see Figs. 7(a) and 7(b)). As pointed out by Schlieder [11], Lewitt and Lawton wrongly claimed that the shaded regions in the figure can be distinguished based on the panoramas. However, the cyclic order of landmarks is identical for all these regions. Schlieder also showed that it is indeed possible to differentiate between these regions by going from the simple panorama representation to one in which the

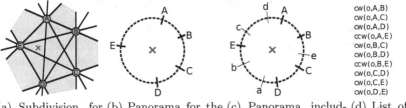

(a) Subdivision for point landmarks

(b) Panorama for the marked position (Levitt & Lawton)

(c) Panorama including opposite directions (Schlieder)

(d) List of triangle orientations

Fig. 7. Subdivision for point landmarks according to [10] and three ways to describe the region containing the agent (cross)

opposite directions of landmarks is included (see Fig. 7(c)) which is equivalent to storing the orientations of the triangles, clockwise (cw) or counterclockwise (ccw), formed by the observer o and every pair of landmarks.

As in our approach certain configurations of objects may lead to similar problems as in the point-based approach by Levitt and Lawton in that we might get large concave regions in the middle of the objects and as this can be problematic for the navigation, we here consider an extension similar to the one proposed by Schlieder for the point case. We do this by going from the visibility-based representation to one which incorporates a suitable left / right distinction to describe the orientation of the configuration formed by three objects. The model we employ is the 5-Intersection model of Billen and Clementini [17] for projective relations between extended objects. We start by briefly describing this model and how it relates to our visibility model. Afterwards, we discuss the resulting finer grained decomposition of space, the identification of the regions, and navigation in this extended version.

6.1 5-Intersection Model

The 5-Intersection model developed by Clementini and Billen in [17,18] defines ternary projective relations between regions, starting from the basic concept of collinearity among points which is an invariant in projective geometry. Clementini and Billen start by introducing a set of relations between points similar to the FlipFlop calculus [21,22]: Given three points $x, y, z \in \mathbb{R}^2$ they build a frame of reference using an oriented line $\overrightarrow{yz}$ passing first through y and then through z. The resulting five relations are summarized and illustrated in Table 1[1].

The really important innovation of the 5-Intersection model is the extension of these relations to configurations of three extended objects or regions A, B, C. In the case of regions, Clementini and Billen subdivided the plane using the internal and external tangents between the reference objects B and C like in our visibility model. Fig. 8(a) depicts the resulting five acceptance areas:

---

[1] We here focus on a subset of the proposed relations for which $y \neq z$ holds and we have slightly adapted the notation to be in line with the rest of this text.

Table 1. Relations of the 5-Intersection model for points

Name	Short Name	Example
rightside	$rs(x,y,z)$	
leftside	$ls(x,y,z)$	
between	$bt(x,y,z)$	
before	$bf(x,y,z)$	
after	$af(x,y,z)$	

$BF(B,C)$ (before), $BT(B,C)$ (between), $AF(B,C)$ (after), $LS(B,C)$ (leftside), and $RS(B,C)$ (rightside).

The relations of the 5-Intersection model for regions $bf(A,B,C)$, $bt(A,B,C)$, $af(A,B,C)$, $ls(A,B,C)$, and $rs(A,B,C)$ are defined based on the acceptance area which contains the primary object A, e.g., $bf(A,B,C) \iff A \subseteq BF(B,C)$.

The comparison between the acceptance areas of our visibility model and the 5-Intersection models in Fig. 8 shows that both complement each other well as each makes finer distinctions in an area where the other one only makes coarse distinctions.

When we employ the 5-Intersection model for deriving a finer grained subdivision of space in the following section, it is important that the relations *ls* (leftside), *rs* (righside), and *bt* (between) are distinguishable from the egocentric perspective of the agent. This is indeed possible with only a minor extension of the agent's sensor abilities: If the agent is located within the acceptance area $LS(B,C)$ of two objects B and C, it will perceive both objects to be completely lying within an angle of less than 180° and C will be entirely to the left of B. If the agent is located within the acceptance area $RS(B,C)$, the angle will be again smaller than 180° but this time C will be to the right of B. For $BT(B,C)$ the perceived angle containing both objects will be 180° or larger. Hence, we need to add the ability of determining whether two objects are completely contained

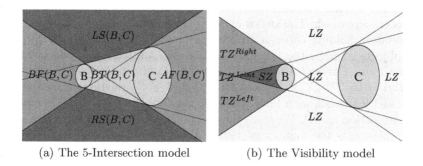

(a) The 5-Intersection model   (b) The Visibility model

Fig. 8. Acceptance areas of the 5-Intersection and the visibility model

in a 180° angle or not to our agent which corresponds to being able to tell apart the different triangle orientation in the approach of Schlieder described above.

6.2 Subdivision, Place Recognition, and Navigation

The decomposition of space works similar to our approach described in Sect. 3. However, instead of only employing the half-tangents for every pair of objects to divide the navigable space into regions, we also include the segments of the middle parts $middle(\lambda)$ for every tangent λ which separate the acceptance areas $BT(X,Y)$, $LS(X,Y)$, and $RS(X,Y)$ from each other. Similar to the case of the half-tangents, we can only use those middle tangent segments which do not intersect another object. Fig. 9(a) shows the resulting subdivision for the example from Fig. 3(a) using dashed lines for the region boundaries of the coarser version and solid lines for the new boundaries leading to additional regions within the convex hull of all three objects. The resulting topological map is shown in Fig. 9(b) and contains 14 more vertices than for the coarser version.

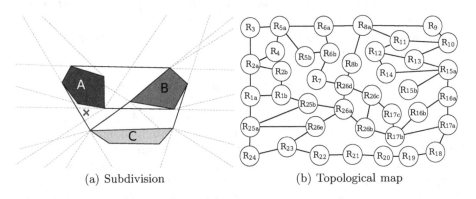

(a) Subdivision   (b) Topological map

Fig. 9. The subdivision and topological map for the finer grained approach based on a combination of the 5-Intersection model and the visibility model

To be able to identify the regions in the finer grained subdivision approach, we need a more complex alternative to the VCOs similar to the triangle orientations in Schlieder's approach. The analog of doing this in our case, is to store the qualitative relation holding between an arbitrary point within the region and each pair of objects whereby the relations used are the 5-Intersection relations combined with the $PartiallyVisible^{Left}(X,Y,Z)$, $PartiallyVisible^{Right}(X,Y,Z)$, $PartiallyVisible^{Joint}(X,Y,Z)$, and $Occluded(X,Y,Z)$ relations from the visibility model which allows a finer distinction within the $bf(X,Y,Z)$ relation of the 5-Intersection model. We store these region descriptions in form of a $n \times n$ matrix where n is the number of visible objects as shown in Fig. 10 for the region marked in Fig. 9(a). The matrix element in row A and column B contains the acceptance area in which any point from the region at hand lies given the reference frame formed by A and B (in this case it is the acceptance area

	A	B	C
A	/	$TZ^{Left}(A,B)$	$BT(C,A)$
B	$AF(B,A)$	/	$LS(C,B)$
C	$BT(A,C)$	$RS(B,C)$	/

Fig. 10. Matrix used as a label for the region marked by the cross in Fig. 9(a)

$TZ^{Left}(A,B))$. In principle, not the entire matrix needs to be stored as corresponding entries mirrored along the diagonal are related in specific ways: For instance, an entry $BT(A,B)$ must result in an entry $BT(B,A)$ and vice versa. If the entry is $AF(A,B)$ instead, the corresponding entry for B and A has to be $TZ^{Left}(B,A)$, $TZ^{Right}(B,A)$, $TZ^{Joint}(B,A)$, or $SZ(B,A)$ and vice versa. In general, it is sufficient to store the more restricting relation for every pair of objects.

The matrix labels are unique most of the time but cases in which two regions have the same label may occur in the same way as for the VCOs in the coarser version. Again these regions can be distinguished by taking a neighboring region into account. For adjacent regions the two matrices vary for exactly one pair of objects while all other relations are identical. In addition, an agent with the sensor abilities described above can easily determine the matrix for the region it is currently in. Hence, the exploration algorithm sketched in Sect. 5.2 can still be applied.

For navigating with a given map, the set of navigation behaviors needs to be extended to handle the cases in which connected regions are separated by the middle segment of an external tangent.

Overall, by applying the finer grained version, a finer level of localization can be achieved while still preserving the qualitative nature of the approach as the different locations can still be distinguished and the approach does not depend on exact metric information about the environment.

7 Conclusions

The approach developed in this paper demonstrates that localization and navigation can be realized in a qualitative way without a need for precise geometric information. By basing our spatial representation on the fundamental concepts of visibility and occlusion we have ensured that it is applicable for agents with very basic sensor abilities and well-suited to be shared between multiple agents. In addition, the approach greatly simplifies path planning, navigation, and map learning. Our model also constitutes a significant improvement over existing approaches as the underlying visibility and 5-Intersection models allow extended objects to be handled instead of only point landmarks.

One limiting factor of our approach currently seems to be the underlying assumption that the objects are uniquely identifiable. Therefore, investigating how well we actually need to be able to distinguish objects while still being able to navigate successfully is a goal for future research. In addition, always using

328 P. Fogliaroni et al.

all visible objects for the space subdivision may often lead to an overly detailed decomposition in practice. A possible solution could be to only take objects near the agent into account in order to achieve an adequate level of granularity. Finally, as already pointed out, we plan to incorporate the exploitation of local geometric information to achieve smoother navigation and want to investigate autonomous exploration and map learning more closely. In particular, reasoning about the existence of regions and connections in the subdivision based on partial map information seems a very promising and interesting approach.

Acknowledgments

The authors would like to thank Christian Freksa and the anonymous reviewers for valuable comments. Funding by the Deutsche Forschungsgemeinschaft (DFG) under grant IRTG Semantic Integration of Geospatial Information and grant SFB/TR 8 Spatial Cognition is gratefully acknowledged.

References

1. Montello, D.: Navigation. In: Shah, P., Miyake, A. (eds.) The Cambridge Handbook of Visuospatial Thinking, pp. 257–294 (2005)
2. Thrun, S.: Robotic mapping: A survey. In: Lakemeyer, G., Nebel, B. (eds.) Exploring Artificial Intelligence in the New Millenium. Morgan Kaufmann, San Francisco (2002)
3. Kuipers, B.: The Spatial Semantic Hierarchy. Artificial Intelligence (119), 191–233 (2000)
4. Franz, M.O., Schölkopf, B., Mallot, H.A., Bülthoff, H.H.: Learning view graphs for robot navigation. Autonomous Robots 5, 111–125 (1998)
5. Remolina, E., Kuipers, B.: Towards a general theory of topological maps. Artificial Intelligence 152(1), 47–104 (2004)
6. Siegel, A.W., White, S.H.: The development of spatial representations of large-scale environments. In: Reese, H.W. (ed.) Advances in Child Development and Behavior, vol. 10, pp. 9–55. Academic Press, London (1975)
7. Tversky, B.: Distortions in cognitive maps. Geoforum 23, 131–138 (1992)
8. Denis, M.: The description of routes: A cognitive approach to the production of spatial discourse. Cahiers Psychologie Cognitive 16(4), 409–458 (1997)
9. Sorrows, M.E., Hirtle, S.C.: The nature of landmarks for real and electronic spaces. In: Freksa, C., Mark, D.M. (eds.) Spatial Information Theory. Cognitive and Computational Foundations of Geopraphic Information Science (COSIT), Berlin, August 1999. Lecture Notes on Computer Science, vol. 1661, pp. 37–50. Springer, Heidelberg (1999)
10. Levitt, T.S., Lawton, D.T.: Qualitative navigation for mobile robots. Artificial Intelligence 44, 305–361 (1990)
11. Schlieder, C.: Representing visible locations for qualitative navigation. In: Carreté, N.P., Singh, M.G. (eds.) Qualitative Reasoning and Decision Technologies, pp. 523–532 (1993)
12. Wagner, T., Visser, U., Herzog, O.: Egocentric qualitative spatial knowledge representation for physical robots. Robotics and Autonomous Systems 49, 25–42 (2004)

13. Tarquini, F., De Felice, G., Fogliaroni, P., Clementini, E.: A qualitative model for visibility relations. In: Hertzberg, J., Beetz, M., Englert, R. (eds.) KI 2007. LNCS, vol. 4667, pp. 510–513. Springer, Heidelberg (2007)

14. Galton, A.: Lines of sight. In: Keane, M., Cunningham, P., Brady, M., Byrne, R. (eds.) AI and Cognitive Science 1994, Proceedings of the Seventh Annual Conference, September 8-9, 1994, Trinity College Dublin, pp. 103–113 (1994)

15. Köhler, C.: The occlusion calculus. In: Proc. Workshop on Cognitive Vision (2002)

16. Randell, D.A., Cui, Z., Cohn, A.: A spatial logic based on regions and connection. In: Nebel, B., Rich, C., Swartout, W. (eds.) Principles of Knowledge Representation and Reasoning: Proceedings of the Third International Conference (KR 1992), pp. 165–176. Morgan Kaufmann, San Mateo (1992)

17. Billen, R., Clementini, E.: A model for ternary projective relations between regions. In: Bertino, E., Christodoulakis, S., Plexousakis, D., Christophides, V., Koubarakis, M., Böhm, K., Ferrari, E. (eds.) EDBT 2004. LNCS, vol. 2992, pp. 310–328. Springer, Heidelberg (2004)

18. Clementini, E., Billen, R.: Modeling and computing ternary projective relations between regions. IEEE Transactions on Knowledge and Data Engineering (2006)

19. Dijkstra, E.W.: A note on two problems in connexion with graphs. Numerische Mathematik 1, 269–271 (1959)

20. Hart, P.E., Nilsson, N.J., Raphael, B.: A formal basis for the heuristic determination of minimum cost paths. IEEE Transactions on Systems Science and Cybernetics 4, 100–107 (1968)

21. Ligozat, G.: Qualitative triangulation for spatial reasoning. In: Campari, I., Frank, A.U. (eds.) COSIT 1993. LNCS, vol. 716, pp. 54–68. Springer, Heidelberg (1993)

22. Scivos, A., Nebel, B.: The finest of its class: The practical natural point-based ternary calculus $\mathcal{LR}$ for qualitative spatial reasoning. In: Freksa, C., Knauff, M., Krieg-Brückner, B., Nebel, B., Barkowsky, T. (eds.) Spatial Cognition IV. LNCS, vol. 3343, pp. 283–303. Springer, Heidelberg (2005)

Showing Where To Go by Maps or Pictures: An Empirical Case Study at Subway Exits

Toru Ishikawa[1] and Tetsuo Yamazaki[2]

[1] Graduate School of Interdisciplinary Information Studies & Center for Spatial Information Science, University of Tokyo, 7-3-1 Hongo, Bunkyo-ku, Tokyo 113-0033, Japan
ishikawa@csis.u-tokyo.ac.jp
[2] Graduate School of Frontier Sciences, University of Tokyo
5-1-5 Kashiwanoha, Kashiwa, Chiba 277-8563, Japan

Abstract. This study empirically examined the effectiveness of different methods of presenting route information on a mobile navigation sysyem, for accurate and effortless orientation at subway exits. Specifically, it compared participants' spatial orientation performance with pictures and maps, in relation to the levels of their spatial ability. Participants identified the directions toward the goals after coming onto the ground faster when viewing pictures than when viewing maps. Spatial orientation with maps was more difficult than that with pictures at exits where body rotation was necessary, especially for people with low mental-rotation ability. In contrast, pictures were equally effective for people with low and high mental-rotation ability. Reasons for the effectiveness of pictures and possibilities of using other presentation formats are discussed.

Keywords: Spatial orientation; Navigational aids; Route information; Presentation formats; Spatial representations.

1 Introduction

Getting oriented in space is a fundamental act for people. For example, when visiting an unfamiliar city to attend a conference, one needs to locate a shuttle bus stop at an airport; to find a booked hotel among other buildings on a street; to identify a recommended restaurant in a downtown area; to move between different meeting rooms in the hotel; and so on. People thus engage in a variety of spatial orientation tasks on a daily basis. But this is not equal to saying that these tasks are straightforward. There are people who have difficulty with knowing where they are, or constructing accurate maps of the surrounding space in their heads.

Such poor sense-of-direction people especially have trouble in a new environment, and one of the locations that frequently pose problems for these people is the exit from a subway station. At subway stations, it is possible, at least theoretically, to move toward desired exits by following signage systems installed at several locations. The difficulty lies in that people need to figure out which way to go after coming onto the ground. In large cities with well developed public transportation systems, such as New

K. Stewart Hornsby et al. (Eds.): COSIT 2009, LNCS 5756, pp. 330–341, 2009.

York, London, Paris, or Tokyo, there are many anecdotes that show that people become disoriented at subway exits. In fact, our preliminary interviews with 58 people (31 men and 27 women, from 21 to 26 years old) showed that 81% of them either frequently or sometimes become lost when coming onto the ground out of a subway station, whereas 29% do so while heading toward an exit in the underground concourse.

To be a potential help for orientation at subway stations, maps are installed at several locations, such as in the underground concourse or at exits. However, there are some well-documented problems concerning the use of maps. First, it has been shown that a significant portion of people, not only children but also adults, have difficulty using maps in the real world, by interrelating the self, the map, and the represented space (Liben et al., 2002). Simply showing maps may hence not be a sufficient solution to disorientation. Also, when showing maps to people, their orientation affects the viewer's understanding (called *map-alignment effects*). People find it easier to understand maps that are aligned with the environment, with the upward direction on the map corresponding to the forward direction in the represented space (e.g., Levine et al., 1982; Warren & Scott, 1993). Despite this fact, one can frequently encounter misaligned maps at many public locations (Levine et al., 1984).

Recently, with the advancement of information and communication technologies, many kinds of navigation systems and location-based services have been developed (Gartner et al., 2007). In addition to mobile devices equipped with Global Positioning System receivers, systems employing various identification or positioning techniques such as IC tags are widely used, enabling indoor as well as outdoor navigation (e.g., Bessho et al., 2008; Hightower & Borriello, 2001).

On such mobile navigation systems, the format for presenting route information to the user can take various forms, and in consideration of the fact that some people have difficulty with maps, formats other than maps have been frequently used. One notable example is pictures or other spatial representations (Radoczky, 2007). There are similarities and differences between maps and pictures. Maps and pictures are both symbols, in the sense that they represent the real space. To understand and use symbols, one needs to comprehend the symbols and interrelate them to their referents (what DeLoache, 2004, calls *dual representation*). On the other hand, maps show two-dimensional, vertical views, while pictures show three-dimensional, horizontal views. Pictures also convey a sort of photographic realism (Plesa & Cartwright, 2008). Due to these characteristics, pictures and other images are often expected as an effective presentation format. There are studies that empirically examined the effectiveness of different types of representations; some indicated effectiveness of 3D representations (e.g., Coors et al., 2005), and others suggested value of schematization or less detail (e.g., Dillemuth, 2005; Plesa & Cartwright, 2008). We believe that the question about what is an effective format for presenting route information requires and deserves more systematic investigation.

Thus, in this study, we selected pictures and maps as major formats typically used on mobile navigation systems, and compared their effectiveness as a presentation format for route information at exits from a subway station. Also, we included spatial ability as an important attribute variable of the user, and examined whether the effectiveness of different types of route information varies depending on the user's spatial ability; and if so, how. Our major research hypotheses were two-fold. First, an effective format for presenting route information should enable the user to identify the locations of goals with ease and confidence. Second, an effective presentation format should help low-spatial and high-spatial people equally, showing little or no relationship between people's orientation performance and their spatial ability.

2 Method

2.1 Participants

Fifty college students (25 men and 25 women) participated in the experiment. Their ages ranged from 21 to 26, with a mean of 22.7 years. These participants were not familiar with the subway station or its neighborhood areas used in this study.

2.2 Materials

2.2.1 Study Area
We chose a subway station in the central Tokyo area (Kayabacho Station) for the study area, and selected five exits from the station (Figure 1). Participants walked toward the five exits starting from different places in the underground concourse of the station.

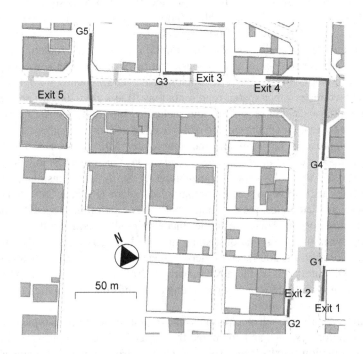

Fig. 1. Map of the study area. Five exits from the subway station (Exits 1-5) were selected. From each exit, participants walked to a goal (denoted by G1-G5) along a route shown by a solid line. Polygons overlaid along streets depict the underground concourse of the station.

These five exits enabled us to examine different patterns of spatial relationships among an exit, street, and goal (schematically shown in Figure 2). At Exit 1, the direction in which participants face when coming onto the ground and the direction in which the goal is located are opposite to each other; in contrast, these two directions are the same at Exit 2 (left panel of Figure 2). At Exit 3, the direction in which participants face when coming onto the ground is perpendicular to the street running in front of it, and they need to make a right turn to go to the goal (right panel of Figure 2). At Exits 4 and 5, participants come onto the ground near a major intersection; at Exit 4 participants turn backward to go to the goal, whereas at Exit 5 participants go forward to the goal.

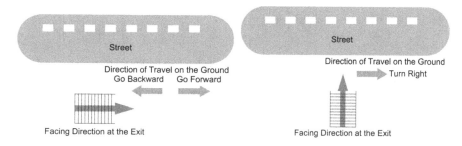

Fig. 2. Different patterns of spatial relationships among an exit, street, and goal. At Exits 1 and 2, people come onto the ground parallel to the street in front of it, and go backward or forward to the goal (*left*). At Exit 3, people come onto the ground perpendicular to the street, and make a right turn to go to the goal (*right*). Exits 4 and 5 are similar to Exits 1 and 2, except that they are located at a major intersection.

2.2.2 Route Information

We showed information about routes in two different formats: maps and pictures. To examine the effectiveness of these formats for mobile navigation systems, we showed maps and pictures in the size of a PDA screen (or a hand-held communication terminal device). To avoid cluttering small maps with too much information, we presented maps in as simple a format as possible. Maps showed the location of an exit (or a starting point on the ground, denoted by S) and a route to the goal (denoted by G). As examples, maps used for Exit 1 are shown in Figure 3.

Pictures showed views that participants would have when coming onto the ground at an exit facing straight ahead, and views along a street to the goal. In the pictures, the direction of travel along the route was indicated by an arrow. Figure 4 shows a picture used for Exit 1 with an arrow indicating a backward turn (left panel) and a picture used for Exit 3 with an arrow indicating a right turn (right panel).

We varied the location where the maps were presented to participants, and the orientation in which the maps were presented to them. Concerning the location, maps were presented either in the underground concourse or at an exit. Concerning the orientation, maps were presented either with the direction of travel corresponding to the upward direction on the map (Figure 3, left), or with the facing direction at the exit corresponding to the upward direction on the map (Figure 3, right). At Exits 1, 3, and 4, maps in the former orientation are misaligned with the surrounding space, whereas

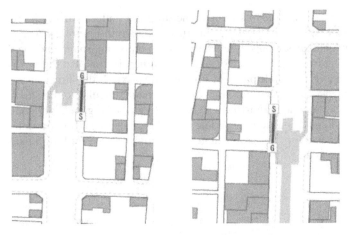

Fig. 3. Maps used for Exit 1, where participants needed to turn backward to go to the goal. The map on the left is shown in misaligned orientation at the exit, with the direction of travel corresponding to the upward direction on the map. The map on the right is shown in aligned orientation, with the facing direction at the exit corresponding to the upward direction on the map.

Fig. 4. Pictures used for Exit 1 (*left*) and Exit 3 (*right*). These pictures show views that participants had at these exits. Arrows indicate the direction of travel after coming onto the ground (turning backward or making a right turn).

maps in the latter orientation are aligned with it. At Exits 2 and 5, the distinction between these two map orientations disappears, because the facing direction and the travel direction coincide (see Figure 2). For pictures, these two properties, location and orientation, were fixed: they were shown at an exit, in aligned orientation. Participants were given no information about the orientations of pictures or maps. They needed to figure out by themselves the relationships between directions in the pictures or maps and directions in the environment.

2.2.3   Mental Rotations Test

As a measure of spatial ability, we asked participants to take the Mental Rotations Test, in which they were to identify figures that are identical to a criterion figure after being rotated into different orientations. This is one of the major psychometric spatial tests, assessing people's ability to mentally rotate imagined pictures, and this mental-rotation ability has been found to correlate with map-reading skills in the field (e.g., Liben & Downs, 1993). We used this test to see whether and how this ability affected participants' orientation and navigation performance with different types of route information. Participants were allowed 6 min to complete this test, consisting of 21 items, and they received plus one point for each correctly marked figure and minus one point for each wrongly marked figure.

2.3 Design

By varying the format (maps or pictures), location (in the underground concourse or at an exit), and orientation (aligned or misaligned), we derived five methods of presenting route information: (a) showing a map in the underground concourse with the direction of travel on the ground upward; (b) showing a map in the underground concourse with the facing direction at the exit upward; (c) showing a map at an exit with the direction of travel upward; (d) showing a map at an exit with the facing direction at the exit upward; and (e) showing a picture at an exit. At Exits 2 and 5, maps (a) and (b), and maps (c) and (d) were the same, because the facing direction and the travel direction coincided.

Participants viewed these five types of route information, one for each exit. The allocation of the five types of route information to the five exits was counterbalanced with a Latin square design, thus five groups being created. Participants were randomly assigned to one of the five groups, with each group ($n = 10$) having equal numbers of men and women. The order in which participants visited the five exits was randomized across participants.

2.4 Procedure

At the beginning of the experiment, participants were individually taken to the starting point for the first exit in the underground concourse. Participants who viewed a map in the underground concourse for the exit were given a map at this point, and asked to learn the map until they thought that they knew how to reach the goal when coming onto the ground. When they indicated that they learned it, the map was removed, and they did not look at it after that. (Participants who viewed a map or picture at the exit were given no information at this point.) Participants were then instructed to go to the exit, by following signage systems installed at several places in the concourse (i.e., following signs showing directions to the designated exit).

When participants reached the exit, participants who viewed a map or picture at the exit were given a map or picture. (Participants who viewed a map in the underground concourse received no information at the exit.) And then they were asked to indicate in which direction the goal was located, by pointing in that direction or verbally describing it. We recorded the time that participants took to indicate the direction and whether the indicated direction was correct. Also, we asked participants to indicate the degree of certainty about their orientation at the exit, on a 7-point scale (1 = *very uncertain*; 7 = *very certain*). After conducting these tasks, participants were instructed to walk to the goal. Then, they were taken to the starting point for the next exit along a circuitous path, and repeated the same procedure as above for a total of five exits.

At the end of the experimental session, participants filled out a questionnaire asking about their age, sex, familiarity with the study area, and so on, and took the Mental Rotations Test. Participants finished all these experimental tasks within 120 min.

3 Results

3.1 Accuracy of Orientation

The numbers of participants who indicated wrong directions toward the goals at the five exits were six (12%) at Exit 1, two (4%) at Exit 2, four (8%) at Exit 3, three (6%) at Exit 4, and zero (0%) at Exit 5. The difference in these numbers for the five exits was not statistically significant in the Friedman test.

We further examined, for each exit, whether the number of wrong responses was different among the five types of route information. We did not find a statistically significant difference for any of the five exits in the Kruskal-Wallis test.

Compared to spatial orientation at exits, following routes guided by the five types of route information did not pose difficulties for participants. Once identifying the directions toward the goals correctly at the exits, participants reached the goals walking along directed routes successfully.

3.2 Response Time

3.2.1 Differences among Information Types
For each of the five exits, we examined whether there was a difference in response time among the five types of route information. (We defined response time as the time that participants took to view the presented picture or map and to indicate the direction toward the goal.) To do that, we analyzed response times for participants who identified the directions toward the goals correctly, by conducting an analysis of variance (ANOVA) with the information type as a between-subject variable. Since the distribution of response time was positively skewed, we normalized it through a log transformation.

For all exits, the main effect of information type was statistically significant, $F(4, 39) = 3.81, p < .05; F(2, 45) = 3.74, p < .05; F(4, 41) = 11.23, p < .001; F(4, 42) = 4.68, p < .01; F(2, 47) = 27.09, p < .001$, respectively for Exits 1-5 (Figure 5).

As can be seen in Figure 5, at all exits, participants who viewed pictures identified the directions of the goals most quickly. In comparison to maps shown at exits, in misaligned or aligned orientation, pictures yielded a significantly shorter response time at Exit 1, $t(17) = 2.46, p < .05; t(15) = 2.01, p < .10$; and at Exit 3, $t(18) = 2.47, p < .05; t(15) = 4.14, p < .01$, respectively. At Exit 4, a picture yielded a shorter response time than a misaligned map, $t(17) = 2.01, p < .10$.

3.2.2 Relationships with Mental Rotation Ability

We next examined the relationships between participants' response times and mental-rotation ability, and whether the relationships vary depending on different information types. To do that, we conducted regression analyses taking interaction effects into account, with a dummy-coded vector for the five types of route information, a continuous variable of the mental-rotation score, and their products representing their interactions (e.g., Aiken & West, 1991).

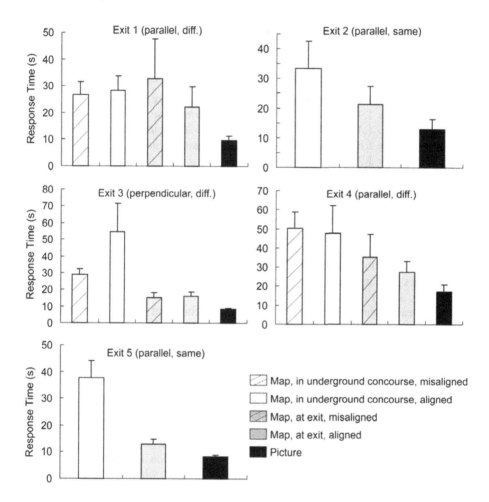

Fig. 5. Comparisons of response times with different types of route information for each exit. Vertical lines depict standard errors of the means. At all exits, pictures yielded the quickest orientation. At Exits 1 and 3, pictures yielded a significantly shorter response time than maps shown at the exits, in misaligned or aligned orientation. At Exit 4, a picture yielded a shorter response time than a map shown at the exit in misaligned orientation. Notes in parentheses for each exit indicate whether people come onto the ground parallel or perpendicular to the street in front of it, and whether the facing direction at the exit and the travel direction are the same or not.

For Exit 1, there was a significant interaction effect of information type and mental-rotation ability, $F(4, 34) = 3.23$, $p < .05$. In light of the existence of interaction, we further conducted a separate regression analysis for each information type, and obtained a significant relationship between response time and mental-rotation ability for maps shown at the exit, indicating a longer response time (RT) for participants with lower mental-rotation ability (MRT).

$$\log RT = 2.51 - 0.04 \times MRT \quad \text{(Map at the exit, misaligned)}$$
$$\log RT = 2.43 - 0.04 \times MRT \quad \text{(Map at the exit, aligned)}$$

For Exit 3, there was a significant main effect of mental-rotation ability, $F(1, 40) = 5.81$, $p < .05$; and the difference among intercepts for the five information types was also statistically significant, $F(4, 40) = 13.00$, $p < .001$. For all information types, participants with lower mental-rotation ability took a longer time.

$$\log RT = 1.82 - 0.01 \times MRT \quad \text{(Map in the concourse, misaligned)}$$
$$\log RT = 1.94 - 0.01 \times MRT \quad \text{(Map in the concourse, aligned)}$$
$$\log RT = 1.46 - 0.01 \times MRT \quad \text{(Map at the exit, misaligned)}$$
$$\log RT = 1.60 - 0.01 \times MRT \quad \text{(Map at the exit, aligned)}$$
$$\log RT = 1.27 - 0.01 \times MRT \quad \text{(Picture)}$$

3.3 Degree of Certainty

At each exit, we asked participants how certain they felt about their responses concerning the direction toward the goal. As a self-report measure of the difficulty of orientation, we examined participants' answers to this question for different types of route information. In the Kruskal-Wallis test, there was a significant difference among the five types of route information for Exits 1, 3, 4, and 5, $\chi^2(4) = 19.57$, $p < .001$; $\chi^2(4) = 19.74$, $p < .001$; $\chi^2(4) = 10.88$, $p < .05$; and $\chi^2(2) = 10.62$, $p < .01$, respectively. At all these four exits, participants felt more certain or confident when viewing pictures (mean rating $= 6.5$) than when viewing maps (mean rating $= 4.8$).

4 Discussion

This study examined the effectiveness of different methods of presenting route information for accurate and effortless orientation at subway exits. At subway exits, people tend to become disoriented and have difficulty figuring out which way to go. Our major concern was to examine what is an effective format for presenting route information to people with different levels of spatial ability in such a situation. In particular, we looked at whether and how much the format of pictures is helpful, which has been popularly used on mobile navigation systems as a potential alternative to traditional maps.

First, our results show that there are people who have difficulty orienting themselves at subway exits. The routes selected for this study were not too difficult, and people encounter situations requiring spatial orientation of this sort in everyday life. Despite that, up to 12% of participants failed to identify the locations of the goals correctly at exits. This finding shows that it is not viable to assume that such orientation tasks pose no difficulties for anybody, and suggests the necessity of providing helpful navigation aids or orientation information at subway exits.

Second, showing maps or pictures at exits was better than showing maps in the underground concourse. Thus, viewing maps in the underground concourse and matching the remembered image to the actual view when coming onto the ground poses time and effort additional to viewing maps or pictures at exits. This finding provides validity to installing maps or pictures at subway exits, preferably as well as installing them in the underground concourse.

Concerning the comparison of pictures and maps, our participants identified the directions toward the goals most rapidly when viewing pictures at all exits, showing the effectiveness of pictures for presenting route information. On average, maps shown at exits yielded 1.9 times as long a response time as did pictures. Pictures provide actual views that the traveler has in the environment, and save mental translation from a two-dimensional, vertical view to a three-dimensional, horizontal view. This ease of orientation with pictures is also reflected by participants' self-ratings of certainty or confidence about their orientation performance.

Our participants followed the routes easily by viewing pictures, as well as identifying the directions toward the goals rapidly, because there were salient buildings or objects along the routes, making it relatively easy to match the pictures and the depicted environment. Conversely, for routes that do not have good landmarks, the effectiveness of pictures observed in this study might not be fully realized. Thus, considerations about formats for route information should take contextual information of routes into account (Klippel & Winter, 2005; Raubal & Winter, 2002; Richter & Klippel, 2005; Winter, 2003).

Superiority of pictures to maps shown at exits was observed at Exits 1, 3, and 4, where the direction in which participants faced when coming onto the ground was different from the direction toward the goal, thus requiring participants either to turn backward or to make a right turn to go to the goal. In cases where body rotation is necessary, quicker orientation is hence possible with pictures than with maps, especially when the maps are misaligned with the surrounding space.

Furthermore, at Exits 1 and 3, ease of orientation was related to participants' spatial ability. At Exit 1, participants with low mental-rotation ability took a longer time to identify the direction toward the goal than those with high mental-rotation ability, when viewing maps at the exit. In contrast, participants with low and high mental-rotation ability did not differ when viewing pictures. At Exit 3, participants with low mental-rotation ability took a longer time for all the five types of route information, with maps yielding a longer response time than pictures. Thus, spatial orientation with maps is more difficult than that with pictures at cognitively demanding exits, especially for people with low mental-rotation ability. And pictures are equally effective for people with low and high mental-rotation ability, which should be a good property for effective navigational aids.

A comment is in order about the maps and pictures used in this study. They were presented in the size of a PDA-screen, thus the amount of information shown on the maps was rather limited. Research has shown that navigation with information shown on a small cellphone screen is difficult (Ishikawa et al., 2008). Thus, effects posed by the screen size of a hand-held navigation system need to be further examined, such as the desired level of detail or photorealism (Nurminen & Oulasvirta, 2008; Plesa & Cartwright, 2008).

Results from this study at least show that there are cases where pictures are an effective format for presenting route information, namely when matching views in the pictures and views in the real space is relatively easy. However, there may not be a single optimal format that works best for all situations. There are other formats that are potentially effective, and speech guidance is one candidate (Bradley & Dunlop, 2005; Denis et al., 2006; Reagan & Baldwin, 2006; Streeter et al., 1985). Essentially the same information that was given by pictures in this study can be conveyed through verbal descriptions. But due to issues inherent in spatial language, such as ontological ambiguity of spatial terms or existence of different types of reference frames, effective verbal route instructions are rather difficult to construct and require careful consideration. Also, aside from route guidance for people with visual impairments, navigation with verbal directions only may provide a lesser degree of excitement or engagement for the traveler than pictures do.

The question about the effectiveness of different formats for presenting route information is an empirical one. We believe that continued systematic studies should lead to theoretically and practically significant insights into this question.

References

Aiken, L.S., West, S.G.: Multiple regression: Testing and interpreting interactions. Sage, Thousand Oaks (1991)

Bessho, M., Kobayashi, S., Koshizuka, N., Sakamura, K.: A space-identifying ubiquitous infrastructure and its application for tour-guiding service. In: SAC 2008: Proceedings of the 23rd Annual ACM Symposium on Applied Computing, pp. 1616–1621 (2008)

Bradley, N.A., Dunlop, M.D.: An experimental investigation into wayfinding directions for visually impaired people. Personal and Ubiquitous Computing 9, 395–403 (2005)

Coors, V., Elting, C., Kray, C., Laakso, K.: Presenting route instructions on mobile devices: From textual directions to 3D visualization. In: Dykes, J., MacEachren, A.M., Kraak, M.-J. (eds.) Exploring geovisualization, pp. 529–550. Elsevier, Amsterdam (2005)

DeLoache, J.S.: Becoming symbol-minded. Trends in Cognitive Sciences 8, 66–70 (2004)

Denis, M., Michon, P.-E., Tom, A.: Assisting pedestrian wayfinding in urban settings: Why references to landmarks are crucial in direction-giving. In: Allen, G.L. (ed.) Applied spatial cognition, pp. 25–51. Erlbaum, Mahwah (2006)

Dillemuth, J.: Map design evaluation for mobile display. Cartography and Geographic Information Science 32, 285–301 (2005)

Gartner, G., Cartwright, W., Peterson, M.P. (eds.): Location based services and telecartography. Springer, Berlin (2007)

Hightower, J., Borriello, G.: Location systems for ubiquitous computing. Computer 34(8), 57–66 (2001)

Ishikawa, T., Fujiwara, H., Imai, O., Okabe, A.: Wayfinding with a GPS-based mobile navigation system: A comparison with maps and direct experience. Journal of Environmental Psychology 28, 74–82 (2008)

Klippel, A., Winter, S.: Structural salience of landmarks for route directions. In: Cohn, A.G., Mark, D.M. (eds.) COSIT 2005. LNCS, vol. 3693, pp. 347–362. Springer, Heidelberg (2005)

Levine, M., Jankovic, I.N., Palij, M.: Principles of spatial problem solving. Journal of Experimental Psychology: General 111, 157–175 (1982)

Levine, M., Marchon, I., Hanley, G.: The placement and misplacement of you-are-here maps. Environment and Behavior 16, 139–157 (1984)

Liben, L.S., Downs, R.M.: Understanding person-space-map relations: Cartographic and developmental perspectives. Developmental Psychology 29, 739–752 (1993)

Liben, L.S., Kastens, K.A., Stevenson, L.M.: Real-world knowledge through real-world maps: A developmental guide for navigating the educational terrain. Developmental Review 22, 267–322 (2002)

Nurminen, A., Oulasvirta, A.: Designing interactions for navigation in 3D mobile maps. In: Meng, L., Zipf, A., Winter, S. (eds.) Map-based mobile services, pp. 198–227. Springer, Berlin (2008)

Plesa, M.A., Cartwright, W.: Evaluating the effectiveness of non-realistic 3D maps for navigation with mobile devices. In: Meng, L., Zipf, A., Winter, S. (eds.) Map-based mobile services, pp. 80–104. Springer, Berlin (2008)

Radoczky, V.: How to design a pedestrian navigation system for indoor and outdoor environments. In: Gartner, G., Cartwright, W., Peterson, M.P. (eds.) Location based services and telecartography, pp. 301–316. Springer, Berlin (2007)

Raubal, M., Winter, S.: Enriching wayfinding instructions with local landmarks. In: Egenhofer, M.J., Mark, D.M. (eds.) GIScience 2002. LNCS, vol. 2478, pp. 243–259. Springer, Heidelberg (2002)

Reagan, I., Baldwin, C.L.: Facilitating route memory with auditory route guidance systems. Journal of Environmental Psychology 26, 146–155 (2006)

Richter, K.-F., Klippel, A.: A model for context-specific route directions. In: Freksa, C., Knauff, M., Krieg-Brückner, B., Nebel, B., Barkowsky, T. (eds.) Spatial Cognition IV. LNCS, vol. 3343, pp. 58–78. Springer, Heidelberg (2005)

Streeter, L.A., Vitello, D., Wonsiewicz, S.A.: How to tell people where to go: Comparing navigational aids. International Journal of Man-Machine Studies 22, 549–562 (1985)

Warren, D.H., Scott, T.E.: Map alignment in traveling multisegment routes. Environment and Behavior 25, 643–666 (1993)

Winter, S.: Route adaptive selection of salient features. In: Kuhn, W., Worboys, M.F., Timpf, S. (eds.) COSIT 2003. LNCS, vol. 2825, pp. 349–361. Springer, Heidelberg (2003)

The Abduction of Geographic Information Science: Transporting Spatial Reasoning to the Realm of Purpose and Design

Helen Couclelis

University of California
Santa Barbara 93106, USA
cook@geog.ucsb.edu

Abstract. People intuitively understand that function and purpose are critical parts of what human-configured entities are about, but these notions have proved difficult to capture formally. Even though most geographical landscapes bear traces of human purposes, visibly expressed in the spatial configurations meant to serve these purposes, the capability of GIS to represent means-ends relationships and to support associated reasoning and queries is currently quite limited. This is because spatial thinking as examined and codified in geographic information science is overwhelmingly of the descriptive, analytic kind that underlies traditional science, where notions of means and ends play a negligible role. This paper argues for the need to expand the reach of formalized spatial thinking to also encompass the normative, synthetic kinds of reasoning characterizing planning, engineering and the design sciences in general. Key elements in a more comprehensive approach to spatial thinking would be the inclusion of abductive modes of inference along with the deductive and inductive ones, and the development of an expanded geographic ontology that integrates analysis and synthesis, form and function, landscape and purpose, description and design.

1 Introduction

For those of us who have ever wondered how to build a good bird house, a book by Halsted [1] is enlightening. In Chapter XVII, which is entirely devoted to this important topic, we read: "It is a mistake to have bird houses too showy and too much exposed. Most birds naturally choose a retired place for their nests, and slip into them quietly, that no enemy may discover where they live. All that is required in a bird house is, a hiding place, with an opening just large enough for the bird, and a watertight roof. There are so very many ways in which these may be provided, any boy can contrive to make all the bird houses that may be needed." (p. 203). An illustration depicting three different bird houses clarifies these principles (Figure 1). We see an old hat nailed on the side of a barn with a hole for an entrance; a three-level pyramid on top of a pole, made of six 'kegs' nailed to planks; and a house-like structure made of wood. Two of these designs actually reuse obsolete objects originally intended for very different purposes. We thus have three very different-looking spatial configurations realized with three very different kinds of materials. Yet 'any boy' can understand that a single functional and spatial logic is giving rise to these three contrasting forms.

K. Stewart Hornsby et al. (Eds.): COSIT 2009, LNCS 5756, pp. 342–356, 2009.
© Springer-Verlag Berlin Heidelberg 2009

place for their nests, and slip into them quietly, that no enemy may discover where they live. All that is required in a bird house is, a hiding place, with an opening just

Fig. 211.—HAT HOUSE. Fig. 212.—KEG HOUSE. Fig. 213.—LARGE HOUSE.

large enough for the bird, and a water-tight roof. There are so very many ways in which these may be provided, any boy can contrive to make all the bird houses that may be needed. An old hat, with a hole for a door,

Fig. 1. Designing a birdhouse: three different spatial arrangements, three different kinds of materials, one set of functional requirements, one purpose. (Source: Halsted 1881, p. 203).

Let us now move on to something more familiar to geographic information science researchers. Figure 2a shows an ordinary urban streetscape. We see buildings, roads, cars, parking spaces, and benches. No natural process has created any of these objects, and no natural principles can explain either their individual shapes or the overall arrangement of the scene. Further, the location of a couple of natural objects visible on the picture – the trees – cannot be explained by any cause known to nature. Like in the case of the birdhouses, some intentional agent must have decided that the space in question needed to be configured in that particular way. And finally, exhibit number three (Figure 2b): Here is a natural landscape. Or is it?... A closer look reveals a number of straight lines crisscrossing the scene that have close to zero chance of having been generated by natural processes. The alert observer immediately realizes that these are the traces of earlier cultivation on a now abandoned landscape. This landscape and the streetscape of Figure 2a thus have something very important in common: even though no human presence is directly visible in either of them, they both reflect human *purposes*, the former through the outlines of old fences and retaining walls that used to support specific agricultural practices, the latter through the urban functions of shopping, resting and circulation served by its constituent parts and overall configuration.

So here is the point: Most landscapes that GIS deals with today are to a greater or lesser extent humanized landscapes, spatially organized so as to support specific functions, and changing over time as human purposes change. People intuitively understand why spaces are configured in particular ways, and they can anticipate what kinds of things *should* or *should not* be there based on explicit or implied purposes and the spatial functions that serve these purposes. In less obvious cases people will ask – but may lack the information to answer – questions such as:

- What is the purpose of this spatial object?
- What is the function of this place?
- What parts should compose this place?
- What else should be next to this place?
- How should this place be connected to other places?
- Is there a reason (*not* cause) for these changes on the map between time *a* and time *b*?

(a)   (b)

Fig. 2. (a) is an ordinary streetscape making no secret of its functions and purposes, including the purpose of the few 'natural' elements visible; whereas (b) is a natural-looking landscape that still bears physical traces from a time when it was used to meet specific agricultural needs. (Sources: (a) http://www.quinn-associates.com/projects/DecorVillage; (b) © J. Howarth.)

- Has the function of this place changed recently?
- How should this place change now that related activities or functions are changing?
- How should this place be configured in order to support the anticipated activities?

Yet the ability of GIS to support these kinds of queries is currently very limited. This is because *the thinking behind the understanding of function and purpose is not analytic but synthetic and normative,* whereas GIS is foremost an analytic tool. It may be argued that about one-half of natural spatial reasoning – the synthetic half – is not properly supported by GIS and is largely ignored in geographic information science. This paper presents a case for expanding the formal reach of the latter and the practical vocabulary of the former by introducing ideas and methods from the normative and synthetic sciences of planning, engineering, and design – more generally, from the disciplines known as the *design sciences*. Herbert Simon's seminal essay *The Sciences of the Artificial* [2] has contributed significantly to the recognition of the design disciplines as a distinct field of systematic intellectual endeavor, and to the understanding of the products of design – whether material or abstract – as belonging to an ontologically distinct class.

This paper focuses not on the activity of design itself but rather, on the question of how to introduce into GIS the concepts and modes of reasoning that will allow users to query and better understand human-configured – that is, designed – spatial entities. These include natural entities adapted for human use, from the vegetable garden in your yard to the Grand Canyon in its role as international tourist attraction, as well as those entities that are created entirely by humans to be university campuses, freeway or sewer networks, or cities. Even though the ultimate objective is practical, the problem of representing designed entities in a form that supports non-trivial automated reasoning and queries raises some very fundamental theoretical issues. Current approaches based on attribution facilitate the classification of such entities but as we will see below, this may not be sufficient. On the other hand, decades of research and software development in architectural and industrial design have yielded important insights into how designs are generated, but not on the inverse problem, more relevant to geographic information science, of how to decipher a 'designed' geographical entity. Still, there is much to be learned from these efforts to formalize the design process. Drawing on the theoretical design literature, the next section discusses the logic of design, which relies on synthetic, normative thinking and uses abductive inference extensively. Section 3, entitled 'The language of design', briefly explores the question of whether synthetic thinking in the geographic domain may require additional spatial concepts. It then introduces the concepts of functions, purposes, and plans that are central to synthetic thinking, and proposes a tentative solution to the problem of expanding the representation of function and purpose in human-configured geographic entities. Inevitably, the Conclusion that follows is brief and open-ended, more geared towards a research agenda than any concrete findings or recommendations.

2   The Logic of Design: Synthetic Thinking and Abduction

2.1   GIS and the Sciences of the Artificial

Spatial thinking as currently represented in geospatial models and software is overwhelmingly of the classic analytic kind that characterizes traditional science. Analytic thinking has been formalized, codified and successfully applied to scientific problems for centuries. However, it is not the only kind of systematic thinking of which humans are capable. Analytic thinking describes the world as it is (or may be), while humans are also very adept at reasoning about the function and purpose of things, easily switching between 'how does it *look*' to 'how does it *work*' to 'what is it *for*'. For example, an experienced engineer can look at a piece of machinery and (a) describe its structure, (b) based on that description, figure out how the machine works, and (c) from its function, infer the purpose for which the machine was built [3] Conversely – and more typically – the engineer will begin with a goal (purpose) to be met and will synthesize a product that functions in desired ways based on specific analytic properties of material and structure. Goal-oriented synthetic thinking of this kind is also known as *normative* because it is concerned with how things *should* or *ought* to be in order to fulfill their intended purpose. While 'normative' is usually contrasted with 'positive', normative reasoning also bears a symmetric relationship with causal reasoning which seeks to derive analytic explanations (Tables 1& 2). Synthetic, normative thinking characterizes not only the engineering sciences but more generally, the *design* disciplines which, in the geospatial domain, range from architecture, landscape architecture, and planning to decision science, spatial optimization, and various forms of spatial decision support. In actual fact, both the traditional sciences and the design disciplines use synthetic as well as analytic thinking, and normative as well as causal reasoning, though in different ways and with different emphases [4]. This paper argues for the need to expand the scientific reach of spatial thinking so as to formally integrate analysis and synthesis and thus also enable normative inferences that involve the functions and purposes of things.

Table 1. Normative versus causal reasoning

Normative	Causal
what *x* <u>in-order-for</u> *y*	*y* <u>because</u> *x*
what *purpose* <u>in-order-for</u> *spatial-pattern*	*spatial pattern* <u>is-caused-by</u> *process*
what *spatial-pattern* <u>in-order-for</u> *purpose*	*process* <u>causes</u> *spatial-pattern*

This integration is critical because the majority of the earth's landscapes now bear the traces of purposeful human intervention as the land is continuously adapted to support specific functions and activities of everyday life. Landscapes adapted by people are not just abstract spaces but *places* rich in personal and cultural meanings. Yet we still lack the formal conceptual frameworks, methodologies, and tools to deal with notions that will help analyze, understand, and anticipate spaces configured for human use and the reasons why these are the way they are, and to understand places as well as spaces. While concern for function and purpose is common in both the social and the life sciences, that interest is not well supported – at least not in the geospatial domain – by an appropriate scientific infrastructure to help formalize and implement modes of thinking that connect the configuration of the land with its functions and purposes. We would like to be able to represent in geospatial databases, and better understand with their help, the human motivations that led to the emergence of particular geographic landscapes and the ways that changing motivations lead to changes in land use and land cover. Such understandings will be essential if we are to model the complex changes now occurring in the ways the earth's surface is used, and their implications for the sustainability of both human populations and biota.

A number of different areas of thought, many with a spatial emphasis, have developed around normative and synthetic thinking. As mentioned earlier, engineers design structures, machines, devices, algorithms, and new materials that function in particular ways to meet specific purposes. So do the planning disciplines, from architecture and urban design to landscape architecture and regional planning, which seek to configure geographic space so that it may better support particular human activities or ecological functions. Decades of efforts have gone into formalizing the design process, resulting in both increased theoretical understanding and the development of software for the support of architectural, industrial, and other design activities [5] [6] [7]. On the formal side, artificial intelligence has contributed considerably to our understanding of synthetic and normative reasoning through the work on plan generation, frames, expert systems and other topics characterized by inferential reasoning that is neither primarily deductive nor inductive [8]. More recently the international DEON[1] conference series, "...designed to promote interdisciplinary cooperation amongst scholars interested in linking the formal-logical study of normative concepts and normative systems with computer science, artificial intelligence, philosophy, organization theory and law", [9] has helped broaden the appeal of normative thinking well beyond its traditional strongholds.

---

[1] 'Deon' is the Greek word for 'what needs to be' and gives rise to 'deontic logic', a notion closely related to the term 'normative' which is derived from the Latin.

Table 2. Contrasting the dominant analytic stance of GIS with the synthetic stance of the design sciences

GIS & traditional sciences	The Design sciences
Analysis	Synthesis
From instances to principles	From principles to instances
Causal	Goal-oriented
Descriptive	Prescriptive
Positive	Normative
IS	*OUGHT*

GIS continues to rely on a predominantly analytic mode of thinking, despite the fact that many of the entities it represents, from roads and cities to rice paddies and ski runs, are such as they appear because they are configured *in order to* support specific human activities and purposes (Table 2). While for years GIS has very successfully supported spatial decisions relating to the allocation and configuration of such entities [10], it cannot yet support queries as to why – say – there is a small structure at the bottom of the ski slope and whether another one should be expected to be at the top, or how that entire configuration of open spaces and installations may change if the ski resort closes for good. This is because GIS databases and operations can provide highly detailed descriptive information on what is out there but they don't normally place entities and relationships in the context of the human activities that require spaces to be configured in particular ways [11]. The difficulty of distinguishing between land cover and land use in GIS provides the archetypal example of what may be missing from analytic descriptions of the geographic world [12]. In traditional representations of land use, residential areas, roads, and buildings are coded no differently than lakes, streams and rock outcrops, with only an item key or map legend indicating that these are actually *artifacts* – artificial things that people made and placed there for a purpose. While the purpose itself is invisible, it is reflected in characteristic functional spatial relations robust enough to be sometimes recognizable not only by human intelligence but also by machines. Thus Ahlqvist and Wästfelt [13] were able to develop a neural net algorithm that could identify summer farms in Sweden from medium resolution satellite imagery. These farms consist of a collection of different land cover patches that stand in specific spatial relations to one another. Their complex spectral signatures defy automated detection at medium resolution, but giving the algorithm some hints about necessary functional relations (here, a couple of distances between patches that belong to different land cover classes) results in highly accurate identification of summer farms. What kind of spatial thinking does that experiment point to? What connects Swedish summer farms, streetscapes, abandoned fields, ski slopes, and birdhouses –geographically speaking? How could we harness the underlying logic so as to expand the range of queries that GIS could support? These are the questions that this paper sets out to address.

2.2 Abduction in Geographic Information Science and GIS

As Chaigneau et al. [14] note, "Function is central to our understanding of artifacts. Understanding what an artifact is used for, encompasses a significant part of what we know about it" (p. 123). The inability of GIS to properly support this kind of understanding is likely what prompted Bibby and Shepherd [15] to write: "the 'objects' represented in GIS are unquestionably assumed to have a prior, unproblematic existence in the external world....a crippingly restrictive conceptualization of objects"(p. 583). Artifacts, which in the general case include both artificial objects like roads and buildings and spatial adaptations like gardens and Swedish summer farms, are always the formal or informal, explicit or implicit products of *design*. Note that 'design' means both drawing and intention. It is the intentional dimension of artifacts that is the key to understanding their nature, and which makes a simple tin can, in Simon's [2] example, an object of an ontologically more complex order than a tree. While humans will immediately recognize the tin can as an instance of an artificial container, no amount of analysis of geometry, topology and attributes can provide a satisfactory understanding of that object.

All complex reasoning, including spatial reasoning, involves three complementary modes of inference: deduction, induction and abduction. The relationship between these modes is shown in Table 3. All three involve, at different stages (a) *rule(s)*, by which we mean the general principle(s), premises or constraints that must hold; (b) *case(s)*, that is, exemplar(s) of phenomena to which the rules do or may apply; and (c) a *result*, or the specific state of affairs to which the rule(s) is or may be applicable. These three modes have different properties. Deduction, induction and abduction yield certain inference, probable inference and plausible inference, respectively. But also, the amount of entropy (information to be obtained) from the inference increases in that order, being minimal for deduction and maximal for abduction. All three modes are present in scientific reasoning, from the deductive power of mathematics to the value of fruitful generalization from a sample, to the inferential leap leading to new discoveries.

Table 3. Symmetries connecting the three basic modes of inference (after Peirce: see [16])

Deduction	Induction	Abduction
Rule	Case	Result
Case	Result	Rule
Result	Rule	Case

Deduction and induction have both been extensively formalized over centuries of mathematical and scientific development and deduction in particular is amply supported in software, including in GIS. Abduction on the other hand, even though it was sketched out (and named) by Aristotle, was only rediscovered in the late 1800s by Peirce [16] and still lacks the recognition and degree of formal support that deduction and induction have enjoyed since antiquity. As Worboys and Duckham [17] note, "In general, computers rely solely on deductive inference processes, although inductive and abductive reasoning are used in some artificial intelligence-based systems. As a consequence, processing in a computer is deductively valid, but this mode of reasoning prevents computers from generating new conclusions and hypotheses" (p.297). Indeed, abduction produces a plausible explanation or hypothesis ('case') for a given state of affairs ('result') such that the explanation satisfies a number of premises or constraints ('rule') that may be theoretical, methodological, empirical or pragmatic.

Formalizations of the abduction problem can be found in the design and artificial intelligence literatures, e.g. [5],[18]. For example, according to [18], "the abduction problem is a tuple $\langle D_{all}, H_{all}, e, pl \rangle$ where D_{all} is a finite set of all the data to be explained; H_{all} is a finite set of all the individual hypotheses; e is a map from subsets of H_{all} to subsets of D_{all} (H explains $(e(H))$; pl is a map from subsets of to a partially ordered set (H has plausibility $pl (H)$). H is complete if $e(H) = D_{all}$; H is a best explanation if there is no H' such that $pl(H') > pl (H)$." (p. 28). The range of $pl (H)$ may be a Bayesian probability or other measure with a partially ordered range and is estimated in the context of the 'rules' (prior knowledge) applicable to the case.

Abduction is often discussed as the logic of medical diagnosis or detective work although its application is much broader, since it underlies constraint satisfaction problems and hypothesis generation of any kind. Abduction has also been recognized as the hallmark of synthetic thinking in general and of design in particular, in the sense that every design problem is a constraint satisfaction problem, and every design solution is a hypothesis in that the design in question is a plausible answer given the facts of the matter. For example, the design of a house must satisfy a number of environmental, legal, social, and resource constraints, along with spatial constraints of minimum area and height, of adjacency, connectivity, occlusion, etc. that derive directly from the domestic functions (cooking, entertaining, sleeping,…) to be supported. While many routine tasks in the geographic information domain involve abductive thinking (e.g., the interpretation of imagery or more generally, of patterns in the data, or the development of models of spatial processes in the absence of general laws), certain equally important tasks relating to artificial entities in particular are not currently supported by GIS and related tools. Examples include: identifying the function and/or purpose of an untypical spatial configuration; reconstructing the plan of a partially preserved archaeological site; identifying the location of spatial parts functionally related to a particular artificial entity; predicting changes in land use and land cover given knowledge of changes in human activities; deciding whether an apparent change on a map relative to an earlier map is real or the result of a mapping error; and designing any land use, watershed, or landscape plan to serve specific human or ecological purposes within existing geographical and resource constraints [19].

The next section argues that the tools to support such tasks are not well developed because the necessary concepts and modes of reasoning, and their relationships to more familiar spatial concepts and analytic modes of reasoning, have not been sufficiently investigated in geographic information science. These are however well established in the synthetic, design sciences, from which we may have much to learn.

3   The Language of Design

3.1 The Vocabulary of Design

What may be the ontological implications of expanding the language of GIS so as to include synthetic thinking and the purpose-orientation of design? A useful place to begin is the investigation of the spatial concepts involved in design and in analytic geospatial science, respectively. Are they the same concepts? If so, what distinguishes the two perspectives? Are there two overlapping but not identical sets of concepts? If so, what are the differences? At least one project exploring these questions is already underway, currently developing a catalog of spatial terms extracted from the literature of both analytic geography and design [20]. For the purposes of the present paper a top-down approach building on the discussion in the previous sections appears more suitable.

As mentioned earlier, both analytic and synthetic reasoning are based on the three complementary inference modes of deduction, induction and abduction, though these are used differently in the two perspectives and to different ends: for describing a present, past, or future state of the world in the analytic case, and for bringing about a different state of the world in the synthetic case. This difference is akin to the distinction made in the philosophy of mind between the two ways, or 'directions of fit', in which intentional mental states can relate to the world[2]. Thus the mind-to-world direction of fit, which includes beliefs, perceptions and hypotheses, concerns actual states of the world (*facta*), whereas the world-to-mind direction, which includes intentions, commands, desires and plans, concerns states of the world that do not yet exist but that one wants to make happen (*facienda*) [21]. It is easily seen that analytic thinking is about *facta* whereas synthetic thinking is about *facienda* – this is precisely the IS-OUGHT distinction highlighted at the bottom of Table 1.

This brief foray into the philosophy of mind reinforces the notion that the critical difference between analytic and synthetic thinking is on the side of the observer-actor's intentional stance rather than the world. It also suggests, though it does not prove, that any differences in vocabulary between analysis and design should also be on the side of intentional rather than spatial terms. One may surmise that there will be differences in emphasis (i.e., some spatial terms will be more prominent in analysis or in design because they relate to concepts that are more important in one or the other tradition), and that there will be qualifiers to spatial concepts commensurate with the objectives of each of the two perspectives. As an example of differences in emphasis, take the concept of 'pattern'. It is central to both analytic geospatial science and design, but many more synonyms of the term are commonly used in the latter: shape, structure, configuration, arrangement, composition, design, motif, form, etc. [22]. This is because 'pattern' and related notions are very critical to design, being in many cases the end result of the design activity itself. Spatial analysis, on the other hand, places considerable emphasis on uncertainty-related spatial concepts such as fuzzy regions, epsilon bands and error ellipses, because its objective is not to change the world but to accurately represent it. (Clearly, these concepts are also very important in some areas of engineering design, though as constraints rather than as objectives). Because of the emphasis on correct representation, accuracy, precision, fuzziness, and so on are also important qualifiers in analytic spatial thinking, whereas the synthetic stance is much more invested in concepts that qualify the fitness-for-use (aesthetic as

---

[2] This distinction is also familiar from the philosophy of language, where the focus is on speech acts rather than intentional mental states. See [27].

well as practical) of the products of design, such as 'efficient', 'functional', 'harmonious', 'pleasant', 'symmetric', 'human-scale', or simply 'good'.

Having found no significant differences in the vocabulary, i.e., in the spatial terms used in analysis and in design, the next question should be about the syntax: How are spatial elements put together to yield arrangements that support specific human purposes? What is it about the resulting forms of artificial entities and spatial configurations that allows them to serve such purposes? How can we understand what these configurations and entities are *for*, and how they relate to human activities? The first question concerns design as a process and is beyond the scope of this paper. The other two, which are about making sense of what humans have designed and built, are explored in the following.

3.2 Purposes, Functions, Activities, and Plans

Purpose is what makes the human world tick and yet there is no place for it in traditional analytic science, whether natural or social. Traditional science is the realm of causes and effects. Purpose, on the other hand, is what the design sciences are about. In the spatial realm, purpose is the interface between the human world of intentions and the world of intentional spatial configurations – the adapted spaces that we call farms, airports, transport networks, or cities. Purpose itself is invisible and immaterial, but it is expressed spatially through *activities* and *functions*. Thus farmers engage in a host of different activities that may include feeding, breeding and moving livestock, growing, harvesting, storing and transporting crops, running a horseback-riding barn or bed-and-breakfast, and so on. Similarly, airports are the places where airplanes take off and land, where aircraft and service vehicles circulate and park, and that people enter and navigate to specific departure gates; and so on. Mirroring the activities, which are temporally bounded occurrences, are the corresponding functions, which are associated with the corresponding artificial entities in a more enduring (though not necessarily permanent) fashion. The barn is still the barn after the animals have left; the airport is still the airport, and the departure gates are still the departure gates during the night hours when there are no rushing passengers or departing flights. Functions reflect the abstract relational structures characterizing human-configured spatial entities of a particular class: Every farm is unique, but all store the hay as close as possible to both a delivery road and to where the cows are kept. Further, the functions themselves are reflected in the *adapted spaces*, the concrete, appropriately configured entities made up of specialized sub-spaces with the required geometrical and other attributes, standing in specific spatial relationships to one another. Finally, these four elements – purpose, activity, function and adapted space come together in (spatial) *plans* (Figure 3). These plans may be implicit or explicit; they may be formal or informal; they may be individual or collective; they may be laid down on clay tablets or on paper, or they may reside in peoples' minds; and they may be finite and immutable or they may be always in flux. No matter in what form, plans always express a desire to adapt geographic space to specific purposes and to the functions and activities these purposes entail.

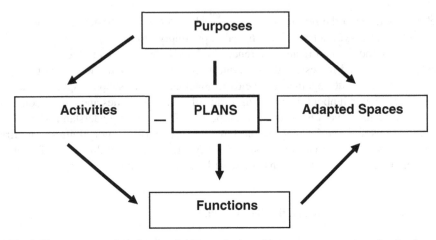

Fig. 3. How purposes, functions, activities, and adapted spaces may come together in plans

The tightly knit nexus of concepts represented in Figure 3, where some of the relations are as important as the entities, presents challenges for current, mostly object-oriented geographic information ontologies. The concept of function is particularly problematic because it cannot be properly specified except in the context of activity on the behavioral side, and of adapted-space on the spatial: functions mirror the former and are expressed in the later. Functions change when activities change and this also (frequently) causes changes in adapted spaces. Functions are thus relational concepts, somewhat like roles, which get their meaning from being associated with both a role-player and a context and which for that reason are not attributes, though both roles and functions are sometimes treated as such [23] [24]. It is indeed often sufficient to assume that naming an artifact is enough to specify its function (a knife is for cutting, a bridge is for crossing), but this ignores the fact that functions are largely in the eye (interests) of the beholder: a knife is also for spreading butter and for prying open lids, and so on. Thus a road bridge IS-A bridge and will normally be classified along with railroad and covered bridges. But if one is interested in the function of allowing vehicles to get to the other side of the river, the road bridge will more likely be classified along with fords and ferries, while from the perspective of the ecologist concerned with wildlife corridors, the function of a road bridge may be to provide a safe underpass for animals This fluidity of function, the endless affordances provided by natural as well as artificial entities that help support human activities and meet goals, is not well supported by current geographic information ontologies. This is also why functional classification is usually considered too problematic to undertake, even though it would often make more sense from the user's viewpoint [25].

Elsewhere I proposed an ontological framework for geographic information that includes the concepts of purposes, functions, and adapted spaces [26] but leaves out activities and plans. Not coincidentally, activities and plans are the only two synthetic concepts in this group. Activities are complexes of individual actions, at different levels of granularity, woven together so as to help realize specific purpose(s); plans are the quintessential examples of synthetic, design-oriented thinking. Purposes, on the other hand are antecedents, adapted spaces are outcomes, and functions are abstract expressions of the properties that make adapted spaces and activities correspond with each other. In this scheme, (user) purposes are at the very top of the hierarchy, thus drastically narrowing and refining the scope of relevant functions to be considered. Along similar lines, Howarth [19] has proposed and implemented a model for mapping activities at different levels of granularity into correspondingly nested adapted spaces by means of plans, and demonstrated the practical utility of that work with case studies from a historic California island ranch. For example, the function of an unlabeled rectangular space represented on an old map could be abductively inferred from a knowledge of the activities constituting a day of rounding up sheep by the ranch cowboys: based on qualitative information on the spatio-temporal pattern of these activities, and map information on distances and site characteristics, it was correctly concluded that the rectangle in question was used as a horse corral for the cowboys' mid-day break.

Searle's theory of Intentionality [27] may help provide a framework for completing the integration of the five elements of Figure 3. As discussed earlier, desires and purposes underlie the 'world-to-mind' direction of fit – changing the world so as to fit what's in the mind – and this notion of 'direction of fit' is closely connected in philosophy to that of 'conditions of satisfaction' What kinds of conditions can satisfy the purpose of making a living on a farm, or of running an efficient airport that is attractive to travelers and safe for everybody? Not all of the answers are of course spatial, but many are, and formalized spatial thinking ought to be able to grasp them. According to Searle the conditions of satisfaction for the world-to-mind direction of fit require world-changing physical *actions* by intentional agents. Two related kinds of intentions are distinguished: *prior intentions*, which are mental, and *intentions in action*, which are involved in carrying out the intended physical act and thus realize the conditions of satisfaction of the original desire or purpose. We are here interested in intentional actions that enfold in, and change the geographic-scale world. Figure 3 suggests two different kinds of such actions: those making up the complex activities that directly realize in whole or in part the conditions of satisfaction of the original desire or purpose (e.g., the daily activities involved in running a horseback-riding barn), and those needed to adapt a space so that it may adequately support these activities (e.g., developing the barns, corrals, riding rings, storage areas, office spaces, access routes, parking areas, etc. in the required sizes, configurations and spatial relations to one another). In addition to these two kinds of activities (one on-going, the other temporally delimited) which reflect two different intentions-in-action, there is a single prior intention expressed in the plan.

An important point in Searle's [27] theory of intentionality is the symmetry between the world-to-mind direction of fit that characterizes world-changing actions by intentional actors, and the more passive mind-to-world direction of fit that results in understanding and interpreting the world and its changes. With this last observation we may now have enough conceptual ammunition to tackle the central question of this paper: How could we expand current formalizations of spatial reasoning so that they may also support the interpretation and understanding of spatial entities purposefully developed or changed by humans. In other words, the challenge is how to get at the hidden underlying plan that ties together the functions and purposes of a human-configured feature in the landscape, its spatial organization, and any activities relating to it that we may be able to observe.

This appears to be a task tailored for abductive treatment. As discussed above (see Table 3), the basic form of abductive inference may be written as: *result* x *rule* → *case*. Here the 'result' may be the spatial configuration on the ground that must be interpreted; the 'rule' would then be the set of known functional organizations that may correspond to the spatial configuration at hand; and the 'case' would be the tentative identification of the artificial spatial entity in question as being very likely - a summer farm, a winery, a sacrificial place, a spa, a secret military installation. As Howarth [19] has shown, abduction may be used in this way for obtaining plausible answers to many different kinds of queries, under both static conditions and in cases where some changes from previous states have been observed, depending on what may be known and what may need to be inferred from among the elements illustrated in Figure 3. It thus appears possible that we may eventually be able to approach the outcomes of normative, synthetic human thinking in geographic space as rigorously and systematically as we do the products of natural processes.

4 Conclusion

Spatial configurations in humanized landscapes realize implicit or explicit spatial plans, which are schemas for promoting specific human purposes related to the land. Purposes are reflected in functions; functions support activities; activities enfold in adapted spaces; and plans connect these elements into normative configurations or designs. I argued that geographic information science should embrace this fundamental insight stemming from the design disciplines – not because it is itself a design discipline, or needs to be, but so as to be able to properly represent and analyze the human-configured landscapes around us. So far, with the exception of activities, the concepts surrounding the notion of (spatial) plan have received scant attention in mainstream geographic information science. It seems that the reasons for this apparent neglect have less to do with a lack of interest in these issues, and more to do with the ways spatial reasoning has been formalized and codified to date.

The paper identified two areas, quite possibly connected, where current approaches to spatial reasoning could be augmented. The first is the facilitation of abductive inference so as to complement the inductive and deductive forms already routinely supported in available models and software. The second is the expansion of current geographic information ontologies so as to encompass and implement the interconnected concepts of purpose, function, activities, adapted spaces, and plans. This will by far be the harder task of the two, requiring us to grapple with culturally contingent issues of means and ends, of needs, wants and choices, as well as with some controversial chapters in the philosophy of mind and social reality. Yet human purpose and its traces on the land, and conversely, the land's role in shaping human purpose, have been for decades the central themes in certain qualitative areas of geography from within the humanistic, cultural, and regional perspectives. We may have something to learn from these old-fashioned approaches also. As geographic information science matures with the passing years, I am reminded of a quote by a now anonymous (to me) researcher from the RAND Corporation: "In our youth we looked more scientific". I would not be the least offended if geographic information science were also to have *looked* more scientific in its youth.

Acknowledgements. An early version of this paper was presented at the *Specialist Meeting on Spatial Concepts in GIS and Design*, NCGIA, Santa Barbara, California, December 15-16, 2008. Jeff Howarth's dissertation research helped rekindle my old professional fascination with design in the new context of geographic information science. Thank you Jeff for this, and for the pictures (Fig. 1 and 2(b)).

References

1. Halsted, B.D.: Barn Plans and Outbuildings. Orange Judd, New York (1881)
2. Simon, H.A.: The Sciences of the Artificial. The MIT Press, Cambridge (1969)
3. Norman, D.A.,: The Psychology of Everyday Things. Basic Books, New York (1988)
4. March, L.: The Logic of Design. In: Cross, N. (ed.) Developments in Design Methodology, pp. 265–276. John Wiley & Sons, Chichester (1984)
5. March, L.: The Logic of Design and the Question of Value. In: March, L. (ed.) The Architecture of Form, pp. 1–40. Cambridge University Press, Cambridge (1976)
6. Cross, N.: Developments in Design Methodology. John Wiley & Sons, Chichester (1984)
7. Tomiyama, T., Takeda, H., Yoshioka, M., Shimomura, Y.: Abduction for Creative Design. AAAI Technical Report SS-03-02 (2003)
8. Cohen, P.R., Feigenbaum, E.A.: The Handbook of Artificial Intelligence. William Kaufmann, Inc, Los Altos (1982)
9. DEON 2008, http://deon2008.uni.lu/cfp.html
10. Aerts, J.: Spatial Optimization Techniques and Decision Support Systems. PhD Dissertation, University of Amsterdam (2002)
11. Kuhn, W.: Semantic Reference Systems. International Journal of Geographical Information Science 17(5), 405–409 (2003)
12. Fisher, P., Comber, A., Wadsworth, R.: Land Use and Land Cover: Contradiction or Complement. In: Fisher, P., Unwin, D. (eds.) Re-Presenting GIS, pp. 85–98. Wiley, Chichester (2005)
13. Ahlqvist, O., Wästfelt, A.: Reconciling Human and Machine Based Epistemologies for Automated Land Use Detection. In: Abstract. Conference on Geographic Information Science 2008, Park City, Utah (2008)
14. Chaigneau, S.E., Castillo, R.D., Martinez, L.: Creator's Intentions Bias Judgments of Function Independently of Causal Inferences. Cognition 109, 123–132 (2008)
15. Bibby, P., Shepherd, J.: GIS, Land Use, and Representation. Environment and Planning B: Planning and Design 27(4), 583–598 (2000)
16. Houser, N., Kloesel, C. (eds.): The Essential Peirce: Selected Philosophical Writings, vol. 1(1867-1893). University of Indiana Press, Indianapolis (2009)
17. Worboys, M., Duckham, M.: GIS: a Computing Perspective. CRC Press, Boca Raton (2004)
18. Bylander, T., Allemang, D., Tanner, M.C., Josephson, J.R.: The Computational Complexity of Abduction. In: Brachman, R.J., Levesque, H.J., Reiter, R. (eds.) Knowledge representation, pp. 25–60. MIT Press, Cambridge (1991)
19. Howarth, J.T.: Landscape and Purpose: Modeling the Functional and Spatial Organization of the Land. PhD Dissertation, Department of Geography, University of California, Santa Barbara, USA (2008)
20. http://teachspatial.org/

21. Velleman, J.D.: The Guise of the Good. Noûs 26(1), 3–26 (1992)
22. Lynch, K.: Good City Form. The MIT Press, Cambridge (1981)
23. Fonseca, F., Egenhofer, M., Agouris, P., Câmara, G.: Using Ontologies for Integrated Geographic Information Systems. Transactions in GIS 6(3), 321–357 (2002)
24. Boella, G., van der Torre, L., Verhagen, H.: Roles, an Interdisciplinary Perspective. Applied Ontology 2, 81–88 (2007)
25. Reitsma, R.: Functional Classification of Space: Aspects of Site Suitability Assessment in a Decision Support Environment. IIASA Research ReportRR-90-2, Laxenburg (1990)
26. Couclelis, H.: Ontology, Epistemology, Teleology: Triangulating Geographic Information Science. In: Navratil, G. (ed.) Research Trends in Geographic Information Science, Springer, Berlin (in press)
27. Searle, J.R.: Intentionality: an Essay in the Philosophy of Mind. Cambridge University Press, Cambridge (1983)

An Algebraic Approach to Image Schemas for Geographic Space

Lisa Walton and Michael Worboys

Department of Spatial Information Science and Engineering, University of Maine,
Orono, ME USA
lisa.walton@spatial.maine.edu, worboys@spatial.maine.edu

Abstract. Formal models of geographic space should support reasoning about its static and dynamic properties, its objects, their behaviors, and the relationships between them. Image schemas, used to embody spatiotemporal experiential abstractions, capture high-level perceptual concepts but do not have generally accepted formalizations. This paper provides a method for formally representing topological and physical image schemas using Milner's bigraphical models. Bigraphs, capable of independently representing mobile locality and connectivity, provide formal algebraic specifications of geographic environments enhanced by intuitive visual representations. Using examples from a built environment, we define topological schemas CONTAINER and LINK as static bigraph components, dynamic schemas INTO and LINKTO as rule-based changes in static components, and more complex schemas REMOVAL_OF_RESTRAINT and BLOCKAGE with sequences of rules. Finally, we demonstrate that bigraphs can be used to describe scenes with incomplete information, and that we can adjust the granularity of scenes by using bigraph composition to provide additional context.

Keywords: Image Schemas, Spatial Relations, Built Environments, Bigraph Models.

1 Introduction

Modern information systems require models that incorporate notions of physical space in ways that move beyond the use of building information models (Laiserin 2002), geographic maps, and wayfinding tools for assisting human navigation though these spaces. Ubiquitous computing environments (Weiser 1993), such as ambient intelligent systems, require software agents that can detect and cause changes in physical environments. Conversely, humans and human aides (e.g., smart phones) in physical space can access and change information objects that exist in virtual space. An additional complication is that bounded regions in both physical and virtual spaces can have either tangible borders (e.g., brick and mortar walls) or borders established by fiat (e.g., the intangible dividing lines between co-worker workspaces in a shared unpartitioned office). In virtual environments systems may also be tangibly bounded (e.g., the confines of a single hard drive) or bounded by an internal software partition or firewall. Communication links in either kind of space can often cross boundaries freely. Although these issues are not new, this hybrid virtual-physical space of interaction continues to provide modeling challenges for spatial information theory.

K. Stewart Hornsby et al. (Eds.): COSIT 2009, LNCS 5756, pp. 357–370, 2009.
© Springer-Verlag Berlin Heidelberg 2009

Image schemas (Johnson 1987), used to embody spatiotemporal experiential abstractions, model conceptual patterns that can be physical or non-physical but do not have generally accepted formalizations despite numerous attempts (Egenhofer and Rodriguez 1999, Frank and Raubal 1999, Raubal, Egenhofer, Pfoser, and Tryfona 1997, Raubal and Worboys 1999). They have a rich history as a support for spatial reasoning and inference, particularly in the use of translating spatial prepositions in language to relations (Mark 1989, Freundschuh and Sharma 1996, Frank 1998), as a basis for describing geographic scenes (Mark and Frank 1989), and as fundamental theories underlying good user interfaces and query languages (Mark 1989, Kuhn and Frank 1991, Kuhn 1993). However, distinct differences exist in the use of schemas in small-scale and large-scale environments (Frank 1998, Frank and Raubal 1998, Egenhofer and Rodriguez 1999), due in part to changes in inference patterns for spatial relations when the scale changes.

Milner (2009) argues that models for spatially-rich systems should provide visualization tools that are tightly coupled with a formal system in order to support the needs of diverse communities including end-users, programmers, system designers, and theoretical analysts. Milner's bigraphical models (Milner 2001) provide a formal method for independently specifying mobile connectivity and mobile locality. Combined with a set of reaction rules that dictate appropriate system transformations, bigraphs provide a unified platform for designing, formally modeling, analyzing, and visualizing ubiquitous systems. Bigraphs were developed for the virtual world of communicating processes and ambient information objects. While it has been argued that agents can be physical and can influence informatics domains, it has not yet been demonstrated that bigraphs are suitable models for mixed virtual and physical environments.

This paper presents an approach for using bigraphs to formally model image schemas for use in built environments. We argue that realizing key image schemas in bigraphs provides a novel and useful means to represent and visualize the static and dynamic relationships and behaviors of entities in built environments. This work arises out of research on a theoretical framework for formally modeling ubiquitous computing environments as part of the multinational Indoor Spatial Awareness Project[1], and is a step towards our goal of developing a more comprehensive spatial theory for built environments.

In the remainder of the paper we provide the background of image schemas and bigraphical models and propose the use of bigraghs to formally represent and visualize key static and dynamic image schemas in built environments. We show how bigraphs can represent scenes with incomplete information, how bigraph composition can be used to increase and decrease the granularity of scenes by providing additional context, present a built environment example utilizing image schemas and granularity shifts, and finally present our conclusions and future work.

2 Background

2.1 Image Schemas

Image schemas are abstractions of spatiotemporal perceptual patterns. In his survey Oakley (2007) describes them as "condensed redescriptions of perceptual experience

---

[1] http://u-indoor.org/

for the purpose of mapping spatial structure onto conceptual structure". Johnson (1987) stated that these patterns "emerge as meaningful structures for us chiefly at the level of our bodily movements through space, our manipulation of objects, and our perceptual interactions" and that schemas can also be applied to "events, states, and abstract entities interpreted as spatially bounded entities". Since our goal is to model the locality and connectivity of entities in built environments, here follow several of the more useful topological and physical schema.

2.1.1 CONTAINER
This schema associates an entity serving as a container with expected relationships and behaviors. These can be either static or dynamic, for example we often associate the spatial relations *inside* and *outside* with some physical entity (e.g., in a room) or conceptual entity (e.g., in a conversation) as well as the dynamic behavior of moving *out-of* or *into* the entity serving as the container. There is also typically an explicit or implicit second entity that participates in the relationship or behavior associated with the container. This schema is often associated with either SURFACE (Rodriguez and Egenhofer 1997) or SUPPORT (Kuhn 2007). However, many place relations, actions and behaviors in built environments can be explained with CONTAINER alone.

2.1.2 LINK
The basic LINK schema (Johnson 1987) consists of two entities connected by a "bonding structure" which could be physical, spatial, temporal, causal, or functional. He also identified potential extensions to this basic schema including relationships between more than two entities or between entities that are spatially or temporally discontiguous. When the link structure is directed, this schema can become a building block for more complex schemas such as PATH (Kuhn 2007).

2.1.3 FORCE
Johnson identified seven basic FORCE schemas, two of which, BLOCKAGE and REMOVAL_OF_RESTRAINT, are particularly relevant. BLOCKAGE involves a force vector encountering a barrier and taking any number of possible directions in response, including the removal of the barrier. REMOVAL_OF_RESTRAINT involves the actions of the force vector once the barrier has been removed by another.

2.1.4 Spatial Relations from Schemas
Many researchers have expanded upon Johnson's original concepts to incorporate aspects of importance for GIS such as spatial cognition and relations (Freundschuh and Sharma 1996, Raubal, Egenhofer, Pfoser, and Tryfona 1997, and Frank and Raubal 1999). For example, Lakoff and Nunez (2000) expand upon Johnson's schemas to derive conceptual schemas for common spatial relations, such as *in, out*. Given a CONTAINER schemas where the entity serving as the container has an associated interior, exterior, and boundary, IN and OUT schemas can be defined where the focus of the container is either on its interior (for IN) or exterior (for OUT).

2.1.5 Algebras for Image Schemas
Egenhofer and Rodríguez (1999) defined a CONTAINER-SURFACE algebra for reasoning with inferences about the spatial relations in/out and on/off. They argue that

these schemas are particularly appropriate for small-scale "Room Space", but demonstrate that there are problems when applying it to large-scale geographic objects and spaces, or mixed environments. In particular, they note that some inference-based reasoning does not directly translate between scales (Rodríguez and Egenhofer 2000).

Kuhn (2007) proposed an algebraic theory built on image schemas for an ontological specification of spatial categories. He focused on topological (e.g., CONTAINMENT, LINK, PATH) and physical (e.g., SUPPORT, BLOCKAGE) schemas, and provided algebraic specifications that associate type classes with universal ontological categories and type membership with universal class instantiation. His model offers a powerful formalism for combining schema behaviors using type class subsumption. Our approach, which formalizes schemas using an algebraic approach, should complement his theory.

2.2 Bigraphical Models

Milner's bigraphs (Milner 2001) provide a formal method for independently specifying mobile connectivity and mobile locality, and are intended to provide an intuitive formal representation of both virtual and physical systems. Bigraphs have a formal definition in category theory (specifically as abstract structures in strict symmetric monoidal categories), but in this paper we use Milner's simpler visual descriptions that are tightly coupled with the underlying algebra. Bigraphs originate in process calculi, especially the calculus of mobile ambients (Cardelli and Gordon 2000) for modeling spatial configurations, and the Pi-calculus (Milner 1999) for modeling connectivity. Ambients, represented as nodes in bigraphs, were originally defined as "bounded places where computation occurs" (Cardelli 1999). In bigraphs nodes have a more general interpretation as bounded physical or virtual entities or regions that can contain other entities or regions. For example, a built environment can be modeled with a bigraph containing nodes representing agents, computers, rooms, and buildings that also represents the connectivity between agents and computers (Milner 2008).

2.2.1 Place Graphs

Containment relations between nodes in bigraphs are visualized by letting nodes contain other nodes. Other spatial relationships between regions such as overlap, meet, or equals are not expressible as place relations. Every bigraph B has an associated place graph B^P, which shows only the containment relations between entities. Figure 1 shows a simple bigraph B of an agent A_1 and a computer C_1 in a room R_1 with its place graph B^P (a tree). The agent is not connected to the computer.

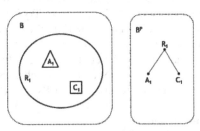

Fig. 1. Bigraph B and place graph B^P

2.2.2 Link Graphs

Connectivity, or linking relations, are represented as hypergraphs, generalizations of graphs in which an edge may join any number of nodes. Each node has a fixed number of *ports* indicating the number of links that it is permitted. Each edge represents a particular connection (relation) between the nodes it links, but does not typically denote a spatial relation such as physical adjacency. Every bigraph B has an associated link graph B^L which shows only the connectivity between nodes. Figure 2 shows the bigraph B for the simple scene when the agent A_1 is connected to the computer C_1. Both the agent and the computer now have one port each that supports the link. The place graph B^P shows only the containment relations in the scene and the link graph B^L shows only the linking relations in the scene.

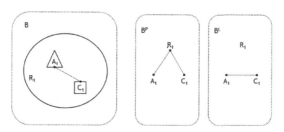

Fig. 2. Bigraph B with place and link graphs

Figure 3, a variant of Milner's example (Milner 2008), shows a more complicated scene with additional agents and an extra room and building.

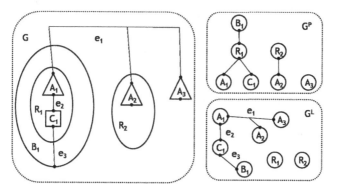

Fig. 3. Built environment G with place and link graphs

Bigraph G describes a scene in which agent A_1 and computer C_1 are inside room R_1 in building B_1, another agent (A_2) is inside a second room (R_2), and a third agent (A_3) is elsewhere. This place configuration is reflected in the place graph G^P. Three agents participate in a conference call (e_1), the first agent is on a computer (e_2), and the computer is connected to a LAN in the building (e_3). This link configuration is reflected in the link graph G^L. In this example rooms have no ports, but buildings have one for a LAN. Agents and computers each have two ports so that any agent can simultaneously connect to a conference call and to a computer. Any computer can simultaneously connect to an agent and network.

3 Image Schemas in Bigraphs

When considering a domain in which the location and connectivity of entities are the most important aspects of the setting, topological spatial schemas for CONTAINMENT and LINK can be used to describe key static topological relationships, and sequences of bigraph transformations can be used to represent changes in connectivity and locality. More complex behaviors that use the FORCE schemas of BLOCKAGE and REMOVAL_OF_RESTRAINT can be used to model scenes in which, for example, an agent has to enter a locked room or building.

3.1 Static Image Schemas in Bigraphs

Bigraphs directly support the visualization of two static spatial schemas, CONTAINMENT and LINK. A node can contain another node (IN) or be connected to another node (LINK). Examples of both schemas are found in Figures 1-3 such as an agent being IN a room or having a LINKTO a computer. These correspondences follow directly from Milner's discussions of his basic bigraph model (Milner 2008).

3.1.1 Reaction Rules for Static Image Schemas

Actions such as going *into* or *out of* a node or *linking/unlinking* can be described using bigraph reaction rules, pairs of bigraph parts indicating permissible atomic changes in bigraphs. For example, the action of an agent logging off a computer can be represented by an UNLINK_FROM rule. Figure 4 shows the visualization of the rule (left) and an example of its application to simple bigraph B to produce a new bigraph B'. No change in containment occurs when applying this rule.

This is only one example of an unlink rule. If we needed a rule to permit agents to leave a conference call we would need a new rule that permitted links between agents to be broken. Another useful rule is one that represents an agent's ability to leave a room. For this we need a reaction rule OUT_OF between an agent and a room. Since the room can contain other entities, our general rule should account for that by

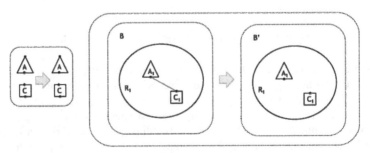

Fig. 4. UNLINK_FROM rule applied to bigraph B

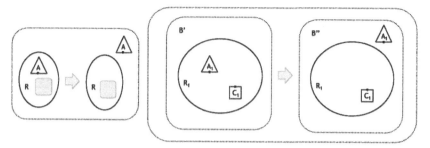

Fig. 5. OUT_OF rule applied to bigraph B′

including a placeholder (illustrated with an empty grey region) for things in a room whose placement doesn't change when the agent leaves the room. This rule and the result of applying it to bigraph B′ to produce bigraph B″ are shown in Figure 5.

To apply the rule we replace the placeholder region with the actual contents of the room besides the agent (e.g., in Figure 5 the computer is also in the room). The converses of these rules (LINKTO and INTO) are defined in the obvious ways by reversing the arrows.

3.2 Dynamic Schemas in Bigraphs

In order to represent scenes in which an agent encounters a locked room physical FORCE schemas are required. For REMOVAL_OF_RESTRAINT Johnson (1987) suggested that a force can move forward when either a barrier is not there or when it is removed by another force. Figure 6 illustrates the simplest example of this schema, an INTO rule (the converse of OUT_OF), in which an agent can enter a region that has no barrier, or had a barrier that was previously removed by another force.

If a region is locked (barricaded) then a BLOCKAGE schema must be invoked first. In the following examples agents are the forces seeking to move forward, "locks" are barriers, and "keys" are the forces that remove or bypass barriers. Given this choice, we must determine the relations between agents and keys, and between the locks and regions. Figure 7 shows three possibilities.

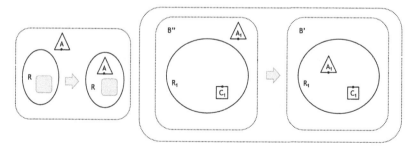

Fig. 6. Simplest REMOVAL_OF_RESTRAINT

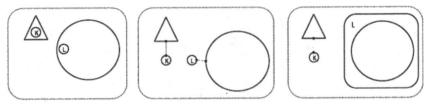

Fig. 7. Three Agent-Key, Lock-Region relations

In the first scene the agent possesses (contains) the key and the lock is in the region. In the second scene the agent and the region are only linked to the key and lock, and in the third the agent is linked to a key but the lock is represented as a barrier surrounding the region. The choice of relationships is dependent on the scene that is to be modeled. According to Johnson BLOCKAGE includes at least three possible reactions of a force when it encounters a barrier – it could go through the barrier (or remove it), go around (bypass) it, or take off in another direction.

Figure 8 illustrates one possible formalization of BLOCKAGE followed by REMOVAL_OF_RESTRAINT. The key is in (possessed by) the agent and the lock is in the room. The key links to the lock (encounters the barrier) and bypasses it. The agent is free to enter the room once the lock is bypassed, and the lock link is released, permitting other keys to engage the lock in future reactions. As before, we include a grey placeholder region in the rules to indicate that other entities can be in the room when the agent enters it. The placeholder would be filled in with the actual contents of the room when the rule was used to modify a bigraph.

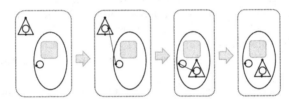

Fig. 8. BLOCKAGE followed by REMOVAL_OF_RESTRAINT

Figure 9 presents an alternative version in which the key is linked to the agent but the lock is a barricade around the region. The key links to the lock (encounters the barrier) and removes it. Being unneeded, it also disappears (this could correspond, for example, to an agent using a pass to gain entry to an area and then discarding it). The agent is free to enter the region once the lock (barrier) has been removed. Again, the empty grey inner region indicates that the rules could be applied to settings in which there are other entities in the region that the agent enters.

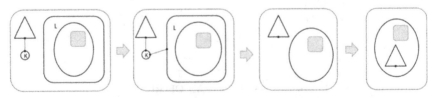

Fig. 9. BLOCKAGE followed by REMOVAL_OF_RESTRAINT (alternate)

The first step in both examples is an instance of a LINKTO reaction rule, and a later step is an instance of an INTO reaction rule. You would not want to have a more general INTO rule in the set of atomic rules allowing an agent to enter a locked room directly, or a rule that permitted the key to link to a lock it was not intended for. The property of a key being the right key to open a lock is not typically shown explicitly in a bigraph visualization but would be part of the algebraic specification. These are only two possible realizations of the schemas. The manner in which particular reaction rules and schemas are defined is dependent on the properties, relationships, and behaviors of the entities in the particular domain being modeled.

Bigraphs can be modified by the application of reaction rules, which typically change a single place or link relationship. Another way to create a new bigraph is by composition with other bigraphs to provide additional contextual information.

4 Modifying Granularity though Bigraph Composition

Bigraph composition typically does not modify existing relationships in a bigraph in the way that reaction rules do, but it may expand or refine them. For example, if there is an open link on a conference call (indicating that additional agents could participate) then composition could add a participant to the call but would not remove any of the current participants. This is similar to the case shown previously when a reaction rule with a room containing a placeholder region was used to modify a specific bigraph by filling in the open place with the actual contents of the room when applying the rule. Enhancing bigraphs in this manner allows us to define settings for other bigraphs that permit decreasing and increasing the level of scene granularity.

4.1 Bigraph Interfaces

Composition between bigraphs is performed by merging *interfaces*. An interface is a minimal specification of the portions of a particular bigraph that support additional *placings* or *linkings* (openings for more containment or linking information). If there are no such openings (e.g., the bigraphs in Figures 1-3) then the bigraph has the trivial empty interface.

A *place* in a bigraph is a *node* or a *root* (outer region) or *site* (inner region). Links can also be expanded with *inner names* and *outer names* for open links that support additional connectivity. Bigraph H in Figure 10 shows a room R_1 containing a computer C_1 and a site (0), a placeholder for other entities that can appear in rooms. The computer also has an open link x_1 for connecting to other nodes.

Bigraphs can only be composed if their interfaces match. When bigraphs H and I are composed their interface (open link x_1 and place 0) merge and disappear, resulting in bigraph J, where the agent appears in the room linked to the computer. In the merged place and link graphs J^P and J^L all the open places and links have disappeared.

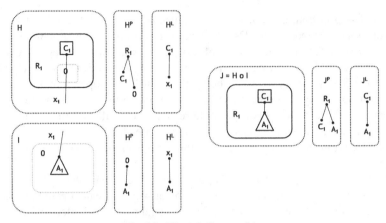

Fig. 10. Bigraph Composition

5 Built Environment Example

Consider a situation in which an agent with a key is confronted by a barrier (lock). Some information is missing, including whether she has the right key or not, whether she can reach the barrier and what is actually inside the barrier. Figure 11 shows one possible representation of this scene, a bigraph with open links and places.

Bigraph I contains a scene in which there are two unspecified outer places (roots 1 and 2). These might not be actual regions; the locality of the agent with respect to the barrier is currently unknown, but possibly different. In the first open place an agent is linked to a key for an unknown lock (indicated by open link x_1). In the second a lock (barrier) has an unknown key (open link x_2) and guards an unknown place (3). Suppose that we also have a coarser-grained view of the scene, a host bigraph H containing only an edge with two connected open links (x_1 and x_2) and open places (sites 1 and 2). Figure 12 shows bigraphs H, I and also J, which describes a scene with a single unspecified place (3) containing a room.

The bigraphs can be composed in any order. Composing H and I adds additional information to the scene. Open links with the same name merge, indicating that the key is linked to the lock. The open outer places 1 and 2 in bigraph I merge with their open place counterparts in bigraph H and disappear, which means that the agent, key, and lock are in the same place. Together, these merges mean that the agent has the right key for the lock, and that she can reach it since she is in the same place it is. Composing I with J adds the information that the entity guarded by the lock is the

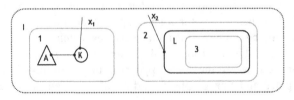

Fig. 11. Bigraph I

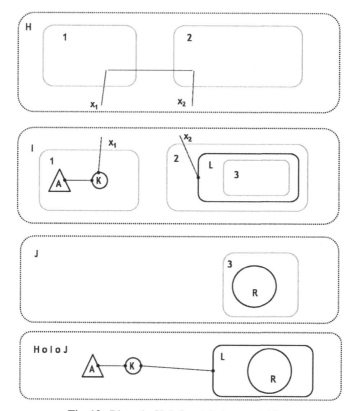

Fig. 12. Bigraphs H, I, J and their composition

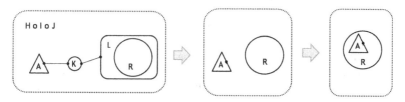

Fig. 13. Invoking BLOCKAGE and REMOVAL_OF_RESTRAINT

room R. The result of the double composition is the bottom bigraph, which provides the finest grained view of the scene.

Now that we have a more complete representation of the scene we can invoke images schemas BLOCKAGE and REMOVAL_OF_RESTRAINT through the application of reaction rules to model removing the barrier and allowing the agent to enter the room. Figure 13 shows an application of the rules from Figure 9.

In the original view of the scene shown in bigraph I, it would have been difficult to model these actions with reaction rules because it was not known which lock (barrier) the key was for, whether the agent was in the same place as the lock, and what exactly the lock was guarding. Bigraphs with open links and places support the description of scenes with incomplete information, and composing bigraphs can increase the granularity of scenes by providing additional contextual information.

6 Conclusions and Future Work

This paper has provided a method for formally representing and visualizing topological and physical image schemas for built environments. Our primary goal was to demonstrate that bigraphical models were appropriate formalization and visualization tools for representing static and dynamic image schemas. Our examples take a subset of existing topological and physical image schemas and describe them using bigraph models in the context of built environments. Static schemas CONTAINER and LINK are represented using individual bigraph subcomponents with natural visual representations. Dynamic schemas (such as INTO and LINKTO) are captured as bigraph reaction rules and visualized as a progression from one static bigraph component to another. More complex dynamic schemas, such as BLOCKAGE and REMOVAL_OF_RESTRAINT, were derived using sequences of bigraph reaction rules for static schemas. These have many possible formalizations using reaction rules, and we demonstrated a few using variants of INTO, LINKTO, and their converses. This suggests the existence of classes of reaction rules for certain schemas, and the possibility that more general specifications involving type classes of entities could be leveraged to form more general theories. We also demonstrated that bigraphs could be used to represent scenes with missing information, and that adjusting the granularity of scenes was possible using bigraph composition to add extra context.

Ontologies for built spaces are of interest, and we are developing one for hybrid outdoor-indoor spaces for built environments based on a typology of the space and the entities it contains. Types for reactive systems are ongoing areas of research (Birkedal, Debois, and Hildebrandt 2006, Bundgaard and Sassone 2006), and could be used to guide an ontological categorization of entities, relations, and behaviors in outdoor-indoor built environments, as Kuhn (2007) did in his ontological categorization of image schemas and affordances (Gibson 1977).

An algebraic theory for bigraphs for built spaces is being developed and realized as algebraic specifications in the functional language Haskell, which provides rich support for defining and reasoning about algebraic concepts such as monoids and their accompanying laws. Milner's bigraphs form a strict symmetric monoidal category (Milner 2009) with the bigraphs serving as the arrows and the interfaces serving as the objects in the category. Given this characterization, bigraph composition (used, for example, to modify scene granularity) is well-defined and avoids some of the issues that arise in the composition of arbitrary graphs.

The possible behaviors of entities in certain specific environmental settings are also of interest, and we believe that modeling affordances in bigraphs would improve behavioral specifications. In the original ambient calculus reaction rules were established based on the abilities of certain ambients (processes) to perform actions in the context they are in. Similar reasoning should be possible in a hybrid virtual-physical environment for the richer classes of ambients used in bigraphs.

Regarding the bigraph formalism itself, future work needs to focus on supporting a larger set of spatial relations (e.g., adjacency and overlap) and the kinds of spatial reasoning that are customarily available in GIS. While it is possible to describe spatial relations using links, it would be desirable to have a formal representation more closely tied to the simple spatial relation visualizations we have become accustomed to. Worboys is currently developing formal spatial enhancements to bigraphs that would support richer spatial reasoning and inference tools.

We would also like to develop support for reasoning about image schematic inference patterns in the bigraph algebra. Since inference patterns have been shown to be different in large and small scale environments (Egenhofer and Rodriguez 1999), once we have established the appropriate pattern formalisms we intend to examine whether or not these differences are still a major factor in our outdoor and indoor built environments.

Another possible area of improvement is adding support for directed movement. We may develop directed bigraphs built on top of Worboys' richer spatial bigraph model, or take advantage of existing directed bigraph extensions (Grohmann and Miculan 2007). An alternative or complementary approach would be to follow Kuhn's strategy (Kuhn 2007) and create PATH from LINK and SUPPORT schemas. We could generate series of bigraph snapshot sequences produced by the application of appropriate LINK and SUPPORT reaction rules. It remains to be seen how best to incorporate direction or more quantitative measures of location and distance should they be necessary.

Acknowledgements. This research was supported by a grant (07KLSGC04) from Cutting-edge Urban Development - Korean Land Spatialization Research Project funded by the Ministry of Land, Transport and Maritime Affairs. This material is also based upon work supported by the National Science Foundation under Grant numbers IIS-0429644, IIS-0534429, and DGE-0504494.

References

Birkedal, L., Debois, S., Hildebrandt, T.: Sortings for reactive systems. In: Baier, C., Hermanns, H. (eds.) CONCUR 2006. LNCS, vol. 4137, pp. 248–262. Springer, Heidelberg (2006)

Bundgaard, M., Sassone, V.: Typed polyadic pi-calculus in bigraphs. In: Proceedings of the 8th ACM SIGPLAN international conference on Principles and practice of declarative programming, Venice, Italy, pp. 1–12 (2006)

Cardelli, L.: Abstractions for Mobile Computation. In: Vitek, J., Jensen, C. (eds.) Secure Internet Programming. LNCS, vol. 1603, pp. 51–94. Springer, Heidelberg (1999)

Cardelli, L., Gordon, A.: Mobile Ambients. In: Le Métayer, D. (ed.) Theoretical Computer Science, Special Issue on Coordination, vol. 240(1), pp. 177–213 (2000)

Egenhofer, M.J., Rodríguez, A.: Relation algebras over containers and surfaces: An ontological study of a room space. Spatial Cognition and Computation 1(2), 155–180 (1999)

Frank, A.U.: Specifications for Interoperability: Formalizing Spatial Relations 'In', 'Auf' and 'An' and the Corresponding Image Schemata 'Container', 'Surface' and 'Link'. Agile-Conference, ITC, Enschede, The Netherlands (1998)

Frank, A.U., Raubal, M.: Specifications for Interoperability: Formalizing Image Schemata for Geographic Space. In: Proceeding from the 8th International Symposium on Spatial Data Handling, pp. 331–348. IGU, Vancouver (1998)

Frank, A.U., Raubal, M.: Formal specification of image schemata – a step towards interoperability in geographic information systems. Spatial Cognition and Computation 1(1) (1999)

Freundschuh, S.M., Sharma, M.: Spatial Image Schemata, Locative Terms, and Geographic Spaces in Children's Narrative: Fostering Spatial Skills in Children. Cartographica: The International Journal for Geographic Information and Geovisualization 32(2), 38–49 (1996)

Gibson, J., Shaw, R., Bransford, J.: The Theory of Affordances. Perceiving, Acting, and Knowing (1977)

Grohmann, D., Miculan, M.: Directed Bigraphs. Electron. Notes Theor. Comput. Sci. 173, 121–137 (2007)

Johnson, M.: The Body in the Mind: The Bodily Basis of Meaning, Imagination, and Reason. University Of Chicago Press (1987)

Kuhn, W.: Metaphors Create Theories for Users. In: Campari, I., Frank, A.U. (eds.) COSIT 1993. LNCS, vol. 716, pp. 366–376. Springer, Heidelberg (1993)

Kuhn, W.: An Image-Schematic Account of Spatial Categories. In: Spatial Information Theory, pp. 152-168 (2007)

Kuhn, W., Frank, A.U.: A Formalization of Metaphors and Image-Schemas in User Interfaces. Cognitive and Linguistic Aspects of Geographic Space (1991)

Lakoff, G., Nuñez, R.: Where Mathematics Comes From: How the Embodied Mind Brings Mathematics into Being. Basic Books (2000)

Laiserin, J.: Comparing Pommes and Naranjas. The Laiserin Letter (15) (2002)

Mark, D.M.: Cognitive image-schemata for geographic information: Relations to user views and GIS interfaces. In: Proceedings, GIS/LIS 1989, Orlando, Florida, vol. 2, pp. 551–560 (1989)

Mark, D.M., Frank, A.U.: Concepts of space and spatial language. In: Proceedings, Ninth International Symposium on Computer-Assisted Cartography (Auto-Carto 9), Baltimore, Maryland, pp. 538–556 (1989)

Milner, R.: Communicating and mobile systems: the π-calculus. Cambridge University Press, Cambridge (1999)

Milner, R.: Bigraphical Reactive Systems. In: Larsen, K.G., Nielsen, M. (eds.) CONCUR 2001. LNCS, vol. 2154, pp. 16–35. Springer, Heidelberg (2001)

Milner, R.: Bigraphs and Their Algebra. Electronic Notes in Theoretical Computer Science 209, 5–19 (2008)

Milner, R.: The Space and Motion of Communicating Agents. Cambridge University Press, Cambridge (2009)

Oakley, T.: Image Schemas. In: Geeraerts, D., Cuyckens, H. (eds.) The Oxford handbook of cognitive linguistics, pp. 214–235. Oxford University Press, Oxford (2007)

Raubal, M., Egenhofer, M.J., Pfoser, D., Tryfona, N.: Structuring space with image schemata: wayfinding in airports as a case study. In: Proceedings of the International Conference on Spatial Information Theory, vol. 1329, pp. 85–102 (1997)

Raubal, M., Worboys, M.: A Formal Model of the Process of Wayfinding in Built Environments. In: Spatial Information Theory. Cognitive and Computational Foundations of Geographic Information Science, p. 748 (1999)

Rodríguez, A., Egenhofer, M.: Image-schemata-based spatial inferences: The container-surface algebra. In: Spatial Information Theory, A Theoretical Basis for GIS, pp. 35–52 (1997)

Rodríguez, A., Egenhofer, M.: A Comparison of Inferences about Containers and Surfaces in Small-Scale and Large-Scale Spaces. Journal of Visual Languages and Computing 11(6), 639–662 (2000)

Weiser, M.: Some Computer Science Problems in Ubiquitous Computing. Communications of the ACM (July 1993)

Spatio-terminological Inference for the Design of Ambient Environments

Mehul Bhatt, Frank Dylla, and Joana Hois

SFB/TR 8 Spatial Cognition
University of Bremen, Germany

Abstract. We present an approach to assist the smart environment *design process* by means of automated validation of work-in-progress designs. The approach facilitates validation of not only the purely structural requirements, but also the functional requirements expected of a smart environment whilst keeping in mind the plethora of sensory and interactive devices embedded within such an environment. The approach, founded in spatio-terminological reasoning, is illustrated in the context of formal ontology modeling constructs and reasoners, industrial architecture data standards and state-of-the-art commercial design software.

Keywords: Spatio-Terminological Reasoning, Ontology, Requirements Modeling, Smart Environment Design, Architecture.

1 Motivation

The field of Ambient Intelligence (AmI) is beginning to manifest itself in everyday application scenarios in public and private spheres. Key domains include security and surveillance applications and other utilitarian purposes in smart homes and office environments, ambient assisted living, and so forth [30, 3]. Notwithstanding the primarily commercial motivations in the field, there has also been active academic (co)engagement and, more importantly, an effort to utilize mainstream artificial intelligence tools and techniques as a foundational basis within the field [26, 4]. For instance, the use of quantitative techniques for sensor data analysis and mining, e.g., to look for patterns in motion-data, and for activity and behavior recognition has found wide acceptability [33, 24].

Shift in Design Perspective. As AmI ventures start to become mainstream and economically viable for a larger consumer base, it is expected that AmI projects involving the design and implementation of smart environments such as smart homes and offices will adopt a radically different approach involving the use of formal knowledge representation and reasoning techniques [4, 6]. It is envisioned that a smart environment will be designed from the initial stages itself in a manner so as to aid and complement the requirements that would characterize its anticipated functional or intelligent behavior [1]. Presently, a crucial element that is missing in smart environment (and architecture) design pertains to the formal modeling – representation and reasoning – of spatial structures

K. Stewart Hornsby et al. (Eds.): COSIT 2009, LNCS 5756, pp. 371–391, 2009.

and artifacts contained therein. Indeed, since AmI systems primarily pertain to a spatial environment, formal representation and reasoning along the ontological (i.e., semantic make-up of the space) and spatial (i.e., configurations and constraints) dimensions can be a useful way to ensure that the designed model satisfies key functional requirements that enable and facilitate its required *smartness*. Broadly, it is this design approach in the initial modeling phase that we operationalize in this research. Although the presented methods can be applied to general architecture as well, they are of specific interest in ambient environments as the number of entities is much higher and thus, keeping track of possible dependencies is more complex and complicated.

Absence of Semantics. Professional architecture design tools are primarily concerned with the ability to develop models of spatial structures at different levels of granularity, e.g., ranging from low-fidelity planar layouts to complex high-resolution $3D$ models that accurately reflect the end-product. For instance, using a CAD tool to design a floor plan for an office, one may model various spatial elements representative of doors, windows, rooms, etc., from primitive geometric entities that collectively reflect the desired configuration. However, such an approach using contemporary design tools lacks the capability to incorporate and utilize the semantic content associated with the structural elements that collectively characterize the model. Furthermore, and partly as a consequence, these tools also lack the ability to exploit the expertise that a designer is equipped with, but unable to communicate to the design tool explicitly in a manner consistent with its inherent human-centered conceptualization, i.e., semantically and qualitatively. Our approach utilizes formal knowledge representation constructs to incorporate semantics at different layers: namely the conceptual or mental space of the designer and a quality space with qualitative abstractions for the representation of quantitatively modeled design data.

Semantic Requirements Constraints. As a result of *absence of semantics*, it is not possible to formulate spatial (and non-spatial) requirement constraints, expressed semantically at the conceptual level, that may be validated against a work-in-progress design (i.e., the realization) at the precise geometric level. For instance, from a purely structural viewpoint, a typical requirement in an arbitrary architectural design scenario would be that the extensions of two rooms need to be in a particular spatial (topological or positional) relationship with each other – it may be stipulated that certain structural elements within a real design that are semantically instances of concepts such as ChemicalLaboratory and Kitchen may not be *next* to each other, or should be separated by a minimum distance. Such spatial constraints are important not because of the level of their inherent complexity from a design viewpoint, which is not too much, but rather because they are semantically specifiable, extensive, and hard to handle for a team of engineers collaboratively designing a large-scale, inter-dependent environment. Our approach formalizes the conceptualization and representation of such constraints in the context of practical state-of-the-art design tools.

1.1   Requirements Constraints in AmI Design

Semantic descriptions of requirement constraints acquires real significance when the spatial constraints are among strictly spatial entities as well as abstract *spatial artifacts*. This is because although *spatial artifacts* may not be spatially extended, but they need to be treated in a real physical sense nevertheless, at least in so far as their relationships with other entities are concerned. Since a conventional working design may only explicitly include purely physical entities, it becomes impossible for a designer to model constraints involving spatial artifacts at the design level, thereby necessitating their specification at a semantic level. For example, in the design of ambient intelligence environments, which is the focus of this paper, it is typical to encounter and model relationships between spatial artifacts (see Section 3.2) such as in (A1–A3):

A1. the *operational space* denotes the region of space that an object requires to perform its intrinsic function that characterizes its utility or purpose
A2. the *functional space* of an object denotes the region of space within which an agent must be located to manipulate or physically interact with a given object
A3. the *range space* denotes the region of space that lies within the scope of a sensory device such as a motion or temperature sensor

Indeed, the characterizations in (A1–A3) are one set of examples relevant for the example scenario presented in this paper. However, from an ontological viewpoint, the range of potential domain-specific characterizations is possibly extensive, if not infinite. Constraints such as in (C1–C3) may potentially need to be satisfied with the limited set of distinctions in (A1–A3):

C1. the functional space of the door of every office should overlap with the range space of one or more motion sensors
C2. there should be no region of space on the floor that does not overlap with the range space of at least one camera
C3. key monitored areas that are connected by doors and/or passages should not have any security blind spots whilst people transition from one room to another

Constraints such as (C1–C3) involve semantic characterizations and spatial relationships among strictly spatial entities as well as other spatial artifacts. Furthermore, albeit being modeled qualitatively at a conceptual level, they also need to be validated against a quantitatively modeled work-in-progress design (e.g., a CAD model) in addition to checking for the consistency of a designer's requirements per se (Section 3.6).

1.2   Key Contribution and Organization

We apply the paradigm of integrated spatio-terminological reasoning for the design and automated validation of smart spaces. The validation encompasses the structural as well as functional requirements expected of a smart environment from the viewpoint of the sensory and interactive devices embedded within such an environment. In essence, a quantity space, modeled accurately using primitive geometric elements, is validated against a domain conceptualization consisting of ontology of spatial entities, artifacts, architectural elements, sensory devices and the relationships among these diverse elements.

Section 2 presents the ontological underpinnings of this work. Here, the concept of integrated spatio-terminological inference is illustrated and the use of spatial ontologies for AmI systems modeling is explained. Section 3 sets up the apparatus for formal requirement constraints modeling with ontologies. This is in turn utilized in the example scenario of Section 3.6, where the proposed approach is demonstrated in the context of an industrial standard for data representation and interchange in the architectural domain, namely the Industry Foundation Classes (IFC) [9], and state-of-the-art commercial architecture design techniques, as enabled by the ArchiCAD [7] design tool. Finally, Section 4 discusses the work in its relationship to existing research, whereas Section 5 concludes with pointers to the outlook of this research.[1]

2   Ontology, Architecture and Ambient Intelligence

Ontologies are defined as "a shared understanding of some domain of interest" [31]. Their structure consists of classes, relations between classes, and axiomatizations of classes and relations (cf. [28] for further information). In the case of AmI systems ontologies can then provide a formalization of entities, relations, and axiomatizations specifically for the AmI environment, as described below.

2.1   Ontologies and Spatio-terminological Reasoning

Although ontologies can be defined in any logic, we focus here on ontologies as theories formulated in description logic (DL), supported by the web ontology language OWL DL [23]. In general, DL distinguishes between TBox and ABox. The TBox specifies all classes and relations, while the ABox specifies all instantiations of them. Even though ontologies may be formulated in more or less expressive logics, DL ontologies provide constructions that are general enough for specifying complex ontologies [17]. Several reasoners are available for DL ontologies, one of them is the reasoning engine RacerPro [13] with its query language nRQL. Here, we use this reasoner for spatio-terminological inference.

Reasoning over the TBox allows, for instance, to check the consistency of the ontology and to determine additional constraints or axioms that are not directly specified in the ontology. Reasoning over the ABox allows, for instance, to classify instances or to determine additional relations among instances. In particular for AmI ontologies, the domain of buildings and their ambient characteristics and constraints have to be specified in order to formalize their requirements.

---

[1] Additional (independent) information in support of the paper is linked at the end of the article.

In addition to reasoning over ontologies (TBox) and their instances (ABox), however, spatial reasoning is of particular interest for the AmI domain, as spatial positions of entities and qualitative spatial relationships between entities are highly important to describe the environment.

A specific feature of the reasoning engine RacerPro is to support region-based spatial reasoning by the so-called SBox [11]. Besides TBox and ABox, this layer provides spatial representation and reasoning based on the REGION CONNEC-TION CALCULUS (RCC) [27]. The SBox can mirror instances of the ABox and specify RCC-relations and consistency among these instances. The separation between spatial and terminological representation and reasoning supports prac-ticability of reasoning, reduces complexity, and benefits ontological modeling in general, as a domain can be described from different perspectives [25], e.g., with a focus on terminological, spatial, temporal, functional, action-oriented, or other thematically different perspectives. The integration of perspectives then allows a comprehensive representation of the domain. While each ontology specifies and axiomatizes its respective view on the domain, alignments across different ontologies provide further axiomatizations [19]. Our particular AmI ontological representation and reasoning are described in Section 3.5 and Section 3.6.

2.2   Industry Foundation Classes and Design Tools

Industry Foundation Classes (IFC) [21] are specific data models to foster inter-operability in the building industry, i.e., a non-proprietary data exchange format reflecting building information. Former models, like 2D or 3D CAD models are based on metric data referring to geometric primitives, e.g., points, lines, etc., without any semantics of these primitives. In contrast, IFC is based on object classes, e.g., *IfcWall* or *IfcWindow*, and their inherent relationships containing metrical data as properties. The advantage of this kind of representation is that data for complex calculations like structural analysis or energy effort can be gen-erated automatically. Within our work we apply the latest stable release IFC2x3 TC1 [21]. Overall, IFC 2x3 defines 653 *building entities* (e.g., *IfcWall*) and ad-ditionally, several *defined types*, *enumerations*, and *select types* for specifying their properties and relationships. Commercial design tools such as Graphisoft's ArchiCad [7] support export capabilities in XML and binary format in a manner that is IFC compliant. Free software tools also exist for modeling, visualizing, syntax checking, etc., of XML and binary IFC data. Note that since our ap-proach utilizes IFC data, datasets from any IFC compliant design tool remain utilizable.

3   Requirements Consistency in AMI Design

In this section, the modular specification of ontologies for AmI that support architectural design processes are presented. We show how the architectural rep-resentation is associated with the ontological representation, how the IFC repre-sentation can be instantiated, and how ontological information can be grounded

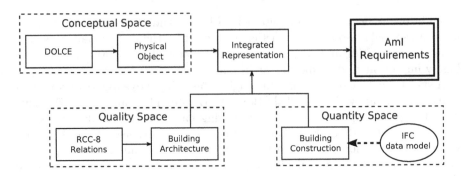

Fig. 1. Spatial ontologies for AmI: three ontological modules for conceptual, quality, and quantity space reflect the different perspectives on the domain

in architectural designs. Subsequently spatial and terminological requirements formulated within and across the ontologies are presented. An example illustrating reasoning and consistency checking is provided in Section 3.6.

3.1   Modular Spatio-terminological Ontologies for AmI Design

As outlined in Section 2.1, space can be seen from different perspectives and ontologies are a method to formalize these perspectives. In order to support the architectural design process for AmI environments, requirement constraints of architectural entities have to be defined from terminological and spatial perspectives. Space is then defined from a *conceptual, qualitative,* and *quantitative* perspective. The resulting three modules consist of different ontologies, illustrated in Fig. 1.

M1. Conceptual Space. This ontological module reflects terminological information of architectural entities. Here, the entities are regarded as such, i.e., they are defined according to their properties without taking into account the context in which they are put. The ontology Physical Object formalizes entities with respect to their attributes, dependencies, functional characteristics, etc. It is based on DOLCE [22], in particular, on the OWL version DOLCE-Lite, and refines DOLCE's Physical Endurants. It defines the entities that occur in the AmI domain, such as Sensor, SlidingDoor, or ChemicalLaboratory.

M2. Quality Space. This ontological module reflects qualitative spatial information of architectural entities. Similar to the previous module, the ontology Building Architecture formalizes entities of the AmI domain, but it specifies their region-based spatial characteristics. In particular, the ontology uses relations as provided by the spatial calculus RCC [27]. A Room, for instance, is necessarily a *proper part* of a Building. Here, we reuse an RCC ontology, that has been introduced in [10], which defines the taxonomy for RCC-8 relations by approximations of the full composition table. RCC-related constraints by the TBox can be directly inferred. Inference given by the composition table for combinations of

RCC relations is then provided by SBox reasoning. Architecture-specific entities in the Building Architecture ontology are further described in Section 3.2.

M3. Quantity Space. This ontological module reflects metrical and geometric information of architectural entities. It is closely related to the IFC data model and partially mirrors the IFC classes. In particular, the ontology Building Structure of the module specifies those entities of the architectural domain that are necessary to describe structural aspects of ambient environments. Especially, information that is available by construction plans of buildings are described here. For example, Door, Wall, and Window are characterized together with their properties length, orientation, placement, etc. Data provided by IFC for a concrete building model can then be instantiated with respect to the Building Structure ontology (cf. Section 3.3). Even though the IFC model itself is not an ontology, parts of it are directly given by their correspondences in the ontological specification in this module.

Integrated Representation. The connection of the three different modules result in formalizing relations across modules. The Integrated Representation defines couplings between classes from different modules, i.e., counterparts and dependencies are defined across modules, based on the theory of $\mathcal{E}$-connections [20]. For example, an instance of Wall from the quantity space is related to an instance of Wall from the quality space or conceptual space. An IFC wall that is illustrated in a construction plan can then be instantiated as a Wall in Building Construction and connected to an instance of Wall in Building Architecture as well as an instance of Wall in Physical Object. It is described by its length and position in the first module, while its counterpart in the second module defines region-based relations to other walls and relations to rooms it constitutes and its counterpart in the third module defines its material and color. Details on modularly specified ontological modules for architectural design are given in [15].

3.2 Space: Objects and Artefacts

We present an informal characterization of the primitive spatial entities within the spatial ontology, or precisely, the modular component as reflected by the *quality space* within the overall spatial ontology (see Section 3.1; Fig. 1). For all ontological characterizations here, precise geometric interpretations are provided in Section 3.3. Here, a high-level overview suffices.

Regions are either the absolute spatial extensions of physical objects, or of spatial artifacts that are not truly physical, but are required to be regarded as such. A *region* of space should be measurable in terms of its area (2D) or volume (3D) and the region space should be of uniform dimensionality, i.e., it is not possible to express a topological relationship between a 2D and and 3D region. The spatial categories in (S1–S4) are identifiable. Spatial relationships between these categories are utilized for modeling structural and functional requirement constraints for a work-in-progress design:

S1. Object Space. The *object space* of a primitive entity refers to the region covered by the physical extent of the respective entity itself. If objects are static, non-deformable, and reconfigurable they cover a well-defined region in the world. In contrast, if an entity is non-static, deformable, or reconfigurable, its spatial extension depends on its specific state s, e.g., the opening angle of a door or window. Let $obspace(o)$ and $obspace(o, s)$ denote the state independent and dependent space covered by such an object. Here, $\mathcal{S}_o$ is the set of all potential states an object may be in and $s \in \mathcal{S}_o$. Since we only deal with static worlds in our modeling, we abstract away from the state parameter s and simply use $obspace(o)$ to denote the space covered by the object in a specific *predefined state* s_p. We assume that the predefined state s_p is consistent with the way how an object is modeled in the design tool, e.g., windows and doors are closed. To really calculate the regions that, for instance, may be covered by a door in a state, the data on the panel extent (*IfcDoorPanelProperties*) and the frame properties (*IfcDoorLiningProperties* and *IfcDoor*) can be applied. In the example scenario of Section 3.6, we only take into account the stable state of an object, as may be modeled within a structural design tool. For instance, the object space does not cover any space occupied by, e.g., deformable or reconfigurable parts of an object.

S2. Operational Space. The operational space of an object, henceforth *operational space*, refers to the region of space that an object requires to perform its intrinsic function that characterizes its utility or purpose. For example, for a door that may be opened in one direction, the operational space characterizes the region of space required to facilitate the free movement of the door between, and including, the fully-opened and fully-closed states. For example, if the operational space of a door and a window would overlap, these two may collide if both are open at the same time, resulting in damages. Similarly, the operational space of a rotating surveillance camera is characterized by the angular degrees of movement, which its controllers are capable of. The operational space comprises all space an object may cover regarding all states it can be potentially in:

$$opspace(o) = \bigcup_{s \in \mathcal{S}_o} obspace(o, s)$$

S3. Functional Space. The functional space of an object, henceforth simply *functional space*, refers to the region of space surrounding an object within which an agent must be located to manipulate or physically interact with a given object. The functional space is not necessarily similar to the object's convex hull; however, in some cases involving arbitrarily shaped concave objects where interaction is limited to a pre-designated intrinsic front, this space could tend to be more or less equivalent to the object's convex hull. We denote the functional space of an object o by the function $fspace(o)$ – the precise geometric interpretation of this function being determinable in domain specific or externally defined ways depending on issues such as object granularity, the scale of the ambient environment being modeled, etc. The functional space of an object is dependent on the object o itself, the current state $s \in \mathcal{S}_o$ an object is in, the capabilities of agent a, and the function f, i.e., an action the agent wants to perform on the object. The set of all objects is given by $\mathcal{O}$, the set of all agents by $\mathcal{A}$, and the set of all available functions $\mathcal{F}_a^o$. Thus, we define functional space in $\mathcal{P} = \mathbb{R}^n$ as:

$$fspace(o, a, f, s) = \{p | p \in \mathcal{P} \wedge a \text{ is within range to perform } f \in \mathcal{F}_a^o \text{ on } o \text{ in } s\}$$

As there are arbitrary numbers of agents, functions, and states, it is in many cases impossible to give metrical definitions for combinations of them. Nevertheless, in cases considered here a detailed description of $fspace(.)$ is not necessary to know. Therefore, we use the union $fspace(o)$ of all functional spaces encompassing functions, agents, and states:

$$fspace(o) = \bigcup_{a \in \mathcal{A}} \bigcup_{f \in \mathcal{F}_a^o} \bigcup_{s \in \mathcal{S}_o} fspace(o, a, f, s).$$

S4. Range Space. The (sensory) range space, henceforth *range space*, refers to the region of space that lies within the scope of sensory devices such as motion, temperature, heat, humidity and fire sensors, infrared and laser scanners, cameras, and so forth. In order to fulfill functional requirements, it might be necessary to either directly or indirectly ensure certain spatial relationships between a sensor's range space and other artifacts and objects. For instance, it is desirable that the range space of temperature sensor should not overlap with that of a heating device, and desirably, the metric distance between the *object space* of the heating device and the range space of the heating sensor be at a certain minimum for a given room layout. The range space of a sensor in a particular state s is denoted by $range(o, s)$. If the sensor is reconfigurable, the maximal range by the sensor can be given by:

$$range_{max}(o) = \bigcup_{s \in \mathcal{S}} range(o, s).$$

Since this paper is restricted to static environments, as they exist on a work-in-progress design, we do not model different states over time. As such, the range of a sensor in the state is given by: $range(o, s) = range(o)$. Simply, for a stationary sensor o, $range(o) = range(o, s) = range_{max}(o)$.

Explicit characterizations of spatial artifacts are necessary to enforce structural and functional constraints (e.g., Section 1.1, C1–C3) during the AmI design process, especially when the environment is intended to consist of a wide-range of sensory apparatus (cameras, motion sensors, etc.).

3.3   Spatial Artifacts: Concrete Goemetric Interpretations in R^2

As illustrated in Section 3.1, fine-grained semantic distinctions at the level of the spatial ontology (i.e., the conceptual space) and the capability to define their precise geometric interpretation by domain-specific parameters (i.e., the quantity space) are both necessary and useful to stipulate spatial and functional constraints during the design phase. Figure 2 provides a detailed view on the different kinds of spaces we introduced in Section 3.2 for R^2. Although all illustrations in this paper deal with 2D projections of 3D information, note that for

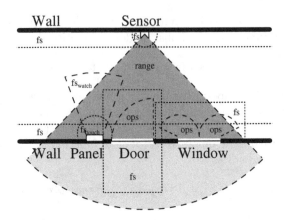

Fig. 2. Differences between functional space (fs, dotted lines), operational space (ops, dashed lines), and range (gray regions surrounded by dashed line)

the general case, there is no difference between the different characterizations in functional, operational, and range spaces – they all refer to a physical spatial extension in R^n. However, these do differ with respect to their ontological characterizations within the modular spatial ontologies (Section 3.1), and in the manner of deriving their respective geometric interpretations in R^n. These interpretations differ and depend on, in addition to an object's inherent spatial characteristics (e.g., size and shape), on one or more additional parameters, as elaborated in (G1–G4):

G1. Object Space. As walls, panels and sensors are, in general, not reconfigurable or deformable, the object space for the spatial entities is just the space occupied by them, which can be derived on the basis of IFC data. In contrast, doors and windows are reconfigurable. The most natural state of these components is assumed to be *closed* and thus, the object space is defined by the space covered in this state. This also complies with the modality by which doors and windows are depicted in architectural drawings.

G2. Operational Space. As doors and windows are reconfigurable they also possess an operational space (dashed line). In Fig. 2 we depict a single panel, right swing door which is defined by its two corner points $t_1 = \begin{pmatrix} x_1 \\ y_1 \end{pmatrix}$ and $t_2 = \begin{pmatrix} x_2 \\ y_2 \end{pmatrix}$ with t_1 denoting the position of the hinge. The respective vectors are denoted by t_1 and t_2. Data necessary for deriving these points are given in the IFC building entities, e.g., *IfcDoorStyle* (type of door) and *IfcDoor* (swing direction). For example, the type of door at hand is referred by *SINGLE_SWING_RIGHT*. We represent the closed door by the vector $\tau = \begin{pmatrix} x_\tau \\ y_\tau \end{pmatrix}$. Currently, opening angles are not represented explicitly in IFC2x3. For reasons of simplicity, we assume a maximum opening angle of 90° for doors. We represent the maximally open door by τ', which is in our specific case equal to the normal vector τ^n of τ. The specific direction of τ^n is defined by the opening direction and hinge position of the door (Fig. 3). Then, the operational space of the door comprises any point

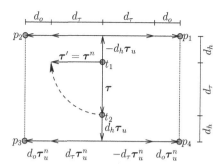

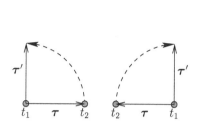
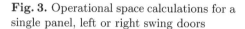

Fig. 3. Operational space calculations for a single panel, left or right swing doors

Fig. 4. Functional space calculations for a single panel, right swing door

between the two vectors τ and τ' starting at t_1, which is a sufficient description to calculate any overlap or containment with other regions. In Fig. 2, this results in the quadrant depicted by ops. The two parts of the window may be opened by 135°, which results in the operational space in the manner as depicted.

G3. Functional Space. Regarding different types of agents the area for possible interactions may vary. For humans, e.g., functional space for touching a wall is within the range of an approximate arm length. For artificial agents like robots or semi-autonomous wheelchairs the concrete value may vary. In Fig. 2 we assumed this range by $40cm$ regarding walls. For doors and windows we approximate the space where they can be opened or closed by rectangular shapes. We denote the length of τ by d_τ and the related unit vector by τ_u. Additionally, we define reachability distances d_h (to the left and right of the door) and d_o (in front and behind the opened door). Based on this information we can derive the four corner points p_i of a rectangle representing *fspace(Door)* by:

$$p_i = t_{\lceil \frac{i}{2} \rceil} + (-1)^{\lceil \frac{i}{2} \rceil} d_h \tau_u + (-1)^{\lfloor \frac{i}{2} \rfloor}((d_\tau + d_o)\tau_u^n) \text{ with } i \in \{1,2,3,4\}.$$

We illustrate this formula in Fig. 4. In Fig. 2, the door panel is one meter wide and we defined $d_h = d_o = 20cm$. For doors or windows where opening angles larger than 90° are possible, the point calculations depend on further aspects. To give an impression we refer to Fig. 5. Note that the shape of the functional spaces may need to be refined depending on the agents and their intended functions with respect to an object at hand (Section 3.2). For example, one must be in close proximity to a panel in order to touch it. In Fig. 2 this is depicted by a semi-circle. In contrast, with respect to an agent who wants to watch the content, this definition is insufficient with respect to the function. The viewing angle (available in technical documentation) and the maximum distance the content can be detected must be considered. This distance may vary with size and the height at which it is fitted to the wall. In Fig. 2, we depicted *fspace(panel, a, watch, s)* assuming a viewing angle of 20° and a visibility distance of 1.5 meters.

G4. Range Space In general, sensors have angles and maximum measurement distances defined which serve as a direct basis for extracting range spaces. Again,

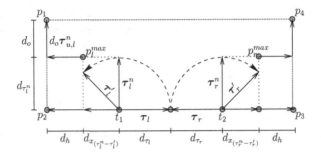

Fig. 5. Operational and functional space calculations for a vertical double panel window with a maximum opening angle of $135°$

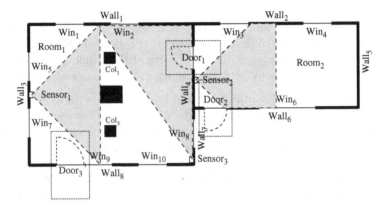

Fig. 6. Exemplary design with Windows, Doors, Columns, and Sensors

two vectors can be determined to calculate the overlap with other spaces with respect to the sensor's position in the environment. For a camera, for example, these values are the field viewing angle and the camera's resolution compared to the visual angle of objects to detect. In our example, a sensor angle of $110°$ and a maximum distance of 4 meters are assumed. The light-gray area at the bottom is also part of the range space but will not be accessible by sensors which rely on visibility, e.g., laser or sonar, because it is behind a wall as seen from the sensor. Nevertheless, other sensors, such as magnetic fields or WLAN, overbear these barriers and interaction with an agent is possible although positioned behind the wall.

3.4   Spatial Objects and Artifacts: An Example Grounding

In our simple example scenario, depicted in Fig. 6, we only use a small subset of the available components. The scenario consists of eight walls, ten windows, three doors, three columns, and three sensors. For to reasons of clarity only the operational and functional spaces of doors as well as the range spaces of the sensors are presented. Due to the connectivity relation between the walls two rooms are constituted ($Room_1$: Walls 1, 4, 7, 8, and 3; $Room_2$: Walls 2, 5, 6, and 4). Additionally, we have to calculate the spatial artifacts defined in Section 3.2. The different artifacts are calculated automatically in the qualitative module (cf. Section 3.1) from the geometrical data based on predefined formulae as presented in Section 3.3. The necessary data is available in the IFC representation. Additionally, the connectivity structure of walls is given in IFC. If open rooms, i.e., rooms not completely enclosed by walls, are given this can be additionally modeled by zones (*IfcZone*) or spaces (*IfcSpaces*) Thus, we can assume that sufficient metric data is given for any object to calculate RCC-8 relations. For example, the metric representation for $Room_1$ and Col_2 is given by:

$$
\begin{aligned}
&(\text{ROOM } Room_1 \ (\text{Wall}_1 \ \text{Wall}_4 \ \text{Wall}_7 \ \text{Wall}_8 \ \text{Wall}_3)) \\
\Rightarrow \ &(\text{ROOM } Room_1 \ ((0\ 0)(6\ 0)(6\ 3)(6\ 5)(0\ 5))) \\
&(\text{COL} \quad Col_2 \quad ((2.6\ 2.2)(2.6\ 2.8)(3.4\ 2.8)(3.4\ 2.2)))
\end{aligned}
$$

Following the formulae and definitions in Section 3.3 (with $d_h = d_o = 0.2$ meters), *opspace*() and *fspace*() can be derived easily. For example, the door panel of $Door_1$ is 0.8 meters long and $t_1 = (6, 0.8)$ ($t_2 = (6, 1.6)$). For *opspace*($Door_1$) follows: $\tau = \binom{0}{0.8}$ and $\tau' = \binom{-0.8}{0}$. For *fspace*($Door_1$) follows: $p_1 = (7, 0.6), p_2 = (5, 0.6), p_3 = (5, 1.8)$ and $p_4 = (7, 1.8)$.

Based on the metrical data, the qualitative model consisting of topological relationship between two regions can be derived. For example, as no point of the region defined by Col_2 (including the boundary) is outside the region or touches the boundary defined by $Room_1$, Col_2 is a *non-tangential proper part* of $Room_1$. Additionally, the functional space of $Door_1$ overlaps with the range spaces of $Sensor_1$ and $Sensor_3$. This is reflected in the SBox by:

```
( rcc-related   Col₁       Room₁       :NTPP  )
( rcc-related   Door₁_fs   Sensor₁_rs  :PO    )
( rcc-related   Door₁_fs   Sensor₃_rs  :PO    )
```

These calculations are performed for all building entities such that a complete qualitative spatial model with RCC-8 relations is available.

3.5   Requirements Constraints for AmI

Given the Integrated Representation of Section 3.1, particular requirements for AmI environments can be defined. The requirements are formalized by constraints within and across the ontologies. The Integrated Representation itself merely defines general relationships across modules. As such, it supports spatial and terminological reasoning by constraints across modules on a general level. This reflects the modular nature of our ontological representation distinguishing spatial perspectives. An example of such a constraint is the requirement that all M2.Door in the Building Architecture ontology are composed of one M1.Door in the quantitative layer (formulated in Manchester Syntax [16], namespaces are added as prefixes):

```
Class:        construction:Door
SubClassOf: compose exactly 1 arch:Door
```

Based on such general relationships, the AmI Requirements ontology is formalized. It specifies particular restrictions for building automation in the architectural design process. Hence, it needs to be adjusted to particular building requirements depending on formal structural, functional and potentially other aspects.[2] Particular requirements for our example scenario are based on smart office environments. An example of such a constraint is the requirement that all functional spaces of doors should be a proper part of some range space of some motion sensors. In more detail, it has to be ensured that different motion sensors properly perform a 'handshake' when monitoring persons changing rooms. Within the Building Architecture ontology, the requirement is specified as follows:

```
Class:        arch:DoorFunctionalSpace
SubClassOf: arch:FunctionalSpace,
              rcc:properPartOf some (arch:MotionSensorRangeSpace)
```

Here, the requirement is defined on the basis of the quality space module (M.2). The Building Architecture ontology defines the classes for functional spaces and range spaces, while the RCC-8 Relations ontology provides the region-based relations. Requirements that take into account different modules, however, are also specified. The requirement that, for instance, all buildings have an intelligent navigation terminal that provides building information for visitors is defined in the following requirement constraint:

```
Class:        arch:Building
SubClassOf: rcc:inverseProperPartOf min 1 (arch:Display
              and (ir:conceptualizedBy some physObj:NavigationTerminal))
```

In this example, the classes Building and Display are defined in the qualitative model and NavigationTerminal is defined in the conceptual model. Their connection conceptualizedBy is defined in the Integrated Representation (ir), while the constraint on the class Building is formalized in the AmI Requirements ontology.

Besides ontology design criteria of such a modular representation, our modeling also shows practically adequate formalizations of spatial entities from the different perspectives. In the quantitative module four walls of a room might actually be modeled by more than four walls. For instance, in the example case in Fig. 6, $Wall_4$ and $Wall_7$ of $Room_1$ are distinguished as two walls in the quantity module. Both walls, however, are mapped to one instance of a Wall in the quality space module. This wall constitute the $Room_1$ counterpart in the quality space. Note that this mapping is only applicable in this example. In general, it is also possible to define more than four wall instances in the quality space module that may constitute a room. The distinction is then directly supported by the modular structure of the ontologies.

---

[2] Section 5 discusses our outlook on other constraints that may be formalized within the requirements ontology.

3.6   Requirements Consistency: An Example Scenario

We use spatio-terminological reasoning to ensure that the requirements specified at the AmI design phase are satisfied. For this purpose, the reasoner proves the consistency of the ABox (terminological instances) according to definitions of the TBox and the consistency of the SBox (spatial instances) according to RCC-8 relations. An actual floor plan representation can then be analyzed whether it fulfills the requirements that are defined by the AmI Requirements ontology. Fig. 7 illustrates two alternatives of a selected part of a floor plan (the floor plan illustrations are reduced for simplicity, e.g., windows are omitted).

The examples are specified in the quantity space module on the basis of their IFC data. The classes Room, Sensor, Wall, and Door of the Building Construction ontology are used to instantiate the rooms $M3.Room_1$ and $M3.Room_2$, the sensors $M3.Sensor_1$ and $M3.Sensor_2$, the door $M3.Door_1$, and several walls (omitted for simplicity). These instances have to be connected to their counterparts in the quality space module. The instances and their relations are then specified in the Building Architecture (M.2) ontology:

> **Individual**: arch:Room1
> **Types**:   arch:Room
> **Facts**:   rcc:externallyConnectedTo   arch:Room2
>
> **Individual**: arch:DoorFunctionalSpace1
> **Types**:   arch:DoorFunctionalSpace
> **Facts**:   rcc:particallyOverlaps   arch:SensorRange1
> **Facts**:   rcc:particallyOverlaps   arch:SensorRange2
> $\vdots$

AmI requirements are then be satisfied by proving consistencies of TBox, ABox, and SBox, outlined herein. Note that the consistency itself is proven by using the DL reasoner RacerPro. For completeness, we also describe an example for a TBox consistency proof, albeit that is not directly connected to our specific example scenarios:

TBox Inconsistency: An inconsistency in the TBox is used to determine whether or not the AmI requirements specified by a designer may possibly be fulfilled by a model per se. In the Integrated Representation ontology, the relation isConceptualizedBy, for instance, may define an injective mapping between M2.Room in the qualitative module and M3.RoomType in the conceptual module (the latter defines specific rooms, such as kitchen, office, laboratory, etc.). Assuming that another relation conceptualize in the AmI Requirements ontology allows several mappings between M3.RoomType and M2.Room and that this relation is defined as the inverse relation of isConceptualizedBy, reasoning over the TBox would then detect an inconsistency in the ontology definition itself.

ABox Inconsistency: Inconsistencies in the ABox arise from instances of the ontology that are not compliant with the ontological constraints, both spatial

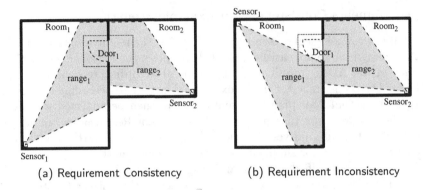

(a) Requirement Consistency   (b) Requirement Inconsistency

Fig. 7. A two room scenario with the requirement that the door must be supervised by sensors, i.e., the functional space must be completely covered by some sensor range (not necessarily only from a single sensor)

and otherwise. In our case, the instances reflect information of a floor plan while the constraints reflect AmI requirements. An inconsistency in the ABox then implies that the floor plan does not fulfill the AmI requirements. A requirement constraint could be that that all rooms should be equipped with doors that satisfy certain (metric) traversibility criteria (e.g., with respect to human, wheel-chair, robot movement). The spatial-centric nature of SBox inconsistency (discussed next) notwithstanding, the reasoning pattern involved there is essentially the same as that for the ABox. Hence further details of ABox inconsistency are excluded herein.

SBox Inconsistency: An inconsistency in the SBox identifies those instantiations that are not compliant with RCC-related constraints. In our scenario, it indicates that a floor plan does not satisfy the qualitative spatial requirements. The AmI Requirements ontology, for instance, constrains that individuals, e.g., humans, can be monitored while they leave or enter rooms at any time (cf. C3 in Section 1.1). For this purpose, all doors have to be monitored and therefore all functional spaces of doors have to be a proper part of some sensor range. In Fig. 7(a), the functional space of Door$_1$ is a proper part of the union of the range spaces of Sensor$_1$ and Sensor$_2$. It therefore satisfies the requirements and is proven to be consistent. This can be verified with RacerPro by proving that no instance exists ('NIL'), that is a functional space of a door and not a non-tangential proper part of the range space of a particular motion sensor. The query in RacerPro infers this result:

```
? (retrieve (?*X ?*Y) (and (?X DoorFunctionalSpace)
                           (?Y MotionSensorRangeSpace)
                           not (?*X ?*Y :ntpp)))
> NIL
```

The same request with the example in Fig. 7(b), however, infers that Door$_1$ is not a non-tangential proper part of some sensor ranges. The example is therefore inconsistent with respect to the AmI requirements. In summary, concrete examples of architectural building plans have to satisfy the requirements given by the AmI Requirements ontology by proving the consistency of their ontological instantiations in the TBox, ABox and SBox.

4   Discussion and Related Work

The field of ambient intelligence has found wide-spread commercial acceptability in the form of applications in the smart environment (e.g., homes, offices) domain [33, 24]. The field has also witnessed considerable inter-disciplinary interactions, hitherto not conceived, from several spheres in artificial intelligence. The design and implementation of this new generation of smart environments demands radically new modeling techniques right from the early design phases. It is necessary for a designer to explicitly communicate the spatial and functional requirement constraints, directly and indirectly related to the perceived smartness of the environment, to the design tool being utilized. Further, it is necessary that such communication accrue in a way that is consistent with the inherent semantic and qualitative manner in which the requirements are conceptualized by the expert. Albeit differing in application and approach, similar sentiments are expressed in [1, 2]. A formal state-based approach for design reasoning that is structures and functions is proposed in [1]. In this approach, structures are defined as states and operations on them are defined as functions. Reasoning is then formulated as an interaction between the two. For the domain of architecture design, this approach has been taken further to create a process by which requirements can be converted into working design solutions through front-end validation [2]. Although the studies in [1, 2] are different in approach, i.e., we utilize formal methods in ontological and spatial reasoning, the motivations of both from a general architectural design viewpoint remain the same, i.e., to reduce design errors and failures by iterative design validation and verification, and from our AmI design perspective, also to ensure that a work-in-progress design fulfills the functional requirements in order for it to be able to deliver the perceived *smartness*. The crux of such a design approach is that it becomes possible to automatically validate the designer's *conceptual space* against the precisely modeled work-in-progress *quantity space* of the design tool. The operationalization of such a design approach and intelligent assistance capability is the objective of our research, and this paper is a foundational contribution in that regard.

Spatio-terminological reasoning is a well-founded approach for integrated reasoning about spatial and descriptive terminological information [11, 12]. Applications of this paradigm in the GIS domain also exist, e.g., [32] and [18]. The first approach follows the idea of the Semantic Web [5] in the context of GIS applications, i.e., users are able to formulate queries with respect to temporal, spatial, and environmental aspects. The approach of BUSTER [32] aims to improve search options by enriching their data with semantic information. Although we use a combination of spatial and terminological reasoning as well, our focus is on the analysis of consistencies in architectural floor plans anchored in the field of AmI environments. We specify architectural data models in a modular ontological way and reason over concrete instantiations of building representations with spatial objects and artifacts.

Closely related to the architecture data interoperability standard utilized in this work, namely the IFC, is the Building Information Model (BIM) [8]. BIM is an emerging and all-encompassing technological framework that *enables* users to integrate and reuse building information and domain knowledge pertaining to the entire life cycle of a building. The concept of Green Building [14], also connected to the BIM, aims at creating structures and utilizing standards and policies that are environmentally responsible and resource-efficient from a sustainability viewpoint. The concept extends throughout a building's complete life-cycle encompassing the design, construction, operation, maintenance, renovation and deconstruction phases. We further touch upon the importance of such emerging standards and connections with our work in whilst positioning our ongoing work in Section 5.

5   Summary and Outlook

In this paper, we propose, formalize, and demonstrate the application of formal knowledge and spatial modeling constructs, and the paradigm of integrated spatio-terminological reasoning in the domain of smart environment design, or more generally AmI design. The proposed application enables the capability to ensure that semantically specified functional requirement constraints by an AmI designer are satisfied by a metrically modeled work-in-progress design such as a floor plan. The paper presented an example scenario in the context of an architectural data interoperability standard, namely the IFC, and the state-of-the-art design tool ArchiCAD. The formal representation and reasoning components utilized the OWL DL fragment of the ontology modeling language and the spatio-terminological reasoner RacerPro, and the Region Connection Calculus as a basis of spatial information representation and reasoning.

There are two main areas for further research that our project has adopted. Along the practical front, we are investigating the integration of our approach with a light-weight indoor-environment design tool, namely Yamamoto [29], which offers built-in capabilities for annotating geometric entities at the quantity space with semantic information. From a theoretical viewpoint, we are extending the approach to serve not only a diagnostic function, but also to provide the capability to explicitly prescribe potential ways to resolve inconsistencies within the work-in-progress design. As another line of work, we note that within an *architecture design tool*, metrical changes in the structural layout, which result in qualitative changes along the *conceptual space* of the designer, directly or indirectly entail differing end-product realizations in terms of building construction costs, human-factors (e.g., traversability, safety, productivity, personal communication), aesthetic aspects, energy efficiency, and long-term maintenance expenses thereof based on present and perceived costs. We propose to extend our approach toward the estimation of such material and non-material costs that arise solely by minor conceptual and spatial variations within a design. It is envisioned that these extensions will be achieved within the framework of emerging standards and frameworks such as BIM, IFC and GreenBuilding, and integrated within a state-of-the-art design tool. In a more general direction, we also have to investigate spatial design considerations that derive from cognitive or practical requirements, i.e., whether the design is cognitively adequate. This could, for instance, be realized by simulations that analyze how people interact with the designed environment. Finally, from a rather long-term viewpoint, we regard it interesting to directly collaborate with professional architects via a dialog interface for the communication of design descriptions and constraints with the system. All suggested extensions remain focused to the domain of *smart space* design.

Acknowledgements. We gratefully acknowledge the financial support of the DFG through the COLLABORATIVE RESEARCH CENTER SFB/TR 8, projects R3-[Q-SHAPE] and I1-[ONTOSPACE]. Additionally, the first author also acknowledges funding by the ALEXANDER VON HUMBOLDT STIFTUNG, GERMANY. We also thank Bernd Krieg-Brückner and John Bateman for fruitful discussions and impulses. Educational licenses have been utilized for ArchiCAD and RacerPro.

References

[1] Akin, Ö.: Architects' reasoning with structures and functions. Environment and Planning B: Planning and Design 20(3), 273–294 (1993)

[2] Akin, Ö., Özkaya, I.: Models of design requirement. In: Proceedings of the 6th International Conference on Design and Decision Support Systems in Architecture and Urban Planning, Landgoed Avegoor Ellecom, The Netherlands (2002)

[3] Augusto, J., Shapiro, D. (eds.): Advances in Ambient Intelligence. Frontiers in Artificial Intelligence and Applications, vol. 164. IOS Press, Amsterdam (2007)

[4] Augusto, J.C., Nugent, C.D. (eds.): Designing Smart Homes. LNCS, vol. 4008. Springer, Heidelberg (2006)

[5] Berners-Lee, T., Hendler, J., Lassila, O.: The semantic web. Scientific American 284(5), 34–43 (2001)

[6] Bhatt, M., Dylla, F.: A qualitative model of dynamic scene analysis and interpretation in ambient intelligence systems. International Journal of Robotics and Automation: Special Issue on Robotics for Ambient Intelligence (to appear, 2009)

[7] Day, M.: The Move to BIM with ArchiCAD 12, 23, AEC Magazine (August 2008)

[8] Eastman, C., Teicholz, P., Sacks, R., Liston, K.: BIM Handbook: A Guide to Building Information Modeling for Owners, Managers, Designers, Engineers and Contractors. In: Frontiers in Artificial Intelligence and Applications. Wiley, Chichester (2008)

[9] Froese, T., Fischer, M., Grobler, F., Ritzenthaler, J., Yu, K., Sutherland, S., Staub, S., Akinci, B., Akbas, R., Koo, B., Barron, A., Kunz, J.: Industry foundation classes for project management - a trial implementation. ITCon 4, 17–36 (1999), http://www.ifcwiki.org/

[10] Grütter, R., Scharrenbach, T., Bauer-Messmer, B.: Improving an RCC-derived geospatial approximation by OWL axioms. In: Sheth, A., Staab, S., Dean, M., Paolucci, M., Maynard, D., Finin, T., Thirunarayan, K. (eds.) ISWC 2008. LNCS, vol. 5318, pp. 293–306. Springer, Heidelberg (2008)

[11] Haarslev, V., Lutz, C., Möller, R.: Foundations of spatioterminological reasoning with description logics. In: Proceedings of Sixth International Conference on Principles of Knowledge Representation and Reasoning (KR 1998), pp. 112–123. Morgan Kaufmann, San Francisco (1998)

[12] Haarslev, V., Möller, R.: Description logic systems with concrete domains: Applications for the semantic web. In: Bry, F., Lutz, C., Sattler, U., Schoop, M. (eds.) Proceedings of the 10th International Workshop on Knowledge Representation meets Databases (KRDB 2003), Hamburg, Germany, September 15-16, 2003. CEUR Workshop Proceedings, vol. 79 (2003)

[13] Haarslev, V., Möller, R., Wessel, M.: Querying the semantic web with Racer + nRQL. In: Proceedings of the KI 2004 International Workshop on Applications of Description Logics, ADL 2004 (2004)

[14] Hall, K. (ed.): Green Building Bible: In Depth Technical Information and Data on the Strategies and the Systems Needed to Create Low Energy, Green Buildings, 4th edn., vol. 2. Green Building Press (2008)

[15] Hois, J., Bhatt, M., Kutz, O.: Modular ontologies for architectural design. In: 4th Workshop on Formal Ontologies Meet Industry (2009)

[16] Horridge, M., Patel-Schneider, P.F.: Manchester OWL syntax for OWL 1.1 (2008); OWL: Experiences and Directions (OWLED 2008 DC), Gaithersberg, Maryland

[17] Horrocks, I., Kutz, O., Sattler, U.: The Even More Irresistible SROIQ. In: Knowledge Representation and Reasoning (KR). AAAI Press, Menlo Park (2006)

[18] Kovacs, K., Dolbear, C., Goodwin, J.: Spatial concepts and OWL issues in a topographic ontology framework. In: Proc. of the GIS (2007)

[19] Kutz, O., Lücke, D., Mossakowski, T.: Heterogeneously Structured Ontologies— Integration, Connection, and Refinement. In: Meyer, T., Orgun, M.A. (eds.) Advances in Ontologies. Proceedings of the Knowledge Representation Ontology Workshop (KROW 2008), pp. 41–50. ACS (2008)

[20] Kutz, O., Lutz, C., Wolter, F., Zakharyaschev, M.: $\mathcal{E}$-Connections of Abstract Description Systems. Artificial Intelligence 156(1), 1–73 (2004)

[21] Liebich, T., Adachi, Y., Forester, J., Hyvarinen, J., Karstila, K., Wix, J.: Industry Foundation Classes: IFC2x Edition 3 TC1. In: International Alliance for Interoperability (Model Support Group) (2006)

[22] Masolo, C., Borgo, S., Gangemi, A., Guarino, N., Oltramari, A.: Ontologies library. WonderWeb Deliverable D18, ISTC-CNR (2003)

[23] Motik, B., Patel-Schneider, P.F., Grau, B.C.: OWL 2 Web Ontology Language: Direct Semantics. Technical report, W3C (2008)

[24] Philipose, M., Fishkin, K.P., Perkowitz, M., Patterson, D.J., Fox, D., Kautz, H., Hahnel, D.: Inferring activities from interactions with objects. IEEE Pervasive Computing 3(4), 50–57 (2004)

[25] Pike, W., Gahegan, M.: Beyond ontologies: Toward situated representations of scientific knowledge. International Journal of Human-Computer Studies 65(7), 659–673 (2007)

[26] Ramos, C., Augusto, J.C., Shapiro, D.: Ambient intelligence: The next step for artificial intelligence. IEEE Intelligent Systems 23(2), 15–18 (2008)

[27] Randell, D.A., Cui, Z., Cohn, A.G.: A spatial logic based on regions and connection. In: Proceedings of the 3rd International Conference on Knowledge Representation and Reasoning (KR 1992), pp. 165–176. Morgan Kaufmann, San Mateo (1992)

[28] Staab, S., Studer, R. (eds.): Handbook on Ontologies. International Handbooks on Information Systems. Springer, Heidelberg (2004)

[29] Stahl, C., Haupert, J.: Taking location modelling to new levels: A map modelling toolkit for intelligent environments. In: Hazas, M., Krumm, J., Krumm, J. (eds.) LoCA 2006. LNCS, vol. 3987, pp. 74–85. Springer, Heidelberg (2006)

[30] Streitz, N.A., Kameas, A.D., Mavrommati, I. (eds.): The Disappearing Computer. LNCS, vol. 4500. Springer, Heidelberg (2007)

[31] Uschold, M., Grüninger, M.: Ontologies: Principles, methods and applications. Knowledge Engineering Review 11, 93–155 (1996)

[32] Visser, U.: Intelligent Information Integration for the Semantic Web. LNCS, vol. 3159. Springer, Heidelberg (2004)

[33] Youngblood, G.M., Cook, D.J.: Data mining for hierarchical model creation. IEEE Transactions on Systems, Man, and Cybernetics, Part C 37(4), 561–572 (2007)

Defining Spatial Entropy from Multivariate Distributions of Co-occurrences

Didier G. Leibovici

Centre for Geospatial Sciences, University of Nottingham, UK
didier.leibovici@nottingham.ac.uk
http://www.nottingham.ac.uk/cgs

Abstract. Finding geographical patterns by analysing the spatial configuration distribution of events, objects or their attributes has a long history in geography, ecology and epidemiology. Measuring the presence of patterns, clusters, or comparing the spatial organisation for different attributes, symbols within the same map or for different maps, is often the basis of analysis. Landscape ecology has provided a long list of interesting indicators, *e.g.* summaries of patch size distribution. Looking at content information, the Shannon entropy is also a measure of a distribution providing insight into the organisation of data, and has been widely used for example in economical geography. Unfortunately, using the Shannon entropy on the bare distribution of categories within the spatial domain does not describe the spatial organisation itself. Particularly in ecology and geography, some authors have proposed integrating some spatial aspects into the entropy: using adjacency properties or distances between and within categories. This paper goes further with adjacency, emphasising the use of co-occurences of categories at multiple orders, the adjacency being seen as a particular co-occurence of order 2 with a distance of collocation null, and proposes a spatial entropy measure framework. The approach allows multivariate data with covariates to be accounted for, and provides the flexibility to design a wide range of spatial interaction models between the attributes. Generating a multivariate multinomial distribution of collocations describing the spatial organisation, allows the interaction to be assessed via an entropy formula. This spatial entropy is dependent on the distance of collocation used, which can be seen as a scale factor in the spatial organisation to be analysed.

Keywords: spatial information, entropy, co-occurrences, spatial statistics, Multivariate data, spatial point process, R programming.

1 Introduction

Like the correlation measure, the entropy is a quasi universal measure used in many different fields, sometimes very far from their initial domain. It is probably after the seminal paper [1] in information theory in the field of signal processing, already quite far from statistical thermodynamics (where entropy concepts were

K. Stewart Hornsby et al. (Eds.): COSIT 2009, LNCS 5756, pp. 392–404, 2009.

initially devised), that the term entropy and the Shannon's formula became popular in applied science. Entropy, information content, uncertainty and diversity are often part of the discourse and the entropy measure is intended to capture the measure of interest, its gain or its loss:

$$H(p_1, p_2, ..., p_i, ..., p_n) = -\sum_{i=1}^{n} p_i \log(p_i) \tag{1}$$

where p_i are observed proportions of the n classes or states within the studied system (with $\sum_{i=1}^{n} p_i = 1$ and if $p_i = 0$, $p_i \log(p_i)$ is set $= 0$). The qualifier of *measure of uncertainty* can then be understood from the fact that if all the $p_i = 1/n$ that is a uniform distribution (the most uncertain situation where every outcome has the same probability), then $H(\{p_i\}) = log(n)$ which is the maximum of entropy. This provides the following normalisation relative to uniformity:

$$0 \leq H_u(p_1, p_2, ..., p_i, ..., p_n) = -1/log(n) \sum_{i=1}^{n} p_i \log(p_i) \leq 1 \tag{2}$$

an interesting measure to observe the data distribution structure and its changes during a monitoring process or any transformation imposed on the dataset. Notice that $H(X)$ and $H_u(X)$ are used when X is a random variable with the n discrete outcomes with probabilities given by or estimated by p_i. Now being interested by a spatial organisation say for example of a binomial variable, it is easy to see that applying the entropy measure given above is independent of the spatial organisation (see the chessboard example on figure 1).

This problem of getting the same entropy for very different spatial configurations, did not preclude the use of Shannon entropy as a measure of diversity for a variable measured over a spatial domain. This is the case for instance in

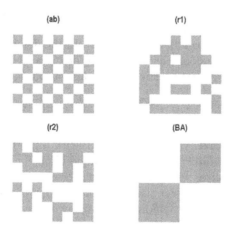

Fig. 1. All chessboard configurations have the same normalised Shannon entropy, $H_u = 1$: (ab) normal Chessboard with a regular spatial structure; (r1 and r2) random permutations of the 64 cells; (BA) another well structured chessboard permutation

ecology, see [2] and related references, also in economy and geographical studies, particularly in urban geography and regional science [3,4,5]. Pointing out the aspatiality of the use of the Shannon entropy in the literature, [6] proposed the use of a weighted matrix describing a spatial configuration related to proximities or contiguities of the classes. The modified Shannon entropy uses this matrix in a similar way as the Moran's statistic does with the correlation measure:

$$H_W(p_1, p_2, ..., p_i, ..., p_n) = -\sum_{i=1}^{n} p_i^W \log(p_i^W) \qquad (3)$$

where $p_i^W = \sum_j W_{ij}p_j$ and the matrix W express the proximities of categories i and j. This nonetheless displaces the problem to that of modelling the spatial proximities, that is the form of a matrix W. In spatial analysis, W is usually associated with all geographical units observed on the map, so the equation 3 could be adapted to consider n as the number of spatial units instead. Therefore W_{ij} would express some spatial proximity of units combined with a category proximity, depending on the purpose of the analysis, and p_j would be replaced by the joint probability of zones and categories, $p_j p_{c(j)}$, considered a priori independent. Nonetheless in Landscape Ecology, proximity of categories evaluated by adjacency has been used in the contagion index to measure clumping aspects of the map, with different normalisation according to the authors [7,8]:

$$D_2 = -H(\{p_{ij}\}) = \sum_{i,j=1}^{n} p_{ij} \log(p_{ij}) \qquad (4)$$

where p_{ij} is the probability of having a pixel of category i adjacent to a pixel of category j. In [9], on similar lines the authors used neighbour counts of different quantities to build entropies related to the metric information (proportion of areas using Voronoi regions of labels), topological information (no entropy but average number of neighbours in the Delaunay graph), and thematic information (using Voronoi regions adjacency counts). In the domain of image analysis, [10,11] used Markov Random Fields theory to define a spatial entropy also emphasising the neighbouring properties and configurations in terms of pixel labels:

$$S_{EV}(\Omega \, \mathcal{N}) = -\sum_{N_v} \sum_{\lambda \in \Omega} p(\lambda, N_v) \log(p(\lambda \, N_v)) \qquad (5)$$

where Ω is the set of pixel values of an image with a neighbourhood system $\mathcal{N}$ from which N_v realises either the neighbourhood of pixel v in terms of pixel values, or, its equivalent class with v representing the class number. The formula is obtained as the expectation over the N_v of the entropy of the conditional probabilities, and so, is in fact the conditional entropy of the random field given the neighbouring system or its transformation in equivalent classes. While conceptually this formulation is appealing, the limitation is that it relies on a Markov Random Field model to be estimated.

Recently in geographical analysis, [12] made an interesting discussion and contribution regarding spatial diversity, complexity, redundancy and their use in spatial information; he proposed a measure of spatial diversity accounting for the spatial configuration as a weight factor in the Shannon entropy. This weight factor is the ratio of the average distance between objects of a particular class (called the intra-distance), to the average distance of a particular class to the other classes (called the extra- distance):

$$H_s(p_1, p_2, ..., p_i, ..., p_n) = -\sum_{i=1}^{n} \frac{d_i^{int}}{d_i^{ext}} p_i \log_2(p_i) \qquad (6)$$

where the weight factor is bounded ≤ 2, see also [13]. This Diversity Index has the advantage of being simple to compute and appears promising as the figure.3 of the paper [13] illustrates. Figure 2 reproduces this result together with our proposal for a Spatial Entropy measure (see next section); the two measures are similarly efficient in discriminating at least the panel (d) from the others. As the two measures are not normalised in the same way, only discriminating power can be really compared.

Building on the diverse approaches developed in the literature, discussed briefly in this introduction, the next section will propose the conceptual basis for a new Spatial Entropy Index. The following sections will (i) explain how the concept of multiple co-occurrences can be used to build some Spatial Entropy measures, (ii) show some results on hypothetical and real datasets, and (iii) describe applications for spatial analysis with multivariate data. A short discussion explores some potential issues and problems related to these concepts.

2   Spatial Entropy from Co-occurrence Counts

The problem of extending the Shannon entropy measure for the spatial domain, inherits the problems encountered when translating probability and statistical methods from identical independent experiments to dependent measurements. The random field framework may provide a satisfying solution, at least in terms of probabilistic coherence, [10,11], but may be restrictive due to regularity properties necessary for the estimation paradigm. The chosen approach here is also deliberately non-parametric. The introduction of constraints from the spatial configuration, expressed by the matrix W in equation 3, seems natural, considering its successful use in correlation analysis with the Moran's Index. Somehow Claramunt's Diversity follows this idea of spatial constraint by using weights expressing proximities and separabilities of the categories of objects. But in equation 6, the question is whether the weighting factor is the ratio of distances or if it is the uncertainty factor $-log(p_i)$ used in Shannon's entropy. In other words is it a mean uncertainty weighted by separability or a mean separability weighted by uncertainty? The index proposed is a combination of these two concepts.

The Markov Random Field approach [10,11] and the adjacency approach [7,8] preserve the spatial dependency of the categories in the formulae. Keeping a generic framework may be useful to incorporate spatial probability theories, but also allows the discrete and continuous cases to be based on the same principles, which is important for sampling aspects. The dependency of categories can be seen as the fact that some categories tend to occur closely to each other or far from each other. So the fundamental event is the occurrence, at the same place or in the same surroundings, of categories or attributes from the same or different processes. The co-occurrence of attributes generalises the adjacency in terms of its spatial concept and its multivariate content. The latter aspect will be explored in more detail in section 5. Based on these considerations one can propose as Spatial Entropy the index:

$$H_{Su}(C_{oo}, d) = -1/log(N_{c_{oo}}) \sum_{c_{oo}}^{Nc_{oo}} p_{c_{oo}} \log(p_{c_{oo}}) \tag{7}$$

where C_{oo} is a multivariate co-occurrence defined from multivariate measurements, categories or attributes of the spatial objects considered in the map or image. One collocation event is dependent on the chosen distance d and on the order of collocation, (see next section). This multivariate co-occurrence C_{oo} generates the collocations c_{oo} at this chosen order, which are counted or estimated. Therefore it defines the distribution of collocations, or the multivariate co-occurrence distribution, that is the $p_{c_{oo}}$. An example of this Spatial Entropy index is given in figure 2, where the collocations of order 3 of the binomial outcome is building the distribution on which the normalised Shannon entropy

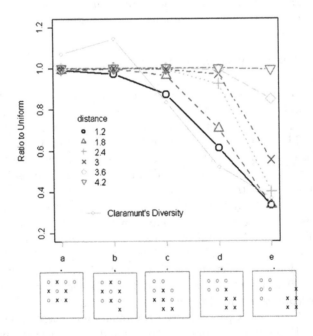

Fig. 2. Spatial Entropy of order 3 on the toy example used in fig.3 in [13]

is computed. For univariate data, the order of collocation gives the possibility to increase the spatial density constraint.

3 Measuring Spatial Co-occurrences

Some aspects of this section have been described in [14]. Depending on the study, the procedure to define the multivariate co-occurrence characteristic will be linked to spatial concepts, noticeably from point pattern analysis theory, [15], also from GIS spatial topology principles, and from image analysis. Let $i = 1,$ $...I$ be categories of a variable vI, $j = 1, ...J$ be categories of a variable vJ, $k = 1, ...K$ be categories of a variable vK, and let $s = 1, ...S$ be the locations where one can record either vI, vJ, vK, or all of them, this for one or more records: depending on the type of geometrical object behind the locations and the meaning given to a location. A location can be a polygon representing a village where different kinds of events can be recorded and summarised as counts or proportions according to, say the population or the area, *e.g.* the same village could be associated to a point but keeping the same meaning. The variables vI, vJ, vK (or more) are associated with a general event, *e.g.* (i) a person of age i with social class level j, diagnosed for a certain disease k and living at location s, or, (ii) a plant species i, on a soil class j, at location s with annual rainfall k, or even (iii) a crime of type i, at time slot j, in a zone of wealth class k of location s. So we are concerned about some possible associations of vI, vJ, vK in terms of their spatial observations, either as multivariate observations on a spatial domain or as already collocated observations of different variables or both. Clearly (i) is multivariate on the persons, (ii) is a collocation of different measurements (iii) is a mixed of both. One could argue that (ii) could be seen as multivariate characteristics of the location.

3.1 Counting Pairs

We shall note, $n_{ii'}$, n_{ij}, and n_{ijk} the number of collocations respectively of events i and i', i and j, i, j and k. When the distance of collocating events needs to be explicit we use n_{ij}^d, where d is the distance of collocating events. A very general definition of collocations is, Cg: **a collocation of marks or labels** $\{i, j, k\}$ **is recorded if, the distance between the locations** $\{s, s', s''\}$, **all together expressing the labels** $\{i, j, k\}$, **is at most** d. Usually the standard way of computing the collocations (here without edge corrections) for two labels takes a form like:

$$n_{ij}^d = \sum_S \#_{C(s_i, d)}\{j\} = \sum_S \#_{C(s_j, d)}\{i\} = n_{ji}^d \qquad (8)$$

where $\#_{C(s_i, d)}\{j\}$ means the number of "events" or marked locations j at maximum distance d from a location s labelled i, that is the number of s_j found in a circle of radius d and centre s_i. With flexibility about the geometry defining the searching area $(\mathcal{G}_{(s_i, d)})$, with d being the buffer size of the geometry s_i instead of the radius of a circle (or even a set deformation with parameter d), and also flexibility about the way ($\mathcal{O}$) the occurrence is recorded "within" the geometry (now also depending on the actual geometry of the j mark), gives a more general formula:

$$n_{ij}^d = \sum_s \mathcal{OG}_{(s_i,d)}(\{s_j\}) \tag{9}$$

An example is given in figure 6 for a different searching area. This approach allows the s_i to be associated with different geometries all through the zone S. Depending on the choices for $\mathcal{O}$ and $\mathcal{G}_{(.,d)}$ the collocation value may not be symmetrical, $e.g$ being in a north angular zone (a triangle or a heart shape).

3.2   Counting Triples

For n_{ijk}, the number of collocations of marks i, j and k, different approaches can be taken. As for two events, variable symmetry should considered as it is implicit in the definition, but writing down the computational aspects could lead naturally to asymmetrical pseudo-similarities, which can also be accounted for, even for two variables. For three or more marked events the standard formula has to be rewritten if symmetry is to be kept:

$$n_{ijk}^d = \sum_S \#_{C(d)}\{i,j,k\} \tag{10}$$

$$\neq \sum_s \#_{C(s_i,d)}\{j,k\} \neq \sum_s \#_{C(s_j,d)}\{i,k\} \neq \sum_s \#_{C(s_k,d)}\{i,j\} \tag{11}$$

$$n_{ijk}^d =_{Cg} \sum_S \#_{C(s_i,d)\cap C(s_j,d)\cap C(s_k,d)}\{i,j,k\} \tag{12}$$

As seen in figure 3, the non-location-centred circle method, giving symmetrical spatial co-occurrences, results in different collocations counts from asymmetrical methods based on marked-location-centred circles.

As will be discussed in section 5, different counting methods and their interpretations introduce a range of methods suitable to discover spatial pattern collocations and assess the associated spatial organisation using the spatial entropy equation formula 7. To differentiate the way a 3-way-collocation is counted one

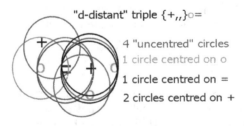

Fig. 3. Symmetrical and asymetrical ways of counting collocations of a triplet with circles of radius d (not necessarily verifying definition Cg)

can write: $n^d_{ijk} = \sum_S \#_{C(d)}\{i,j,k\}$ and $n^d_{s_i jk} = \sum_s \#_{C(s_i,d)}\{j,k\}$. Marginals of these counts are to be used to analyse either different profiles or whole proportions, such as:

$$p^d_{jk/i} = n^d_{ijk}/n^d.jk \quad \text{and} \quad p^d_{ijk} = n^d_{ijk}/n^d... \tag{13}$$

where "." means a summation over the index it is replacing: $n^d.jk = \sum_i n^d_{ijk}$. For the following examples order 2, 3, 4 collocations have been used on only one categorical variable which provides interesting insight into the use of the multiple collocation power.

4 Some Examples

Reusing the toy dataset from Li and Claramunt's paper the figure 4, bottom panel, shows that as the order increases the consistency across the scale effect is maintained. But as the data "gets" more structured the discriminating power

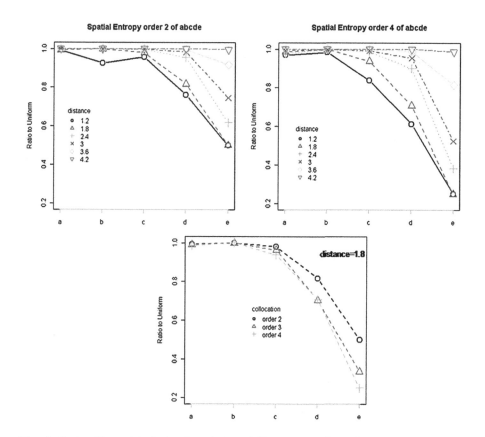

Fig. 4. Scale effect with Order 2 and 4; and Variation of Spatial Entropy with order 2, 3, 4 at distance 1.8 unit on the toy example used in fig.3 in [13]

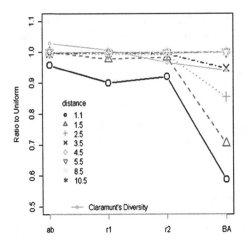

Fig. 5. Scale effect with Order 3 on the chessboard example of fig.1

among the different distances is improved for higher collocation orders. When comparing orders with the same collocation distance, higher order also improves the relative gain in negentropy or loss of entropy.

4.1   The Chessboard Example

For this second toy example, more comparable to a situation of image analysis or remote sensing data, where every pixel can be classified, the spatial entropy of order 3 still performs well comparatively to the Claramunt's Diversity index, (see figure 5). This confirms the statement proposed in the introduction about the separability property characterising this index, that is the ability to discover if the categories are apart. Perhaps with more categories, this index could be valuable in detecting separated categories, for example in remote sensing imaging.

4.2   The Lansing Data

Now an example still using one categorical variable, but with many categories, here representing 6 different tree species of this well known dataset used in point pattern analysis, figure 6. First steps of co-occurrence design with increasing distances are represented on the figure, the second step being an intersection with another circle (with the chosen radius) centred on a point of the first one, the third step being just a counting process within the delimited zone. The heart shape buffer zones represent a potential co-occurrence design if the interest is on angular co-occurrences which can make sense in ecology. The collocation entropy of order 3 shown on figure 7 exhibits some local scale spatial organisation. This organisation is up to 0.2 units radius from a given point. The nonparametric envelope of 5% is built from 19 random permutations of the labels of the points. Claramunt's index was 2.2309 and the 5% envelope was [2.33588 − 2.34531],

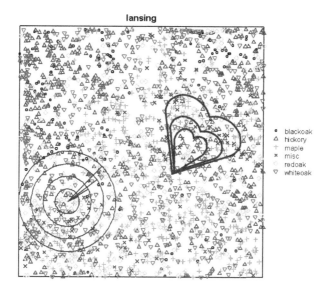

Fig. 6. Tree Dataset in a 1 unit x 1 unit window from the R package Spatstat [16]: collocation radii 0.05, 0.1, 0.15, 0.2, and shaped collocation examples

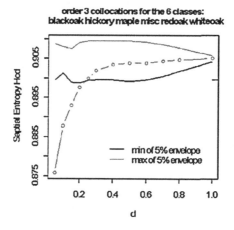

Fig. 7. Scale effect of Spatial entropy on the Lansing data with random permutation envelope

so outside this interval. This is 95.50% in percent of the lowest bound, with the interval becoming [100.00% − 100.40%]. With H_{Su}, one gets at $d = 0.1$ unit: 98.50% with the interval [100% − 100.75%]. Claramunt's index appears more than 11 times the spread of the interval and H_{Su} is 2 times, but both are outside the 5% confidence envelope.

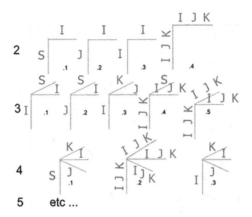

Fig. 8. k-way Tables for collocation discovery analysis: all 2-modes, 3-modes, 4-modes possibilities with cell values as counted collocations as explained in the text: I, J, K categorical variables, S spatial locations

5   Multiway Co-occurrence Analyses

Coming back to the original problem outlined in the abstract, spatial interactions among a collection of variables can be associated with the same process or different spatial processes. One can now argue that the general framework described in the previous sections can accommodate many analysis questions. The way of measuring the co-occurrence, the order of co-occurrence, and finally the variables involved, are all parameters to play with in defining an appropriate exploratory analysis. Figure 8 expresses some collocations tables targeted when computing the co-occurrences, here with three categorical variables vI, vJ, vK. Notice that S, the spatial locations are also represented, but usually before computing the Spatial Entropy, the table is collapsed or aggregated according to a chosen attribute (say vI categories) or a geographical partition (a covariate part of the sampling design). This increases the flexibility of this multiway methodology. The distance d can also be integrated as another "dimension" of the multiway tables in order to perform multiscale entropies. Some other types of analysis has been done using these multiway tables of co-occurrences in [14]. The context was also related to finding spatial organisation of the categories of few variables, but using a different measure of the distributions of co-occurrences. Instead of the entropy, the chi-squared of lack of independence was used, as conveniently, it is linked to correspondence analysis. A generalisation of correspondence analysis to multiway tables was then described using a tensorial framework, [17,18].

6   Conclusion and Further Work

This paper discussed a possible extension of the Shannon's entropy to the spatial domain in a simple way. Looking at the multivariate distribution of co-occurrences and the way it is built or estimated allows us to investigate various spatial multiway interactions to be analysed on a range of scales. Normalising this entropy and studying its variations under hypothesis of random labelling enables the existence of a particular spatial interaction effect encompassed in the studied spatial distribution to be tested. Using this framework with useful entropy formula such as the cross-entropy and the conditional entropy can further extend the type of analysis to be used when studying spatial interactions, [19]. Aiming to be applied to fuzzy information, Wuest et al. [20] debated probabilistic versus non-probabilistic information for a map but this development does not seem to take into account spatial proximities. This fuzzy aspect of the spatial data either in position or attribute value was explored also in [21], but the focus was on fuzziness rather than on the entropy concept itself, then still remaining an important challenge for spatial entropy.

Acknowledgments

As part of the Bridging the Gap awarded project: "Geospatial mash up technologies in Health and disease mapping", this work was supported by the Engineering and Physical Sciences Research Council [grant number EP/E018580/1] and the University of Nottingham.

References

1. Shannon, C.: A mathematical theory of communication. Bell System Technical Journal 27, 379–423, 623–656 (1948)
2. Bogaert, J., Farina, A., Ceulemans, R.: Entropy increase of fragmented habitats: A sign of human impact? Ecological Indicators 5(3), 207–212 (2005)
3. Batty, M.: Spatial entropy. Geographical Analysis 6, 1–31 (1974)
4. Snickars, F., Weibull, J.: A minimum information principle: Theory and practice. Regional Science and Urban Economics 7, 137–168 (1977)
5. Cutrini, E.: Using entropy measures to disentangle regional from national localization patterns. regional science and urban economics. Regional Science and Urban Economics 39(2), 243–250 (2009)
6. Karlström, A., Ceccato, V.: A new information theoretical measure of global and local spatial association: S. The Review of Regional Research 22, 13–40 (2002)
7. O'Neill, R., Krummel, J., Gardner, R., Sugihara, G., Jackson, B., DeAngelis, D., Milne, B., Turner, M., Zygmunt, B., Christensen, S., Dale, V., Graham, R.: Indices of landscape pattern. Landscape Ecology 1(3), 153–162 (1988)
8. Li, H., Reynolds, J.F.: A new contagion index to quantify spatial patterns of landscapes. Landscape Ecology 8, 155–162 (1993)
9. Li, Z., Huang, P.: Quantitative measures for spatial information of maps. International Journal of Geographical Information Science 16(7), 699–709 (2002)
10. Maitre, H., Bloch, I., Sigelle, M.: Spatial entropy: a tool for controlling contextual classification convergence. In: Proceedings of IEEE International Conference on Image Processing, ICIP 1994, vol. 2, pp. 212–216 (1994)
11. Tupin, F., Sigelle, M., Maitre, H.: Definition of a spatial entropy and its use for texture discrimination. In: International Conference on Image Processing, vol. 1, pp. 725–728 (2000)

12. Claramunt, C.: A spatial form of diversity. In: Cohn, A.G., Mark, D.M. (eds.) COSIT 2005. LNCS, vol. 3693, pp. 218–231. Springer, Heidelberg (2005)
13. Li, X., Claramunt, C.: A spatial entropy-based decision tree for classification of geographical information. Transactions in GIS 10(3), 451–467 (2006)
14. Leibovici, D., Bastin, L., Jackson, M.: Discovering spatially multiway collocations. In: GISRUK Conference 2008, Manchester, UK, April 2-4, 2008, pp. 66–71 (2008)
15. Diggle, P.J.: Statistical Analysis of Spatial Point Patterns. Hodder Arnold, London (2003)
16. Baddeley, A., Turner, R.: spatstat: An r package for analyzing spatial point patterns. Journal of Statistical Software 12(6) 1, 1–42 (2005)
17. Leibovici, D., Sabatier, R.: A Singular Value Decomposition of k-Way Array for a Principal Component Analysis of Multiway Data. PTA-k. Linear Algebra and Its Applications 269, 307–329 (1998)
18. Leibovici, D.: PTAk: Principal Tensor Analysis on k modes. Contributing R-package version 1.1-16 (2007)
19. Bhati, A.S.: A generalized cross-entropy approach for modeling spatially correlated counts. Econometric Reviews 27(4), 574–595 (2008)
20. Wuest, L.J., Nickerson, B., Mureika, R.: Information entropy of non-probabilistic processes. Geographical Analysis 35(3), 215–248 (2003)
21. Shi, Y., Jin, F., Li, M.: A total entropy model of spatial data uncertainty. Journal of Information Science 32(4), 316–323 (2006)

Case-Based Reasoning for Eliciting the Evolution of Geospatial Objects

Joice Seleme Mota, Gilberto Câmara, Maria Isabel Sobral Escada,
Olga Bittencourt, Leila Maria Garcia Fonseca, and Lúbia Vinas

INPE – National Institute for Space Research,
Av. dos Astronautas 1758,
12227-001 São José dos Campos, Brazil
{joice,gilberto,isabel,olga,leila,lubia}@dpi.inpe.br

Abstract. This paper proposes an automated approach for describing how geospatial objects evolve. We consider geospatial objects whose boundaries and properties change in the time, and refer to them as *evolving objects*. Our approach is to provide a set of rules that describe how objects change, referred to as *rule-based evolution*. We consider the case where we are given a series of snapshots, each of which contains the status of the objects at a given time. Given this data, we would like to extract the rules that describe how these objects changed. We use the technique of case-based reasoning (CBR) to extract the rules of object evolution, given a few representatives examples. The resulting rules are used to elicit the full history of all changes in these objects. This allows finding out how objects evolved, recovering their history. As an example of our proposed approach, we include a case study of how deforestation evolves in Brazilian Amazonia Tropical Forest.

Keywords: Spatio-temporal data, evolving objects, Case-Based Reasoning.

1 Introduction

The computational modelling of geospatial information continues to be, after decades of research, a problem which defies a definitive solution. Since computer models assign human-conceived geographical entities to data types, matching geospatial data to types and classes has been the focus of intense research. Recently, there has been much interest on modelling and representation of geospatial objects whose properties change [1-5]. Such interest has a strong practical motivation. A new generation of mobile devices has enabled new forms of communication and spatial information processing. Remote sensing data is becoming widespread, and more and more images are available to describe changes in the landscape. As new data sources grow, we are overwhelmed with streams of data that provide information about change.

Representing *change* in a GIS (Geographical Information System) is not only an issue of handling time-varying data. It also concerns how objects gain or lose their identity, how their properties change, which changes happen simultaneously, and what causes change. As Goodchild et al.[5] point out, the distinction between geospatial entities as continuous fields or discrete objects also applies in the temporal domain. In this paper, we deal with computational models for time-varying discrete

K. Stewart Hornsby et al. (Eds.): COSIT 2009, LNCS 5756, pp. 405–420, 2009.

geospatial entities. We refer to those as *geospatial objects* and distinguish two broad categories. The first category concerns objects whose position and extent change continuously, referred to as *moving objects*. The second type concerns objects bound to specific locations, but whose geometry, topology and properties change but at least part of its position is not altered. We refer to these as *evolving objects*, which arise in urban cadastre and in land cover change. For an alternative characterization of spatio-temporal objects, see Goodchild et al.[5].

This paper describes a computational model for *evolving objects*, which tracks changes that occurred during an object's lifetime. The proposed model aims to answer questions such as *"What changes took place for each object?"*, *"When did these changes occur?"* and *"How did the changes take place?"*. We aim to extract the history of an object from its creation to its disappearance, including references to other objects involved. Eliciting the history of each object helps us to understand the underlying causes of change. To be able to record the complete history of each object, we need a model that uses previous cases as well as knowledge obtained from an expert as the main sources of knowledge used to solve new problems. This leads to the subject of this paper. We propose a computational method that contains a set of rules that describe how *geospatial objects* evolve, based on a sample of existing cases. For this task we have used the Case-Based Reasoning (CBR) technique, which defines a set of rules that arise from knowledge about the application domain.

In what follows, we review previous work on section 2 and describe our proposal in Section 3. In section 4, we describe an experiment where we applied our method to a spatiotemporal study of deforestation evolution. This paper builds on previous work by the authors [6-8].

2   Challenges in Describing How Spatial Objects Evolve

In this section, we consider previous work on models for evolving objects and introduce the challenges in describing how these objects change. Evolving objects are typical of cadastral and land change applications. Computational models for describing such objects are also referred to as *lifeline models*. Lifeline models use three ideas: *identity*, *life*, and *genealogy*. Identity is the characteristic that distinguishes each object from others during all its life. Life is the time period from the object's creation until its elimination. Genealogy implies managing the changes that occur to an object has during its life. Hornsby and Egenhofer [9] stress the need to preserve an object's identity when its geometry, topology, or attributes change, a view supported by Grenon and Smith [3]. Consider parcels in an urban cadastre. A parcel can have its owner changed, be merged with another, or split into two. A possible approach is to describe an object's history based on operations such as creation, splitting and merging [9, 10]. However, these authors do not consider the problem of extracting evolution rules from the objects themselves. They consider objects of a single type. In this paper, we consider objects of different types and we provide ways to extract their evolution rules.

To take a simple motivational example, consider Figure 1, where there are three objects: S1 of type *'Street'* and P1 and P2 of type *'Parcel'*. *Given the geometries of these objects at times T1 and T2, how can we find out how these objects evolved?* To

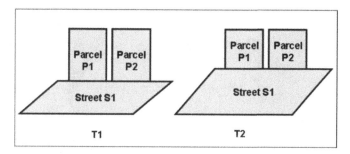

Fig. 1. Example of objects evolution

model this example, we need to consider different rules for spatial operations. Consider the case of the *'merge'* geometric operation, which joins the geometries of two objects. When the objects have different types, merging two objects can produce different results. When the object types are *'Street'* and *'Parcel'*, there should be different rules for the result of the merging two objects. One possible set of rules is: (a) "merging two *Parcels* results in a *Parcel*"; (b) "merging a *Street* with a *Parcel* results in expanding the *Street*".

As a second example, consider how the internal and external borders of Brazil changed, as shown in Figure 2. Each polygon in Figure 2 is a Brazilian state. The Brazilian borders have changed significantly since the 18th Century, both because of internal division (creation of new states from existing ones) and inclusion of external areas (through international treaties). Suppose we want to devise a procedure that, given the snapshots shown in Figure 2, tries to extract the history of Brazil´s internal and external borders. Such method would have to distinguish at least three data types (*'Country'*, *'State'*, *'ExternalArea'*) and would need a set of type-dependent rules for object merging and splitting. As a first guess, this set would have these rules:

R1. Splitting an existing State produces two States: a new State and the existing State with a smaller area.
R2. An existing State can be converted into a new State with the same borders.
R3. Merging a State with an existing State produces a State with larger area. The new area is assigned to an existing State.
R4. Merging a Country with an External Area produces a Country with larger area. The new area is assigned to an existing State.
R5. Splitting a State from a Country produces a Country with smaller area and a new part of the External Area.

These rules are not the only possible set. They may be able to rebuild a believable history of the Brazilian states, but may fail to be historically accurate. Given a set of snapshots which show that state of spatial objects in different times, we are not always able to remake their precise history. However, often snapshots are all we have, and we need to devise ways to make a likely guess about the objects' evolution.

These examples and similar cases lead us to propose the idea of *rule-based evolution of typed geospatial objects*. Our view of types comes from Computer Science, where types are tools for expressing abstractions in a computer language [11]. On a

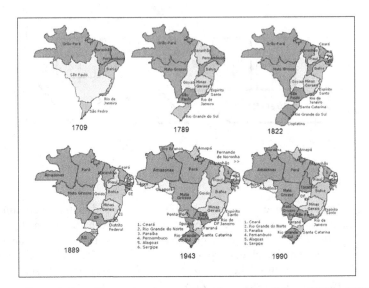

Fig. 2. Evolution of internal and external borders of Brazil from 1709 to 1990

theoretical level, a type is a set of elements in a mathematical domain that satisfy certain restrictions. A *typed object* is an object whose evolution is subject to constraints that are specific to its type. Thus, in the Brazilian borders example, for objects of type '*Country*' and those of type '*State*' we need different rules to describe their evolution. Models in which objects have different types and evolution cases are richer and more powerful than typeless ones.

3   Extracting the Evolution Rules Using Case-Based Reasoning

In this section, we describe the use of Case-Based Reasoning (CBR) to extract the evolution rules for a set of geospatial objects. CBR is a method for problem solving that relies on previously used solutions to solve similar problems [12]. In contrast to techniques that rely solely on general knowledge of a problem domain, CBR is able to use the *specific* knowledge of previously experienced, concrete problems [13]. This is a recurring technique in human-problem solving. To solve a new problem, we recall how we handled a similar one in the past and try to reuse it. A new problem is solved by finding a similar past case and reusing it in the new problem. CBR is an incremental technique, since experience learned in a case is applied for solving future problems.

Traditional rule-based inferences have no memory of previous cases, and use the same set of rules for solving all problems. Instead of doing inferences using a large group of rules, a CBR-based software keeps track of previous cases. When confronted with a new problem, it tries to adapt rules that were useful in similar cases, thus increasing continually its knowledge base [13]. The similarity between two problems can often ensure the interpretation adopted for a previous case can also adopted for the new one.

There are two types of CBR implementations: automatic procedures and information recovery methods [12]. Automatic systems solve the problem in an autonomous way and provide methods to evaluate the results of their decisions. Information recovery methods use human experts to set up the problem-solving rules based on well-known examples. These rules are then used to perform the desired task. The current work uses a CBR system of the second type. Following [13], our proposed CBR technique has the following main steps: 1) Select a set of exemplary cases in the database; 2) Use these cases to set up a set of evolution rules with the help of a domain expert; 3) Test the proposed solution and, if necessary, revise it; and 4) Store the experience represented in the current set of rules for future reuse. The steps to model and to represent how spatiotemporal objects evolve (shown in Figure 3) are:

1. *Retrieval of snapshots of the area that contains a set of geospatial objects whose history we want to describe.*
2. *Selection of a subset of this data that allows the human expert to find out the different types of geospatial objects and set up their evolution rules.*
3. *Represent these evolution rules using CBR.*
4. *Recover all objects from the database and compute their history based on the evolution rules.*

The domain expert defines two types of rules to characterize the objects' evolution: *description rules* and *evolution rules*. The *description rules* define the types of geospatial objects. The *evolution rules* define how objects evolve under spatial operations such as 'split' and 'merge'. The expert defines the *description rules* considering the objects' properties and their spatial relationship, including topological predicates such as 'cross', 'close to' and 'touch'. Consider Figure 4, where some prototypical land change objects are portrayed. Figure 4(a) shows three objects at time T1. At time T2, three new objects appear as shown in Figure 4(b). After applying the description and evolution rules described below, the resulting objects are shown in Figure 4(c).

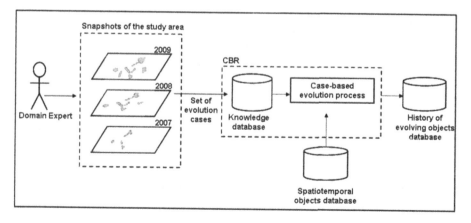

Fig. 3. General view of CBR method for eliciting geospatial objects evolution

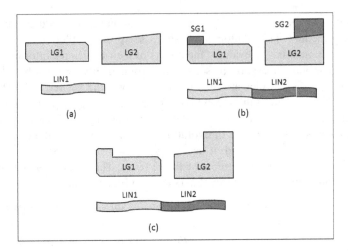

Fig. 4. Evolution of prototypical land change objects: (a) Time T1; (b) Time T2 before application of description rules; (c) Time T2 after application of evolution rules

In this example, the description rules define three types of objects: *LargeGeometric (LG), Linear (LIN)* and *Small Geometric (SG),* according to the following rules:

> *DR1. An object with perimeter/area ratio smaller than 10 hectares^{-1} is a Linear object.*
>
> *DR2. An object with perimeter/area ratio greater than 10 hectares^{-1} and whose area is less than 50 hectares is a Small Geometric object.*
>
> *DR3. An object with perimeter/area ratio greater than 10 hectares^{-1} and whose area is more than 50 hectares is a Large Geometric object.*

These rules allow us to identify the objects in Figure 4, as shown in the labels assigned to each object. For this case, a possible set of evolution rules would be:

> *ER1. A Small Geometric object that touches a Large Geometric object is merged with the Large Geometric object.*
> *ER2. Two adjacent Small Geometric objects are merged.*
> *ER3. Two Linear objects that are adjacent are not merged.*

Applying these rules, the *SmallGeometric* objects shown in Figure 4(b) are merged with the adjacent *LargeGeometric* objects, thus resulting in a spatial expansion of the latter. This example shows the need for a system that is able to represent the *description* and *evolution rules* and to apply them to extract the history of a set of objects. This system architecture is described in the next section.

4   CBR-Based Geospatial Object History Extractor

This section describes the architecture of a geospatial history extractor based on Case-Based Reasoning (CBR) technique. A CBR system stores knowledge as a set of cases. Each case contains data about a specific episode, with its description and the context in which it can be used. The contents of each case include a set of rules set up by the domain expert. Among the several existent techniques for knowledge acquisition for CBR, we used unstructured interviews, where the information is obtained through direct conversation with the specialist. In these interviews, he gives his perspective of the problem, and a computer specialist records these cases. The expert elicits the knowledge domain in two steps:

1. Describing the objects in their environment (*description rules*).

2. Analyzing this outcome of spatial operations between the objects (*evolution rules*).

After the expert produces the rules, the CBR system stores a set of rules for each case, as shown in Figure 5.

The knowledge base consists of a series of cases, indexed by the object´s attributes. Based on the problem´s description, the indexes point out which attributes should be compared, finding out the case that can be useful for the solution. Each attribute receives a weight (among 0 and 1) according to their degree of importance in the solution of the case. In our model we built the indexes using an explanation-based technique, where the specialist points out which attributes are relevant for the solution of the problem. Figure 5 shows the indexes for the cases that described the problem described in Figure 4. The indexes for the *description rules* are area and perimeter/area ratio; the indexes for *evolution cases* are the *objects types* and their *spatial relationship*.

After creating and indexing the knowledge base, we can then create the history of all objects. Each object is considered as a new problem and processed separately in two phases. Processing starts by taking the objects from the *Geospatial Objects* Database that contains snapshots of the geospatial objects at different periods of time. For the example shown in Figure 2 (evolution of Brazil's borders), the database would contain six snapshots for the years 1709, 1789, 1822, 1889, 1943 and 1990. The CBR system starts at the earliest snapshot. For each object in each snapshot, the CBR tries to find out its type based on the Description Rules, defined by a domain expert. The CBR system measures the similarity between each case stored in the database and the new object, according to their attribute values. Expressed as a real number between 0.0 (no similarity) and 1 (equality), similarity is calculated for each case in the database according to the attribute values. The software recovers the best match, shows it to the expert for confirmation, and stores the confirmed solutions in *the Typed Geospatial Objects Database*. After processing all the information from the first snapshot, the system recovers all objects from the next snapshot in the *Geospatial Objects Database*. It describes them according to the Description Rules and stores them in the *Typed Geospatial Objects Database*.

The second phase of the CBR-based system takes the objects from two consecutive snapshots of the Typed Geospatial Objects Database to describe their evolution. The specialist verifies if the objects are neighbors and the system compares the objects from the two consecutive snapshots according to the rules of the Evolution Cases Database. These rules consider the objects' spatial relationships to find out if two

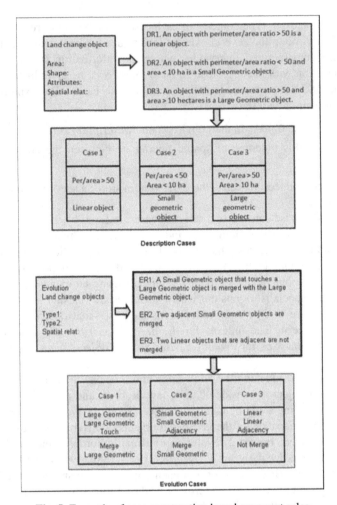

Fig. 5. Example of case construction based on expert rules

objects should be merged in agreement with their description. The system creates the history of each object and stores it in the History Objects Database. The attributes of the History Objects Database are:

1. New Object: the new identification for the object created.

2. FatherObject1: identification of the first object that generated the new.

3. YearObject1: year of the first object that generated the new.

4. FatherObject2: identification of the second object that generated the new.

5. YearObject2: year of the second object that generated the new.

6. Result: the new description for the new object.

7. Year result: the year in which the new object was created.

8. New area: the area of the new object given by the sum of the areas of the objects that were merged.

These attributes keep the origin of the new object allowing the recovery of its history. Considering that the snapshots are stored in increasing temporal order, taking time as a sequence $T = \{1, ..., n\}$ the evolving process can be described in the following steps:

1. Let $T = 1$.

2. Take the objects from time T from the Geospatial Objects Database. Describe these objects according to the Description Cases Database. Store the results in the Typed Geospatial Objects Database.

3. Take the objects from time T+1 from the Geospatial Objects Database. Describe these objects according to the Description Cases Database. Store the results in the Typed Geospatial Objects Database.

4. Compare the objects of times T and T+1 using the Evolution Cases Database. Evolve the objects if possible. Store the results in a History Objects Database.

5. If there are further snapshots in the Geospatial Objects Database, make $T = T+1$ and go to step 2 above. Otherwise, exit the program.

The example shown in Figure 5 is a simple one, where the cases are mutually exclusive and simple ad-hoc rules could solve the problem without resorting to CBR. Many cases are not mutually exclusive and require reasoning techniques which are more complex. To better explain the possible uses of the proposed technique, we present a case study using real data in the next section.

5 Land Change Objects in Brazilian Amazonia: A Case Study

This section presents a case study about extraction of the history of geospatial objects associated to deforestation areas in the Brazilian Amazonia rainforest. The motivation is the work carried out by the National Institute for Space Research (INPE). Using remote sensing images, INPE makes yearly assessments of the deforestation in Amazonia region. INPE's data show that around 250,000 km^2 of forest were cut in Amazonia from 1995 to 2007 [14].

Given the extent of deforestation in Amazonia, it is important to figure out the agents of deforestation. We need to assess the role and the spatial organization of the different agents involved in land change. Our idea is to associate each land change patch, detected in a remote sensing image, to one of the agents of change. Extensive fieldwork points out the different agents involved in land use change (small-scale farmers, large plantations, cattle ranchers) can be distinguished by their different spatial patterns of land use [15] [8]. These patterns evolve in time; new small rural settlements emerge and large farms increase their agricultural area at the expense of the forest. Farmers also buy land from small settlers to increase their property for large-scale agriculture and extensive cattle ranching. Therefore, CBR system will aim to distinguish land change objects based on their shapes and spatial arrangements.

We selected a government-planned rural settlement called Vale do Anari, located in Rondônia State, in Brazilian Amazonia Tropical Forest. This settlement was

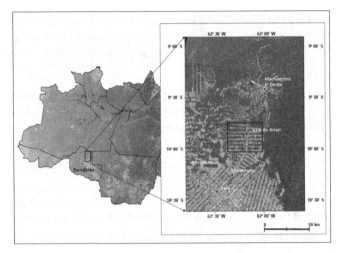

Fig. 6. Location of the study area. The Brazilian Amazonia is on the left, and the Vale do Anari area in the state of Rondônia is on the right. Light-coloured areas indicate deforestation.

established by INCRA (Colonization and Land Reform National Institute), in 1982, with lots of approximately 50 ha (see Figure 6). The third author had carried out fieldwork on the area [16, 17]. In their work, Silva et al [8] used a decision-tree classifier to describe shapes found in land use maps extracted from remote sensing images. They associated these shape descriptions to different types of social agents involved in land use change. Silva et al [8] work did not find out how individual objects evolved, but presented their results comparing the overall types of objects found in each snapshot. They classified deforestation patterns as Linear (LIN), Irregular (IRR) and Geometric (GEO) in *Vale do Anari* region [8]. These objects associated a deforestation patterns were the input of our system. In our study, we distinguish three types of land change objects: *Small Lot* (LOTS), *Along Road Occupation* (AR) and *Concentration Areas* (CON). The main characteristics of those objects are:

1. *Along Road Occupation*: Small settlement household colonists living on subsistence agriculture or small cattle ranching. Their spatial patterns show up as *linear* patterns following planned roads built during earlier stages of colonization.
2. *Small Lot*: Small household colonists associated to settlement schemes living on subsistence agriculture or small cattle ranching. Their spatial patterns show up as *irregular* clearings near roads, following parcels defined by the planned settlement.
3. *Concentration*: Medium to large farmers, associated to cattle ranches larger than 50 ha. This pattern results from the selling of several 50 ha lots to a farmer aiming to enlarge his property. Their spatial patterns is *geometric*, close to roads or population nuclei.

The *Description Rules* (DR) for deforestation objects in our case study are:

DR1. "A geometric spatial pattern is an object of type land concentration".

DR2. "An irregularly shaped pattern that touches a road is an object of type along road occupation".

DR3. "An irregular spatial pattern doesn't touch a road is an object of type small lot".

DR4. "A linear spatial pattern that touches a road is an object of type along road occupation".

DR5. "A linear spatial pattern that doesn't touch a road is an object of type small lot".

A subset of the deforestation objects in the Vale do Anari is shown in Figure 7. The sequence starts with objects representing 1982-1985 deforestation on the right side.

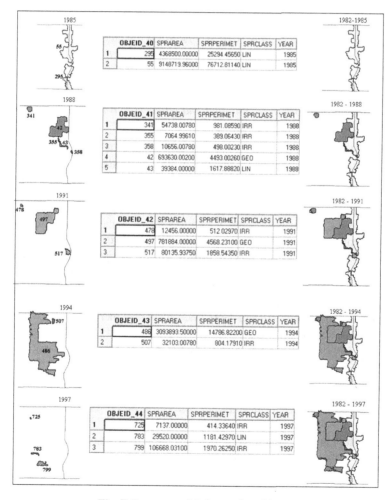

Fig. 7. Sequence of deforestation objects

The next set of *deforestation objects* represents new deforested areas detected during the 1982-1985 period and so forth. These three year snapshot show how deforestation occurred; the objects' labelling was confirmed by experts on deforestation domain. On the left side of Figure 7 the deforestation objects detected in the intervals of three years are shown and linked to an attribute table by an identification number.

The next step is defining the evolution rules. These rules depend on the object´s type and its relation with adjacent objects. An object of type *along road occupation* does not evolve, since it signals the start of the occupation. When two *small lots* touch each other, they are merged into a new *small lot*. When a *concentration* touches a *concentration* or a *small lot*, they are merged and the result is a new *concentration*. A small lot with area greater than 50 ha represents the results of small lots that were merged. If a *concentration* touches a *small lot* with area greater than 50 ha, it doesn't evolve. The evolution rules are:

ER1 – *"Two adjacent land concentration are merged into a land concentration"*.

ER2 – *"An object of type along road occupation is not merged with other objects"*.

ER3 – *"Two adjacent small lots are merged into a small lot"*.

ER4 – *"A small lot with area < 50 ha adjacent to a land concentration is merged with it and the result is a land concentration"*.

ER5 – *"A small lot with area >= 50ha and adjacent to a land concentration is not merged with other object."*

The CBR system builds the *Description Cases Database* using the *description rules* and the *Evolution Cases Database* using the *evolution rules*. Then, it considers all deforestation objects using the procedure described in Section 4 above. For each new object, it looks for a similar case in *Description Cases Database* to define its type. The next step is applying the evolution rules. Given an object´s type and spatial arrangements, the CBR software looks for similar cases in the *Evolution Cases Database*. Based on these cases, it establishes the history of each object, including the originating objects (if the new objects results from a merge operation). The results produced for a sample of the *deforestation objects* are presented in Figure 8.

The report of the CBR system shows how deforestation objects evolved. Until 1991, no objects evolved due to rule ER2: "*An object of type along road occupation is not merged with other objects*". In 1991, the object 478 merged with the object 341 following rule ER3 ("Two adjacent *small lot* objects are merged and the new object is a *small lot*") and the result is the object 1. Also in 1991, object 497 merges with object 42 according to rule ER1, (*"Two adjacent land concentration objects are merged and the new object is a land concentration"*), creating object 2. In 1994, land concentration object 486 appears and merges with object 43 following rule ER4 (*"A small lot with area < 50 ha adjacent to a land concentration object is merged with it and the result is a land concentration object".*), creating object 3. In the same year, object 2 merges with object 517 again following rule ER4, creating object 4. In 1994 object 3 merges with object 355, following rule ER4, creating object 5. Still 1994, object 4 merges with object 5, following rule ER1, creating object 1, that is again expanded, producing object 7, which merges with object 1. In 1997, object 7 merges with objects 725, creating object 8. Then it merges with object 783, creating object 9, and finally merges with object 799, producing object 10. The CBR system was thus able to show

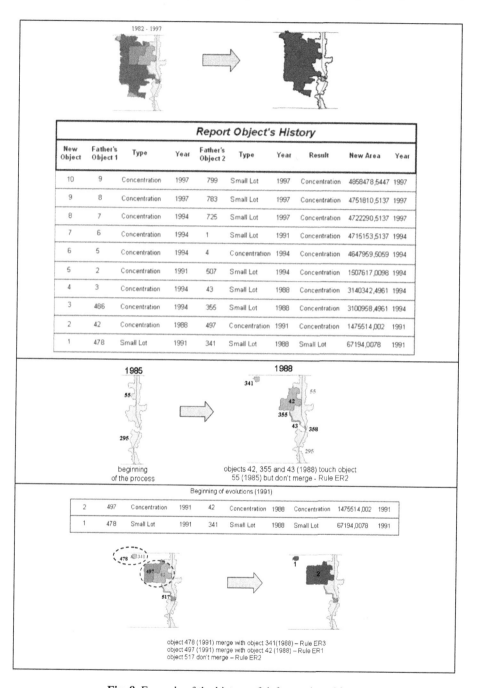

Fig. 8. Example of the history of deforestation objects

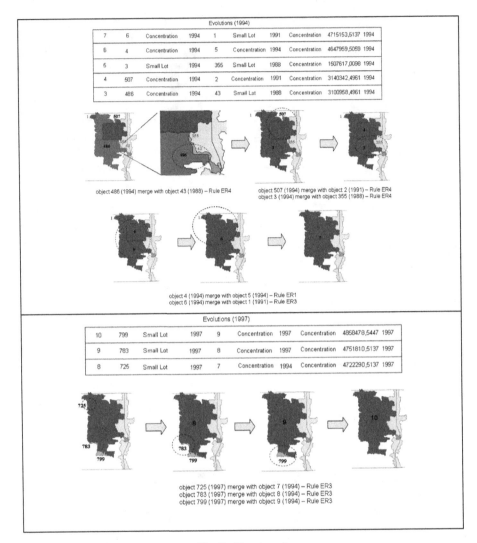

Fig. 8. (*Continued*)

how land concentration occurred in the region, showing that the government plan for settling many colonists in the area has been largely frustrated [16]. The process of land concentration in the *Vale do Anari* settlement described by the CBR system matches what was noted in the interviews performed during fieldwork [16].

6   Conclusions

In this paper, we dealt with evolving objects. We are interested in cases where the simple rules of merging and splitting are not enough to describe their evolution, since such evolution depends on the object´s types. We propose a method that uses previous cases as well as knowledge elicited from a specialist as the main sources of knowledge used to solve new problems. A contribution of our research is the definition a case-based reasoning (CBR) method to describe the object´s type and find out how geospatial objects evolve. Experimental results for the Brazilian Amazonia Tropical Forest corroborate the effectiveness of our proposal. The approach of using typed geospatial objects and evolution rules contributes to solve the problem of automatically modelling and describing the history of evolving geospatial objects.

Use of the CBR software for describing object evolution follows form the work of Silva et al. [13] that developed a method for distinguishing patterns of land use change based on their shapes in static timestamps. Their work did not discuss how spatial patterns evolve in time. The current work addresses the problem of tracking changes during an object's lifetime, based on type-specific evolution rules. In our experiments using case-based reasoning (CBR), we were able to obtain the rules for object evolution and to describe how geospatial objects evolve. The CBR technique proved to be a simple and useful approach to set up the rules for land change trajectories.

In our application domain, CBR presented a satisfactory result, since the knowledge base had only a few cases, which were presented to the expert in an organized way. When there are many data types and different cases, the knowledge base should be generated carefully to avoid conflicting and inconsistent interpretations. Additionally, despite advanced techniques for case indexing and retrieval (neural networks, genetic algorithms), a knowledge base with many cases can have a slow and bad performance. In such cases, the CBR software needs to include adaptation and learning techniques, which also detect inconsistencies in the rules. In this case, the rules would be changed according to the expert´s reaction to examples being presented to him. Adaptation and learning are complex and error-prone techniques that, if not done properly, may result in further inconsistencies in the knowledge base. Therefore, many CBR softwares do not provide an adaptation and learning facility. They simply recover the most similar case and make the solution available for the specialist to determine if it solves his matching problem.

Our experience shows that CBR-based techniques are useful and simple to set up for recovering the history of evolving geospatial objects when there are few types and clear-cut rules for object description and evolution. When there are many types and the evolution rules are complex, the CBR software needs to be carefully designed, and should include a learning phase and techniques for detecting inconsistencies and conflicts. In our work, we consider the CBR software we designed in promising enough; so we plan to extend these CBR techniques for improving the study of land cover change evolution in the Brazilian Amazonia region.

References

1. Frank, A.: Ontology for Spatio-temporal Databases. In: Koubarakis, M., Sellis, T. (eds.) Spatio-Temporal Databases: The Chorochronos Approach, pp. 9–78. Springer, Berlin (2003)
2. Worboys, M.: Event-oriented approaches to geographic phenomena. International Journal of Geographical Information Science 19(1), 1–28 (2005)
3. Grenon, P., Smith, B.: SNAP and SPAN: Towards Dynamic Spatial Ontology. Spatial Cognition & Computation 4(1), 69–104 (2003)

4. Galton, A.: Fields and Objects in Space, Time, and Space-time. Spatial Cognition and Computation 4(1), 39–68 (2004)
5. Goodchild, M.F., Yuan, M., Cova, T.J.: Towards a general theory of geographic representation in GIS. International Journal of Geographical Information Science 21(3), 239–260 (2007)
6. Bittencourt, O., et al.: Rule-based evolution of typed spatiotemporal objects. In: VIII Brazilian Symposium in Geoinformatics, GeoInfo 2007, Campos do Jordão, Brazil (2007)
7. Mota, J.S., et al.: Applying case-based reasoning in the evolution of deforestation patterns in the Brazilian Amazonia. In: Proceedings of the 2008 ACM Symposium on Applied Computing, SAC 2008, Fortaleza, Ceara, Brazil, March 16 - 20, 2008. ACM, New York (2008)
8. Silva, M.P.S., et al.: Remote Sensing Image Mining: Detecting Agents of Land Use Change in Tropical Forest Areas. International Journal of Remote Sensing 29, 4803–4822 (2008)
9. Hornsby, K., Egenhofer, M.: Identity-Based Change: A Foundation for Spatio-Temporal Knowledge Representation. International Journal of Geographical Information Science 14(3), 207–224 (2000)
10. Medak, D.: Lifestyles. In: Frank, A.U., Raper, J., Cheylan, J.-P. (eds.) Life and Motion of Socio-Economic Units. ESF Series. Taylor & Francis, London (2001)
11. Cardelli, L., Wegner, P.: On Understanding Type, Data Abstraction, and Polymorphism. ACM Computing Surveys 17(4), 471–552 (1985)
12. Kolodner, J.L., Jona, M.Y.: Case-based reasoning: An overview: Technical Report 15, Northwestern University (June 1991)
13. Aamodt, A., Plaza, E.: Case-based reasoning: foundational issues, methodological variations and system approaches. AI Communications 7(1), 39–59 (1994)
14. INPE, Monitoramento da Floresta Amazônica Brasileira por Satélite (Monitoring the Brazilian Amazon Forest by Satellite), Report, São José dos Campos: INPE (2005)
15. Lambin, E.F., Geist, H.J., Lepers, E.: Dynamics of land-use and land-cover change in Tropical Regions. Annual Review of Environment and Resources 28, 205–241 (2003)
16. Escada, M.I.S.: Evolução de Padrões da Terra na Região Centro-Norte de Rondônia, Instituto Nacional de Pesquisas Espaciais: São José dos Campos, p. 164 (2003)
17. Souza, L.S., Escada, M.I.S., Verbug, P.: Quantifying deforestation and secondary forest determinants for different spatial extents in an Amazonian colonization frontier (Rondonia). Applied Geography (Sevenoaks) (in press)

Composing Models of Geographic Physical Processes

Barbara Hofer and Andrew U. Frank

Department of Geoinformation and Cartography,
Vienna University of Technology, 1040 Vienna, Austria
{hofer,frank}@geoinfo.tuwien.ac.at

Abstract. Processes are central for geographic information science; yet geographic information systems (GIS) lack capabilities to represent process related information. A prerequisite to including processes in GIS software is a general method to describe geographic processes independently of application disciplines. This paper presents such a method, namely a *process description language*. The vocabulary of the process description language is derived formally from mathematical models. Physical processes in geography can be described in two equivalent languages: partial differential equations or partial difference equations, where the latter can be shown graphically and used as a method for application specialists to enter their process models. The vocabulary of the process description language comprises components for describing the general behavior of prototypical geographic physical processes. These process components can be composed by basic models of geographic physical processes, which is shown by means of an example.

Keywords: geographic physical processes, process modeling, process language, GIS.

1 Introduction

Geography is investigating distributions of objects, people, goods, etc. in space [2]. Distributions are subject to change, which is caused by processes. Getis and Boots define spatial processes as "tendencies for objects to come together in space (agglomeration) and to spread in space (diffusion)" [11, p.1]. Examples for processes of interest in geography are the migration of people, the spread of diseases, the movements of goods, the flow of water, the transport of sediments, etc. [2,5]. Spatial processes are also studied in disciplines like biology, ecology, hydrology, economics, etc. Numerous approaches for process modeling have been developed and implemented in specialized modeling tools. The knowledge of professional modelers is considerable and models are becoming increasingly realistic and complex. The goal of modeling applications is generally a prediction of effects of processes.

Geographic information systems (GIS), the tools used in geography, play a supportive role for process modeling; they are mostly used for data management

K. Stewart Hornsby et al. (Eds.): COSIT 2009, LNCS 5756, pp. 421–435, 2009.

and for visualization of results calculated by modeling tools. The interoperation between GIS and modeling tools is hindered by a lack of capabilities to represent process related information on the GIS side. The integration of time and process in GIS is one of the unsolved issues of geographic information science [20,7,12].

The integration of sophisticated process modeling capabilities that address the specialized methods of the different models is impossible. Instead, GIS need to be complemented with a basic ability to represent processes in order to enhance the interoperability with process modeling tools. A first step towards GIS with process handling capabilities, requires the identification of prototypical process behaviors across disciplines. A second requirement is the development of a general method to describe geographic processes and to compose process components to models. Our proposal to meeting these requirements is a *process description language* for geographic physical processes (section 3). The hypothesis of our work is that mathematical languages can be used for describing and composing models of the general behavior of geographic physical processes.

A fundamental requirement for a process description language is the identification of a vocabulary that allows entering the particularities of processes. The focus of the proposed process description language is on the description of the general behaviors of geographic physical processes. The restriction to geographic physical processes, which are a subset of physical processes, allows the reuse of existing knowledge on physical process modeling (section 2). Physical models can be formulated in mathematical languages such as (partial) differential equations. An analysis of basic partial differential equations lead to the identification of prototypical processes (section 4). The contribution of the process description language is the linkage between mathematical formulations of prototypical process behaviors and concise examples of geographic physical processes. For this purpose processes are conceptualized as block models, which are expressed with difference equations (section 5). Block models have the advantage that they can be graphically depicted for supporting the modelers at their task. The application of the process description language to modeling and composing process models is shown by means of an example (section 6).

2   Geographic Physical Processes and GIS

Processes are sequences of events that are connected by a mechanism; they lead to a recognizable pattern [11,28]. The mechanism may be initiated by different kinds of forces such as physical, social, or political forces [11]. *Geographic* or *spatial* processes are processes that shape distributions of elements in space; they create spatial structures.

Geographic processes can be social or physical processes, which are two ontologically different groups of processes [10]. Modeling applications using geographic information systems (GIS) show an emphasis on environmental, i.e., physical processes [4]. We consider physical processes that are of interest in geography, geographic physical processes, a good start for developing a process description language.

An example for a geographic physical process is the dispersion of exhaust fumes of a factory. The question of interest in relation to this process is where the areas are that are most affected by the exhaust fumes [32]. In order to answer this question, the spreading of the exhaust fumes in the atmosphere has to be modeled. From the moment the exhaust fumes are released to the air from the factory's chimney, they spread continuously and are moved by air currents. The exhaust fumes affect the air quality in a region surrounding the factory. A model of this process is discussed in section 6. Other examples for geographic physical processes are water runoff, groundwater flow, sediment transport, and hill slope erosion.

Geographic physical processes are a subset of physical processes and share their characteristics. From an ontological point of view, physical processes are subject to material causation in contrast to social processes that are following information causation [9]. Information causation acting in social processes is not limited to temporal or spatial neighborhood. In case of material causation the energy transmitted from one unit to the next corresponds to the gain of energy in the receiving unit. This conservation of energy or some other property like mass is a common principle of physical processes. Physical processes establish the physical reality, which is continuous [13]. In addition, physical processes are considered to be local processes; this means that their influences are restricted to the neighborhood. The spectrum of interest in geography "excludes quantum or relativistic effects and is thus rigidly Newtonian" [12, p.1].

The integration of processes in GIS is a longstanding question of geographic information science [20,7,12]. A key difficulty regarding the integration of GIS and process models is the static nature of GIS. Kavouras [19, p.50] recognizes in GIS a "... lack of a concrete theoretical foundation, which among others, has not found acceptable ways to represent generically data, processes, and data on flows and interactions associated with socio-economic applications". There are reasons to integrate process models and GIS, despite all conceptual differences [4].

Various approaches have been presented aiming either at extending GIS with time [20,17,35] or at integrating process models in GIS [33,36,29,34]. Time-oriented approaches generally focus on objects and their change [17,35]. Theories developed from this point of view are not generally applicable to geographic physical processes, because they are continuous processes with field-like characteristics. Work by Yuan [36] refers to phenomena with both, object and field characteristics, and applies to the analysis of rainstorms. General approaches for integrating geographic phenomena in GIS are, e.g., PCRaster [33], the vector map algebra [34], and the *nen* data model [28,29].

The integration of process models in GIS is referred to as embedded modeling in GIS. Three other levels of integration of modeling tools and GIS are differentiated, which are: loose coupling, tight coupling, and modeling tools integrating GIS functionality [30,4,6]. Loose coupling means that GIS are used for generating the input for modeling software and for visualizing the calculated results. The option of tight coupling connects GIS and modeling software through a common interface; this is only achieved for single models. Highly specialized modeling tools may integrate the GIS functionality they require in their system. Mitasova and Mitas [24] additionally mention the group of GIS and web-based models, where widely used models are provided as web applications together with required input data and parameters.

A series of applications are implementing the integration of GIS and process models on the different levels. The applications are often successful at integrating a particular model of a process from the viewpoint of a certain discipline. In contrast to the focus of these applications, we aim at identifying general functionality required to extend GIS with general process modeling capabilities.

3   Why Develop a Process Description Language?

The potential of GIS regarding the spatial aspects of process modeling and analysis is not yet exploited. We take a look at the requirements of process modeling for identifying areas of improvement on the GIS side. Improving the capabilities of GIS regarding process modeling can lead to a better interoperability between GIS and process modeling tools in the long run.

In the Virtual GIS project [3] criteria for the development of spatial modeling systems have been identified. These criteria include a graphical user interface, a component for the interactive development of scenarios, functionality for spatial analysis and visualization, and "a generic system that operates as a toolbox independent of a specific domain" [8, p.3]. To achieve such a system, we have to abstract from specifics of different disciplines and from details of the quantitative analysis. Our proposal is the development of a *language* to describe and to compose models of prototypical geographic processes; with a restriction to geographic physical processes.

The process description language contributes two things: a) the identification of prototypical processes and b) a method to generally describe and compose models of geographic physical processes. Composition of model components is an important feature for process modeling as previously mentioned by [22,26]. The qualitative description focuses on general principles of processes, which of course does not replace quantitative process modeling. The process description language allows the generation of *qualitative sketch models* of processes describing the general behavior of the processes. These sketch models are created with the language not requiring deep mathematical knowledge. The models can serve as input for existing modeling tools and facilitate process modeling for non-expert modelers.

We see the proposed process description language as a layer on top of existing modeling tools for spatial processes such as PCRaster [33] and the vector map algebra [34]; a layer with a more rigorous mathematical foundation in partial differential equations. The motivation of developing the process description language is an improvement of GIS functionality, which is a distinguishing feature from modeling tools such as the Spatial Modeling Environment (SME) [22], SIMILE [25], or the 5D environment [23] developed in other disciplines than the GIS discipline. In addition, modeling tools developed for modeling primarily non-spatial processes that are based on STELLA can solve differential equations but no partial differential equations.

4   A Mathematical Model and Prototypical Processes

"Process models generally express theories predicting the nature of the exchange of energy and mass within systems, over time" [27, p.361]. A model of a physical process comprises a configuration space, interactions between the elements of the configuration space, governing equations and constitutive relations [15]. The configuration space contains information about the elements of the modeled system together with physical parameters, initial and boundary conditions, etc. The interactions between the elements of a system are related to governing equations. The governing equations of physical models are based on natural laws. Constitutive relations are required to fully describe the model of the process. The practice of process modeling is well summarized in the following statement: "Because of the complexity of the Earth systems, process-based modeling of geospatial phenomena relies in practice on the best possible combination of physical models, empirical evidence, intuition and available measured data" [15, p.2].

The governing equations of physical phenomena are based on natural laws that generally refer to the conservation of a property such as mass or energy [16]. Fundamental conservation laws state that the total amount of, e.g., mass in a system remains unchanged. The amounts of a quantity going in, going out, and being created or destroyed in a region, have to correspond to the amount of change in a certain region [21]. The general conservation law consists of three main components: the component specifying the change of the concentration of a substance $u(x, t)$ over time, the flow of the substance $\phi(x, t)$, and sources or sinks in the system $f(x, t)$. The specification of these three components is sufficient for describing the general behavior of a series of physical phenomena [21]:

$$\frac{\partial u(x, t)}{\partial t} + \frac{\partial \phi(x, t)}{\partial x} = f(x, t). \tag{1}$$

Equation 1 is a general conservation law in one spatial dimension formulated as a partial differential equation (PDE). PDEs are one possible mathematical language for formulating models of physical and thus geographic physical processes. A PDE is widely applicable, because "it can be read as a statement about how a process evolves without specifying the formula defining the process" [1].

The terms $u(x, t)$ and $\phi(x, t)$ in Equation 1 are unknowns, when assuming that sources and sinks are given [21]. For describing the unknowns, an additional equation is required: a constitutive relation [21]. These constitutive relations generally define the flow term of the conservation equation. They give a rule to link flow $\phi(x, t)$ and concentration $u(x, t)$ of a substance [16]. The definition of the constitutive relations is based on physical characteristics of the system and often founded on empirical evidence.

Two commonly differentiated kinds of flow, which are described by constitutive relations, are advective flow ϕ_A and diffusive flow ϕ_D. Advection (transport by flow) and diffusion (random spread of particles) are important kinds of processes also for geography. The flow of a substance or object due to advection is specified by Equation 2; the flow of a substance due to diffusion is modeled by Equation 3.

$$\phi_A = u(x,t) * v, \text{ with } v...\text{flow velocity}. \tag{2}$$

$$\phi_D = -k * \frac{\partial u(x,t)}{\partial x}, \text{ with } k...\text{diffusion constant}. \tag{3}$$

The flow term ϕ in the conservation equation (Equation 1) is composed of the advective component ϕ_A and the diffusive component ϕ_D:

$$\phi = \phi_A + \phi_D. \tag{4}$$

Depending on the process, advective and diffusive components of flow can be present or either of them can be zero. Inserting the constitutive relations describing the flow terms in the conservation law, under consideration of the possible combinations of terms, leads to a series of equations: the advection equation, the diffusion equation, and the advection-diffusion equation.

Advection equation: The advection equation (Equation 5) describes the bulk movement or flow of a substance in a transporting medium [21]. The direction and velocity of flow are determined by the flow direction and velocity v of the transporting medium. The equation is an evolution equation that shows how the process evolves over time. An example for an advection process is the transport of pollen by wind.

$$\frac{\partial u(x,t)}{\partial t} + v * \frac{\partial (u(x,t))}{\partial x} = f(x,t) \tag{5}$$

Diffusion equation: The diffusion equation (Equation 6) describes a process where a substance spreads from areas of higher concentrations of the substance or areas with higher pressure to areas with lower concentrations or pressure [21]. Diffusive flow ϕ_D is specified by flow down the concentration gradient, which is expressed by the minus sign in Equation 3. The motion of the particles is random. The diffusion equation is again an evolution equation. An example for a diffusion process is a contaminant diffusing in standing water.

$$\frac{\partial u(x,t)}{\partial t} - k * \frac{\partial^2 u(x,t)}{\partial x^2} = f(x,t) \tag{6}$$

Advection-diffusion equation: In the case of an advection-diffusion process, diffusive and advective flows take place. An example is the spread of a toxic liquid in a lake: the liquid diffuses from areas of higher to those of lower concentrations of toxins and in addition, the water current of the lake moves the toxic liquid. In the case of this process, both types of flows are present in the conservation equation and lead to an advection-diffusion equation (Equation 7).

$$\frac{\partial u(x,t)}{\partial t} + v * \frac{\partial \left(u(x,t)\right)}{\partial x} - k * \frac{\partial^2 u(x,t)}{\partial x^2} = f(x,t) \qquad (7)$$

Steady-state equation: Important are also the steady-state versions of the conservation equation, where the term including time is zero ($\frac{\partial u(x,t)}{\partial t} = 0$) and the source is a function of space $f(x)$ [16]. In a steady-state process the available amount of a quantity remains unchanged; what changes is the amount of the quantity occupying a position in space over time. The equation models steady-state flow in fields. This kind of equations is known as equilibrium equation. The following equation (Equation 8) shows the steady-state Poisson equation in two dimensions. The Poisson equation contains a source or sink term, which the second steady-state equation, the Laplace equation, does not contain.

$$\frac{\partial^2 u(x,y,t)}{\partial x^2} + \frac{\partial^2 u(x,y,t)}{\partial y^2} = f(x) \qquad (8)$$

The list of equations is usually complemented by the wave equation, which describes the propagation of waves such as sound waves or water waves. The discussion of the wave equation would exceed the scope of this paper and is left for a future report on the topic.

PDEs allow an analysis of process behavior from a theoretical point of view. The identified PDEs provide the core for the establishment of the vocabulary of the process description language. We have to show that the equations modeling prototypical processes can be linked to geographic physical processes. This task is achieved by the process description language.

5   A Process Description Language

The fundamental laws governing the models of geographic physical processes are conservation laws and flow laws. The specification of these laws describes the qualitative behavior of a process. A summary on kinds of processes described by these laws was given in section 4. Additional requirements for the specification of a model of a physical process are the definition of the configuration space including parameters, boundary and initial conditions etc. [15, c.f. section 4]. These aspects of process models are left aside in the work presented in this paper. The focus is on the key task of the process description language: the assignment of the equations modeling general process behavior to a certain geographic physical process.

For the purpose of selecting an appropriate equation for modeling a certain process, we conceptualize geographic physical processes with deterministic block models [32]. Blocks can be aligned next to each other, on top of each other, or on top and besides each other, just as required to represent a process in 1D, 2D, or 3D in the model. This kind of representation is comparable to raster and voxel representations in GIS. Block models are useful for conceptualizing geographic physical processes; they describe the behavior of a process with respect to blocks of finite size. For the specification of the process behavior we define the storage of a substance, the flow of the substance and sources and sinks in the system. We will see that there are geographic physical processes, whose behavior corresponds to the behavior of processes described with the equations from Section 4.

From a mathematical point of view the approach using block models is closely related to finite difference methods. The formulation of the block models is done mathematically with difference equations. Differential and difference equations are seen as equivalent languages for expressing process models. Differential equations are continuous representations of a phenomenon; difference equations are a discretization of differential equations. Linking PDEs to difference equations and vice versa is always possible for the basic, linear equations we are working with. A previous account of the connection between PDEs and difference equations has been given in [14]. The advantage of using blocks for establishing a model is that the model can be visualized.

General conservation laws apply to block models as to continuous representations of phenomena. The formulation of the conservation equation as a difference equation in three dimensions is ($\triangle$ labels a difference):

$$\frac{\triangle F}{\triangle t} + \frac{\triangle \phi}{\triangle x} + \frac{\triangle \phi}{\triangle y} + \frac{\triangle \phi}{\triangle z} = f \,. \tag{9}$$

The change in the density or concentration of a substance F stored in a block over a time interval is expressed by $\frac{\triangle F}{\triangle t}$. This change is caused by flows across the boundaries of a block (e.g., flow in x-direction $\frac{\triangle \phi}{\triangle x}$) and sources or sinks ($f$) in the system [18]. The flow in a model is captured by describing the flow between two neighboring blocks; it is specified by the gradients of the flux terms ϕ in x, y, and z direction. The gradients of the flux terms can either be negative and refer to flows out of a block or positive and refer to flows into a block. Figure 1 shows a block and the flux terms in all directions of the block. $\phi_{x|x}$ refers to the flow across the left border of a block and $\phi_{x|x+\triangle x}$ refers to the flow across the right border of a block in positive x, respectively y and z, direction. The gradient of the flux terms in x-direction, which gives the flow in x-direction, is defined by (Equation 10):

$$\frac{\triangle \phi}{\triangle x} = \frac{\phi_{x|x+\triangle x} - \phi_{x|x}}{\triangle x} \,. \tag{10}$$

The general definition of the flow across a face has to be extended by the specification of the type of flow taking place, which depends on the ongoing process. This means that in Equation 9, the flux terms ϕ have to be replaced by the

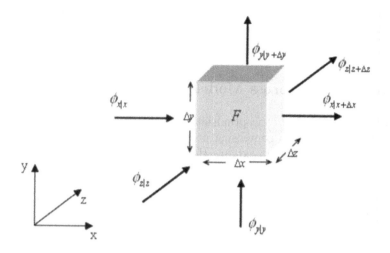

Fig. 1. Flux terms in three dimensions in respect to a block [31]

specific type of flow going on. We discuss here three possible types of flow (c.f. section 4):

- advective flow: $\phi_A = F * v$,
- diffusive flow: $\phi_D = -k * \frac{\triangle F}{\triangle x}$,
- and advective-diffusive flow: $\phi = \phi_A + \phi_D = F * v - k * \frac{\triangle F}{\triangle x}$.

The outcome of this procedure is a difference equation that describes the general behavior of an ongoing process. The kinds of equations considered in this paper were discussed in section 4. The presented approach produces a language to describe the processes, which so far are missing in a GIS. The language consists of:

1. the state variable (e.g., F) referring to data in a GIS,
2. operations applicable to the state variables (primarily partial differences like $\frac{\triangle F}{\triangle x}$),
3. and multiplicative constants to form
4. equations.

The language has further the important composability property: simpler descriptions can be composed to more complex descriptions. The example in section 6 shows that first, a simple model can be created that considers only diffusive flows; this model can be extended for a component referring to advective flows. Adding diffusive and advective flows gives the total amount of flow in the system.

   The description of geographic physical processes resulting from the use of the process description language is qualitative; it can be seen as a *sketch-model* of a process. A detailed quantitative evaluation of processes is not in the foreground of our work. However, because of the formulation of the models with difference equations and the possibility to change to a representation as differential equations, the output of the process description language can serve as input for a quantitative analysis of processes.

6   Composing Process Models - An Example

A strength of the chosen conceptualization of the processes based on blocks is the intuitive approach to composing process models. Adding components and links to existing components extends the model. The illustrative example of a geographic physical process that we model with the process description language is the dispersion of exhaust fumes of a factory (c.f. section 2).

First, we build a simple model that considers the spread of exhaust fumes with density $F(x, y, z, t)$ in the atmosphere and a source $f(x, y, z, t)$ alone. The source term is zero except for the location of the smoke stacks. We omit the influence of gravity.

For the block model we need to specify the storage equation and the flow equation that describe how fumes spread in the atmosphere. We assume that the fumes are homogeneously distributed in each block; the density of fumes in a block is an indicator of the ratio between fumes and air. The storage equation defines the change in the density of fumes in a cell over a time period. The rate of change in the concentration of fumes $\triangle F$ over a time interval $\triangle t$, depends on how much exhaust fumes come in and go out of a block. The flow of fumes in and out of a block is defined by a flow law. We assume that the amount of flow between two blocks depends on the difference in fume concentration between the two blocks; fumes spread from areas of high fume density to areas of lower fume density. This characterization corresponds to the behavior of a diffusion

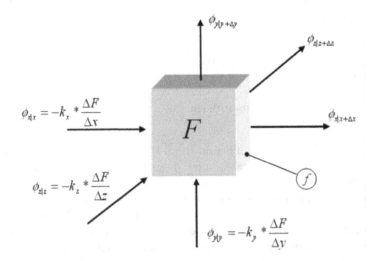

Fig. 2. Diffusive flux terms in three dimensions and source f of fumes

process. The flow taking place is diffusive, the advective component of flow is zero. Figure 2 shows the diffusive flux terms in three dimensions. The diffusive flux term ϕ_D is defined by:

$$\phi_D = -k_x \frac{\Delta F}{\Delta x} - k_y \frac{\Delta F}{\Delta y} - k_z \frac{\Delta F}{\Delta z}. \tag{11}$$

Inserting the specification of ϕ_D (Equation 11) in the general conservation law (Equation 9), leads to a diffusion equation (Equation 12). The diffusion equation, therefore, is a model of the spreading of fumes in the atmosphere.

$$\frac{\Delta F}{\Delta t} - k_x \frac{\Delta^2 F}{\Delta x^2} - k_y \frac{\Delta^2 F}{\Delta y^2} - k_z \frac{\Delta^2 F}{\Delta z^2} = f. \tag{12}$$

The simple model above omits the influence of air currents that have effects on the distribution of exhaust fumes in the air. Therefore, we add a component referring to air currents with velocities in three dimensions $v = (v_x, v_y, v_z)$ to the model.

The change in the concentration of fumes in a block consists again of the amount of fumes coming in and going out of a block. Now we have to consider two kinds of flows; the diffusive flow ϕ_D as described above and the movement of fumes by air currents. The movement of fumes by wind depends on the concentration of fumes in a block and the velocity of the wind in x, y, and z direction: v_x, v_y, and v_z. The fumes are transported by the flow field of air currents; this kind of flow is known as advective flow ϕ_A (c.f. Equation 2). Figure 3 depicts the advective flux terms in reference to a block; the mathematical formulation is given below (Equation 13):

$$\phi_A = F v_x + F v_y + F v_z. \tag{13}$$

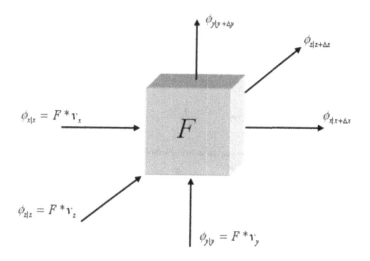

Fig. 3. Advective flux terms in three dimensions describing air currents

We insert the advective flux term ϕ_A (Equation 13) in the conservation law (Equation 9) to get a model of the effects of air currents on fumes (Equation 14):

$$\frac{\Delta F}{\Delta t} + \frac{\Delta(Fv_x)}{\Delta x} + \frac{\Delta(Fv_y)}{\Delta y} + \frac{\Delta(Fv_z)}{\Delta z} = 0. \tag{14}$$

The right side of Equation 14 is zero, because we assume that there are no sources or sinks of air currents in our system. The resulting equation is an advection equation. This equation models the movement of fumes by air currents. To get the complete description of our model that considers the diffusion of fumes and the advection of fumes by wind, we have to compose Equation 12 and 14. We know from Equation 4 that the complete flow in a model consists of the sum of advective flux terms ϕ_A and diffusive flux terms ϕ_D. Combining both types of flows in one equation, leads to an advection-diffusion equation (Equation 15):

$$\frac{\Delta F}{\Delta t} + \frac{\Delta(Fv_x)}{\Delta x} + \frac{\Delta(Fv_y)}{\Delta y} + \frac{\Delta(Fv_z)}{\Delta z} - k_x \frac{\Delta^2 F}{\Delta x^2} - k_y \frac{\Delta^2 F}{\Delta y^2} - k_z \frac{\Delta^2 F}{\Delta z^2} = f. \tag{15}$$

The advection-diffusion equation describes the change in the concentration of fumes in a block under consideration of air currents that move the fumes. Figure 4 shows the total flux terms acting on a block. The modeling process showed that process components can be composed to extend the model of the process. The two kinds of flows taking place in our model of the dispersion of exhaust fumes, can simply be added to give the total amount of flow in the system.

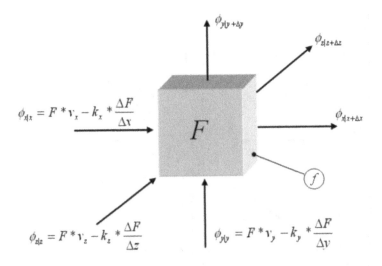

Fig. 4. Total flow in the model of the dispersion of exhaust fumes

The specification of complete models of geographic physical processes requires the definition of a configuration space, boundary and initial conditions, parameters, etc. (c.f. section 4). Adding this information allows the quantitative analysis of the models with, for example, a finite difference or finite element solver. The focus in this paper was on the storage and flow laws describing the behavior of processes; the definition of the configuration space, required data sets, boundary conditions, etc. that complement a process model, yet have to be added. This additional information about the model is, however, implicitly contained in the model description.

In the future, values for parameters, boundary conditions, and initial conditions could come from the data stored in a GIS. In addition, the solution of the difference equation resulting from modeling the process, could be a GIS layer, which gives the distribution of exhaust fumes over time.

7 Results and Conclusions

Mathematical languages can be used for describing and composing qualitative models of geographic physical processes. Based on an analysis of differential equations we identified equations that model prototypical process behavior. By means of deterministic models based on blocks, the process description language links geographic physical processes with these prototypical process equations. The illustrative example of a model for the dispersion of exhaust fumes from a factory showed how a process is qualitatively described and how process components are composed.

The description of processes on a general level, as achieved by the process description language, can enhance the usability of existing modeling tools for non-expert modelers. The sketch model established with the process description language can serve as input for modeling tools that require the description of the process closer to mathematical details of the model equations.

The presented work lays the foundation for a process enriched GIS. The establishment of a process model in a process enriched GIS, could work as follows: The process description language allows the specification of the general behavior of a process. Data on parameters, initial conditions, boundary conditions, etc. can be derived from data available in the GIS. The process description together with data on initial and boundary conditions, parameters, etc. serve as input for, for example, a finite element solver. The quantitative analysis of the process done with the finite element solver results in a layer in the GIS that represents the ongoing process. A series of questions regarding the usage of process modeling in GIS, the assessment of the suitability of the models, the technical realization, etc. are raised by the idea of a process enriched GIS. In the long run such a GIS could become a comprehensive simulation tool.

Acknowledgments. We gratefully acknowledge thoughtful comments of Waldo Tobler and four anonymous reviewers.

References

1. Partial Differential Equation, Encyclopædia Britannica (retrieved April 18, 2007), http://www.britannica.com/eb/article?tocId=9374634
2. Abler, R., Adams, J.S., Gould, P.: Spatial Organization: the Geographer's View of the World, International edn. (1st edn.) Prentice-Hall International, London (1977)
3. Albrecht, J., Jung, S., Mann, S.: VGIS: a GIS shell for the conceptual design of environmental models. In: Kemp, Z. (ed.) Innovations in GIS 4, pp. 154–165. Taylor & Francis, Bristol (1997)
4. Bivand, R.S., Lucas, A.E.: Integrating Models and Geographical Information Systems. In: Openshaw, S., Abrahart, R.J. (eds.) GeoComputation, pp. 331–363. Taylor & Francis, London (2000)
5. Briggs, D., Smithson, P.: Fundamentals of physical phenomena. Routledge (1993); reprinted 1993 edn. first published 1985 by Hutchinson Education
6. Brown, D.G., Riolo, R., Robinson, D.T., North, M., Rand, W.: Spatial Process and Data Models: Toward Integration of Agent-Based Models and GIS. Journal of Geographical Systems 7(1), 25–47 (2005)
7. Burrough, P.A., Frank, A.U.: Concepts and Paradigms in Spatial Information: Are Current Geographic Information Systems Truly Generic? International Journal of Geographical Information Systems 9, 101–116 (1995)
8. Crow, S.: Spatial modeling environments: Integration of GIS and conceptual modeling frameworks. In: 4th International Conference on Integrating GIS and Environmental Modeling (GIS/EM4): Problems, Prospects, and Research Needs, Banff, Alberta, Canada, pp. 1–11 (2000)
9. Frank, A.U.: Material vs. information causation - An ontological clarification for the information society. In: Wittgenstein Syposium, Kirchberg, Austria, pp. 5–11 (2007)
10. Frank, A.U.: Tiers of ontology and consistency constraints in geographic information systems. International Journal of Geographical Information Science (IJGIS) 75(5), 667–678 (2001)
11. Getis, A., Boots, B.: Models of Spatial Processes - An approach to the study of point, line and area patterns. Cambridge Geographical Studies. Cambridge University Press, Cambridge (1978)
12. Goodchild, M.F.: A Geographer Looks at Spatial Information Theory. In: Montello, D.R. (ed.) COSIT 2001. LNCS, vol. 2205, pp. 1–13. Springer, Heidelberg (2001)
13. Hayes, P.J.: The Second Naive Physics Manifesto. In: Hobbs, J.R., Moore, R.C. (eds.) Formal Theories of the Commonsense World. Ablex Series in Artificial Intelligence, pp. 1–36. Ablex Publishing Corp, Norwood (1985)
14. Hofer, B., Frank, A.U.: Towards a Method to Generally Describe Physical Spatial Processes. In: Ruas, A., Gold, C. (eds.) Headway in Spatial Data Handling, Spatial Data Handling 2008, Montpellier France. LNG&C, p. 217 (2008) ISBN: 978-3-540-68565-4
15. Hofierka, J., Mitasova, H., Mitas, L.: GRASS and modeling landscape processes using duality between particles and fields. In: Ciolli, M., Zatelli, P. (eds.) Proceedings of the Open Source GIS - GRASS users conference, Trento, Italy, September 11-13 (2002)
16. Holzbecher, E.: Environmental Modeling Using MATLAB. In: Zannetti, P. (ed.) Environmental Modeling. Springer, Heidelberg (2007)
17. Hornsby, K., Egenhofer, M.J.: Qualitative Representation of Change. In: Hirtle, S.C., Frank, A.U. (eds.) COSIT 1997. LNCS, vol. 1329, pp. 15–33. Springer, Heidelberg (1997)

18. Huggett, R.J.: Modelling the human impact on nature: systems analysis of environmental problems, p. 202. Oxford University Press, Oxford (1993)
19. Kavouras, M.: Understanding and Modelling Spatial Change. In: Frank, A.U., Raper, J., Cheylan, J.P. (eds.) Life and Motion of Socio-Economic Units, ch. 4. GISDATA Series, vol. 8, pp. 49–61. Taylor & Francis, Abington (2001)
20. Langran, G., Chrisman, N.: A Framework for Temporal Geographic Information. Cartographica 25(3), 1–14 (1988)
21. Logan, J.D.: Applied Partial Differential Equations, 2nd edn. Undergraduate Texts in Mathematics. Springer, Heidelberg (2004)
22. Maxwell, T., Constanza, R.: A language for modular spatio-temporal simulation. Ecological Modelling 103, 105–113 (1997)
23. Mazzoleni, S., Giannino, F., Mulligan, M., Heathfield, D., Colandrea, M., Nicolazzo, M., Aquino, M.D.: A new raster-based spatial modelling system: 5D environment. In: Voinov, A., Jakeman, A., Rizzoli, A. (eds.) Proceedings of the iEMSs Third Biennial Meeting: Summit on Environmental Modelling and Software, Burlington, USA (2006)
24. Mitasova, H., Mitas, L.: Modeling Physical Systems. In: Clarke, K., Parks, B., Crane, M. (eds.) Geographic Information Systems and Environmental Modeling. Prentice Hall series in geographic information science, pp. 189–210. Prentice Hall, Upper Saddle River (2002)
25. Muetzelfeldt, R., Massheder, J.: The Simile visual modelling environment. European Journal of Agronomy 18, 345–358 (2003)
26. Pullar, D.: A Modelling Framework Incorporating a Map Algebra Programming Language. In: Rizzoli, A., Jakeman, A.R. (ed.) International Environmental Modelling and Software Society Conference, vol. 3 (2002)
27. Raper, J., Livingstone, D.: Development of a Geomorphological Spatial Model Using Object-Oriented Design. International Journal of Geographical Information Systems 9(4), 359–383 (1995)
28. Reitsma, F.: A New Geographic Process Data Model. Ph.D. thesis, Faculty of the Graduate School of the University of Maryland (2004)
29. Reitsma, F., Albrecht, J.: Implementing a new data model for simulation processes. International Journal of Geographical Information Science (IJGIS) 19(10), 1073–1090 (2005)
30. Sui, D., Maggio, R.: Integrating GIS with Hydrological Modeling: Practices, Problems, and Prospects. Computers, Environment and Urban Systems 23, 33–51 (1999)
31. Sukop, M.: Transient Ground Water Flow (2009), http://www.fiu.edu/sukopm/ENVE279/TransientGW.doc
32. Thomas, R.W., Huggett, R.J.: Modelling in Geography - A Mathematical Approach, 1st edn. Harper & Row, London (1980)
33. Van Deursen, W.P.A.: Geographical Information Systens and Dynamic Models. Phd-thesis, University of Utrecht, NGS Publication 190 (1995)
34. Wang, X., Pullar, D.: Describing dynamic modeling for landscapes with vector map algebra in GIS. Computers & Geosciences 31, 956–967 (2005)
35. Worboys, M.F.: Event-oriented approaches to geographic phenomena. International Journal of Geographical Information Science (IJGIS) 19(1), 1–28 (2005)
36. Yuan, M.: Representing complex geographic phenomena in GIS. Cartography and Geographic Information Science 28(2), 83–96 (2001)

Decentralized Time Geography for Ad-Hoc Collaborative Planning

Martin Raubal[1], Stephan Winter[2], and Christopher Dorr[1]

[1] Department of Geography, University of California, Santa Barbara, USA
raubal@geog.ucsb.edu, chdorr@gmail.com
[2] Department of Geomatics, The University of Melbourne, Australia
winter@unimelb.edu.au

Abstract. For an autonomous physical agent, such as a moving robot or a person with their mobile device, performing a task in a spatio-temporal environment often requires interaction with other agents. In this paper we study ad-hoc collaborative planning between these autonomous peers. We introduce the notion of *decentralized time geography*, which differs from the traditional time-geographic framework by taking into account limited local knowledge. This allows agents to perform a space-time analysis within a time-geographic framework that represents local knowledge in a distributed environment as required for ad-hoc coordinated action between agents in physical space. More specifically, we investigate the impact of general agent movement, replacement seeking, and location and goal-directed behavior of the initiating agent on the outcome of the collaborative planning. Empirical tests in a multi-agent simulation framework provide both a proof of concept and specific results for different combinations of agent density and communication radius.

1 Introduction

This paper studies collaboration between mobile social agents, in particular forms of collaboration that require a physical encounter at some point in space and time. Agents communicate in an ad-hoc, peer-to-peer manner, such that each agent has easy access to local knowledge, but no access to global knowledge. There is no central instance in this network that collects data nor does it provide for centralized time-geographic analysis; analysis, where required, has to be done locally by the distributed agents.

Such agents may be mobile social robots, mobile sensor network nodes, or persons with smart mobile devices. Consider, for example, the following problem: A major accident has happened at an industrial site, leading to a spread of chemicals into the surrounding environment. In order to take measurements of several critical variables, a number of agents with different types of sensors are needed at the site within a specified time frame. Traditionally, these agents are activated and managed centrally, but what if the central emergency command center, or the centralized communication infrastructure, was destroyed by the accident? Let us assume that these agents exist somewhere in the environment, potentially off-site, and have the ability to communicate and collaborate in a peer-to-peer manner. Let us further assume that one agent

K. Stewart Hornsby et al. (Eds.): COSIT 2009, LNCS 5756, pp. 436–452, 2009.

discovers the accident and decides that help is needed. The following questions then arise for this initiating agent: How many agents are needed, and how many can be expected at the site by the upper time limit? How large must the search radius for help be to get those agents together? How difficult is it to reach agents within this search radius? What is the risk of limiting a call for help within these boundaries?

Problems such as the chemical spill require *decentralized collaboration* with local interaction of agents to achieve a common goal. Such collaboration takes place under specific space-time constraints in a heterogeneous environment. These space-time constraints limit theoretically which agents can participate in a collaboration requiring a physical encounter. In addition, communication bandwidth and energy resources (in the worst case agents run out of power) are often restricted. Therefore, an efficient management of the collaboration will consider these constraints in limiting the communication to those agents that are relevant for a given task, thus saving communication bandwidth, battery power of mobile agents, and unnecessary message processing overhead of agents out of collaboration range. For this purpose we propose a novel framework based on the *decentralization of time geography*.

Time geography provides a means for accessibility analysis by considering spatio-temporal constraints. However, classical time-geographic analysis [1, 2] is centralized with complete global knowledge and direct communication between the center and individual agents. An emergency management center would identify agents nearby the spill that can reach the site in time, and contact only those. In contrast, decentralized time-geographic analysis operates on limited local knowledge of the mobile social agents. For example, the agent initiating the chemical spill response collaboration does not know where other agents currently are. Our hypothesis is that (1) time geography can be used locally and decentralized to optimize search and (2) its local application makes communications and computations more efficient compared to centralized problem-solving from a global perspective.

Section 2 reviews related work from time geography and decentralized cooperation. Section 3 develops the decentralized time-geographic framework that agents can use to cooperate locally about message spreading and collaboration planning. Section 4 presents an implementation of the framework in a multi-agent simulation, and Section 5 discusses the results of the simulation experiments. The paper concludes with a summary and open questions.

2 Related Work

In the following we introduce the major ideas behind time geography, provide an overview of peer-to-peer communication and agent collaboration, and describe how these areas relate to our research.

2.1 Time Geography

Agents and resources are available at a limited number of locations for a limited amount of time. Time geography defines the space-time mechanics of locational presence by considering different constraints [1]. The possibility of being present at a specific location and time is determined by the agent's ability to trade time for space,

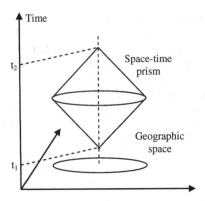

Fig. 1. Space-time prism as intersecting cones

supported by transportation and communication services. *Space-time paths* depict the movement of individual agents in space over time. Such paths are available at various spatial and temporal granularities, and can be represented through different dimensions. All space-time paths must lie within *space-time prisms* (STP). These are geometrical constructs of two intersecting cones [3]. Their boundaries limit the possible locations a path can take (Figure 1). The time budget is defined by $\Delta t = t_2 - t_1$ in which an agent can move away from the origin (the apex at t_1), limited only by the maximum travel velocity and the fact that at t_2 it will be at a certain location (the apex at t_2).

Time geography has been applied to model and measure space-time accessibility in transportation networks [4, 5]. Space-time paths of individuals in networks are limited to movement along edges. The geometry of the STP in a network forms an irregular shape because movement is limited and travel velocity may vary for each edge. Algorithms for calculating the *network time prism* (NTP) can be found in [6] and [7].

2.2 Peer-to-Peer Communication

Peer-to-peer (P2P) communication is ad-hoc communication between distributed agents, without involvement of a dedicated server providing communication services to its clients, or any other hierarchic communication infrastructure. It enables mobile agents to collaborate in an ad-hoc manner provided that they agree on a communication channel and protocol. In a P2P communication network each node is of equal importance. Nodes can take the role of a communication client, receiving messages from, or sending them to other nodes, but they can also provide message forwarding services for the other nodes. P2P communication networks are transient in nature, with nodes entering and leaving the network freely.

A special class of P2P communication is characterized by mobile nodes. For this class the communication is realized wirelessly by radio, which is short-range, due to typically limited bandwidth and on-board energy resources. This means connectivity in mobile networks depends on the physical distance of nodes, which is constantly changing. Communication over larger distances relies on message forwarding and routing. Routing strategies have been studied extensively for static and mobile sensor networks [8, 9], and also for social networks [10].

Parameters defining the connectivity and spread of messages in a peer-to-peer communication network are the communication range and the message forwarding policies. In contrast to classical wireless sensor networks [9], which apply synchronized communication and delayed message forwarding, the types of agents considered in this paper are considered to be 'always-on': broadcasting is possible at any time and message forwarding can happen instantaneously. However, instantaneous (unlimited) message forwarding ('flooding') still reaches only nodes that belong to the connected component of the original sender at the time of broadcasting. Therefore, in mobile networks repeated or opportunistic forwarding are strategies to bridge potential gaps of connectivity, e.g., [11].

Peer-to-peer communication between mobile agents has limited resources, especially bandwidth and on-board battery energy. Accordingly, decentralized algorithms try to minimize the required number of messages to be broadcasted, often at the cost of accuracy.

In the context of the current paper, peer-to-peer communication between mobile agents is complemented by an awareness of the agents of their own position. Agents such as mobile social robots, mobile sensor network nodes, or persons with smart mobile devices are increasingly equipped with positioning technology supporting active or passive locomotion and wayfinding. Tracking positions is an essential factor for time geography, also for decentralized time geography.

2.3 Agent Collaboration

Previous research has focused on technical aspects of P2P collaboration [12] and collaborative, spatio-temporal decision support systems [13]. In [14], a game-theoretic model for the canonical problem of spatio-temporal collaboration was presented with the goal of optimizing individual benefits. Bowman and Hexmoor [15] implemented a simplified agent collaboration system, with boxes that had to be pushed into holes, to investigate the effect of the social network topology on agent collaboration. Agents had to decide whether to collaborate or not based on a payoff criterion. A collaborative, hierarchical multi-agent model integrating knowledge-based communication was implemented in [16] for the RoboCupRescue[1] competition. Global task planning is done in an administrant layer and mobile agents can make their own choices to seek advice from the central agent through an autonomous layer. In general, these robot competitions focus on search and rescue applications: robot capabilities in mobility, sensory perception, planning, mapping, and practical operator interfaces, while searching for simulated victims in unstructured environments.

None of this work explicitly considered the decentralized interaction of mobile agents with both spatio-temporal and communication constraints. These constraints play a role in collaboration within geosensor networks. However, in geosensor networks collaboration has been mostly studied between distributed *immobile* nodes [17, 18], and only recently were mobile nodes allowed to track local movement patterns [19]. Instead we focus here on the coordination of collaboration between different *mobile* agents at a static location and within a given time frame, in some sense generalizing the application-specific approach in [7]. Such spatio-temporal accessibility has been one of the core issues in time-geographic analyses.

---

[1] http://www.robocup.org/

3   A Decentralized Time Geography Framework

The general problem studied in the following is defined as follows: An agent a_l at location l_l and time t_0 needs other agents $a_2 \ldots a_i$, with $i \geq 2$, for collaboration at location l_j, with l_j not necessarily l_l, and time t_k, $k > 0$. The agents can only communicate in a peer-to-peer manner—there is neither a central coordinating instance, nor any hierarchic or centralized communication available. The agents' communication efforts required to facilitate the collaboration can be optimized by multiple ways, e.g., by strategies of movement or message routing. Here we are primarily interested in the local application of time-geographic concepts and their impact on limiting the communication and solving the given problem.

The aim of this section is to explore the general dimensions of this problem. These dimensions will form a framework (a set of variables) of decentralized time geography that allows (a) the initiating agent a_l to specify a particular communication need, and (b) to communicate this need together with the messages to other agents $a_2 \ldots a_i$. In this sense, the framework will form an ontology, or a shared vocabulary between the agents, of decentralized time geography.

3.1   Location of the Collaboration

In the above problem, most likely an agent a_l will initiate a collaboration at its current location l_l, but in principle the location of the collaboration may be different from l_l. From a time-geographic perspective, two other cases can be distinguished. An agent a_l can invite to a location l_j that is reachable for a_l before t_k, i.e., a_l can participate in the collaboration. Alternatively, agent a_l can invite to a location $l_j \neq l_l$ that is not reachable by a_l before t_k, i.e., a_l calls for a collaboration it cannot participate in. As shown in Figure 2, the location l_j and time t_k of the collaboration define a cone, with (l_j, t_k) forming the apex, and the aperture defined by a maximum velocity. Since a_l knows its maximum velocity, i.e., knows the cone, it can compute the base b_0 of the cone at current time t_0, specifying one out of the following three alternatives:

1. a_l is at the center of the base of the cone,
2. a_l is within the base but not at the center, or
3. a_l is outside of the base.

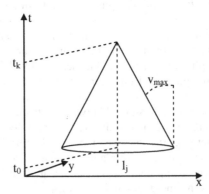

Fig. 2. The cone defined by the location of the collaboration l_j, the time of the collaboration t_k, the current time t_0, and the maximum velocity of agents, v_{max}

The location l_j may even be an extended region instead of a position, $l_j = \int p_k \, dx$. For example, in the introductory scenario a_l is calling to the spread chemical spill. Such an extension does not change the following principles; it only changes the form of a cone (Figure 2) to a frustum.

Agent a_l will communicate the cone parameters together with its call for help, such that every agent receiving the call can locally make use of it in two ways: first, it can decide whether it can help, and second, it can decide whether it should forward the message.

3.2 Agent Travel Capabilities

For any agent in the environment confronted with the problem of reconstructing the cone, three of the four parameters are fixed: l_j and t_k are specified in the request message, and the current time t_i, with $0 \le i \le k$, can be observed from an on-board clock. For the last one, v_{max}, however, three cases must be distinguished:

1. The agents in the environment are homogeneous. In this case, any agent knowing its own capabilities can safely assume the same capabilities for all other agents. Hence, v_{max} is both constant and globally known.
2. The agents in the environment are heterogeneous, but the environment itself constrains the maximum velocity, e.g., by a traffic code. Such an external behavioral code makes v_{max} again constant and globally known. Still, agents in the environment can have v_{max} below the velocity limit. By including them in the set of relevant agents a communication strategy accepts commissions: these agents may not be able to make it to the meeting point in time.
3. The agents in the environment are heterogeneous, such that an individual agent cannot determine the maximum velocity of any other agent from local knowledge. In this case, an agent can only make an estimate of v_{max}, leading to two potential types of errors:

   - Omissions—the agent's analysis will miss some candidates that have higher velocities than the assumed maximum velocity and are currently outside of the calculated base, but would be inside of the base when applying their true v_{max}.
   - Commissions—the agent's analysis will include some candidates that have lower maximum velocities than the assumed maximum velocity and are currently inside of the calculated space-time prism but would be outside of the base when applying their true v_{max}.

If movements in the environments are restricted to a travel network, the same distinction applies, although the cone becomes discrete and distances are computed on the network [7].

3.3 Message Distribution

Depending on the three location alternatives of agent a_l in relation to the base (Section 3.1), a_l may wish to choose different message distribution strategies. The request for collaboration should, according to time geography, be sent to all agents within the

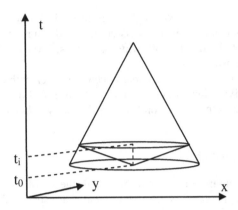

Fig. 3. A call distributed between t_0 and t_i

base b_0 of the cone[2] (l_j, t_k, v_{max}) at t_0. If agent a_1 is inside the base a limited flooding strategy can be applied: all agents receiving the request message, forward (re-broadcast) this message if their radio range intersects with the base of the cone, i.e., if reasonable hope exists to reach relevant agents (agents in b_0). If agent a_1 is outside the base it may wish to geographically route the message first to agents within the base [20] before flooding starts.

The types of agents considered here forward messages instantaneously (in contrast to some geosensor networks that delay each message forwarding by periodical, synchronized communication windows). But this alone does not guarantee that all agents within the base receive the request at t_0. It only guarantees that all agents that are at t_0 also within a_1's connected component of the communication network receive the request at t_0. Since no agent can know the coverage of its own connected component from local knowledge, a_1 can only exploit the dynamically changing connectivity of all agents in its connected component at t_0 by asking these agents to repeat the broadcasting from time to time, $t_i < t_k$, as long as their own radio range intersects with the base b_i of the cone at t_i (the shrinking cone base over time, see Figure 3, can be computed locally). Repeated forwarding increases the probability to reach other relevant agents, although it cannot guarantee reaching all agents that were relevant at t_0.

3.4 Categories of Communication Needs

The type of collaboration may require different communication patterns.

1. In the simplest case, a_1 may only *call* for collaboration, not expecting any other response than agents deciding and physically coming to help. Agent a_1 can construct a cone and determine its base b_0 at t_0 to identify the location of all relevant agents. Then a_1 sends the request for collaboration attaching the cone parameters to the message to enable the recipients of the message to decide whether they can contribute to solving the task.

---

[2] Note though that the computation of the cone base is an (optimistic) approximation for a space cluttered with features. Real travel distances may be longer than straight lines, but the agents do not know better from local information.

2. Agent a_1 may further specify to receive *offers* for help. Collecting offers might be useful to count and assess the help to be expected as soon as possible, such that if necessary additional measures can be taken (such as relaxing the cone parameters, in particular t_k). This requires suited and prepared agents to respond to a received request, which, if again forwarded repeatedly, adds an additional cone to Figure 3. Since a_1, the target of an offer, is moving, an offer must be sent to the base of a_1's own space-time cone of (l_1, t_0, v_{max}) at the current time. Note that agent a_1 cannot decide whether it has received all offers. Two strategies can be applied: (a) a_1 stops listening after a reasonable time, or (b) a_1 specified in the call for help a time t_i by which it will stop accepting further offers.
3. Alternatively, agent a_1 may have specified that no action should be taken without a specific *booking* of collaboration services. Agent a_1 may want to select a specific number of collaborating agents from incoming offers, or the set of collaborating agents best suited for the task at hand. This third type of message, a booking message, is again addressed to specific but mobile agents, and forms a third cone if forwarded repeatedly. Fragile connectivity means again that booking messages may fail to reach an agent.

3.5 Broadcasting Range

A last distinction is made regarding the communication radius, which is technically a radio range.

1. The radio range can be fixed and the same for all agents, e.g., given by a particular wireless technology.
2. The radio range can be fixed for every agent, but different between individual agents.
3. The radio range can be variable. This case allows broadcasting agents to vary the radio range according to the urgency of the collaboration or the size of the base of the collaboration cone.

4 Simulation

This section presents the implementation of the decentralized time-geographic framework and the specific experiments to evaluate the impact of different parameters on the collaborative task performance. The focus of the current implementation is on the use of the framework for collaborative planning and messages are only forwarded to seek replacements.

4.1 Implementation and Data

NetLogo[3], a cross-platform multi-agent modeling environment, was used for the simulation. It is a programmable environment for simulating natural and social phenomena, and is well suited to model complex systems developing over time. The goal of the following experiments was to demonstrate that time geography can be utilized

---

[3] http://ccl.northwestern.edu/netlogo/

locally and in a decentralized framework to optimize search, and that this leads to an enhanced efficiency regarding communications and computations compared to centralized problem solving.

In the simulation, one initiating agent a_l broadcasts a message with a request for help consisting of the task location l_j, upper time limit t_k, and the distribution of different agent types needed. In our case, the simulated environment consists of three different agent types (orange, lime, and magenta) representing different sensing or other capabilities. Every agent that receives the message checks whether it can contribute (i.e., right type of sensor) and reach the location within the specified time limit (i.e., checking whether it is currently located within the space-time cone—compared to the general framework described in Section 3, this only shifts the decision about temporal usefulness from the initiating agent to each individual agent). In addition, there exists a replacement mechanism based on directed messages and 2-way communication within the communication radius, which allows an agent that finds another agent with the same capabilities but closer to the goal to be replaced. For reasons of better comparability the environment is kept constant for all experiments: it

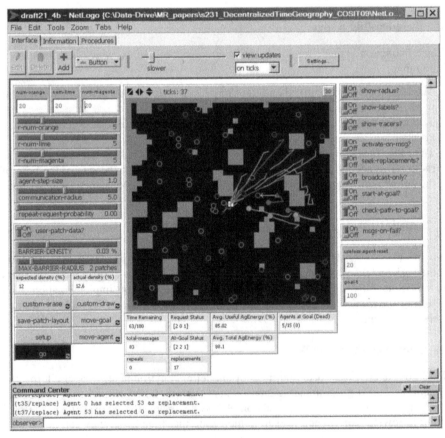

Fig. 4. Screenshot of the NetLogo simulation environment

consists of a 41 x 41 raster grid with a constant barrier structure and density, and the task location l_j is in the center. Furthermore, all agents move with the same velocity. Figure 4 shows the NetLogo simulation environment.

4.2 Experiments

In addition to the general questions of decentralized time geography and its effect on communication and computation, we were particularly interested in the impact of certain variables on solving the given problem. More specifically, we investigated the influence of the following situations:

1. All agents wander randomly in the environment at all times (*AOM false*) vs. agents do not wander without having received a call for help (*AOM true*).
2. Agents seek replacements (*SKR true*) vs. agents do not seek replacements (*SKR false*).
3. Initiating agent a_I starts at the task location l_j (*SAG true*) vs. initiating agent's location is random within the environment (*SAG false*).
4. Initiating agent a_I is not at the task location l_j but checks path to the goal first for potential helping agents (*CGP true*) vs. initiating agent wanders randomly in the environment looking for helping agents (*CGP false*).

The experiments were performed for two different numbers of agents (60, 12) randomly placed for each run, with an equal distribution of the three different agent types (20/20/20, 4/4/4), and three different communication radii (5%, 10%, and 20% of total field coverage of the environment). 25% of agents of each type were required at the goal for solving the task, i.e., 5 and 1 of each respectively. Energy consumption was held constant for different activities, i.e., 1.5 to broadcast, 0.2 to seek message, 0.5 to respond, 0.05 to update, and 0.75 to move; the maximum energy was 200 per agent[4]. Every experiment consisted of 100 runs and the individual results were averaged.

4.2.1 Control Conditions
The different behavior mechanisms were compared against control conditions for every combination of agent number and communication radius. This resulted in 6 controls and for each of them the following variables were kept constant: *AOM false, SKR true, SAG false, CGP false*.

4.2.2 Test Conditions
For each test condition we recorded the following variables:

- *Total success rate*: the number of times out of 100 that the simulation was successful, i.e., all agents required to solve the task existed simultaneously at the task location before the upper time limit.
- *Broadcaster success rate*: the number of times out of 100 that the initiating broadcaster succeeded in completing its request, i.e., all agents required started moving towards the task location.
- *Time*: the average time (in ticks) of completion for the successful runs (excludes unsuccessful runs).

---

[4] These values are arbitrary and can be changed at any time to reflect actual values for a real-time scenario.

- *Total number of messages*: the average number of total messages over all successful runs (excludes unsuccessful runs).
- *Replacements*: the average number of replacement actions per run.
- *Average remaining energy*: the average remaining energy (in %) of all agents.
- *Average used energy*: the average amount of energy used (in units) by all agents. The individual results are given in Table 1 and Table 2.

Table 1. Results of simulation runs for 60 agents and communication radii 5%, 10%, and 20%. Best and worst values (if applicable) within a set of runs are indicated by $^+$ and $^-$ respectively, overall best values for each variable are marked bold.

Scenario (# agents, comm. radius)	Succ.	BCast Succ.	Time	Mess.	Repl.	Remain Energy	Used Energy
control60_10pct	*77*	*91*	*60*	*199.0*	*38.0*	*78.0*	*2639*
E1_60_10_AOMtrue	51$^-$	76$^-$	67$^-$	137.0	18.0	87.0$^+$	1557$^+$
E2_60_10_SKRfalse	64	88	61	57.7$^+$	0.0	81.2	2251
E3_60_10_SAGtrue	79$^+$	93$^+$	55$^+$	160.6	24.1	80.2	2373
E4_60_10_CGPtrue	66	87	64	204.7$^-$	41.0	76.9$^-$	2776$^-$
control60_5pct	*45*	*64*	*72$^+$*	*178.2$^-$*	*43.0*	*73.4$^-$*	*3186$^-$*
E5_60_5_AOMtrue	30$^-$	46$^-$	77$^-$	95.3	13.0	87.1$^+$	1550$^+$
E6_60_5_SKRfalse	46	68$^+$	75	57.5$^+$	0.0	77.9	2650
E7_60_5_SAGtrue	55$^+$	66	72$^+$	155.8	33.6	74.5	3056
E8_60_5_CGPtrue	52	68$^+$	74	174.8	40.5	73.7	3155
control60_20pct	*83$^+$*	*98*	*49*	*248.8$^-$*	*34.5*	*81.7$^-$*	*2192$^-$*
E9_60_20_AOMtrue	58$^-$	95$^-$	57$^-$	172.3	18.8	86.9$^+$	1568$^+$
E10_60_20_SKRfalse	71	96	54	59.7$^+$	0.0	83.7	1957
E11_60_20_SAGtrue	83$^+$	100$^+$	38$^+$	135.0	12.2	86.0	1683
E12_60_20_CGPtrue	75	97	48	238.2	29.2	82.0	2160

Table 2. Results of simulation runs for 12 agents and communication radii 5%, 10%, and 20%. Best and worst values (if applicable) within a set of runs are indicated by $^+$ and $^-$ respectively, overall best values for each variable are marked bold.

Scenario (# agents, comm. radius)	Succ.	BCast Succ.	Time	Mess.	Repl.	Remain Energy	Used Energy
control12_10pct	*65*	*75*	*57$^-$*	*20.0*	*1.6*	*79.4$^-$*	*494$^-$*
E13_12_10_AOMtrue	58$^-$	69$^-$	51	16.4	0.7	90.4$^+$	230$^+$
E14_12_10_SKRfalse	73$^+$	77	52	13.3$^+$	0.0	83.3	400
E15_12_10_SAGtrue	72	83$^+$	48$^+$	19.7	1.3	82.4	424
E16_12_10_CGPtrue	64	76	57$^-$	22.4$^-$	1.9	79.4$^-$	494$^-$
control12_5pct	*56$^+$*	*63$^+$*	*61$^-$*	*15.5*	*1.3*	*77.9$^-$*	*531$^-$*
E17_12_5_AOMtrue	41$^-$	51$^-$	61$^-$	12.8	0.6	89.1$^+$	262$^+$
E18_12_5_SKRfalse	54	62	55$^+$	12.0$^+$	0.0	80.5	468
E19_12_5_SAGtrue	52	58	57	16.2$^-$	1.4	78.1	526
E20_12_5_CGPtrue	52	60	55$^+$	15.3	1.2	78.2	523
control12_20pct	*74*	*87*	*48*	*28.3$^-$*	*2.0*	*82.1$^-$*	*430$^-$*
E21_12_20_AOMtrue	72$^-$	78$^-$	43	21.8	0.8	91.3$^+$	208$^+$
E22_12_20_SKRfalse	81	90	51$^-$	14.2$^+$	0.0	84.8	366
E23_12_20_SAGtrue	90$^+$	97$^+$	39$^+$	22.6	0.7	86.7	319
E24_12_20_CGPtrue	77	89	44	27.7	2.3	84.0	383

5 Analysis and Discussion of Simulation Results

The implementation of the decentralized time-geographic framework demonstrates that time geography can be used locally and optimizes search by selecting only relevant agents, i.e., those agents that can contribute to solving the task and are able to reach the task location within the specified time limit. Furthermore, the replacement mechanism enhances temporal efficiency by substituting helpful agents with other helpful agents of the same type but closer to the goal.

In order to compare our decentralized framework to one with a central instance we calculated an optimal centralized case where the central command has complete knowledge, therefore resulting in the highest success probability. It is based on the assumption that the communication radius of the central instance covers the whole environment and can therefore reach every agent with 1 message. In reality, assuming a flooding strategy utilizing several hops between agents, these numbers will be much higher, and in addition it cannot be guaranteed that every agent will receive the message due to non-connected parts of the network. The total message numbers for the centralized cases are therefore derived by 1 (initiating agent to central instance) + 2 (send from / reply to central instance) * number of agents (60 / 12) + number of requested agents (15 / 3) = 136 and 28 messages respectively. Note that in the centralized case the central instance needs to evaluate whether an agent will make it to the task location on time, whereas in the decentralized framework agents determine this themselves. The results demonstrate that even when compared to the best centralized case, in more than half of the experiments, decentralized time geography leads to a higher communication efficiency (and this will only improve further when compared to flooding strategies). There is a clear dependency between number of agents in the environment and communication efficiency when comparing centralized and decentralized strategies: the lower the agent density the higher the efficiency gain in the decentralized framework. In the case of no replacements, the decentralized time-geographic framework is always superior and there are even scenarios for a large number of agents when this is the case (*60_5_AOMtrue*, *60_20_SAGtrue*).

In the following we discuss the overall results for the control conditions and the individual results for the four test scenarios.

5.1 Analysis of Control Scenarios

To demonstrate the overall picture of the decentralized time-geographic framework, we investigated for each control condition the results for the following variables: total success rate, time of completion, number of messages, number of replacements, and remaining energy. Due to the different units the original values were standardized by calculating their z-values, i.e., the standard deviation functions as the unit of measurement for describing the distances from the mean [21].

The visualization in Figure 5 shows that the total success rate is positively correlated with the length of communication radius. This is independent of the total number of agents. It is important to note that a larger number of agents does not automatically lead to a higher success rate as the *60_5* and *12_5* conditions demonstrate. A significant result is the fact that success rate is inversely related to time: for high success rates the average time of completion for the successful runs is a

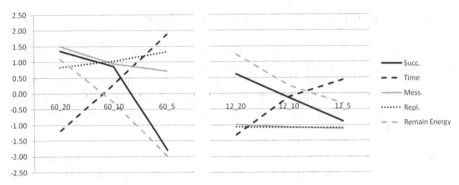

Fig. 5. Comparison of z-transformed values for total success rate, time of completion, number of messages, number of replacements, and remaining energy for the 6 control conditions (numberOfAgents_communicationRadius)

minimum. As expected, more agents in the environment lead to more messages being exchanged and the smaller the communication radius the fewer the messages. The number of replacements stays relatively constant within the 60 and 12 agents conditions and there seem to be no obvious significant overall correlations to any of the other variables. This is surprising because we expected that a large number of replacements will result in more efficiency (in terms of saving time). The most significant counter-example for this is condition *60_5*, which has the largest number of replacements but also uses the most amount of time. Remaining energy correlates positively with success rate and negatively with time, leading to the assertion that high success rates are due to energy-efficient problem-solving.

5.2 Agent Movement

The results clearly demonstrate that (random) agent movement has a large impact on the total and broadcaster success rates, and also on the amount of time needed to reach the task location. Both the total success rate and the broadcaster success rate are lower for all of the scenarios where agents do not wander in the environment without having the message. Except for the *12_10* and *12_20* conditions, such behavior also takes the most time for solving the task. There are fewer messages being exchanged and fewer replacements occur compared to the control conditions. Due to the fact that most of the agents do not move most of the time, the remaining energy is highest for all scenarios.

5.3 Agent Replacements

In the conditions without replacements one could argue both ways, i.e., for a higher success rate because fewer agents are expected to die due to energy loss, and on the other hand for a lower success rate because agent replacements save time (the replacing agent is closer to the goal) resulting in fewer cases where total time runs out. The results indeed demonstrate a mixed effect: On average the conditions for 12 agents show a higher success rate, whereas the conditions for 60 agents show a lower average success rate. The differences to the control conditions are also smaller for 12 agents. The reason for this is that fewer agents imply fewer replacement possibilities and therefore a smaller visible effect. This effect can also be seen regarding time: All scenarios with 60 agents need more time for solving the task without replacements. As expected, without replacements there are fewer messages being sent and these savings in communication costs lead to savings in energy.

5.4 Location of Initiating Agent

Locating the initiating broadcaster at the task location has a significant effect on the outcome of the simulations. Except for the *12_5* condition, the total and broadcaster success rates are on average (*60_20* is a tie for total success rate) significantly higher than for the control conditions. This is not surprising because the broadcaster immediately gets a portion of the requested agents to the goal within a short period of time. A small number of agents and a small communication radius lead to fewer useful agents around the goal and this explains the outlier *12_5*. The condition of having the initiating agent at the goal location also results in the fastest task performance overall (see bold numbers in Table 1 and Table 2), again with *12_5* being the exception. As expected there are fewer total messages (for all but *12_5*) and on average agents use less energy because there is lower cost for travel. Note that for 60 agents there is a significant reduction in replacements because the first portion of useful agents is closest to the goal and therefore does not need replacements. For 12 agents this effect is barely visible.

5.5 Goal-Directed Behavior of Initiating Agent

In this condition, the initiating agent broadcasts the message while moving to the goal itself. While we expected an increase in success rate compared to the control condition, where the initiating agent moves randomly in the environment, the results are mixed. Only the *60_5* and *12_20* conditions result in higher success rates whereas in the other 4 conditions success rates decrease. One possible explanation for this is the extra way of going back from the goal when some of the required agents are still missing. An interesting case is condition *12_10* where the total success rate is lower although the broadcaster success rate is slightly higher. For all other cases total success rate and broadcaster success rate have a positive correlation. Another surprising result is that compared to the control conditions, more time is needed on average for 60 agents to solve the task, whereas for 12 agents this result is reversed. The number of total messages is similar to the control conditions, slightly fewer on average but more for 10% communication radii. There are also only minor differences with regard to remaining energy and number of replacements.

Overall, the simulation results indicate that a high success rate and using as little time as possible, can be achieved with a large communication radius and the initiating agent located at the goal. Scenarios with static agents are very energy-efficient but show the worst results regarding success rates and time. Minimizing the number of messages can be achieved in scenarios without replacement mechanisms but this comes at the cost of lower (though not worst) success rates.

6 Conclusions and Future Work

In this paper we have developed a theoretical framework for *decentralized time geography*. Traditional time-geographic analysis is centralized and assumes complete global knowledge, whereas the move to decentralized time geography allows for the consideration of limited local knowledge. Such framework is important for networks with peer-to-peer collaboration of mobile social agents, where there is no central instance and time-geographic analyses must be performed locally by the distributed agents.

In particular, we investigated the problem of one agent initiating ad-hoc collaborative planning and decision-making among several agents, which eventually leads to physical support at a specific site. We identified as the major components of the decentralized time-geographic framework the location of the collaboration, the agents' travel capabilities, the message distribution strategy, the communication pattern, and the broadcasting range. In order to demonstrate the functioning of decentralized time geography, we performed experiments in a multi-agent simulation framework for different combinations of agent density and communication radius. The results demonstrated that time geography can be used locally and decentralized to optimize search. For more than half of our experiments its local application made communications and computations more efficient compared to problem-solving from a global perspective, i.e., with a central command having complete knowledge of the environment and its agents. We further showed the impact of specific conditions with regard to agent movement, replacement seeking, and location and goal-directed behavior of the initiating agent on the simulation outcomes.

The framework of decentralized time geography has many potential application areas, such as physical robot interaction, search and rescue, environmental analyses, transportation problems, and mobile location-based gaming. Our work serves as a foundation for the theory of decentralized time geography and leads to many directions for future research:

- The theoretical framework developed here needs to be thoroughly tested through extended simulations covering various application domains and additional aspects such as heterogeneous agents with different capabilities and maximum velocities, and different message distribution strategies, such as geographical routing and flooding. A major question that arises for these distribution strategies concerns the communication of the current request state to other agents in the network, i.e., how many agents of different types are still needed at a given point in time.

- In addition to evaluating whether they make it to the task location on time, the spatio-temporal constraints of time geography should enable agents to plan their own communication behavior, such as, in a given situation, deciding how far messages should travel, or where or how often messages should be spread.

- Limited local knowledge, especially about dynamically changing connectivity, loses some agents but this is compensated for by lower energy consumption. It is important to investigate the correlation between such lower consumption and total success rates, as well as how repetitive forwarding strategies may alleviate the problem of disconnected network components.

- In our simulation we have kept the environment fixed, but variations of environmental structure, barrier density, and task location will most likely have a large impact on the outcome of the experiments. Future work should utilize representations of real-world environments, in order to better investigate the impact of such factors, also with regard to choosing optimal spatial search and travel strategies. This may include the possibility of agents joining and leaving a network at any given point in time.
- Extending our scenario to situations with multiple collaborations at different locations or times will help in generalizing the proposed framework for decentralized time geography.

Acknowledgments

The comments from three anonymous reviewers provided useful suggestions to improve the content and clarity of the paper. This work is supported under the Australian Research Council's Discovery Projects funding scheme (project number 0878119).

References

1. Hägerstrand, T.: What about people in regional science? Papers of the Regional Science Association 24, 7–21 (1970)
2. Miller, H.: A Measurement Theory for Time Geography. Geographical Analysis 37(1), 17–45 (2005)
3. Lenntorp, B.: Paths in Space-Time Environments: A Time-Geographic Study of the Movement Possibilities of Individuals. Lund Studies in Geography, Series B (44) (1976)
4. Miller, H.: Measuring space-time accessibility benefits within transportation networks: Basic theory and computational methods. Geographical Analysis 31(2), 187–212 (1999)
5. Kwan, M.-P.: GIS Methods in Time-Geographic Research: Geocomputation and Geovisualization of Human Activity Patterns. Geografiska Annaler B 86(4), 267–280 (2004)
6. Miller, H.: Modeling accessibility using space-time prism concepts within geographical information systems. International Journal of Geographical Information Systems 5(3), 287–301 (1991)
7. Raubal, M., Winter, S., Teßmann, S., Gaisbauer, C.: Time geography for ad-hoc shared-ride trip planning in mobile geosensor networks. ISPRS Journal of Photogrammetry and Remote Sensing 62(5), 366–381 (2007)
8. Nittel, S., Duckham, M., Kulik, L.: Information dissemination in mobile ad-hoc geosensor networks. In: Egenhofer, M., Freksa, C., Miller, H. (eds.) Geographic Information Science - Third International Conference, GIScience, pp. 206–222. Springer, Berlin (2004)
9. Zhao, F., Guibas, L.: Wireless Sensor Networks. Elsevier, Amsterdam (2004)
10. Marti, S., Ganesan, P., Garcia-Molina, H.: SPROUT: P2P Routing with Social Networks. In: Lindner, W., et al. (eds.) Current Trends in Database Technology - EDBT, Workshops, pp. 425–435. Springer, Berlin (2004)
11. Xu, B., Ouksel, A., Wolfson, O.: Opportunistic Resource Exchange in Inter-Vehicle Ad-Hoc Networks. In: Fifth IEEE International Conference on Mobile Data Management, pp. 4–12. IEEE, Berkeley (2004)

12. Schoder, D., Fischbach, K., Schmitt, C.: Core Concepts in Peer-to-Peer Networking. In: Subramanian, R., Goodman, B. (eds.) Peer-to-Peer Computing: The Evolution of a Disruptive Technology, pp. 1–27. Idea Group Inc., Hershey (2005)

13. Jankowski, P., Robischon, S., Tuthill, D., Nyerges, T., Ramsey, K.: Design Considerations and Evaluation of a Collaborative, Spatio-Temporal Decision Support System. Transactions in GIS 10(3), 335–354 (2006)

14. Luo, Y., Bölöni, L.: Children in the Forest: Towards a Canonical Problem of Spatio-Temporal Collaboration. In: AAMAS 2007, Int. Conference on Autonomous Agents and Multiagent Systems, pp. 990–997. IFAAMAS, Honolulu (2007)

15. Bowman, R., Hexmoor, H.: Agent Collaboration and Social Networks. In: International Conference on Integration of Knowledge Intensive Multi-Agent Systems KIMAS 2005: Modeling, EVOLUTION and Engineering, pp. 211–214. IEEE, Waltham (2005)

16. Peng, J., Wu, M., Zhang, X., Xie, Y., Jiang, F., Liu, Y.: A Collaborative Multi-Agent Model with Knowledge-Based Communication for the RoboCupRescue Simulation. In: International Symposium on Collaborative Technologies and Systems (CTS 2006), pp. 341–348 (2006)

17. Duckham, M., Nittel, S., Worboys, M.: Monitoring Dynamic Spatial Fields Using Responsive Geosensor Networks. In: Shahabi, C., Boucelma, O. (eds.) ACM GIS 2005, pp. 51–60. ACM Press, New York (2005)

18. Farah, C., Zhong, C., Worboys, M., Nittel, S.: Detecting Topological Change Using a Wireless Sensor Network. In: Cova, T.J., Miller, H.J., Beard, K., Frank, A.U., Goodchild, M.F. (eds.) GIScience 2008. LNCS, vol. 5266, pp. 55–69. Springer, Heidelberg (2008)

19. Laube, P., Duckham, M., Wolle, T.: Decentralized Movement Pattern Detection amongst Mobile Geosensor Nodes. In: Cova, T., et al. (eds.) GIScience 2008. LNCS, vol. 5266, pp. 199–216. Springer, Heidelberg (2008)

20. Ko, Y.-B., Vaidya, N.: Location-aided routing (LAR) in mobile ad hoc networks. In: Proceedings of the 4th Annual ACM/IEEE International Conference on Mobile Computing and Networking (MobiCom 1998), pp. 66–75. ACM Press, Dallas (1998)

21. Devore, J., Peck, R.: Statistics - The Exploration and Analysis of Data, 4th edn. Duxbury, Pacific Grove (2001)

Adaptable Path Planning in Regionalized Environments

Kai-Florian Richter

Transregional Collaborative Research Center SFB/TR 8 Spatial Cognition
Universität Bremen, Germany
`richter@sfbtr8.uni-bremen.de`

Abstract. Human path planning relies on several more aspects than only geometric distance between two locations. These additional aspects mostly relate to the complexity of the traveled path. Accordingly, in recent years several cognitively motivated path search algorithms have been developed that try to minimize wayfinding complexity. However, the calculated paths may result in large detours as geometric properties of the network wayfinding occurs in are ignored. Simply adding distance as an additional factor to the cost function is a possible, but insufficient way of dealing with this problem. Instead, taking a global view on an environment by accounting for the heterogeneity of its structure allows for adapting the path search strategy. This heterogeneity can be used to regionalize the environment; each emerging region may require a different strategy for path planning. This paper presents such an approach to *regionalized path planning*. It argues for the advantages of the chosen approach, develops a measure for calculating wayfinding complexity that accounts for structural and functional aspects of wayfinding, and states a generic algorithm for regionalization. Finally, regionalized path planning is demonstrated in a sample scenario.

Keywords: Path planning, regionalization, wayfinding complexity, cognitive ergonomics.

1 Introduction

Wayfinding is defined to be a purposive, directed, motivated activity to follow a route from origin to destination that reflects the cognitive processes going on during navigation [1]. It is based on a wayfinder's mental representation of an environment or, in case of not well known environments, on external representations delivered as a means of wayfinding assistance. In urban environments, planning one's route through the environment is largely determined by the network of streets [2]. This network is the predominant structure movement occurs in, i.e., the structure of the environment drives planning and execution of the wayfinding task—at least on the large-scale, environmental level [3,4].

Automatically planning a path through such networks classically only accounts for geometric properties of the network, predominantly path length and

K. Stewart Hornsby et al. (Eds.): COSIT 2009, LNCS 5756, pp. 453–470, 2009.

the distance between two nodes. In recent years, several approaches have been presented that instead account for human principles of path planning and the cognitive complexity of traveling through a network. Their calculated paths may result in large detours. This is because they focus only on one aspect of path planning, namely the reduction of cognitive complexity. The approaches ignore geometric aspects and the overall structure of an environment. They only account for the neighborhood of the current location for deciding on how to proceed. This paper presents an approach to regionalized path planning in network environments where the strategy employed for path search can be adapted to properties of the part of the environment planning currently occurs in.

In the following section, some background on human wayfinding in urban environments, wayfinding complexity and cognitively motivated path search is presented. Section 3 argues for a regionalized approach to path planning that acknowledges possible heterogeneity of an environment. Section 4, then, develops a measure for wayfinding complexity and a generic algorithm for performing regionalization of an environment, followed by an illustration of regionalized path planning in Section 5. Section 6, finally, concludes the paper with a summary and an outline of some future work.

2   Wayfinding in Urban Environments

In the following, human path planning and execution (Section 2.1) and its relation to environmental complexity (Section 2.2) is discussed. Different approaches allow for measuring environmental complexity (Section 2.3). Finally, different path search algorithms accounting for these findings are presented in Section 2.4.

2.1   Path Planning in Urban Environments

In their planning, humans often do not take the shortest or fastest path, which today's automatic assistance systems (e.g., internet route planners or car navigation systems) usually calculate. Instead, their conceptualization or perception of the environment determines path choices. Golledge [5] lists a number of factors that influence human path choice. Next to distance and time these include number of turns, shortest or longest leg first, many curves or turns, first noticed and most scenic route. Others (e.g., [6]) have explored how the angle with which the direction of the current or initial segment deviates from the direction to the destination influences path choice. Since these factors are based on cognitive and perceptual aspects and do not rely on distance (or at least just on perceived, not actual distance), routes may differ between traveling to a place and back from that place again [5].

In [7], Wiener et al. found that there is a difference between the chosen path depending on whether it is communicated to others, planned for being traveled by oneself, or actually traveled along. There have been significant differences between the routes that participants (who knew the environment well) had chosen to communicate to others and those they actually traveled themselves. While this may be due to communication considerations, i.e., selecting a route that is easier to describe and remember, there are also differences between the route planned to travel and the actually traveled one. This seems to indicate that routes are not fully planned ahead in every detail, but some options can only be verified and, thus, chosen in situ. This supports the concept of hierarchical planning and execution in wayfinding as argued for by Timpf [2,3], for instance.

2.2 Environmental Complexity

The structure of an environment has a strong influence on people's wayfinding behavior. The complexity of this structure to a large part determines the complexity of wayfinding. In complex environments, people have more difficulties building up a mental representation, i.e., learning the environment (e.g., [8]). Wayfinding itself is also more difficult. People take longer and make more mistakes [9,10]. Different factors influence the complexity of an environment's structure. Heye and Timpf [11] elicited some of these factors for traveling in public transport. Here, complexity mostly depends on the structure of the stations and the paths between places to be traveled there. Generally, structural complexity of built environments depends on architectural differentiation, the degree of visual access, and the complexity of the layout [12,13]. In urban environments, a layout's complexity is determined by the types and orientation of streets [14] or competing spatial reference systems [15], among others.

2.3 Measuring Environmental Complexity

There are different approaches to calculating measures of the structural complexity of an environment. One prominent example is the approach of space syntax [16], which uses largely topological measures to capture the influence of a space's structure on human (social) behavior. A key concept is intelligibility that describes how predictable (understandable) the global structure of space is from observations of local properties. The structure of an environment may further be described using methods of integration or choice, among others.

Others have focused on the structure of the underlying network. O'Neill [17] introduced interconnection density (ICD) as a measure. The ICD value is the average number of nodes connected to every node. In that it captures the connectedness of an environment. It relates to measures of integration (as in space syntax). With increasing ICD, wayfinding performance decreases [17]. This indicates that many possible ways through an environment aggravate constructing mental representations of that environment (there is more to be stored) and increase the chance of making errors (there are more options to choose a wrong turn). In a similar line of thinking, Mark [18] characterized intersections of an environment according to the number of possible choices at them. For each type of intersection, different costs are assigned that reflect the difficulty of navigating it. The simplest action to perform is to go along a straight segment, followed by turning around a corner, which requires a mental update of one's heading. Intersections, i.e., points in a network where there are several options to continue the path, are weighted according to their number of branches. Coming to the dead end of a T-intersection, which forces a decision on how to continue, is treated as a special case with lower costs than other intersections.

In his theory of wayfinding choremes, Klippel [4] elicited human conceptualizations of turning actions in networks. One of the main results shows that humans have prototypical mental representations of the actions performed in the network, rather than prototypical representations of an intersection's structure. Taking this into account, an intersection's complexity increases if it offers several turns into the same conceptual direction (cf. also [19]).

2.4 Cognitively Motivated Path Search Algorithms

To capture wayfinding complexity as discussed above in automatically planning a path through an environment, in recent years different path search algorithms have been proposed that account for principles of cognitive ergonomics. These principles are human-centered, in that the optimization criterion used to determine a path is based on aspects that emerge from the conceptualization of wayfinding situations, for example, the likelihood of going wrong or the ease of describing the path. The criteria are not directly dependent on geometric properties of the path network, such as distances between nodes. All the different approaches can be implemented as a variant of Dijkstra's shortest path algorithm [20]. In the following, these approaches will be presented.

Duckham and Kulik [21] extend standard shortest path search by a heuristic that reflects the complexity of negotiating the decision point represented by the two adjacent edges (e.g., turning from one edge onto another). The specific weighting used is based on an adaptation of Mark's measure [18]. Duckham and Kulik term their algorithm *simplest path* algorithm. In a simulation experiment, they show that their algorithm generally results in paths that are only slightly longer than the shortest path. While the costs employed account for structural differences of intersections, they do not account for functional aspects, for example, (possible) ambiguity in the direction to take at an intersection, nor landmarks or other environmental characteristics that might be exploited in instructions. Like shortest paths, the simplest path finds the cheapest path according to a cost function. Unlike shortest paths, the cost function used applies to the complexity of navigation decisions rather than travel distance or time.

The *landmark spider* [22] is an approach that acknowledges landmarks as crucial elements in wayfinding and performs routing along (point-like) landmarks located in the environment. For these landmarks, a salience value is assumed to be known (e.g., [23]). The landmark spider accounts for three parameters in determining edge weights: the salience of a landmark, its orientation with respect to the wayfinder and its distance to the edge. To compute orientation, a simple homogeneous four sector model is used, dividing the plane in "front," "left," "right" and "back" direction. The landmark spider approach calculates the 'clearest' path, i.e., the path that leads a wayfinder along the most suitable landmarks. The algorithm does not account for any distance information.

The approach to *most reliable paths* presented by [24] aims at minimizing the possibility of choosing a wrong turn at an intersection. As in simplest paths, the structure of an intersection influences path choice. Here, the assumption is that while instructions, such as "turn left," are easy to understand, they may still result in wrong decisions given that there are several options to turn left at an intersection. Accordingly, paths in this approach are optimized with respect to the ambiguity of instructions describing the decision to be taken at intersections. For each turn, i.e., for passing from an edge e to an edge e' that are both connected by a middle vertex, the unreliability measure is calculated. This measure is defined as the number of turns that are instruction equivalent. Two turns are instruction equivalent if they are described using the same linguistic variable. Geometrically, both turns head approximately in the same direction. To compute the linguistic variable, a qualitative direction model is used (cf. [25,26] for a discussion of the adequacy of different direction models).

In its original implementation, geometric length of edges is used as a secondary criterion to distinguish equally unreliable paths—the shorter one is selected. Consequently, Haque and coworkers extended this original algorithm to allow for a weighting between unreliability (r) and path length (w):

$$op(e, e') = \lambda_d w(e') + \lambda_u r(e, e')$$

This way, the trade-off between reliability and length can be shifted in one direction or the other. In simulation experiments, Haque and coworkers have been able to show that an agent using most reliable paths to navigate through a street network performs better than an agent relying on shortest paths. This also holds for the optimized most reliable paths.

Richter and Duckham [27] combine the reasoning behind simplest paths and context-specific route directions [19]. Just as simplest paths, the algorithm for *simplest instructions paths* finds the best route, i.e., the route associated with the lowest costs in terms of instruction complexity. However, the algorithm utilizes instruction principles and optimization criteria that are related to functional aspects of human direction giving and avoid ambiguity in instructions. The algorithm makes use of the systematics of route direction elements developed in [28]. Consequently, it allows for multiple alternative instructions (labels) to navigate a pair of edges. Furthermore, it realizes spatial chunking [29]. When selecting the node with the currently lowest cost, the algorithm tries to spread all instructions that node has been reached with forward through the graph. To this end, it checks whether neighboring nodes are chunkable using these instructions as well. Using superordinate chunking rules that check for cognitive and structural plausibility, this spreading of instructions allows for traversing several nodes with a single instruction, thus reducing the travel costs, or, in terms of route directions, the number of instructions that need to be communicated.

3   Path Planning in Regionalized Environments

An analysis of the different cognitively motivated path search algorithms shows that compared to the shortest path each approach results in detours in the path from origin to destination [30]. The length of the detour largely depends on the structure of the environment. Since the algorithms are designed to account for complexity in wayfinding and aim at minimizing this complexity one way or other, all avoid complex parts of the environment by navigating around them. Depending on the environment, the resulting detours may turn the calculated paths nearly useless as the increased effort (time and distance) is much greater than the decreased complexity and, thus, users would not accept this path as the solution to their wayfinding problem.

As a consequence, cognitively motivated path search algorithms need to counteract these detours somehow in order to render them useful in any given situation. One way to achieve this is to take the geometric aspects that have been excluded previously into account again. In most reliable paths [24], for example, a weighting is possible between unreliability (a cognitive aspect) and path length (a geometric aspect). Setting the weights is an optimization problem. Favoring the cognitive aspects too much bears the danger of getting large detours, while a strong favor of the geometric aspects may counter the original intention of the algorithms, namely accounting for the complexity of wayfinding.

Introducing this second parameter into the path planning algorithm tries to mediate between the ease of following a route and the length of paths by weighting the parameters. This has two fundamental drawbacks, which are related. First, using a weighting between two factors is a bottom-up approach that tries to fix the detour problem on a uniform level. Everywhere across the environment the same weighting is used. This ignores that the problem domain at hand (the environment) may be heterogenous and may require different strategies in different areas. Second, because of this ignorance of the global structure of an environment the adaptation of the weights has to be done for each new environment individually and mostly from scratch in order to find a sensible balance. It is largely a trial-and-error process that runs until the results "look good."

Instead, since the structure of an environment plays such a crucial role in wayfinding complexity and the performance of the path search algorithms, an obvious approach is to directly exploit environmental structure in countering the detour behavior of cognitively motivated path search algorithms. This is a top-down approach, acknowledging that the environment at hand may be heterogenous with respect to its structure and, as a consequence, its wayfinding complexity. To enable this approach, the environment needs to be analyzed to identify crucial differences within its structure. If no such differences exist, the path search algorithms behave uniformly across the environment. In this case emerging detours will either be small or the environment is so complex that it would be hard to find a sensible weighting that counters the detours anyway.

There are different aspects and different methods to identify the structural differences within an environment (see Section 4). Independent of the chosen method, the analysis results in the identification of different regions, i.e., a regionalization of the environment. Those parts of a network that are in the same region share the same properties according to the chosen regionalization method. They belong to the same *class*. The number of different classes may vary depending on both the environment and the chosen method. For example, there may be just two classes of regions—simple and complex, or the differentiation may be more elaborate, for example, "highway system," "major streets," "suburb," "downtown," and so on. Having a representation of an environment's network that contains information about the regionalization, it is possible to use different path search strategies for each region class. This adapts searching for an optimal path to the environment's structure at hand. The general algorithm and its properties are discussed next, Sections 4 and 5 then illustrate this approach with some examples.

The Regionalized Path Planning Algorithm

The *regionalized path planning algorithm* takes as its input a graph G representing the environment's network of streets and a node o that represents the origin of the route to be taken. The algorithm works on both the original graph and the complete linegraph [21]. The original graph reflects the geometry of the environment, i.e., each node has a position coordinate. Also, each node is annotated with the region class it belongs to. The algorithm also takes a list of region-tuples (class,function); class is a value for the region class at hand, function represents the cost function to be used for path search for this region class. The algorithm's result is a path from origin to destination, represented as a sequence of nodes that need to be traversed. Since these nodes have a position, this sequence can be directly mapped back to the geographic data the graph is derived from.

Path planning is performed as it is done in Dijkstra's shortest path algorithm [20]. To account for the heterogeneous structure of the environment that is identified by regionalization, the algorithm is adapted to allow for using different cost functions depending on the region a node is in. To this end, when expanding a new node it is checked to which region class it belongs. According to this, the costs to reach the neighboring nodes is calculated using the corresponding cost function. Algorithm 1 summarizes the regionalized path planning algorithm. In this notation, the algorithm works on the linegraph [21], which is used by most cognitively motivated path search algorithms. Searching on the original graph can be realized accordingly.

Algorithm 1 is very similar to the well known Dijkstra algorithm. The only addition is the dependence of the cost function on the region class. This reflects that regionalized path planning is a top-down approach. Once the representation of the environment—the graph—is set up with region information, path planning itself is a matter of correctly adapting to the region of the environment search currently occurs in. In that, Algorithm 1 can be seen as a kind of meta-level path search algorithm. It selects the right strategy depending on the current situation. Note that it is not possible to stop search the first time the destination is reached. This is because cost functions, such as the one used in simplest instruction paths [27], might not only depend on local information (the current edge), but on global information (e.g., previously traversed edges or the current region) as well. This may result in situations where a path that reaches the destination later on in the search process may be better than the one that has reached the destination first.

Algorithm 1. The path planning algorithm for regionalized environments.

Data: $G = (V, E)$ is a connected, simple, directed graph; $G' = (E', \mathcal{E})$ is the complete linegraph; $o \in E$ is the origin (starting) edge; $w : R \times E \to \mathbb{R}^+$ is the edge weighting function, which depends on the region class R.

Result: Function $p : E \to E$ that stores for each edge the preceding edge in the least cost path.

1 // *Initialize values*;
2 **forall** $e \in E$ **do**
3   Initialize $c : E \to \mathbb{R}^+$ such that $c(e) \leftarrow \infty$;

4 Set $S \leftarrow \{\}$, a set of visited edges;
5 Set $p(o) \leftarrow o$;
6 Set $c(o) = 0$;
    // *Process lowest cost edge until all edges are visited*
7 **while** $|E \backslash S| > 0$ **do**
8   Find $e \in E \backslash S$ such that $c(e)$ is minimized;
9   Add e to S;
10   **forall** $e' \in E \backslash S$ such that $(e, e') \in \mathcal{E}$ **do**
       // *Update costs to e' based on region's cost functon*
11   Set $r \leftarrow$ **region**(e), the region class of e;
12   **if** $c(e') > c(e) + w(r, e')$ **then**
13   Set $c(e') \leftarrow c(e) + w(r, e')$;
14   Set $p(e') \leftarrow e$;

---

4 Regionalization of Environments

As pointed out in Section 3, there are different aspects and methods that may cause a regionalization of an environment. It may be possible to use existing regions, for example, districts of a city to structure the environment. However, this does not tell anything about differences in environmental complexity since these districts are just administrative boundaries that are drawn for historical or census reasons. Another approach could be to divide the streets represented in the network in different classes according to their road status hierarchy. It can be assumed that traveling along highways is less complex than traveling through small streets in a downtown area. However, this default assumption does not always hold. Further, using this simple approach there is no distinction possible within hierarchy levels, i.e., the actual complexity of the configurations of streets on the same hierarchy level cannot be judged.

Thus, for the purpose of path planning that accounts for environmental complexity, regionalization must be performed on the network based on (some of) its properties. For example, it is possible to use the measures explained in Section 2.3 or those underlying the cognitively motivated path search algorithms (Section 2.4). Both ICD [17] and Mark's complexity measure [18] can be used to determine the complexity of the nodes in the network. These two measures, as well as the derived measure used in simplest paths [21], rely on topological and ordering information only. And they account for local information only. They are restricted to information concerning the number of branches at a given node.

In the following, regionalization is performed by, first, calculating a complexity value for every node of the network and, second, clustering these nodes based on their values to form region classes. Based on the discussion in Section 2, three parameters that are derived from the network are used in the process:

1. *Number of branches*:according to ICD and Mark's measure, the more branches there are at a node, the more difficult it is to correctly navigate the corresponding intersection. While this parameter ignores function in wayfinding (cf. [4]), it reflects the fact that a high number of branches provides more opportunity to take the wrong turn, i.e., increases the chances for wayfinding errors. Further, intersections with many streets meeting there become more complex as there is more information to parse before taking a decision.

2. *Average deviation from prototypical angles*: people have prototypical conceptualizations of turning actions in wayfinding [4]. Street configurations that do not adhere to the prototypical angles are hard to conceptualize and to correctly integrate in the environment's mental representation [14]. Thus, covering a functional aspect of wayfinding, intersections with oblique turns are deemed more complex than those following the prototypical angles of 90 and/or 45 degree turns. As wayfinding complexity is calculated for individual nodes (not the edges), the average deviation for all combinations between two of the node's branches needs to be calculated.

3. *Average segment length*: decision points are the crucial parts of route following [31]. In street networks, decision points correspond to intersections. These are the spots along a route a wayfinder needs to decide on the further way to take. They are the spots where wayfinding errors occur. Therefore, along long segments there are fewer possibilities to make errors since there are fewer decision points. Areas with short segments, on the other hand, have a higher density of decision points. Thus, longer segments indicate an area that is less complex for wayfinding, simply because there are fewer options to decide from. Again, since complexity is calculated for nodes, the average segment length over all branches of a node is used.

These parameters capture both structural and functional aspects of wayfinding. They aim for reflecting human assessment of wayfinding complexity.

4.1   The Combined Wayfinding Complexity Measure

Using the three parameters *number of branches*, *average deviation*, and *average segment length* we can define a combined wayfinding complexity measure *CWC*. The complexity *cwc* of an individual node n_k is the sum of the values for the three parameters. The individual parameters can be weighted to account for differences in the relevance of each parameter. The order of magnitude of the different parameters depends on the environment's structure and the geographic data at hand. The number of nodes can be expected to be in the range of lower positive integers—there will hardly be intersections with more than 8 branches. The length of street segments, however, may range from a few meters to several kilometers. Further, when calculating distances, the values depend on the coordinate system's representation. For Gauss-Krüger, for example, coordinates are represented as 7 digit numbers. Therefore, values of the parameters are normalized to be in the range $[0, 1]$ first before adding them up. To this end, the maximum value for each parameter needs to be known, i.e., the maximum number of branches nb_{max}, the maximum average deviation ad_{max}, and the maximum average segment length al_{max}. As discussed above, for number of branches and deviation, small values denote low complexity, for segment length, however, high values correspond to low complexity. Therefore, the length value is subtracted from 1 to account for this difference in the value's semantics.

Equation 1 states how to calculate the wayfinding complexity value of an individual node. Here, nb_k is the number of branches of the node n_k (its degree), the different λ's are the weighting factors, $d(\gamma)$ is the deviation of an angle γ from 90 (45) degrees, $\gamma(b_1, b_2)$ is the angle between two branches b_1 and b_2, and $l(b)$ gives the length of a branch b.

$$cwc(n_k) = \lambda_{nb}\frac{nb_k}{nb_{max}} + \lambda_d\frac{\frac{\sum_{j=0}^{nb_k}\sum_{i=0,i\neq j}^{nb_k}d(\gamma(b_i,b_j))}{nb_k}}{ad_{max}} + \lambda_l(1 - \frac{\frac{\sum_{i=0}^{nb_k}l(b_i)}{nb_k}}{al_{max}}) \quad (1)$$

A low CWC value for a node corresponds to an intersection that can be expected to be easy to navigate, a high value to a complex intersection.

4.2   Regionalization by Clustering

Applying Equation 1 to all nodes, i.e., calculating the CWC value for each node, results in a distribution of complexity values across the environment. This can now be used to form region classes and, based on the classes, to perform regionalization of an environment.

It can be assumed that the distribution of complexity values is not random, i.e., that there are clusters of nodes which have similar complexity values. However, the values will not necessarily change gradually between neighboring nodes. While this will often be the case, there will also be leaps in values, for example, at those points where local, dense parts of the network connect to main streets that define the global structure of an environment. These considerations can be exploited in the regionalization process.

As a first step, region classes need to be defined based on the CWC values. In principle, there can be an arbitrary number of classes. Usually, a small number of classes will suffice to reflect the environment's structure, though. Each class corresponds to a mutual exclusive interval in the range of $[0, 1]$, i.e. the interval $[0, 1]$ is divided into sub-intervals to form the region classes. According to these intervals, each node belongs to a region class. In order to form regions from the individual nodes, nodes of the same class that are spatially near need to be subsumed into clusters.

In general, neighboring nodes that are in the same region class are subsumed. This way, all nodes that are connected by at least one path that contains only nodes of the same region class form a cluster of that region class. Two further aspects need to be considered: 1) the size of clusters; 2) nodes surrounded by nodes of a different region class. The first aspect relates to the willingness of calling a collection of nodes a "region." Clusters should have a minimum size in order to count as a region. Having two neighbored nodes belonging to the same class hardly can be considered a region. Nodes correspond to intersections in the real world (or to streets when looking at the linegraph). Even if two neigh-bored intersections are much more complex than their neighbors, they may be remarkable for humans, but most probably are not considered to be a region in their own right. Rather, they will be considered to be special in their neighbor-hood. Further, the path search algorithms presented in Section 2.4 will avoid such small clusters without much of a detour.

Regarding nodes surrounded by nodes of another class the reverse argument holds. Even though they belong to a different region class, these nodes should be subsumed within the same cluster. They can be considered to be outliers within a cluster. Often, these outliers will belong to a neighboring region class. They will have a range value one below or above the cluster's class and, thus, can be considered to be similar to the other nodes. Using the same reasoning as above, such a node would correspond to a single intersection that differs in complexity to its surrounding intersections within a part of the street network. While being remarkable, conceptually the node will still belong to this part of the network.

The regionalization process consists of two steps. In a first step, connected nodes belonging to the same region class are subsumed (this may also be done using a minimum spanning tree of the street network with similarity of CWC values as edge weight [32]). This results in a set of potential clusters. Each of these clusters is checked against the size threshold for forming a region. If they are smaller the cluster is disbanded again. In a second step, for each node that does not belong to a cluster (anymore), it is checked whether it can be subsumed with a neighboring cluster of a different region class (see Figure 1). To this end, for all neighbors of the node it is checked which cluster they belong to. If two or more of these neigh-bors belong to the same cluster, the node is added to that cluster. This threshold of two neighbors is used to avoid meandering clusters that consist of a single line of nodes. At the same time clusters are less likely to occur that show large differ-ences in CWC values across the cluster. Using only a single neighbor that needs to belong to a cluster, it becomes more likely that a chain of nodes with slowly in-creasing (or decreasing) CWC values will be added to that cluster. If a node has more than three neighbors and there are two or more potential clusters a node can be subsumed in, the cluster with the most connections to the node is chosen. The second step is repeated until no more nodes can be subsumed to neighboring clus-ters. Algorithm 2 summarizes the regionalization by clustering process.

In principle, using this algorithm nodes may end up belonging to two or more regions at the same time. This reflects the fuzziness of many real world's regions [33]. For the application of the regionalized path planning algorithm (Algorithm 1) this would result in ambiguous situations. Therefore, only one

---

Algorithm 2. The regionalization by clustering algorithm

Data: $G(V, E)$ is the graph representing the street network; for each node $v \in V$ the CWC value cwc(v) has been calculated using Equation 1; a list of tuples (regioninterval,regionclass) that defines which sub-interval of $[0, 1]$ corresponds to which region class; cts is the threshold size of clusters.

Result: regions, a set of regions

1 Set regions ← ∅;
2 Set tempclusters ← ∅;
3 Set unclustered ← ∅;
    // Set the region class for each node.
4 **forall** $v \in V$ **do**
5   **forall** (regioninterval,regionclass) **do**
6   **if** cwc(v) $\in$ regioninterval **then**
7   Set rc(v) ← regionclass;

    // Construct potential clusters.
8 **forall** $v \in V$ **do**
9   **forall** $n \in neighbors(v)$ **do**
10   **if** rc(v) $=$ rc(n) **then**
11   **if** $\exists c : c \in$ tempclusters $\wedge n \in c$ **then**
12   Add v to n's cluster;
13   **else**
                // Add $\{v,n\}$ as new cluster.
14   Set tempclusters ← tempclusters $\cup \{v, n\}$;

    // Check for size threshold.
15 **forall** cluster $\in$ tempclusters **do**
16   **if** size(cluster) $< cts$ **then**
17   Set unclustered ← unclustered $\cup$ cluster;
18   Set tempclusters ← tempclusters\cluster;

    // Add unclustered nodes.
19 **repeat**
20   **forall** $v \in$ unclustered **do**
21   **if** $\exists n_0, .., n_{k-1} : n_0.., n_{k-1} \in neighbors(v) \wedge n_0, ., n_{k-1} \in$ cluster, $k \geq 2$, cluster $\in$ tempclusters **then**
22   Add v to cluster;
23   Set unclustered ← unclustered\$\{v\}$;

24 **until** *no node has been added to any cluster* ;
    // Setting regions to the set of clusters joined with the set of remaining unclustered nodes as default region.
25 Set regions ← tempclusters $\cup$ unclustered;

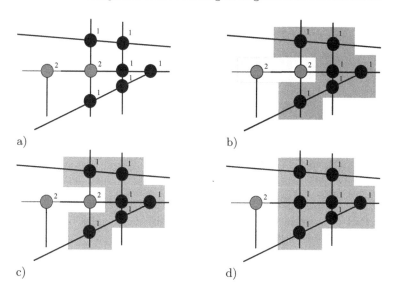

Fig. 1. The clustering process: a) each node is assigned to a region class; b) nodes of the same class are clustered if they are all on the same path; c) clusters smaller than the region threshold (here 3) are removed; d) nodes not belonging to a cluster may be assigned to a neighboring cluster if two or more of their neighbors are within this cluster

of the clusters is kept. Different strategies can be used to decide this: the first cluster the node has been added to may be kept, or the largest or smallest it belongs to. Likewise, some nodes may end up not belonging to any region. Therefore, there always needs to be a default "region" that subsumes all these nodes.

4.3 Discussion

The wayfinding complexity measure CWC has three factors λ_{nb}, λ_d, and λ_l, which are used for weighting the three measure's components. At a first glance, this seems to result in the same effect discussed above for most reliable paths [24]. Setting different weights changes the outcome of the regionalization algorithm. In most reliable paths, setting different weights changes the ratio between accounting for geometric and for cognitive aspects. As argued above, this has drawbacks related to the need to adapt the weights to each environment individually. However, in Equation 1 the weights are used to determine the influence each parameter has on wayfinding complexity. It reflects the influence each parameter has in general; it is not bound to a specific environmental setting. The weights used do not influence the path search strategies employed, but rather determine the regionalized structure path search is performed on.

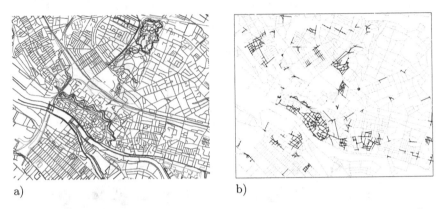

a)   b)

Fig. 2. a) Part of Bremen, the test environment; b) the complex (dark colored) and simple (light) regions identified for this environment

5   Example

The approach to regionalized path planning as discussed above has been tested using different scenarios. Here, one scenario is used for illustration: path planning in the inner-city area of Bremen. A map of that area can be seen in Figure 2a. The area comprises of some big streets connecting different districts of the city, some parts with regular street patterns, and some dense, old parts of the city, which are especially located along the river.

Figure 2b shows a regionalization of the environment. In this example, only two kinds of regions are distinguished: those regions with complex intersections and regions that are easy to navigate. The region threshold in Algorithm 2 is set to 5. This threshold is chosen for illustration purposes to ensure that regions emerge which will result in a clear detour if circumvented by a path search algorithm. Any node that is not part of a region after clustering is assigned to the "easy" region, which is used as default region. A complex region is a region with nodes having a CWC value greater than 0.6 (this value is chosen ad hoc for illustration purposes only—the CWC value which actually delineates easy and complex intersections would need to be determined empirically; cf. Section 6). As can be seen, the 'complex' regions are mostly located along the river in the dense old-town parts where streets meet at odd angles.

To illustrate regionalized path planning in this environment, for the 'simple' regions the chosen path search strategy is 'shortest path,' for the 'complex' regions the 'simplest paths' [21] is chosen as strategy. The reasoning behind this combination is that in those parts with low wayfinding complexity people will have few problems in understanding the shortest path. Often, shortest and simplest path do not differ much here anyway. It is in the complex parts of the environment that people need instructions that take away the complexity of the environment. Forcing the simplest path algorithm to find a path through complex parts of the environment achieves just that. The resulting paths get wayfinders out of these parts in the simplest way available without resulting in large

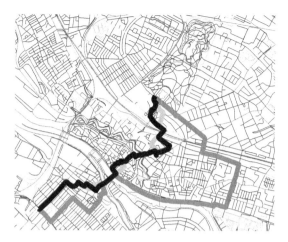

Fig. 3. Planning a path between a sample origin / destination pair. Illustrated are the differences between the shortest path (light colored line), the simplest path (medium colored line taking the bow to the right), and the regionalized path (dark line).

deviations from the course to the destination. In an actual assistance scenario, where verbal or graphical instructions are provided to wayfinders, understanding these paths can be made easier by employing strategies for generating cognitively ergonomic route instructions (cf. [34,29,28]).

Figure 3 shows an example of the shortest, simplest and regionalized path between the same origin / destination pair. As can be seen, the shortest path runs through complex and simple parts of the environment as it only relies on distances between nodes as a geometric property of the network. The simplest path avoids the complex parts, thus, taking a large detour to reach the destination in this example. The regionalized path, finally, lowers complexity in the complex parts of the environment according to the simplest path approach and gets through simple parts of the environment as fast as possible, as predicted.

6   Conclusions

Human path planning relies on several more aspects than only geometric properties of an environment, i.e., the shortest distance between two locations. For wayfinding and path planning in network environments, such as urban street systems, in recent years several cognitively motivated path search algorithms have been presented. These approaches cover different aspects of principles of human path planning. Essentially, all the approaches are extensions of Dijkstra's classical shortest path algorithm. By taking into account cognitive aspects of path planning, the approaches largely ignore geometric aspects. They try to minimize wayfinding complexity, not the length of the traveled path. Accordingly, they avoid complex parts of an environment and, thus, may result in large detours. This renders the calculated paths unacceptable.

In this paper, an approach to regionalized path planning has been presented that acknowledges that environments may be heterogeneous, i.e., that environmental complexity usually is not uniformly distributed across an environment, which is an assumption other cognitively motivated approaches implicitly make. This approach employs adaptable strategies depending on the region path search currently occurs in. This way, the strategy that allows for optimal assistance in a given region of the environment may be chosen. The paper presented the overall approach to regionalized path planning arguing for its advantages, developed a measure for an intersection's wayfinding complexity and a generic algorithm that allows for regionalizing an environment based on this measure, and presented an example applying regionalized path planning to adequately cope with changing complexity in an environment's structure.

Regionalized path planning takes a global approach when it comes to accounting for complexity of an environment. This is different to other approaches to cognitively ergonomic path planning that in each step only account for the immediate surrounding of a single node. These approaches never leave their local perspective and, thus, can only take local decisions in their planning. However, path planning is known to be hierarchic [3] and to be based on regions as well [35]. Accordingly, future work on regionalized path planning comprises path search on different levels—the level of regions and the level of individual nodes—and the implementation of more global path search strategies, for example, region-based ones. The global view will also be more strongly reflected in the complexity analysis where aspects of monotony of an environment and similarity of neighboring intersections will be integrated, in line with [36]. Further, both for the weighting of the different factors in calculating the CWC values and for setting a threshold value for complex intersections future (empirical) work is required to elicit values that reflect human conceptualization and behavior.

Acknowledgments

This work is supported by the Transregional Collaborative Research Center SFB/TR 8 Spatial Cognition which is funded by the Deutsche Forschungsgemeinschaft (DFG). The author likes to thank Jan Oliver Wallgrün and four anonymous reviewers for valuable input.

References

1. Montello, D.R.: Navigation. In: Shah, P., Miyake, A. (eds.) Handbook of Visuospatial Thinking, pp. 257–294. Cambridge University Press, Cambridge (2005)
2. Timpf, S., Volta, G.S., Pollock, D.W., Frank, A.U., Egenhofer, M.J.: A conceptual model of wayfinding using multiple levels of abstraction. In: Frank, A.U., Formentini, U., Campari, I. (eds.) GIS 1992. LNCS, vol. 639, pp. 348–367. Springer, Heidelberg (1992)
3. Timpf, S., Kuhn, W.: Granularity transformations in wayfinding. In: Freksa, C., Brauer, W., Habel, C., Wender, K.F. (eds.) Spatial Cognition III. LNCS (LNAI), vol. 2685, pp. 77–88. Springer, Heidelberg (2003)

4. Klippel, A.: Wayfinding choremes. In: Kuhn, W., Worboys, M.F., Timpf, S. (eds.) COSIT 2003. LNCS, vol. 2825, pp. 320–334. Springer, Heidelberg (2003)
5. Golledge, R.G.: Path selection and route preference in human navigation: A progress report. In: Kuhn, W., Frank, A.U. (eds.) COSIT 1995. LNCS, vol. 988, pp. 207–222. Springer, Heidelberg (1995)
6. Dalton, R.C.: The secret is to follow your nose — route path selection and angularity. In: Peponis, J., Wineman, J., Bafna, S. (eds.) Proceedings of the Third International Space Syntax Symposium, Ann Arbor, University of Michigan, pp. 47.1–47.14 (2001)
7. Wiener, J.M., Tenbrink, T., Henschel, J., Hölscher, C.: Situated and prospective path planning: Route choice in an urban environment. In: CogSci 2008: 30th Annual Conference of the Cognitive Science Society, Washington, D.C. (2008)
8. Kim, Y.O.: The role of spatial configuration in spatial cognition. In: Peponis, J., Wineman, J., Bafna, S. (eds.) Proceedings of the Third International Space Syntax Symposium, Ann Arbor, University of Michigan, pp. 49.1–49.21 (2001)
9. Butler, D.L., Acquino, A.L., Hissong, A.A., Scott, P.A.: Wayfinding by newcomers in a complex building. Human Factors 35(1), 159–173 (1993)
10. Dogu, U., Erkip, F.: Spatial factors affecting wayfinding and orientation — a case study in a shopping mall. Environment and Behavior 32(6), 731–755 (2000)
11. Heye, C., Timpf, S.: Factors influencing the physical complexity of routes in public transportation networks. In: Axhausen, K. (ed.) 10th International Conference on Travel Behaviour Research, Lucerne, Switzerland (2003)
12. Weisman, J.: Evaluating architectural legibility: Way-finding in the built environment. Environment and Behaviour 13(2), 189–204 (1981)
13. Gärling, T., Böök, A., Linberg, E.: Spatial orientation and wayfinding in the designed environment – a conceptual analysis and some suggestions for postoccupancy evaluation. Journal of Architectural and Planning Research 3, 55–64 (1986)
14. Montello, D.R.: Spatial orientation and the angularity of urban routes — a field study. Environment and Behavior 23(1), 47–69 (1991)
15. Werner, S., Long, P.: Cognition meets Le Corbusier – cognitive principles of architectural design. In: Freksa, C., Brauer, W., Habel, C., Wender, K.F. (eds.) Spatial Cognition III. LNCS, vol. 2685, pp. 112–126. Springer, Heidelberg (2003)
16. Hillier, B., Hanson, J.: The Social Logic of Space. Cambridge University Press, Cambridge (1984)
17. O'Neill, M.J.: Evaluation of a conceptual model of architectural legibility. Environment and Behaviour 23(3), 259–284 (1991)
18. Mark, D.: Automated route selection for navigation. IEEE Aerospace and Electronic Systems Magazine 1, 2–5 (1986)
19. Richter, K.-F., Klippel, A.: A model for context-specific route directions. In: Freksa, C., Knauff, M., Krieg-Brückner, B., Nebel, B., Barkowsky, T. (eds.) Spatial Cognition IV. LNCS, vol. 3343, pp. 58–78. Springer, Heidelberg (2005)
20. Dijkstra, E.W.: A note on two problems in connexion with graphs. Numerische Mathematik 1, 269–271 (1959)
21. Duckham, M., Kulik, L.: "Simplest" paths: Automated route selection for navigation. In: Kuhn, W., Worboys, M.F., Timpf, S. (eds.) COSIT 2003. LNCS, vol. 2825, pp. 169–185. Springer, Heidelberg (2003)
22. Caduff, D., Timpf, S.: The landmark spider: Representing landmark knowledge for wayfinding tasks. In: Barkowsky, T., Freksa, C., Hegarty, M., Lowe, R. (eds.) Reasoning with mental and external diagrams: computational modeling and spatial assistance - Papers from the 2005 AAAI Spring Symposium, Menlo Park, CA, pp. 30–35. AAAI Press, Menlo Park (2005)

23. Raubal, M., Winter, S.: Enriching wayfinding instructions with local landmarks. In: Egenhofer, M.J., Mark, D.M. (eds.) GIScience 2002. LNCS, vol. 2478, pp. 243–259. Springer, Heidelberg (2002)

24. Haque, S., Kulik, L., Klippel, A.: Algorithms for reliable navigation and wayfinding. In: Barkowsky, T., Knauff, M., Ligozat, G., Montello, D.R. (eds.) Spatial Cognition 2007. LNCS, vol. 4387, pp. 308–326. Springer, Heidelberg (2007)

25. Montello, D.R., Frank, A.U.: Modeling directional knowledge and reasoning in environmental space: Testing qualitative metrics. In: Portugali, J. (ed.) The Construction of Cognitive Maps, pp. 321–344. Kluwer Academic Publishers, Dordrecht (1996)

26. Klippel, A., Montello, D.R.: Linguistic and nonlinguistic turn direction concepts. In: Winter, S., Duckham, M., Kulik, L., Kuipers, B. (eds.) COSIT 2007. LNCS, vol. 4736, pp. 373–389. Springer, Heidelberg (2007)

27. Richter, K.-F., Duckham, M.: Simplest instructions: Finding easy-to-describe routes for navigation. In: Cova, T.J., Miller, H.J., Beard, K., Frank, A.U., Goodchild, M.F. (eds.) GIScience 2008. LNCS, vol. 5266, pp. 274–289. Springer, Heidelberg (2008)

28. Richter, K.F.: Context-Specific Route Directions - Generation of Cognitively Motivated Wayfinding Instructions. DisKi 314 / SFB/TR 8 Monographs, vol. 3. IOS Press, Amsterdam (2008)

29. Klippel, A., Tappe, H., Kulik, L., Lee, P.U.: Wayfinding choremes — a language for modeling conceptual route knowledge. Journal of Visual Languages and Computing 16(4), 311–329 (2005)

30. Richter, K.F.: Properties and performance of cognitively motivated path search algorithms (under revision)

31. Daniel, M.-P., Denis, M.: Spatial descriptions as navigational aids: A cognitive analysis of route directions. Kognitionswissenschaft 7, 45–52 (1998)

32. Assunção, R.M., Neves, M.C., Câmara, G., Freitas, C.D.C.: Efficient regionalization techniques for socio-economic geographical units using minimum spanning trees. International Journal of Geographical Information Science 20, 797–811 (2006)

33. Kulik, L.: A geometric theory of vague boundaries based on supervaluation. In: Montello, D.R. (ed.) COSIT 2001. LNCS, vol. 2205, pp. 44–59. Springer, Heidelberg (2001)

34. Dale, R., Geldof, S., Prost, J.-P.: Using natural language generation in automatic route description. Journal of Research and Practice in Information Technology 37(1), 89–105 (2005)

35. Wiener, J.M., Mallot, H.A.: 'Fine to coarse' route planning and navigation in regionalized environments. Spatial Cognition and Computation 3(4), 331–358 (2003)

36. Schmid, F., Peters, D., Richter, K.-F.: You are not lost - you are somewhere here. In: Klippel, A., Hirtle, S. (eds.) You-Are-Here-Maps: Creating a Sense of Place through Map-like Representations, Workshop at Spatial Cognition (2008)

An Analysis of Direction and Motion Concepts in Verbal Descriptions of Route Choices

Karl Rehrl[1], Sven Leitinger[1], Georg Gartner[2], and Felix Ortag[2]

[1] Salzburg Research, Jakob Haringer-Straße 5/III, A-5020 Salzburg, Austria
{karl.rehrl,sven.leitinger}@salzburgresearch.at
[2] Institute for Geoinformation and Cartography, Vienna University of Technology
Erzherzog-Johann-Platz 1/127-2, A-1040 Vienna, Austria
{georg.gartner,felix.ortag}@cartography.tuwien.ac.at

Abstract. This paper reports on a study analyzing verbal descriptions of route choices collected in the context of two in situ experiments in the cities of Salzburg and Vienna. In the study 7151 propositions from 20 participants describing route choices along four routes directly at decision points (100 decision points in total) are classified and compared to existing studies. Direction and motion concepts are extracted, semantically grouped and ranked by their overall occurrence frequency. A cross-classification of direction and motion concepts exposes frequently used combinations. The paper contributes to a more detailed understanding of situational spatial discourse (primarily in German) by participants being unfamiliar with a way-finding environment. Results contribute to cognitively-motivated spatial decision support systems, especially in the context of pedestrian navigation.

1 Introduction

Navigation is one of the most common spatial activities of human beings involving a number of spatial abilities and cognitive processes [25]. Navigation requires continuous sense-making of the proximal surrounds ([12], [25]) which is considered to be a challenging task in unfamiliar spatial environments. With the advent of electronic navigation systems [1] better human decision making in such environments seems to be in reach. One of the major research questions is how the process of human spatial decision making can be supported effectively by such systems. Today's electronic navigation systems provide three types of user interfaces: (1) maps, (2) visual turn instructions and (3) textual or voice-based turn instructions. In the domain of car driving voice-based instructions have gained considerable attention in the last years whereas in other navigation domains (e.g. pedestrian navigation) map-based guiding systems are predominant [1]. Guiding pedestrians by textual or voice-based turn instructions is still in its infancies [26].

Among the open questions are the following: Are textual or voice-based turn instructions useful for pedestrian navigation and if so, how should these instructions be structured? Manufacturers of navigation systems currently tend to disable voice instructions upon switching to pedestrian mode and reduce decision support to

K. Stewart Hornsby et al. (Eds.): COSIT 2009, LNCS 5756, pp. 471–488, 2009.
© Springer-Verlag Berlin Heidelberg 2009

map-based interfaces[1]. Map-based interfaces on small screens lead to problems with map reading performance [7] which gives motivation to explore voice-based interfaces. However, empirical studies testing the performance of voice-based guidance are rare [26]. Several authors explore textual descriptions and come to the conclusion that turn instructions for pedestrians are useful ([4],[24]). In addition elements for good descriptions are suggested [21]. In this context also the importance of landmarks for improved decision support is stressed [5] and the positive impact of landmarks on navigation performance is confirmed [30]. In recent work models for generating turn instructions from combinations of spatial direction concepts and landmarks have been proposed [2]. Also the role of direction concepts in way-finding assistance has been explored [15], resulting in cognitively motivated models for cardinal directions [16]. Striving towards voice-based instructions the role of language in spatial decision support is subject to an ongoing debate (e.g. [16],[35]). One of the established methods to explore spatial discourse are empirical studies ([3],[5],[34]). In experiments participants are asked to verbally describe well-known routes. A review of these studies has identified two major gaps: (1) the lack of in situ studies exploring verbal descriptions of route choices in the context of real world decision situations and (2) the lack of in situ studies involving participants being unfamiliar with the way-finding environment. Due to these gaps we consider the ad-hoc interpretation of spatial decision situations by people being unfamiliar with the way-finding environment as not studied adequately. Studying language use in such situations is crucial for empirically founded turn instructions which do not only reflect expert views (as provided by citizens) but consider perception of space by people being unfamiliar with an environment and thus being spatially challenged. We think that a better understanding of this non-expert view [11] will contribute to user-centered turn instructions and is therefore the main motivation for our study on which we report in this paper.

The structure of the paper is as follows. Section 2 describes the set-up of experiments. Section 3 classifies propositions (following Denis' classification) and discusses results. Section 4 proposes a method for further analysis of spatial direction and motion concepts. In Section 5 and 6 direction and motion concepts are structured in ranked taxonomies. Section 7 discusses cross-classification results and sketches a model for composing turn instructions. Section 8 summarizes and concludes the paper. Since the study was organized in two Austrian cities with German-speaking participants, the primary results are in German language. In order to ensure the validity of the results we primarily refer to German language concepts and try to provide the most suitable English translation (which may differ from the original meaning and should primarily improve readability).

2   Collecting Situational Verbal Descriptions of Route Choices

The analysis of verbal descriptions in spatial context has a long tradition. In one of the first empirical studies [22] Kevin Lynch explored the mental structuring of city-scale spaces [10] in language. The study is one of the first examples of using the method

---

[1] This was a finding in tests with Smartphone-based navigation systems Route 66, Wayfinder 8 and Nokia Maps 2.0 and 3.0.

later referred to as "think aloud" [20] for gaining insight into human conceptions of space [10]. The main contribution concerning the analysis of verbal route descriptions comes from Denis ([4],[5],[6]). In [5], Denis grounds a general framework for analyzing route descriptions through spatial discourse and uses the framework for several experiments with participants familiar with the environment ([5],[6]). Although the work of Denis lacks the aspect of the situational use of language, the general framework is well suited to be adopted in similar studies. Other empirical studies explore the use of spatial language in various settings ([27],[34]). Raubal's case study of way-finding in airports contributes to a better understanding of spatial discourse in indoor settings, but lacks the situational context. A recent study by Brosset [2] explores the spatial discourse of orienteering runners in rural environments. This study also does not answer the question about the situational spatial discourse in decision situations. Another recent study by Ishikawa [14] compares the effect of navigation instructions using different frames of reference (absolute and relative). Since the study uses predefined turn instructions questions about situational spatial discourse are not answered. Some other contributions deal with landmarks in route directions (e.g. in the context of route descriptions as navigational aids [4] or as enrichment for way-finding instructions [28]) but do not address direction and motion concepts or situational use. Concluding from related work our motivation for the study was twofold:

1. Obviously, there is a lack of in situ studies analyzing situational spatial discourse. However, since human spatial decision making is highly situational, the only way to study human cognitive processes in this context is the situation itself. One of the main motivations for our study is to narrow this gap.
2. Most of the existing studies explore route descriptions by asking participants to recall memorized route knowledge. Until now we do not have empirical evidence how people walking a route for the first time and thus unfamiliar with the spatial surrounds describe spatial decision situations. By studying spatial discourse in such situations we will get verbal descriptions of route choices coming closest to situations where turn instructions by way-finding assistance systems are typically given. Since our experimental set-up fosters such situations we think we can learn from these for composing empirically-founded turn instructions.

2.1 Experimental Set-Up

Selection of routes: The experiments were organized in the two Austrian cities Salzburg and Vienna. In each city we pre-defined two routes, one in the inner city and one in a peripheral district. Routes were composed of 22 and 27 decision points. Different cities and environments were chosen in order to address the question to which extend situational spatial discourse is depending on the physical environment. In order to avoid learning effects the sequence of routes was changed between participants.

Participants: We selected two test groups consisting of 10 participants in each city. Participants in each city were half female and half male. Participants were first term students and stayed in the city for no longer than 3 weeks before the experiment. All participants confirmed to have no or very limited spatial knowledge of the cities and to be unfamiliar with the test routes. Each participant had to complete both routes in one city. In order to avoid problems with the think aloud method [20] we did a pre-test consisting of 3 decision points. The route for the pre-test was separated from the test routes in order to avoid prior acquisition of route and environmental knowledge of participants. Participants were paid (Salzburg) or got credits for a course at the University (Vienna).

Implementation: A test instructor accompanied participants to the starting point of each test route and supervised the tasks defined by the experiment along the pre-defined route to the end point. At decision points test candidates were asked to (1) describe the surrounding environment and (2) to describe all their possible choices at this decision point as they would explain to another person. Since participants were unfamiliar with the route they described all possible choices at each decision point. We asked participants to refer to the visible spatial environment (the so-called vista space [25]) in their descriptions and to provide unambiguous descriptions of route choices. If the description of a route choice was not considered unambiguous for any other person (e.g. the use of spatial-dimensional terms such as "left", "right" or "straight" without any reference to a fixed spatial entity was not accepted as unambiguous), the participants were asked to continue the description process. Afterwards the test instructor told participants which choice to take. The test instructor used the same wording as participants for giving route instructions in order to avoid any influences on perception as well as language. Decision points were identified by participants themselves (by stopping their walk). If a participant did not identify a decision point, this point was skipped and the test person was directed in the right direction (by gestures). At the end of the experiment each test person was asked to reproduce a summarizing route description. A final survey about demographic data and self-assessment of spatial abilities completed the experiment. Verbal protocols were recorded and transcribed afterwards.

3   Classification of Propositions

Firstly, we classified propositions using Denis' classification [5] (Table 1).

Table 1. Examples from the protocols showing the classification method

Class	German	English
1 Actions	rechts abbiegen	turn right
2 Actions with references to landmarks	in die Straße abbiegen	turn into the street
3 Landmarks without actions	ich bin an einer Kreuzung	I'm at a crossing
4 Landmark descriptions	gelbes Haus	yellow house
5 Others	Überqueren ist hier verboten	crossing is not allowed here

Due to the situational set-up of our experiments participants were able to refer to any kind of spatial entities from the visible environment. Thus, for the classification we treated all kinds of entities as landmarks (since entities were perceived by participants as visually salient, which satisfies the commonly used definition of the term landmark [31]). Nevertheless we are aware that this use of the term landmark may be in contrast to other definitions (e.g. [22]).

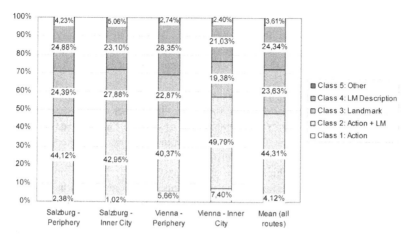

Fig. 1. Comparison of proposition types used throughout the four test routes

We considered 7151 propositions (expressed by 10 participants on 2 routes with 23 and 27 decision points in each city) describing spatial scenery and route choices at decision points for classification (we have not considered 798 additional propositions from route summaries since the focus of this analysis is on the situational use). Fig. 1 compares results from the Denis classification between the four test routes.

The comparison shows that about half of all propositions were used by participants to describe actions (actions without references to landmarks or actions with references to landmarks); the other half of propositions was used by participants to describe landmarks and their characteristics. In the landmark description group (classes 3 and 4, 47.97% mean value) 23.63% propositions are used to introduce landmarks (landmarks which were not referenced before) and 24.34% are used to describe landmarks; most of the propositions in the action group (44.31% out of 48.43%) relate actions to landmarks. Fig. 1 confirms that the overall distribution of propositions in the 5 classes is widely stable between test groups and test routes. Minor deviations occur between test groups in the use of class 1 and class 2 propositions and between the Vienna – Inner City route and the other routes. The increased use of class 1 propositions by the test group in Vienna is considered a characteristic of participants (environmental planning students vs. students of social studies in Salzburg). The increased use of class 2 propositions throughout the Vienna – Inner City route has its origin in differences of the physical environment (more salient landmarks, which do not need further description). All in all the general distribution confirms the extensive use of landmarks in the situational description of route choices (in approximately 90% of all propositions landmarks are referenced), which has been additionally fostered by our experimental set-up.

The comparison with previous studies by Denis [5] and Brosset [2] shows that our results fit in the overall picture. The distribution of action- and landmark-related propositions is half-and-half in all three studies. As expected, our experimental set-up resulted in a higher number of action-landmark relations (44.31% compared to 35 % in the Brosset study and 33.5% in the Denis study). Another noticeable variance is the fact that similar to the Brosset study our study leads to an increased number of land-mark descriptions (26.5% in the Brosset study, 24.34% in our study compared to 11.3% in the Denis study). Although Brosset argues that this aspect is influenced by the difference between natural and urban environments our results clearly show that the increased use of landmark descriptions is not only appearing in natural environ-ments, but also in urban environments. Thus we assume that the increased use of landmark descriptions mainly depends on the in situ aspect of the studies, which is common in Brosset's and our experiments, but is missing in the Denis experiment.

To summarize, Denis' classification leads to four findings: (1) the overall distribu-tion of proposition types in verbal descriptions shows a low variance between differ-ent routes and test groups which allows for the assumption, that personal or physical variances have only minor influence on used proposition types, (2) the half-and-half distribution between action related propositions and landmark description related propositions is stable for verbal descriptions of routes or route choices in different studies, (3) complementing previous studies our study shows a high-relevancy of route descriptions relating actions to landmarks and (4) also shows a clear tendency towards an increased use of detailed landmark descriptions in situ, which has accord-ing to [3] not been sufficiently answered by existing studies.

In the further analysis we focus on propositions classified in the action group (class 1 and 2) and leave the analysis of the landmark description group (class 3 and 4) to future work.

4   Extraction of Direction and Motion Concepts from Propositions

Most of the propositions in class 1 and 2 are so-called directionals expressing a change in the localization of an object ([8],[37]). Directional expressions consist of three main particles which are of further interest to us [2]: spatial relations, motion verbs and landmarks. While we leave landmarks to future analyses, the reported analysis is motivated by the following questions: (1) Which set of spatial direction concepts is used in the propositions, (2) which set of motion concepts is used in the propositions and (3) how frequently are these concepts used by the test groups.

To answer these questions the overall question how to deal with semantics of spa-tial relations in language has to be considered. According to several authors ([489],[32],[37]) spatial prepositions (and adverbs in the German language) are the main language concepts to express spatial relations. Direction concepts given from the view point of the speaker and thus of interest in the context of our study are de-noted as projective relations [13]. According to Herskovits spatial relations can be either spatial-dimensional (e.g. "in front of", "behind"), topological (e.g. "in", "on") path-related (e.g. "across", "through"), distance-related (e.g. "near", "far") or belong to some other category (e.g. "between" or "opposite"). One of the newer accounts contributing to the understanding of semantics of prepositions comes from Tyler and Evans [36]. In their work they integrate previous accounts and propose a theory called principled polysemy as a foundation for analyzing the semantics of English preposi-tions. According to their account semantics of prepositions is not only determined by the preposition itself but distinct senses rely on the context. They call the semantic nucleus of prepositions a "proto-scene", building on our daily experiences with our physical surrounds and describing concrete spatial scenes as highly abstract and schematic relationships. Thus "proto-scenes" allow mapping of different spatial scenes on one schematic concept which can be considered as good foundation for a semantic reference system [18].

Our analysis is based on the proposed breakup of spatial propositions into several particles of a "proto-scene". Tyler and Evans denote the "schematic trajectory" (TR), a "schematic background element" (the landmark or short LM), the "spatial relation" between the TR and the LM and "functional elements" of the landmarks determining a spatial relation. Whereas the nature of the TR and LM particles follows closely the primary breakup of a spatial scene proposed by Talmy [32] (the trajectory acts as the primary object and the landmark acts as the secondary object), the distinct sense of the spatial relation is determined by the spatial preposition and the functional elements of the landmark (functional elements specify the functional role a landmark takes in a spatial relation). Following this approach our method for extracting the particles is structured as follows (repeated for each proposition):

1. Extract the TR (primary and often moving object, in most cases a person)
2. Extract the LM (secondary object(s), static reference in the spatial relation)
3. Extract the spatial relation (spatial preposition or adverb in the German language)
4. Extract the motion verb

The resulting particles are tagged with a unique identifier of the participant (PID), the number of the decision point (DP) and a unique identification of route choices at decision points (C). If one of the particles could not be unambiguously identified from the context the corresponding entry in the result table was left empty (Table 2).

If one proposition contained more than one spatial relation, the proposition was split up and each spatial relation ran separately through the extraction process. Verbs were translated to their infinitive. Helper verbs (such as "can") were not considered. In order to adapt the method to German language spatial adverbs were treated like spatial prepositions.

Table 2. Example propositions and the related break-up

Propositions
and I can walk towards the church (und ich kann in die Richtung der Kirche gehen)
there's the possibility to move along the street (es gibt die Möglichkeit sich die Straße entlang zu bewegen)
first cross the zebra crossing (zuerst den Zebrastreifen überqueren)
through a small archway (durch einen kleinen Bogen)

TR	LM	Motion Verb	Spatial Relation	PID	DP	C
I (ich)	church (Kirche)	walk (gehen)	towards (Richtung)	S6	1	A
	street (Straße)	move (bewegen)	along (entlang)	S3	2	B
	zebra crossing (Zebrastreifen)	cross (queren)	across (über)	S4	2	B
	archway (Bogen)		through (durch)	S3	3	C

5 Results from the Analysis of Direction Concepts

The main goal of the analysis was to identify the set of re-occurring direction concepts used by participants. In total 3940 propositions (3652 of them containing spatial particles) of class 1 and 2 have been analyzed. Extracting the spatial particles resulted in a set of 103 different spatial prepositions and adverbs. Since some of the 103 particles belong to a subcategory of a more general direction concept or are synonyms these concepts could be semantically grouped. The rules for semantic grouping rely on related work ([8],[13],[36]) and on a German semantic dictionary[2].

85 particles could be grouped to one of 26 main direction concepts. The remaining 18 particles were not considered for further analysis since the particles were only used in singular propositions. Fig. 2 shows the taxonomy of spatial relations. The classification is adopted from related work ([13],[36]). Main classes are structured by *orientation*, *goal*, *path*, *topology* and *distance*. The mean occurrence frequency of concepts is shown in brackets.

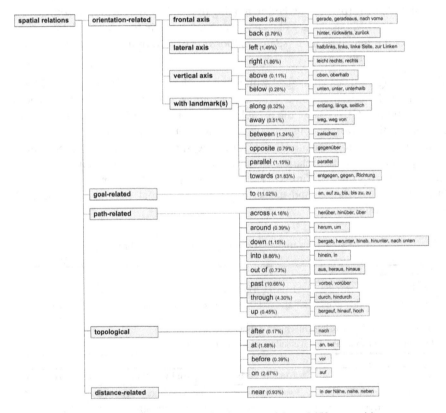

Fig. 2. Taxonomy of spatial relations generated from 3652 propositions

---

[2] Online German Semantic Dictionary based on data provided by the University of Thübingen (Germany). Accessible as http://canoo.net.

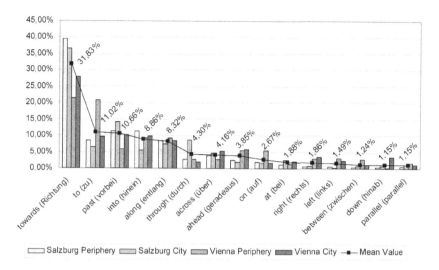

Fig. 3. Occurrence frequency of 15 direction concepts used in more than 1% of all propositions

In the further analysis we compare the 15 most frequently used spatial relations (each class is used in more than 1% of all propositions). We assume that the remaining set of 11 spatial prepositions only plays a minor role in the situational use of verbal descriptions. Fig. 3 compares the usage between the four test routes.

The predominantly used spatial relation in both test groups is "towards" (31.83% mean value). In combination with a landmark or spatial entity "towards" is one of the simplest possibilities to describe route choices in spatial decision situations. The frequent use of this direction concept is in accordance with the frequent use of landmark descriptions (Denis class 4) since landmarks or spatial entities used with "towards" have to be identified unambiguously. Noticeable is a variance in the use of "towards" between the two test groups in Salzburg and Vienna. Considering the whole distribution one will recognize that the test group in Vienna makes more use of "to" as well as the *spatial-dimensional* concepts "ahead", "left", "right" and "between". It seems that instead of the predominant concept "towards" the group in Vienna used a slightly broader range of concepts although variances are not very significant. One interesting question is whether differences in the physical environment influence the use of spatial direction concepts or not. One noticeable variance is the increased use of the preposition "to" along the Vienna – Periphery route (with 20.85% three times higher compared to the mean value of the other routes 8.2%). Since salient landmarks are missing along this route participants frequently referred to structural spatial entities like "streets" or "pathways" and transitively described where these entities "lead them to". Another noticeable variance is the increased use of "through" in the Salzburg – City route. Due to the medieval environment in the old town of Salzburg the number of archways and passages is higher compared to the other test routes. We assume that this aspect of the physical environment is directly reflected in the results, since the occurrence of the concept is more than three times higher (8.56%) compared to the mean of the other routes (2.4%). A similar argumentation is valid for the concept "down", which is more frequently used in the Vienna – City route. This could be explained with differences in the scenery where some streets are leading "down". From the whole distribution we conclude that the use of direction concepts by participants follows a distribution with low variances between the four test routes. We further conclude that the situational use of direction concepts by participants reflects the scenery only in very special cases. We also discovered slight differences in the use of direction concepts between the two test groups. We explain the variances with slightly different skills in the expression of spatial relations (the test group in Vienna used a slightly broader range of direction concepts). Since our study is one of the first studies exploring spatial discourse in real world scenarios we have not found valid data to compare the reported results.

6 Results from the Analysis of Motion Concepts

For the analysis of motion concepts the same 3940 propositions in class 1 and 2 were used (2309 of them containing motion verbs). Extracting the motion particles resulted in 133 different motion verbs (including verbs with different adverbs). We classified motion verb stems into motion verb classes (following Levin classification [19]). Fig. 4 shows relative occurrences of motion verbs in 10 matching Levin classes.

The classification was highly effective since a large number of verbs in German language differ from their stem only because of the attached adverb. 113 verbs could be mapped to one of 10 verb classes. The remaining 20 verbs only occurred in singular propositions or did not express motion and thus were not further considered. The resulting taxonomy (Fig. 5) shows the complete list of used German verbs classified in 10 matching Levin classes.

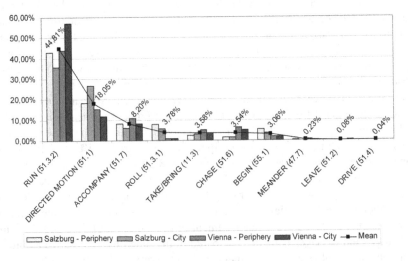

Fig. 4. Relative occurrences of motion verbs in 10 matching Levin classes

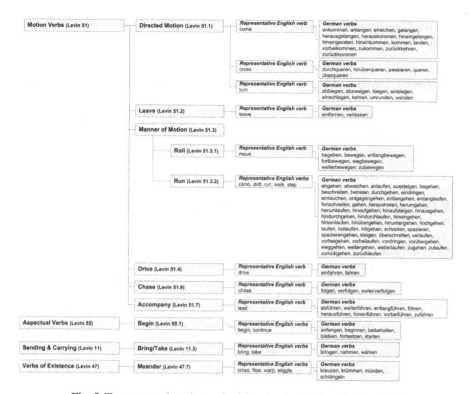

Fig. 5. Taxonomy of motion verbs following Levin classification [19]

The most frequently used verb class is the manner of motion class *run* (Levin class 51.3.2, 44.81% relative occurrences). Since the German language is considered to be a manner-typed language [33] (manner of motion is predominantly expressed in the verb) in most of the propositions the *run* verb "gehen" (walk) in combination with a spatial adverb is used. Other frequently used manner of motion verbs are "laufen" – "run" and "schreiten" – "step". The second largest class is the class of *inherently directed motion* verbs (e.g. the verbs "queren" - "cross", "kommen" – "come" or "passieren" – "pass" are according to Levin considered to be inherently directed). We also classified "turn" in this class since it does not express a manner of motion but is used to express directions. Including "turn" 18.05% of used motion verbs are inherently directed. This percentage could be significantly higher with other languages (e.g. English) since paths are more often directly expressed in verbs and not in adverbs [33]. Worth to mention are also verbs in the *accompany* class (e.g. "führen" – "lead") with 8.2% occurrences. In most of these occurrences the verb is used transitively where the *subject* is not the participant but a spatial entity (e.g. "the path leads us towards the building"). Also the verbs "bring" and "take" are used transitively (e.g. "the path takes me to the building"). The German verb "bewegen" – "move" is classified as *roll* verb. Although the overall use of motion verbs between test groups is very similar, "move" was mainly used by the test group in Salzburg (6.6%). In general, the overall usage of motion verbs shows the predominance of manner type motion verbs in German language. Although experiments with 20 participants are not representative for general conclusions this language characteristic is also confirmed by other authors [33] and thus should be considered in the composition of turn instructions.

7 Cross-Classification Analysis of Direction and Motion Concepts

Results from the distinct analyses of direction and motion concepts lead to the question of combined usage. To explore combined usage we use a cross-classified table revealing occurrence frequencies of 26 direction concepts and 10 motion verb classes (Fig. 6).

In the cross-classification we focus on *orientation-*, *goal-*, *path-* and *topology-*related combinations occurring more than 50 times in all propositions. Combinations with frequencies less than 50 (except the borderline "on" with 49 occurrences) are not considered as frequently used.

The most frequently (in combinations with spatial relations) used verb class is the class of *run* verbs with 948 occurrences. Verbs in this class are predominantly used in combination with the *orientation* relations *towards* (213, e.g. "in Richtung gehen" – "walk towards"), *along* (145, e.g. "entlang gehen" – "walk along") and *ahead* (60, e.g. "geradeaus gehen" – "walk ahead"). Furthermore verbs are used in the *goal* relation *to* (94, e.g. "gehen zu" – "walk to") and in the *path* relations *into* (94, e.g. "hineingehen" – "walk into"), *past* (68, e.g. "vorbeigehen" – "walk past") and *through* (51, e.g. "durchgehen" – "walk through"). There is only one topological relation which is frequently used with a *run* verb: *on* (49, e.g. "gehen auf" –"walk on"). As obvious from the examples above some of the combinations of "walk" with a spatial preposition can also be expressed by path-related verbs in English, e.g. pass, cross or enter. However, in German some of these *directed motion* verbs do not have counterparts and thus have to be expressed with manner type verbs (e.g. walk) and spatial relations as adverbs (e.g. into, out of, through).

The second largest class of motion verbs is the *directed motion* class (e.g. come, cross, turn) with 558 occurrences. Predominant combinations are the *orientation* relation *towards* (94, e.g. "abbiegen in Richtung" – "turn towards"), the *goal* relation *to* (85, e.g. "kommen zu" – "come to") and the *path* relations *into* (108, e.g. "abbiegen in" – "turn into") and *across* (68, e.g. "überqueren" – "cross"). In this class we only

	FRONTAL AXIS		LATERAL AXIS		VERTICAL AXIS		LANDMARK						GOAL	PATH								TOPOLOGICAL				DISTANCE	
	AHEAD	BACK	LEFT	RIGHT	ABOVE	BELOW	ALONG	AWAY	BETWEEN	OPPOSITE	PARALLEL	TOWARDS	TO	ACROSS	AROUND	DOWN	INTO	OUT OF	PAST	THROUGH	UP	AFTER	AT	BEFORE	ON	NEAR	SUMS
DIRECTED MOTION (51)	3	5	23	30			5	3	4	10	3	94	85	68	3		108	10	18	14		1	29	8	24	10	558
LEAVE (51.2)					1	1	1	1	1			1															6
ROLL (51.3.1)			1	2	2		16			1	6	22	12		1		6		1	1	1				4		76
RUN (51.3.2)	60	8	16	11	5	8	145	4	11	5	12	213	94	26	4	24	94	4	68	51	6	1	18	3	49	8	948
DRIVE (51.4)							2										1										3
CHASE (51.6)	1	1					3	1				4													3		13
ACCOMPANY (51.7)	3	3		1	6	8	11	4	3		5	19	58	7		3	23		26	8			1		11	1	201
BEGIN (55.1)	2								2	1	1	29			3		3			2			1		5		49
TAKE/BRING (11.3)			4	9	1							6				3	1		2	1			2				29
MEANDER (47.7)				1											1	0	2										4
SUMS	69	16	40	48	14	28	184	12	21	18	27	388	249	104	9	30	238	14	115	77	7	2	51	11	96	19	1887

Fig. 6. Cross-tabulation of used direction and motion concepts

classified inherently directed German verbs. Thus the occurrence frequency would likely be varying with other languages.

The third largest verb class is the *accompany* class with 201 occurrences. There is only one combination above the threshold of 58 occurrences, namely the transitively used verb *lead* with the preposition *to* (58, e.g. "führen zu" - "lead to").

All other combinations are less frequently used. Since the cross-classification is based on empirical data the extracted combinations do not result in a complete set of prototypical combinations of motion verbs and spatial relations. However, the cross-classification clearly indicates combinations which are frequently used by participants throughout all four experiments (despite different environments, different test routes and two different test groups). Based on the empirical data we classify 13 combinations as the relevant ones, noting that the complete set will certainly include further combinations being necessary for the description of different route tasks (good additional candidates may be combinations from the cross-classification with 20 to 30 occurrences). The following table (Table 3) summarizes the 13 relevant combinations ranked by their occurrence frequency. Furthermore we add the usage context (*orientation*, *goal*, *path* or *topology*) and whether the combination is used in context with a spatial entity or landmark.

Table 3. Frequently (>50) used combinations of motion and spatial relation concepts

NO	VERB	RELATION	USEAGE	LANDMARK	FREQ
1	WALK	TOWARDS	ORIENTATION	Yes	213
2	WALK	ALONG	ORIENTATION	Yes	145
3	TURN	INTO	PATH	Yes	108
4	TURN	TOWARDS	ORIENTATION	Yes	94
5	WALK	TO	GOAL	Yes	94
6	WALK	INTO	PATH	Yes	94
7	COME	TO	GOAL	Yes	85
8	WALK	PAST	PATH	Yes	68
9	CROSS	ACROSS	PATH	Yes	68
10	WALK	AHEAD	ORIENTATION	No	60
11	LEAD	TO	GOAL	Yes	58
12	WALK	THROUGH	PATH	Yes	51
13	WALK	ON	TOPOLOGY	Yes	49

The 13 predominantly used combinations shape a set of re-occurring and prototypical *action schemes* [29] relating prototypical actions to prototypical spatial scenes (following the definition of Tyler and Evans [36]). Whereas actions either specify *manner* or *path* [33] of a moving trajector *proto-scenes* express the *spatial orientation* or the *spatial relation*. *Action schemes* combine both particles and are thus considered well suited for the specification of ontologies for route tasks ([17], [34]).

7.1 Modeling Route Tasks with Action Schemes

As proposed in related work ([9],[29]) a route can be topologically described as view-graph (basically a topological network of interconnected local views). View graphs are

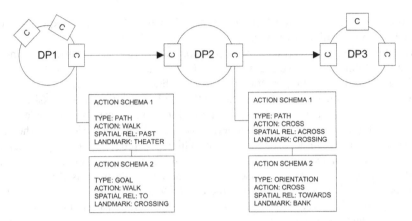

Fig. 7. Example of modeling route choices with *action schemes*

composed from local views and actions linking these views. By executing an *action schema* a trajector can move from one decision point to the next. In our approach we propose a similar model specifying actions and spatial relations in prototypical *action schemes*. A route is modeled by an ordered sequence of decision points where each decision point offers different route choices to trajectors. Decision points are logically linked by a sequence of *action schemes*, allowing trajectors to transit from one decision point to another. *Action schemes* specify the path by relating actions to spatial entities via spatial relations. We differentiate between *orientation*, *goal* and *path schemes*, depending on involved action and relation types. Fig. 7 shows an example of modeling route choices (C) and transitions between decision points (DP). *Action schemes* are attached as ordered sequences to route choices and directed transitions.

In order to complete the route task between two decision points a trajector has to interpret the *action schemes* in their specified order. The task is composed of identifying spatial entities (views), relating actions to entities and performing the actions. Thus *action schemes* have to unambiguously describe the path from one decision point to the next. Instances of *action schemes* may be translated to natural language instructions. Therefore the empirical data from the experiments, the extracted concepts and the *action schemes* model provide the foundation.

8   Summary and Conclusion

In this paper we pursue a user-centered approach analyzing language use in spatial decision situations. We analyzed direction and motion concepts in verbal descriptions of route choices gathered from two in situ experiments in the Austrian cities Salzburg and Vienna. The experiments differ from previous experiments ([2],[5],[27],[34]) in the aspect that the in situ use of spatial language by people being unfamiliar with a spatial environment is explored. In the analysis of verbal descriptions we focus on three questions: (1) How do the results differ from previous studies, (2) which direction and motion concepts are predominantly used by participants and (3) what can we learn for composing turn instruction by analyzing combined usage?

How do the results differ from previous studies?

From the comparison of classified propositions (Denis' classification) with previous studies by Brosset [3] and Denis [5] we conclude that similarities overbalance differences. One interesting finding is that the increased use of landmark descriptions (Denis Class 4) is not a matter of natural environments (as Brosset argues), but likely results from the in situ aspect of the experiments. In this aspect the Brosset study and our study reveal similar results whereas the study by Denis, which does not cover the in situ aspect, shows a clear difference. Additionally, results from both studies provide some evidence that the increased use of propositions relating actions to landmarks is another characteristic of in situ descriptions. Our experimental set-up fostered the use of propositions relating actions to landmarks and increased the overall usage by 10% (compared to the studies by Brosset and Denis). Since the cross-classification in Section 7 clearly benefits from a higher number of class 2 propositions (an increased quantity of actions relating a trajector to spatial entities results in a richer set of qualitative spatial relations) we consider the experiment set-up as successful. Further comparisons of direction and motion concepts have to be postponed to future work due to the lack of comparable studies.

Which direction and motion concepts are predominantly used by participants?

The analysis revealed a set of 15 direction concepts (use in more than 1% of all propositions) and 6 verb classes (use in more than 1% of all propositions) which were predominantly used by participants throughout two experiments and four test routes. Due to a careful experimental set-up (two different test groups, two different cities and two different test routes in both cities) we assume that the overall distribution is only marginally depending on environmental or personal differences. Although test groups with 20 participants are not considered representative for general conclusions, we consider the number of participants as suited for getting first insights into language use in situational descriptions of route choices. Additional experiments exploring cultural as well as environmental influences are needed. However, as a first step towards empirically founded ontologies [17] our results contribute to the task-perspective as to situational aspects. From the experiments we get a bunch of verbal descriptions of concrete route tasks in decision situations. The reported analysis reveals the most frequently used direction and motion concepts. The composition of prototypical *action schemes* has been sketched in Section 7. In contrast to expert ontologies our approach results in empirically founded concepts. We think that the strive towards natural language instructions in electronic pedestrian guidance can benefit from these results. Scholars are encouraged to set up similar experiments for widening the empirical foundation.

What can we learn for composing turn instruction by analyzing combined usage?

The final cross-classification further narrows the set of direction and motion concepts to 13 predominantly used combinations (more than 50 occurrences in all propositions). From the cross classification only 5 motion concepts (come, cross, lead, turn, walk) and 9 direction concepts (across, ahead, along, into, on, past, through, to, towards) remain. We consider the 13 combinations as a good, empirically founded starter set for the composition of semantically enhanced turn instructions. Since the set is based on empirical data it is not likely that it is complete (first attempts to unambiguously describe test routes with the 13 concepts have revealed some missing ones), however the cross classification also reveals further combinations which are used less by participants but may be essential for describing certain route choices (some combinations with frequencies between 20 and 30 have turned out to be among the missing ones). Extending the set is ongoing work. Revealed *action schemes* are tested by describing route tasks along the four test routes. At each decision point we select the most frequently used concepts for the composition of *action schemes*. If the concepts are not among the 13 we further extend the set. Completing all route tasks will lead to a revised set which is considered as a good candidate set for a more generalized model well-suited to describe most of the route tasks in built environments. To complete the model an empirically founded taxonomy of spatial entities and landmarks and their usage in the proposed *action schemes* has to be added. Such taxonomy can be deduced by further analyzing propositions in Denis' class 2.

In future work a second iteration of experiments guiding a new set of participants unfamiliar with the environment along the test routes will show whether the user-centered approach (closing the loop from user descriptions to concepts and back to natural language instructions) is worth to be pursued in the future. If so, the approach complements existing work on route descriptions ([2],[4],[21]) as well as empirical studies on the performance of verbal turn instructions ([14],[30]) and contributes to a foundation for semantically enhanced decision support in future electronic pedestrian navigation systems.

Acknowledgements

The work was mainly accomplished in SemWay, a project partly funded by the Austrian Ministry for Transport, Innovation and Technology in the thematic research program FIT-IT Semantic Systems.

References

1. Baus, J., Cheverst, K., Kray, C.: A survey of map-based mobile guides. In: Meng, L., Zipf, A., Winter, S. (eds.) Map-based mobile services - Theories, Methods, and Implementations, pp. 197–216. Springer, Heidelberg (2005)
2. Brosset, D., Claramunt, C., Saux, É.: A Location and Action-Based Model for Route Descriptions. In: Fonseca, F., Rodríguez, M.A., Levashkin, S. (eds.) GeoS 2007. LNCS, vol. 4853, pp. 146–159. Springer, Heidelberg (2007)
3. Brosset, D., Claramunt, C., Saux, E.: Wayfinding in natural and urban environments: a comparative study. Cartographica: The International Journal for Geographic Information and Geovisualization 43(1), 21–30 (2008)
4. Daniel, M.P., Denis, M.: Spatial descriptions as navigational aids: A cognitive analysis of route directions. Kognitionswissenschaft 7, 45–52 (1998)
5. Denis, M.: The description of routes: A cognitive approach to the production of spatial discourse. Cahiers de Psychologie Cognitive 16, 409–458 (1997)
6. Denis, M., Pazzaglia, F., Cornoldi, C., Bertolo, L.: Spatial discourse and navigation: an analysis of route directions in the City of Venice. Applied Cognitive Psychology 13, 145–174 (1999)

7. Dillemuth, J.: Map Size Matters: Difficulties of Small-Display Map Use. In: Proceedings of the 4th International Symposium on LBS & TeleCartography, Hong Kong, November 8-10 (2007)

8. Eschenbach, C.: Contextual, Functional, and Geometric Components in the Semantics of Projective Terms. In: Carlson, L., van der Zee, E. (eds.) Functional features in language and space: Insights from perception, categorization and development. Oxford University Press, Oxford (2004)

9. Franz, M., Schölkopf, B., Mallot, H.A., Bülthoff, H.H.: Learning view graphs for robot navigation. Autonomous Robots 5, 111–125 (1998)

10. Freundschuh, S., Egenhofer, M.: Human Conceptions of Space: Implications for GIS. Transactions in GIS 2(4), 361–375 (1997)

11. Fontaine, S., Edwards, G., Tversky, D., Denis, M.: Expert and Non-expert Knowledge of Loosely Structured Environments. In: Cohn, A.G., Mark, D.M. (eds.) COSIT 2005. LNCS, vol. 3693, pp. 363–378. Springer, Heidelberg (2005)

12. Gluck, M.: Making Sense of Human Wayfinding: Review of cognitive and linguistic knowledge for personal navigation with a new research direction. In: Cognitive and Linguistic Aspects of Geographic Space, pp. 117–135. Kluwer Academic Press, Dordrecht (1991)

13. Herskovits, A., Bird, S., Branimir, B., Hindle, D.: Language and Spatial Cognition: An Interdisciplinary Study of the Prepositions in English. Cambridge University Press, Cambridge (1986)

14. Ishikawa, T., Kiyomoto, M.: Turn to the left or to the west: Verbal navigational Directions in relative and absolute frames of reference. In: Cova, T.J., Miller, H.J., Beard, K., Frank, A.U., Goodchild, M.F. (eds.) GIScience 2008. LNCS, vol. 5266, pp. 119–132. Springer, Heidelberg (2008)

15. Klippel, A., et al.: Direction concepts in wayfinding assistance. In: Baus, J., Kray, C., Porzel, R. (eds.) Workshop on artificial intelligence in mobile systems (AIMS 2004), pp. 1–8 (2004)

16. Klippel, A., Montello, D.R.: Linguistic and Non-Linguistic Turn Direction Concepts. In: Winter, S., Duckham, M., Kulik, L., Kuipers, B. (eds.) COSIT 2007. LNCS, vol. 4736, pp. 373–389. Springer, Heidelberg (2007)

17. Kuhn, W.: Ontologies in support of activities in geographical space. International Journal of Geographical Information Science 15, 613–631 (2001)

18. Kuhn, W.: Geospatial semantics: Why, of what, and how? In: Spaccapietra, S., Zimányi, E. (eds.) Journal on Data Semantics III. LNCS, vol. 3534, pp. 1–24. Springer, Heidelberg (2005)

19. Levin, B.: English Verb Classes and Alternations. University of Chicago Press (1993)

20. Lewis, C., Rieman, J.: Task-Centered User Interface Design. Clayton Lewis and John Rieman, Boulder, USA (1994)

21. Lovelace, K.L., Hegarty, M., Montello, D.R.: Elements of Good Route Directions in Familiar and Unfamiliar Environments. In: Freksa, C., Mark, D.M. (eds.) COSIT 1999. LNCS, vol. 1661, pp. 65–82. Springer, Heidelberg (1999)

22. Lynch, K.: The Image of the City. MIT Press, Cambridge (1960)

23. May, A.J., Ross, T., Bayer, S.H., Tarkiainen, M.J.: Pedestrian navigation aids: information requirements and design implications. Personal und Ubiquitous Computing 7(6) (2003)

24. Michon, P., Denis, M.: When and Why Are Visual Landmarks Used in Giving Directions? In: Montello, D.R. (ed.) COSIT 2001. LNCS, vol. 2205, pp. 292–305. Springer, Heidelberg (2001)

25. Montello, D.R.: Navigation. In: Miyake, A., Shah, P. (eds.) Cambridge handbook of visuospatial thinking. Cambridge University Press, Cambridge (2005)
26. Ortag, F.: Sprachausgabe vs. Kartendarstellung in der Fußgängernavigation. Diplomarbeit, Institut für Geoinformation und Kartographie, Forschungsgruppe Kartographie, Technische Universität Wien (2005)
27. Raubal, M., Egenhofer, M.J.: Comparing the complexity of wayfinding tasks in built environments. Environment & Planning B: Planning and Design 25, 895–913 (1998)
28. Raubal, M., Winter, S.: Enriching Wayfinding Instructions with Local Landmarks. In: Egenhofer, M.J., Mark, D.M. (eds.) GIScience 2002. LNCS, vol. 2478, pp. 243–259. Springer, Heidelberg (2002)
29. Remolina, E., Kuipers, B.: Towards a general theory of topological maps. Artificial Intelligence 152(1), 47–104 (2004)
30. Ross, T., May, A., Thompson, S.: The use of landmarks in pedestrian navigation instructions and the effects of context. In: Brewster, S., Dunlop, M.D. (eds.) Mobile HCI 2004. LNCS, vol. 3160, pp. 300–304. Springer, Heidelberg (2004)
31. Sorrows, M.E., Hirtle, S.C.: The Nature of Landmarks for Real and Electronic Spaces. In: Freksa, C., Mark, D.M. (eds.) COSIT 1999. LNCS, vol. 1661, pp. 37–50. Springer, Heidelberg (1999)
32. Talmy, L.: How language structures space. In: Pick, H.L., Acredolo, L.P. (eds.) Spatial Orientation: Theory, Research, and Application, pp. 225–282. Plenum Press, New York (1983)
33. Talmy, L.: Toward a Cognitive Semantics. Concept Structuring Systems, vol. II. MIT Press, Cambridge (2000)
34. Timpf, S.: Geographic Task Models for geographic information processing. In: Duckham, M., Worboys, M.F. (eds.) Meeting on Fundamental Questions in Geographic Information Science, Manchester, UK, pp. 217–229 (2001)
35. Tversky, B., Lee, P.U.: How Space Structures Language. In: Freksa, C., Habel, C., Wender, K.F. (eds.) Spatial Cognition 1998. LNCS, vol. 1404. Springer, Heidelberg (1998)
36. Tyler, A., Evans, V.: The Semantics of English Prepositions: Spatial Sciences, Embodied Meaning, and Cognition. Cambridge University Press, Cambridge (2003)
37. Wunderlich, D., Herweg, M.: Lokale und Direktionale. In: von Stechow, A., Wunderlich, D. (eds.) Semantik: Ein internationales Handbuch der zeitgenössischen Forschung. de Gruyter, Berlin (1991)

The Role of Angularity in Route Choice
An Analysis of Motorcycle Courier GPS Traces

Alasdair Turner

Bartlett School of Graduate Studies, UCL, London
a.turner@ucl.ac.uk

Abstract. The paths of 2425 individual motorcycle trips made in London were analyzed in order to uncover the route choice decisions made by drivers. The paths were derived from global positioning system (GPS) data collected by a courier company for each of their drivers, using algorithms developed for the purpose of this paper. Motorcycle couriers were chosen due to the fact that they both know streets very well and that they do not rely on the GPS to guide their navigation. Each trace was mapped to the underlying road network, and two competing hypotheses for route choice decisions were compared: (a) that riders attempt to minimize the Manhattan distance between locations and (b) that they attempt to minimize the angular distance. In each case, the distance actually traveled was compared to the minimum possible either block or angular distance through the road network. It is usually believed that drivers who know streets well will navigate trips that reduce Manhattan distance; however, here it is shown that angularity appears to play an important role in route choice. 63% of trips made took the minimum possible angular distance between origin and destination, while 51% of trips followed the minimum possible block distance. This implies that impact of turns on cognitive distance plays an important role in decision making, even when a driver has good knowledge of the spatial network.

Keywords: Navigation and wayfinding, spatial cognition, distance estimation, route choice, spatial network analysis.

1 Introduction

It is known that route angularity has an effect on the perception of distance. Sadalla and Magel show that, within laboratory conditions, a path with more turns is perceived as being longer than a path with fewer turns [1]. However, where navigation of a learned environment is concerned, we tend to assume that people will optimize the cost of their journey in terms of energy expended when walking, and in terms of time-efficiency when driving. That is, we would expect the physical cost constraint to overcome the cognitive cost of moving from origin to destination, and therefore observe experienced drivers taking the shortest block or Manhattan distance between origin and destination. In Bachelder and Waxman's proposed functional hierarchy of navigation strategies, it is unsurprising that the highest learned behavior is route optimization through distance

K. Stewart Hornsby et al. (Eds.): COSIT 2009, LNCS 5756, pp. 489–504, 2009.

minimization [2]. 'Distance' being the physically shortest distance. Below this comes navigation using the topological network, navigation by landmark, and only at the lowest level is angle considered an important factor, through path integration, or minimizing the angle to the destination.

That angle is a low level navigational strategy is reinforced by how we mentally process angles. Sadalla and Montello show that memory of turns tends to be categorized into distinct units of orthogonal directions rather than a continuous understanding of angle between locations [3]. This grouping of angles together would appear to affect our ability to navigate by using a least-angle heuristic, making it efficient in unknown environments but too susceptible to error to be usable in learned environments [4]. Although Montello shows that the physically shortest path is not always preferred within an urban environment [5], this effect again appears to be due to path integration rather than a higher level navigational strategy. Montello also shows that the direction of travel is important to the route choice. This finding indicates path integration, as a true minimal path would be the same in both directions. Furthermore, even the extent of the angular effect on perception of distance seems to be susceptible to experimental conditions. Heft shows that, if a non-laboratory set up is used and there are distractions in the environment, then the route angularity effect is not necessarily replicable [6].

Given this body of evidence, the proposal that minimization of route angularity is employed as a high-level navigation strategy might seem an unusual direction to take. However, the angularity of a route can work as a useful proxy for the number turns, and therefore as a mechanism to underpin any strategy that minimizes cognitive load for memory. Although turns may be discretized in our heads following Sadalla and Montello [3], the angle gives useful indicator of whether or not we might perceive the turn, without having to work out a schema for quantization. It also allows us to circumvent problems to do with decisions about when a turn becomes a turn. For example, when several small turns over a short distance turn into a perceived change of direction. As an overriding factor, though, the route angularity effect itself is important, as it shows that our conception of distance is mutable. That is, if we are to minimize distance, then we must form a plan of action in response to our cognitive map. In this situation, we may well plan to reduce the complexity of the route in favor of a reduction in the physical distance that it presents us, especially where the means of transport is mechanical rather than corporeal. Indeed, distance memory is notoriously difficult to pin down, but it appears to be more dependent on information or events encountered during a journey rather than the time taken or effort expended through it [7]. In this respect, a path that minimizes turning also minimizes the number of views along it, and so it can be stored in a compact manner, and once again the minimization angle may be thought of as a proxy for the minimization of cognitive distance.

In transportation analysis, however, the mechanism for route choice is almost universally assumed to be the shortest time between locations, and it is often assumed that the participant has access to a perfect map. Nevertheless, there is limited acknowledgment of different route choice within traffic analysis through the field of space syntax. Space syntax originated as a theory of the interaction of space and society, in which twin processes of movement and socialization occurred [8]. Movement was posited to be along continuous lines and socialization within convex areas. This theory led to the construction of an 'axial map' representing the linear movement paths, which was then analyzed to construct minimum turn paths between spaces. One problem with early models was that a turn constituted any move from one line to another, regardless of intersection angle, even a slight bend in the road. In response, various researchers turned either to joining sets of lines into continuous paths [9,10] or analyzing the angle between them [11,12]. In particular, a form of angular analysis which calculates the *betweenness* [13] has been shown to correlate very well with both pedestrian and vehicular aggregate movement levels [14]. In betweenness, the shortest (angular) path is calculated between all pairs of street segments in the system. For each segment that lies on one of the shortest paths, a count is incremented, so that a segment that lies on many shortest paths has a high betweenness rating. Further refinements to the method have moved away from axial lines altogether, and instead use road-center lines complete with topological constraints (such as no-left turn restrictions or one-way streets), again with a demonstration of high correlation with observed movement using a range of measures either angularly or turn-based [15,16].

Despite much correlative evidence, no direct causal link between movement strategy and observed movement levels has established within the field of space syntax[1]. It is suggested that the construction of paths from all segments to all other segments represents an approximation to all possible origins and destinations, and that paths taken by people tend to be the shortest angular paths rather than the shortest block distance paths. However, the observations can only be linked to aggregate numbers. While space syntax maps of angular betweenness identify main roads [16,18], those showing metric betweenness show 'taxi-driver' paths ferreting through back streets [15]. This result in itself leaves the methodology open to criticism. It is suggested that once people learn an environment, they will take these back street rat-runs. That is, rather than the good correlations indicating some underlying cognitive feature of the environment, it is argued that they merely indicate that most people are still learning the environment using path integration, and once they have attained full knowledge, they will move to the taxi-driver routes.

This paper – inspired by the debate around angular versus block distance within space syntax – examines the routes taken by a set of people knowledgeable about the available paths, who would be most expected to minimize physical distance covered. Motorcycle couriers make many deliveries every day within a confined area. This paper examines routes taken by couriers belonging to

---

[1] There is indirect evidence, such as experiments by Conroy Dalton into the route decisions people make within complex building environments [17], but this does not tie directly back to the angular paths used at the urban level to construct betweenness measures.

a single company during the month of October 2008, from global positioning system (GPS) tracking data for each courier. Approximately 50 couriers are observed, for a total of 2425 delivery trips. The courier company is based in London, and almost all the journeys take place within the central West End and City areas, with a typical journey length of 1.2km (more details follow in the coming sections). Motorcycle couriers are used for two reasons. Firstly, due to the number of deliveries they make in a constrained area, they have good knowledge of the road network. Secondly, they do not rely on the GPS data to navigate, so the observed trace represents the rider's own navigation rather than a route suggested by the GPS system. Instead, the GPS system is used by the company to allocate drivers to nearby pickups for delivery. As delivery is often urgent, the riders should be taking time-critical decisions to inform their chosen path, and we would expect them to take the minimum physical distance between origin and destination. However, in this paper, it is shown that the path chosen by the couriers is more often the minimum angular path than the minimum block distance path.

2   Background

There is potential to use global positioning system (GPS) in a range of cognitive contexts, and Dara-Abrams sets out a framework for spatial surveying and analysis using GPS within spatial cognition studies [19]. Furthermore, traditional travel diaries may be replaced with GPS trackers giving more accurate recordings of participants' movements [20]. Yet GPS data has been rarely used for explicit determination of route choice, despite at first seeming the ideal option for data gathering about wayfinding decisions. There may be a number of reasons for this: GPS can only reliably be used in outdoor environments; purchasing units for individual studies can be relatively expensive; and, unless dedicated units are used, typically the GPS will have been used to guide the participant, making navigation data redundant. Thus, on the whole, information mining does not seem to take place at the level of the road unit. GPS information can be used to give population statistics, such as the diurnal movement of commuters, or to derive location-based data from clusters. For example, Jiang et al. have used GPS traces to characterize patterns of movement by taxi in terms of power law distribution and relate this to a Lévy flight pattern [21]. At a more detailed level, Ashbrook and Starner use GPS both to suggest areas of interest, and to predict movement between them using a Markov transition-probability model [22]. Individual studies also exist in an attempt to link angle with behavior. For example, Mackett et al. give GPS units to children to follow their movement. However, the angular analysis they make is simply in terms of the number of turns made, not compared to potential route choices [23].

There have been studies from the transportation research field that do explicitly consider the geometry of routes that are taken. For example, Jan et al. have made an analysis of the routes taken by 216 drivers within the Lexington, Kentucky area over the period of a week [24]. Their main observations concern the difference between the shortest block distance path and the participants' actual routes. As the study is from the field of transportation analysis, Jan et al. are principally concerned with route allocation, and therefore alternative hypotheses concerning other cognitive models of route choice are not explicitly considered. Li et al. make a similar study for 182 drivers in Atlanta, Georgia, but once again route allocation is the paramount consideration [25]. Therefore, the regularity of commuters using one or more routes is examined, and a logit model prepared in order to predict which routes will be taken according to the probabilistic observation, rather than an investigation into cognitive hypotheses about how the drivers are making decisions.

Despite the difference in motivation for their study, Jan et al. do make the observation from individual instances within the data that it appears that some road users stick to the freeway as long as possible, but do not present a quantitative analysis of the routes in these terms. As Jan et al. point out, this seems a reasonable heuristic for movement between locations. From a cognitive perspective, we might add that the freeway path involves relatively fewer decision points, and is relatively free-flowing, although it is frequently not the shortest angular path. The driver must double back or make another diversion to get to their final destination, and so the case of commuters may well differ from those of motorcycle couriers. In particular, it would seem that sticking to freeways is the province of a less knowledgeable user of the road system, and when we look at traces in the following sections, we find tentative support for this hypothesis. More particularly, though, the paper is concerned with angularity as a possible navigation strategy, and this will form the major part of the analysis presented here.

3   Methodology

The methodology for this research follows three stages. Firstly, a subset of credible GPS data is extracted from raw data about courier locations at specific times, and split into individual trips from origins to destinations. Secondly, the trips are attached to a base map, and topological constraints on the possible routes are enforced in order to generate a set of sample trips, for which actual road distance followed and angle turned can be calculated. Thirdly, two forms of analysis are performed. In one, the block distance and the angular distance traveled are compared against a minimal trip using the same starting and destination road segment. In the other, the aggregate number of trips is compared to a space syntax analysis of the road network using betweenness. All the import and analysis algorithms reported were coded as a plug-in to UCL's Depthmap program [26].

3.1   Trip Segmentation

The GPS data used for the study are in the form of *xml* files downloaded from a courier company who track their deliveries[2]. Each *xml* file contains a series of

---

[2] The data are free and may be obtained from http://api.ecourier.co.uk

rows containing a courier identifier (a short alphanumeric code), a courier type (pushbike, motorbike, van and so on), the location in WGS84 format (longitude and latitude), a notional speed on an arbitrary scale, a heading and a timestamp. For the purposes of this paper, we took the courier types 'motorbike' and 'large motorbike'. While pushbike traces would have been interesting to examine, there were too few in the sampled data set. The data were collected for every day through October 2008 between 10am and 4pm. The choice of month was to take a working month where weather is typically not extreme. The time period is chosen to exclude rush hours, which might bias the drivers away from congested streets.

Once the lines for motorbikes were extracted, the data were then subdivided into trips for each courier. The data contain a timestamped location approximately every 10 seconds for each courier, although this may vary considerably according to ability to pick up a signal within a high rise area or in a tunnel. For most of the time, the timestamp in combination with the speed attribute, can be used to pick up when the courier has stopped or is moving, although GPS typically drifts around, and erroneous speed, heading and location data is frequently encountered. However, in relation to stops it can be difficult to tell the difference between a delivery or pickup and a stop at a traffic light. The couriers are incredibly efficient, and can often pick up and drop within a couple of minutes. For the purposes of this paper, we took 90 seconds stationary (speed below '2.5' – roughly equivalent to 2.5mph though not exactly), to be indicative of a stop. Due to GPS drift, starting up again is only registered when two consecutive rows show movement away from the stop point. The short stop requirement may accidentally pick up inter-traffic light journeys, but this is perhaps not too far removed from the journeys undertaken: what is not immediately obvious is that motorcycle courier trips are typically very short, both physically and temporally. This is due to the nature of the job. A motorcycle courier picks up an urgently needed document and takes it to another company, typically still within the center of London. In any case, to prevent overly short or overly long journeys, only trips between 500m and 5000m in length were recorded. Furthermore, the journeys take place in a very restricted area. Almost all the trips (well over 95%) were within the central area of the West End and City of London (see figure 1). However, a few journeys do go outside the central zone, well into the surrounding counties. In order to capture the majority of the sort of trips being undertaken, only trips within a 20km × 20km region containing the center of London were retained[3] Other trips were removed due to clearly erroneous GPS data: locations can and often do completely misread. As there was a significant amount of data available, trips were simply excluded altogether if a reading suggested a speed over 20ms^{-1} (approximately 45mph – the urban limit in the UK is 30mph). After paring down the trips, 2425 individual origin-destination trips were identified, comprising location points joined through a continuous time period.

---

[3] Ordnance Survey grid coordinates 520 000, 170 000 to 540 000, 190 000.

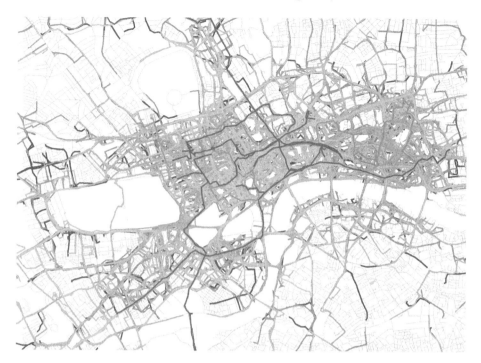

Fig. 1. All GPS traces within the central London area (colored by trip duration), with the base map shown in the background. The area shown is approximately 8km × 6km ≈ 50km^2. (Map data: OpenStreetMap)

3.2   Trace Attachment

The GPS traces obtained in the previous section are very approximate records of journeys taken. In order to ensure an accurate path is taken, the traces were next attached to an 'Integrated Transport Network (ITN) layer' base map supplied by Ordnance Survey. The map includes the topological information such as one way streets and no-right turns. Thus it is possible to derive the path that must actually have been driven from the data. Path reconstruction such as this is relatively unusual to attempt with GPS traces. Typically, algorithms need to find out where a vehicle is at the current time and where it is heading, and if a mistake is made it can be corrected quickly. In our case however, we only have points approximately every 10 seconds, and we must be careful to construct the most plausible path between locations. Figure 2 shows an actual GPS trace and its mapping to the base map. The driver starts on the Western side, drives North around a roundabout before heading South and then East. The traces were attached by what might at first seem a very lax method. All road segments within a 50m buffer of each location point are considered as possible routing segments provided they are within 30 degrees of the heading. However, if a tighter radius is used, then, due to inaccuracies in the location data, dead ends can easily be included, or the actual path missed.

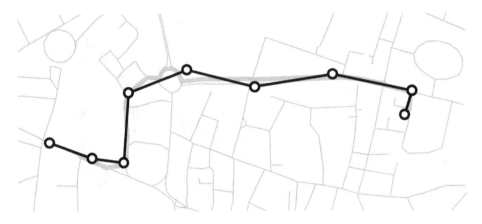

Fig. 2. Sample GPS trace (black) and the path as attached to the topological base map data (thick gray). This example uses OpenStreetMap data rather than the OS ITN layer used for the actual experiments.

Once the candidate segments have been identified, the path finding algorithm works by connecting up the shortest path between one candidate segment and the next, including at least one candidate segment for each location node along the path. The 'shortest path' is of course problematic for the purposes of this paper. Although usually we might just say the shortest block distance path, in order to be fair to both hypotheses we construct two variations: one using the shortest angular path between consecutive candidate segments and the other using the shortest block distance path. All measures concerned are then applied to the path identified with the appropriate distance rule.

There are a couple of caveats to using this method. Firstly, that the trip is truncated somewhat by the 50m buffer distance selecting shorter paths at the beginning an end (see figure 2 for an example). Secondly, and more importantly, in order to impose a route that follows the general behavior of riders, no immediate doubling back along roads is allowed. This is the equivalent of a no U-turn rule. Of course, in certain instances, drivers may well have made a U-turn. In addition, it is to be expected that riders may bend the rules occasionally or perform unexpected behaviors. A few paths generated seemed to take very tortuous paths to reach their destinations, and these would seem to occur when such anomalies have occurred. As these are very rare, they do not make a significant impact on the results.

3.3   Trip Analysis

Once plausible origin to destination routes were constructed, the routes taken were analyzed fairly simply. Firstly, the 'actual' path taken was compared to the shortest possible path between the origin segment for the trip and the destination segment for the trip. For each of these cases, the shortest path was compared to the path constructed using the same metric between candidate nodes to eliminate bias in the construction of the 'actual' path. That is, an 'actual' path constructed using angular shortest path between location nodes along its length was compared to the shortest angular path possible between the origin location and destination location, and similarly for the block distance path.

In addition, space syntax analysis using the *betweenness* measure was performed, which counts the number of times a segment is on the shortest path (either angular or block distance) between all pairs of origins and destinations in the systems. We can constrain this measure to look at just trips of length up to 1500m, based on finding that the length of trace attachment trips in the previous section averages about 1500m (more details about the statistics of the trip lengths are given in the next section). This test cuts in the opposite direction through the data to the per courier analysis. It is intended to discover if couriers on average make use of streets which are well connected to others, either angularly or based on block distance, following findings that angular betweenness appears to correspond to road designation, with main roads having high betweenness [16,18]. This is somewhat different to what the courier does in response to a particular origin and destination, and is compared to the overall numbers of couriers using each segment of road, shown in figure 3.

Fig. 3. Counts of the number of couriers that used each road segment for the central area of London. The inner London ring road (part of which is visible in the top left corner) is largely avoided. (Map data: OpenStreetMap).

4   Results

4.1   Individual Trip Measures

2425 trips were analyzed. The average block path length was 1380m with a standard deviation of 700m. Angle was measured using the space syntax convention of 90 degrees being '1.0' [14]. So a 180 degree turn is '2.0', and continuing straight on is '0.0'. This makes it easier to count 'numbers of 90 degree turns', or effectively give a number of 'turns', if turns approximate 90 degrees, as they would on a grid system. Using this convention, the average angular path length was 5.34 with a standard deviation of 3.64.

We then compared the actual path length against the shortest possible path length for both angular and block distance paths. The results are shown in table 1. In order to further elucidate these results, they are displayed in figure 4 using a logarithmic scale for the fractional ratio of actual path length to possible path length. The optimal fraction is 1.00, where the path taken is the shortest possible route. The couriers take this optimal path a surprising number of times, either taking the shortest block distance path or the shortest angular path in 71% of cases. It should be noted that in 43% of cases, they take a path which is both the shortest angular and shortest block distance path, so in these cases we are unable to distinguish which strategy they are using to guide their path. Of the remainder though, 70% of the time, when the courier takes a shortest path, it is the angular shortest path rather than the block distance shortest path. It has to be noted that in all cases the differences are very small. A lot of the time, the courier takes a trip within a small percentage difference of the actual minimum possible path, be it angular or metric.

The results are for the most part unsurprising: these are short trips and we would expect them to show good optimization of the route taken. However, it is interesting that, when they take a shortest path, it tends to be the angular shortest path rather than the metric shortest path. This suggests that even for short trips where the route is well known, the angular shortest path is the preferred option. This may be less unexpected than at first thought. Making a turn is a time consuming business. One must slow down, perhaps wait for

Table 1. Cumulative number of paths within a factor of the minimum possible path distance for angular and block measurements of the path ($n = 2425$)

Fraction of minimum	Angular	Metric
1.00	1531 (63%)	1232 (51%)
1.01	1574 (65%)	1332 (60%)
1.05	1637 (68%)	1643 (68%)
1.10	1738 (72%)	1863 (77%)
1.20	2010 (83%)	2143 (88%)
1.50	2237 (93%)	2258 (93%)
2.00	2419 (100%)	2424 (100%)

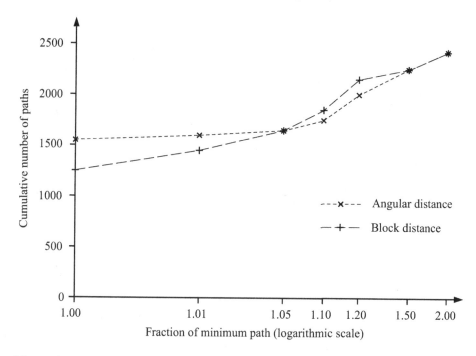

Fig. 4. Cumulative numbers of paths within a factor of the minimum possible distance paths for angular and block measurements of the path

oncoming vehicles to clear, and, on a motorbike lean from side to side. A straight path involves significantly less effort.

From a space syntax perspective, it is immaterial whether the effect is due to a cognitive distance difference or due to a physical effort different. However, it is useful to consider the evidence available to find the difference between the two hypotheses. If the effort hypothesis were true, then we would expect to see an appreciable decrease in journey speed as the angular turn increases[4] In order to attempt to detect this effect, the average speed was calculated for each journey. This was then compared to the total path angle of the the journey. If the effort hypothesis were true, we would expect to see a correlation between the speed and turn, as turning would slow down the journey speed. In fact, no correlation between speed and total angular turning for the path is observed ($R^2 < 0.01$). This is further backed up by a large difference between correlation of trip duration and angle ($R^2 = 0.31$) as opposed to the correlation of trip duration and path block-distance length ($R^2 = 0.66$). That is, traveling further

---

[4] Ideally, we might extract from the data a time-cost table for turns from each segment to each other segment, to determine the exact cost of any turn. However, a number of assumptions have been made already with regards to route, and time and speed are sampled coarsely. Therefore, the argument in the text uses the average journey speed instead.

Fig. 5. Space syntax measures of the central London area (angular betweenness radius 1.5km). At this radius, which reflects local journeys, the inner London ring road (part of which is visible in the top left corner) is not highlighted. (Map data: OpenStreetmap).

physical distance is correlated with increased trip duration, but making more turns is much more weakly correlated with increased trip duration, and hence the trip duration is less dependent on angle turned. Therefore, given the cost of turning seems to have little effect, it is more likely that angle is reduced due to an increase in cognitive difficulty in order to process more turns. As discussed in the introduction, we might suggest that the views along a street segment are relative stable, and thus each turn will lead to a significant increase in memory required to learn the route, and so higher turn journeys will increase cognitive distance.

4.2   Aggregate Trip Measures

By contrast to the individual trip measures, the aggregate trip measures were not found to be informative. If we refer to figure 3, it can be seen that the couriers are confined to, for the most part, a very small subset of streets within London. In particular, it is clear that couriers do not tend to use main ring roads, and their journeys would therefore be unrelated angular betweenness as we have described it: a measure which relates to main roads [18]. However, at such small distances, an average of 1.5km, the main roads are not actually picked out by angular betweenness, as shown in figure 5.

This observed difference to the literature on angular betweenness is due to the fact that in the literature examples, angular betweenness is run of radii up to 80km rather than just 1.5km. In the high-radius case, many trips from outside the city center take place and are transferred through the ring road structure [18]. Hence, in the low radius case much more local possibilities of movement are picked up, spread more evenly across the center. However, if figure 3 and figure 5 are compared, it is clear that angular betweenness is still not correlated to the movements of the couriers. While the angular betweenness does not favor any particular origins and destinations, the paths that the couriers take are very much biased by the contingencies of the specific pick-up and drop locations. Most of the courier routes cut directly through the middle of the map, presumably reflecting jobs transferring documents between companies located at the heart of the City and the West End of London. This is further backed up by our previous findings: the couriers *are* taking, for the most part, the shortest angular and block distance paths between origins and destinations, so this result must be biased by the actual origin and destinations. Otherwise, as the betweenness measure is simply a sum of shortest paths, the sum of courier paths would *have* to be correlated to betweenness. To confirm the results, linear regression tests were performed, and a weak positive correlation with betweenness was observed. For angular betweenness the figure was $R^2 = 0.09$, and for block-distance betweenness the value was $R^2 = 0.07$. This result is again indicative of a preference to angular rather than block-distance once again, but in the aggregate case it is too weak to form definitive evidence.

5   Conclusion

Results from the field of space syntax suggests that, on aggregate, vehicular numbers correlate with measures of the road network based on minimum angle rather than minimum block distance as might initially be expected [14,15]. This tends to imply that navigation may well be based on minimizing angular rather than block distance. However, there has been no established link between cognition and this strategy. It could equally be a physical phenomenon of the properties of road networks that results in the observed movement within them, as it could be a cognitive one. As a response, this paper attempts to uncover if there is a cognitive aspect by looking at the route choice decisions of motorcycle couriers. Motorcycle couriers know the street network very well, and hence they will be expected to optimize the block distance between origin and destination. We find, however, that despite often taking the shortest block distance path, couriers prefer the shortest angular path. Where it is possible to make a preference, they choose the shortest angular path 70% of the time.

It is of course easy to conflate angular choice with other strategies such as path integration. However, given the courier data, the proportion of exact angular routes is so high that it leaves little room for consideration of other factors. For example, it might be expected that turns will be taken toward the beginning of a journey [17], but as the couriers are, for much of the time, choosing an exactly minimum path, any turns that are made are in fact necessary for the minimal path to be followed, rather than because they are early or late in the journey. Alternatively, the opposite may be the case: that the minimal angular path is a result of a path integration decision at the beginning of the journey, that then impacts on automatically following the minimum angular path later.

However, as evidence for space syntax, and that angular routes play a strong part in route choice, the evidence that links individual route choice to a preference for the shortest angular route in secion 4.1 is compelling. That said, there remains much to be done. Such a restricted data set tells us little about other factors in general route choice decisions. Factors such as landmarks and other informational structures need to be added to the analysis, as do many other components that can help bring the cognitive map into the routine study of urban travel, including investigating the decisions made by drivers with different levels of experience of the system [27].

Acknowledgements

The courier data used in this paper is generously supplied free by eCourier. For more details, please see http://api.ecourier.co.uk/. The road-center line and topological connection data used for the study were derived from the Ordnance Survey ITN Layer data, © Crown Copyright 2005. Due to Ordnance Survey licensing restrictions, the base maps depicted in the paper are derived from data supplied free of charge by OpenStreetMap. For details, please see http://www.openstreetmap.org/. OpenStreetMap data is licensed under the Creative Commons Attribution–Share Alike 2.0 license. OpenStreetMap data accessed 06-Jun-2009. Finally, thank you to the reviewers who suggested examining trip time in more detail to resolve the difference between time-cost and cognitive-cost of trips.

References

1. Sadalla, E.K., Magel, S.G.: The perception of traversed distance. Environment and Behavior 12(1), 65–79 (1980)
2. Bachelder, I.A., Waxman, A.M.: Mobile robot visual mapping and localization: A view-based neurocomputational architecture that emulates hippocampal place learning. Neural Networks 7(6/7), 1083–1099 (1994)
3. Sadalla, E.K., Montello, D.R.: Remembering changes in direction. Environment and Behavior 21, 346–363 (1989)
4. Hochmair, H., Frank, A.U.: Influence of estimation errors on wayfinding decisions in unknown street networks analyzing the least-angle strategy. Spatial Cognition and Computation 2, 283–313 (2002)
5. Montello, D.R.: Spatial orientation and the angularity of urban routes. Environment and Behavior 23(1), 47–69 (1991)
6. Heft, H.: The vicissitudes of ecological phenomena in environment-behavior research: On the failure to replicate the "angularity effect". Environment and Behavior 20(1), 92–99 (1988)
7. Montello, D.R.: The perception and cognition of environmental distance: Direct sources of information. In: Frank, A.U. (ed.) COSIT 1997. LNCS, vol. 1329, pp. 297–311. Springer, Heidelberg (1997)

8. Hillier, B., Hanson, J.: The Social Logic of Space. Cambridge University Press, Cambridge (1984)
9. Figueiredo, L., Amorim, L.: Continuity lines in the axial system. In: van Nes, A. (ed.) Proceedings of the 5th International Symposium on Space Syntax, Delft, Netherlands, TU Delft, pp. 163–174 (2005)
10. Thomson, R.C.: Bending the axial line: Smoothly continuous road centre-line segments as a basis for road network analysis. In: Hanson, J. (ed.) Proceedings of the 4th International Symposium on Space Syntax, London, UK, UCL, pp. 50.1–50.10 (2003)
11. Dalton, N.: Fractional configurational analysis and a solution to the Manhattan problem. In: Peponis, J., Wineman, J., Bafna, S. (eds.) Proceedings of the 3rd International Symposium on Space Syntax, Atlanta, Georgia, Georgia Institute of Technology, pp. 26.1–26.13 (2001)
12. Turner, A.: Angular analysis. In: Peponis, J., Wineman, J., Bafna, S. (eds.) Proceedings of the 3rd International Symposium on Space Syntax, Atlanta, Georgia, Georgia Institute of Technology, pp. 30.1–30.11 (2001)
13. Freeman, L.C.: A set of measures of centrality based on betweenness. Sociometry 40, 35–41 (1977)
14. Hillier, B., Iida, S.: Network effects and psychological effects: A theory of urban movement. In: van Nes, A. (ed.) Proceedings of the 5th International Symposium on Space Syntax, Delft, Netherlands, TU Delft, vol. I, pp. 553–564 (2005)
15. Turner, A.: From axial to road-centre lines: A new representation for space syntax and a new model of route choice for transport network analysis. Environment and Planning B: Planning and Design 34(3), 539–555 (2007)
16. Jiang, B., Zhao, S., Yin, J.: Self-organized natural roads for predicting traffic flow: A sensitivity study. Journal of Statistical Mechanics: Theory and Experiment (July 2008) P07008
17. Conroy Dalton, R.: The secret is to follow your nose. In: Peponis, J., Wineman, J., Bafna, S. (eds.) Proceedings of the 3rd International Symposium on Space Syntax, Atlanta, Georgia, Georgia Institute of Technology, pp. 47.1–47.14 (2001)
18. Turner, A.: Stitching together the fabric of space and society: An investigation into the linkage of the local to regional continuum. In: Proceedings of the 7th International Symposium on Space Syntax, Stockholm, Kungliga Tekniska Högskolan (KTH), pp. 116.1–116.12 (2009)
19. Dara-Abrams, D.: Cognitive surveying: A framework for mobile data collection, analysis, and visualization of spatial knowledge and navigation practices. In: Freksa, C., Newcombe, N.S., Gärdenfors, P., Wölfl, S. (eds.) Spatial Cognition VI. LNCS, vol. 5248, pp. 138–153. Springer, Heidelberg (2008)
20. Wolf, J., Guensler, R., Bachman, W.: Elimination of the travel diary: Experiment to derive trip purpose from global positioning system travel data. Transportation Research Record: Journal of the Transportation Research Board 1768, 125–134 (2001)
21. Jiang, B., Yin, J., Zhao, S.: Characterising human mobility patterns over a large street network (2008) Arxiv preprint 0809.5001(accessed, March 1 2009)
22. Ashbrook, D., Starner, T.: Using GPS to learn significant locations and predict movement across multiple users. Personal and Ubiquitous Computing 7(5), 275–286 (2004)

23. Mackett, R.L., Gong, Y., Kitazawa, K., Paskins, J.: Children's local travel behaviour — how the environment influences, controls and facilitates it. In: 11th World Conference on Transport Research, California, Berkeley (2007)
24. Jan, O., Horowitz, A.J., Peng, Z.R.: Using global positioning system data to understand variations in path choice. Transportation Research Record: Journal of the Transportation Research Board 1725, 37–44 (2000)
25. Li, H., Guensler, R., Ogle, J.: An analysis of morning commute route choice patterns using gps based vehicle activity data. Transportation Research Record: Journal of the Transportation Research Board 1926, 162–170 (2005)
26. Turner, A.: Depthmap. In: Computer Program, 9th edn., UCL, London (2009)
27. Golledge, R.G., Gärling, T.: Cognitive maps and urban travel. In: Hensher, D.A., Button, K.J., Haynes, K.E., Stopher, P. (eds.) Handbook of Transport Geography and Spatial Systems, pp. 501–512. Elsevier Science, Amsterdam (2004)

Author Index